Solid-State Spectroscopy

Springer
Berlin
Heidelberg
New York
Barcelona
Budapest
Hong Kong
London
Milan
Paris
Santa Clara
Singapore
Tokyo

Hans Kuzmany

Solid-State Spectroscopy

An Introduction

With 238 Figures, 34 Tables and 110 Problems

Professor Dr. Hans Kuzmany
Institut für Festkörperphysik, Universität Wien
Strudlhofgasse 4
A-1090 Wien, Austria
and
Ludwig-Boltzmann-Institut für Festkörperphysik
Kopernikusgasse 15
A-1060 Wien, Austria

ISBN 3-540-63913-6 Springer-Verlag Berlin Heidelberg New York

Library of Congress Cataloging-in-Publication Data. Kuzmany, H. (Hans), 1940– Solid-State spectroscopy: an introduction / Hans Kuzmany. p. cm. Includes bibliographical references and index. ISBN 3-540-63913-6 (hardcover: alk. paper) 1. Solids–Spectra. 2. Materials–Spectra. 3. Spectrum analysis. I. Title. QC176.8.06K89 1998 530.4'12–dc21 97-48782

Printed in Germany

Typesetting: Data conversion by Satztechnik Katharina Steingraeber, Heidelberg
Cover design: *design & production* GmbH, Heidelberg

SPIN 10530950 54/3144 – 5 4 3 2 1 0 – Printed on acid-free paper

Preface

The dramatic increase in our knowledge of the solid-state over the last 10–20 years has come in great part from new spectroscopic experimental techniques. In this context spectroscopy is used in a broad sense and covers various experiments where energy analysis of a particle or electromagnetic radiation is crucial. Accordingly, spectroscopic methods extend in the electromagnetic spectrum from radio waves to γ radiation but also include particles after their interaction with a solid. Each spectral range has its characteristic technique and addresses particular properties of the solid. A fundamental knowledge of the various methods is therefore a prerequisite for a successful investigation of problems in a particular area.

The actual motivation for writing this textbook was the lack of any didactic review summarizing the methods and applications of spectroscopy with regards to solids. Many of the methods were well known from molecular physics but, even there, reliable and comprehensive textbooks are not available. Also, spectroscopic problems can be characteristically different in molecules and crystalline material because of the periodic arrangements of the atoms and molecules in the latter.

The material presented here is a result of several postgraduate courses on "Solid-State Spectroscopy" given by the author over the last few years. The goal of these lectures was to supply a representative selection of spectroscopic techniques and to describe their field of application. This goal has been retained as the concept of the current textbook. Accordingly, the intention is to provide a broad knowledge of the basic concepts, sufficient to follow specialized lectures or specialized literature later on.

Another source of the subject to be discussed is a textbook written by the author in 1989 in German and edited by Springer-Verlag in 1989. In this textbook the concept of presentation and didactic strategy was developed but the elaboration of the material has been performed in much more detail in the current version and the volume of subjects presented has been substantially increased.

During the formulation of the text particular attention was paid to a physical understanding of the spectroscopic problems rather than to their formal description. To improve information transfer from the text the most important results have been framed. This should be of particular help for the application-oriented reader. To simplify presentation no vector or tensor notation is employed in general. Only in cases where the physical meaning requires the specification of the rank of the variables, bulk letters, or indexed symbols in script are used for vectors and tensors and bulk letters in Roman for operators.

The bulk of the book deals with a description of current spectroscopic techniques and their applications on an introductory level. This is backed up by extensive appendices which contain several useful tables and a considerable number of further details, including some mathematical formulations on an advanced level. In this way a better link could be established with standard textbooks and to formulations used in spectroscopic research. The book is constructed, however, to allow reading and understanding without a study of the appendices. In this context the latter can be used either as a source of additional information for the lecturers or as part of the course work.

The first part of the textbook describes electromagnetic radiation, light sources such as lasers and synchrotron radiation, and general concepts of experimental techniques. The second part concentrates on individual spectroscopic methods using electromagnetic radiation and particles. The problems collected at the end of each chapter are designed to further the understanding of the text. Each of them is flagged for its instructive value. Discussion and solution of the problems is highly recommended. Problems with an asterisk are more difficult and may be considered as an extension of the subject covered by the book. Problems labeled with a superscript *a* require input from the appendix.

During the specification of the problems I strongly benefited from valuable discussions with colleagues and former students in my group. In this context I am particularly grateful to Mag. J. Winter, Mag. R. Winkler, and Mag. M. Hulman for their engagement in the discussion of the problems. A booklet with solutions will be available from the author for interested readers.

The current textbook may be useful as a first text for senior undergraduate students. However, it is particularly designed for postgraduates in physics, chemistry, and material science, before they start to work in a special research field. With the inclusion of the appendices the value of the book is extended to a more knowledgeable audience such as students working on a thesis, academic lecturers who intend to set up a similar course in solid-state spectroscopy, or even researchers in the field.

Two years of education in general physics are a prerequisite for understanding this book. In addition, a basic knowledge of solid-state physics and some background in the concepts of quantum mechanics will be very helpful. At the end of each chapter references are given for readers who need additional information on specific subjects and recent developments.

The subject covered by this textbook extends over a very broad field of material science. Extended discussions with many specialists were therefore extremely important. In this context I would like to acknowledge in particular Prof. G. Vogl and Prof. H. Grosse from the Universität Wien, Prof. M. Mehring from the Universität Stuttgart, Prof. J. Fink from the Institut für Festkörperphysik und Werkstofforschung in Dresden, and Prof. J. Kürti from the Eötvös University in Budapest.

For critical reading and correction of special chapters I acknowledge Doz. B. Sepiol from Wien, and Prof. Kürti and Prof. Mehring from Budapest and Stuttgart, respectively. Also, I am particularly grateful to T. Leitner for his continuous efforts to get the graphics of the textbook into a computer-compatible shape and for designing the coverplate.

Finally, I acknowledge the Springer-Verlag in Heidelberg, in particular Dr. Lotsch and Mr. C.-D. Bachem, for their support and efforts during the preparation of the manuscript.

It has been the idea of this book to provide an overview and an aid for newcomers to the rapidly emerging and colorful field of solid-state spectroscopy.

Wien, January 1998 *Hans Kuzmany*

Contents

1. Introduction

Spectroscopy as a special technique of scientific research is almost as old as the scientific research itself. Originally it consisted of an analysis of the wavelength and intensity of light after absorption or emission from atoms, molecules, or even from condensed matter. This analysis was, however, restricted basically to the visible spectral range. Today the range of a spectrum is much wider. Any analysis of the energy of radiation is considered as spectroscopy, no matter whether the radiation is electromagnetic, mechanical, or embodied in particles. The spectral range of the electromagnetic radiation extends from radio frequencies to γ radiation and the particle radiation includes electrons, neutrons, positrons, muons, and even neutral or charged atoms. The energy range to be considered covers at least ten orders of magnitude. The importance of such an extensive energy range to condensed matter and, in particular, to solids needs to be investigated. Information on the matter is obtained from the radiation spectrum modified by its interaction with the electronic and magnetic configuration of molecules or crystals. This interaction can proceed either by two-particle interaction, such as the processes of absorption or emission of radiation, or by three-particle interaction as in the process of scattering.[1] In fact, many electronic transitions in condensed matter are in the energy range of 10^{-7} eV which allows absorption of radio waves in the MHz region. Transitions of spin states of nuclei or atoms in a magnetic field are examples of such low-energy processes. In the infrared, vibrational transitions or selected transitions between electronic orbitals can be investigated. At the highest energies transitions at the nuclei level score. Information on the condensed material can still be obtained due to the influence of the crystalline environment on the nuclear energy levels. This is the basis of Mößbauer spectroscopy.

This textbook deals primarily with spectra from solid-state materials. The transition energy between two states in the solid is determined from the position of a structure in a spectrum. The magnitude of the structure is related to the rate of the transitions between the two states. This rate is always determined by the product of probability for the transition and number of appropriate configurations for its realization. As such all spectroscopic meth-

[1] In a quantum-mechanical notation absorption is a three-particle process and scattering is a four-particle process.

ods rely on the same principle, not only from a descriptive point of view, but also from a quantitative and mathematical point of view. However, the solids under consideration cannot be described by a single and general enough theory, meaning that each spectral range has its own special approximation.

Not only the theoretical background but also the experimental procedures for the various spectroscopic techniques are strongly related. All of them are based on the use of:

- a proper source for the probe beam,
- the sample which interacts with the probe,
- instrumentation for the analysis of the probe, and
- a detection system for the probe.

Each of these items is subjected to highly advanced technology or, in the case of the samples, to an advanced theoretical description.

In the following chapters the various spectroscopic processes will be discussed. We will start by reviewing the concepts of electromagnetic radiation and then turn to the experimental frame of the classical and fundamental spectroscopic methods. A description of the nature of the "response function" of material to radiation follows. The next set of chapters introduce special spectroscopic methods like Raman scattering, infrared spectroscopy, and magnetic resonance spectroscopy. The concept of a linear response to a radiative perturbation holds, in principle, for all these techniques. For scattering experiments with electrons and neutrons an extended description of the linear response is required, as presented in Chap. 14.

Regarding the history of solid-state spectroscopy some techniques date back to the beginning of scientific work. Others such as pulsed nuclear magnetic resonance or experiments which use synchrotron radiation for excitation are quite new. One characteristic feature is common to many techniques. They start out as an exotic experiment, become part of a pool of experimental techniques used for the analysis of physical properties, and ultimately represent the dominant technique for the determination of these properties in solids. A good example of such a progression is Fourier spectroscopy. It started off as an exotic experiment by *Michelson* [1.1] in 1891. Then, in 1951, *Fellgett* showed the great advantages of this technique for the spectral range of the far infrared [1.2]. In the 1970s the relative value of Fourier spectroscopy and grating spectroscopy for the full infrared spectral range was the subject of extended discussions. Today this technique dominates not only the infrared but also the visible and even the near-ultraviolet spectral ranges. Similar developments can be observed for the positron annihilation or nuclear magnetic resonance spectroscopy.

One recurring problem in solid-state spectroscopy is its relation to spectroscopy with molecules. Since solid-state physics is a relative newcomer in science, many of the spectroscopic techniques have been applied previously to molecular systems. Some of the results from these studies also apply to solids. However, the periodic arrangement of atoms in crystals and the high density of electronic states for certain energies play an important role in solid-state spectroscopy and, in many cases have required the development of new concepts. The difference between molecular and solid-state responses is remarkable in many spectroscopic techniques such as optical absorption, Raman scattering, luminescence, or photoemission. In fact, in solid-state spectroscopy both the molecular and the solid-state points of view may be important since very often properties of molecules are well retained as, for example, in molecular crystals.

A final remark should be made regarding the formal presentations in this text. Since it emphasizes the experimental point of view international units are used, even for formulas which were derived from purely theoretical considerations. A list of the most important fundamental constants is given in these units in Table A.1. On the other hand, different energy units are quite common for the different spectroscopic techniques. Joules, electron volts, wave numbers, frequencies, temperature, and even atomic energy units (Hartree) are used. The values for the different units are correlated to each other in Table A.2.

2. Electromagnetic Radiation

The main part of this textbook deals with spectroscopy utilizing electromagnetic (EM) radiation. Thus, we will first review the most important properties of this particular probe. We will start with the idealized description of a plane wave within Maxwell's theory and continue with an explicit description of more realistic fields, such as radiation from a dipole and from an arbitrarily accelerated charge. In Sects. 2.3 and 2.4 Fourier transforms are discussed and applied to the important case of radiation from sources with limited emission time. These sections are rather short since the readers should be familiar with such subjects. For those who need a tutorial or a review, a more detailed treatment of the subject can be found in Appendix B.

2.1 Electromagnetic Waves and Maxwell's Theory

As long as its wavelength is not too short radiation can be characterized by the classical description of a plane EM wave in the form

$$E = E_0 \cos(kx - \omega t) , \tag{2.1}$$

or

$$E = E_0 \mathrm{e}^{\mathrm{i}(kx - \omega t)} . \tag{2.2}$$

E is the electric field of the wave with the amplitude E_0, wave vector k, and angular frequency $\omega = 2\pi f$ where f is the frequency. The complex form of the electric field (2.2) is often very convenient, but it should be kept in mind that only its real part has a physical meaning. Thus, in order to be correct, the complex conjugate must be added in all calculations. The real field is then obtained as $(E + E^*)/2$. The sign of the imaginary symbol i is arbitrary. It serves to describe the actual phase of the field. Only a consistent sign convention is needed. Different sign conventions can lead to different signs in relations derived from the field. This is the reason why formulas found in the literature often deviate slightly from each other. By convention, a positive sign is used throughout this book. The correlation between the wave vector k, the wavelength λ, the wave number ν, and the quantum energy ϵ of the radiation are given as follows:

$$k = \frac{2\pi}{\lambda} = 2\pi\nu = \frac{n\omega}{c_0} = \frac{n\epsilon}{\hbar c_0} . \tag{2.3}$$

Here, $n = c_0/c$ is the (real) index of refraction expressed as the ratio between the velocity of light in vacuum and in the solid. Using these relations the calculated value of the wave vector for visible light is of the order of $10^5\,\mathrm{cm}^{-1}$. This value is very small compared to typical wave vectors of quasi-particles excited in the first Brillouin zone of real crystals, or as compared to wave vectors of this zone in general. Typical values for the latter two quantities are $q \approx 10^8\ \mathrm{cm}^{-1}$. This fact is of fundamental importance for solid-state spectroscopy and often plays a dominating role in the selection rules of spectroscopic transitions.

Occasionally, general complex numbers $A = A_\mathrm{c} + \mathrm{i}A_\mathrm{s}$ are used to describe EM waves. A_c and A_s are called the components of a *phasor*, and A is the complex amplitude of the wave. A_c and A_s are defined as the coefficients of the cosine term and the sine term if a harmonic oscillation of the general form

$$E(x, y, z, t) = E(x, y, z)\cos(\omega t - \phi)$$

is separated in its cosine and sine component. Phasors are useful if, for instance, a linear superposition of waves is studied. The resulting wave is then obtained as a summation of complex numbers.

The energy of EM waves is characterized by different quantities depending on the spectral ranges. Often used are the wavelength λ given in Å, nm or µm, the frequency f given in Hz, (the angular frequency in s^{-1}), the wave number ν given in cm^{-1}, or the quantum energy $\hbar\omega$ given in eV. For example, lasers are usually characterized by their wavelength, electronic transitions by their energy in eV, and vibrational excitations by wave numbers. The use of different units is not as confusing as it may appear at a first glance since usually only one type of them appears within a particular field of spectroscopy. Thus, in this book the traditional units for the description of radiation and transition energies will be used. It is nevertheless important to keep in mind the quantitative relations between the various units for the description of the energy as they are given in (2.3) or in Table A.2. For practical use it is convenient to remember that

1 meV corresponds to about $8\,\mathrm{cm}^{-1}$ and 1 µm wavelength corresponds to 1.24 eV.

A summary of the energy units commonly used in the various ranges of the EM radiation is given in Table 2.1. This table also lists the spectroscopic techniques applied in the various spectral ranges. Abbreviations are explained in the corresponding sections of this book.

The other characteristic quantities of EM radiation, the electric field E, the magnetic excitation H, the induction B, the vector potential A, etc. can be expressed most conveniently in the SI units V, A, m, and s. These

Table 2.1. The electromagnetic spectrum.

	Wavelength	Wave number [cm^{-1}]	Frequency [s^{-1}]	Energy [eV]	Spectroscopic techniques
Electric waves	∞–0.03 cm		0–E12		ESR, EPR, NMR
Far infrared	3000–40 µm	3–400		(0.4–50)E–3	FTIR, abs., refl.
Infrared	40–0.8 µm	250–12 500	(7–400)E12	0.03–1.6	IR, FTIR
Visible light	0.8–0.4 µm	(12–25)E3		1.6–3	abs., refl., ellipsom.
Ultraviolet	400–10 nm			3–120	abs., UPS
X rays	10–0.01 nm			50–120E3	XPS, XAFS
γ radiation	10–0.1 pm			(2–1200)E4	MB, PAC

quantities are listed in Table 2.2. For the sake of generality the older cgs units are also given.

In addition to the above quantities, which were derived from the definition of the EM wave, photometric (or physiological) quantities such as *candela*, *lumen*, and *lux* are important. The photometric quantities are based on the unit of the lumen (lm) defined as the light flux emitted (and detected by the human eye) from $6 \times 10^{-3}\,\mathrm{m}^2$ of a black body at the temperature of melting Pt which is 2042 K. The photometric radiant intensity is measured in candela (cd); 1 cd is 1 lm/ster[1]. The irradiant intensity is measured in lux (lx); 1 lx is $1\,\mathrm{lm/m}^2$.

Table 2.2. Characteristic quantities of the electromagnetic field.

Quantity	SI [VAms]	cgs [cm g s]	Relations
Electric field (E)	Vm^{-1}	g$^{1/2}$cm$^{-1/2}$s^{-1}	1Vm^{-1} = (1/3)cgs
Magnetic excitation (H)	Am^{-1}	g$^{1/2}$cm$^{-1/2}$s^{-1}	1Am^{-1}=0.0256Oe
Magnetic induction (B)	Vsm^{-2}	g$^{1/2}$cm$^{-1/2}$s^{-1}	1Vsm^{-2} =10^{4}G
Vector potential (A)	Vsm^{-1}	g$^{1/2}$cm$^{1/2}$s^{-1}	
(Radiant) power (P)	VA	erg s^{-1}	
Energy density (W)	VAsm^{-3}	erg cm^{-3}	1VAm^{-2}=10^{3}cgs
Intensity (irradiance) (I)[a]	VAm^{-2}	erg cm^{-2}s^{-1}	1VAm^{-2}=10^{3}cgs
Intensity (radiant) (Φ)[a]	VA ster^{-1}	erg s^{-1} ster^{-1}	
Radiance (brightness) (L)[a]	VAm^{-2} ster^{-1}		

[a] I, L and Φ can be normalized to unit band width. In this case the symbols are supplied with an index labeling the variable used for the normalization.

[1] ster (from *steradian*) is the unit for the solid angle $\overline{\Omega}$; 1 ster = $1/4\pi$; $\mathrm{d}\overline{\Omega} = \sin\theta\mathrm{d}\theta\mathrm{d}\phi$.

The relationships between the quantities listed in Table 2.2 are obtained from Maxwell's equations, and the well known relationship between the vector potential and the field B. For plane waves (2.2) and (B.1) yield

$$\boldsymbol{B} = \frac{1}{\omega}(\boldsymbol{k} \times \boldsymbol{E}_0)\mathrm{e}^{\mathrm{i}(\boldsymbol{kr}-\omega t)} \qquad \text{with} \qquad \boldsymbol{H} = \frac{\boldsymbol{B}}{\mu\mu_0} \,. \tag{2.4}$$

For example, for a wave propagating in the x direction with $\boldsymbol{E} \parallel \boldsymbol{z}$, the vector $\boldsymbol{B}$ is $\parallel -\boldsymbol{y}$ and has the form

$$\boldsymbol{B} = -\frac{k}{\omega}E_0\mathrm{e}^{\mathrm{i}(kx-\omega t)}\boldsymbol{e}_y \,.$$

Similarly, the vector potential (Appendix B.2) is given by

$$\boldsymbol{A} = \frac{\mathrm{i}}{\omega}\boldsymbol{E}_0\mathrm{e}^{\mathrm{i}(\boldsymbol{kr}-\omega t)} \tag{2.5}$$

and, for the wave propagating in the x direction, it has the form

$$\boldsymbol{A} = \frac{\mathrm{i}}{\omega}E_0\mathrm{e}^{\mathrm{i}(kx-\omega t)}\boldsymbol{e}_z \,.$$

From these results the energy density of the radiation becomes

$$W = \frac{1}{2}(\boldsymbol{ED} + \boldsymbol{HB}) = \frac{\varepsilon\varepsilon_0 E^2}{2} + \frac{\mu\mu_0 H^2}{2} \,, \tag{2.6}$$

where ε_0 and μ_0 are the dielectric constant and permeability of vacuum, and E, D, H, and B have been assumed as real. ε_0 and μ_0 are related to the light velocity and to the impedance of vacuum by

$$\frac{1}{\sqrt{\varepsilon_0\mu_0}} = c_0, \qquad \sqrt{\frac{\mu_0}{\varepsilon_0}} = 377\,\text{Ohm.}$$

The numerical values for ε_0 and μ_0 are compiled in Table A.1.

Since the energy spreads perpendicular to $\boldsymbol{E}$ and $\boldsymbol{H}$ and is equally distributed to the electric and the magnetic field, the intensity I is obtained from

$$I = Wc = \varepsilon\varepsilon_0 E^2 \frac{c_0}{\sqrt{\varepsilon\mu}} = \sqrt{\varepsilon\varepsilon_0}E\sqrt{\mu\mu_0}H \cdot \frac{c_0}{\sqrt{\varepsilon\mu}} = EH \,. \tag{2.7}$$

Written as the Poynting vector $\boldsymbol{I}$ and for $\mu = 1$

$$\boldsymbol{I} = \boldsymbol{E} \times \boldsymbol{H} = \boldsymbol{E} \times (\boldsymbol{k} \times \boldsymbol{E})\frac{1}{\omega\mu_0} = \frac{E^2\boldsymbol{k}}{\omega\mu_0} = E^2 c_0\sqrt{\varepsilon}\varepsilon_0\boldsymbol{e}_k \,, \tag{2.8}$$

where $\boldsymbol{e}_k$ is the unit vector in the direction of the wave propagation.

In the above equations W, I, and $\boldsymbol{I}$ are time-dependent (with a term $\cos^2(\boldsymbol{kr}-\omega t)$ for plane waves). To obtain the time average a proper averaging procedure must be performed. In the simple case of plane waves the values of E^2 and B^2 are then replaced by their time average $E_0^2/2$ and $B_0^2/2$.

Using a complex notation for the fields, products of the vectors have to be replaced by the product of one vector with the complex conjugate of the other vector. In this way the sum of the squared components of the vectors represent the square of the magnitude of the field.

The intensity of radiation is very often evaluated for complex fields E from $I = c_0\varepsilon_0 E E^*$. Since in this case the time dependence is lost only the time-average intensity is obtained and its magnitude is two times the magnitude of a real field with the same amplitude.

2.2 Radiation from Accelerated Charges

Even though plane waves are a good example to illustrate electromagnetic theory and are indeed very often useful to describe the EM field locally, they are not a very realistic form of the radiation. In reality EM radiation always originates from accelerated electric charges. The mode of acceleration determines the wave field. Vibrating electric dipoles, such as a vibrating molecule with a finite electric dipole moment, or excited molecular or solid systems provide realistic acceleration and emission patterns. We will therefore first discuss the basic properties of the Hertzian dipole and then make some general remarks about radiation from arbitrarily accelerated charges.

2.2.1 The Hertzian Dipole

The radiation from an oscillating dipole is emitted by moving charges in a pattern shown schematically in Fig. 2.1. A dipole with the length l, charge $\pm Q$, and oscillation amplitude Δl emits radiation in directions $\boldsymbol{r}$ defined by the unit vector $\boldsymbol{e}_r$. Thus, at an arbitrary point $\boldsymbol{r}$ a field $\boldsymbol{E}(\boldsymbol{r})$ will be observed. If the oscillation is harmonic with angular frequency ω the dipole has the form

$$P_\mathrm{D} = P_{\mathrm{D}0} \cos \omega t = Q \Delta l \cos \omega t \, . \tag{2.9}$$

A Hertzian dipole corresponds to the approximation $\Delta l \approx l$ and $\lambda = 2\pi c_0/\omega \gg l$. Its amplitude is therefore $P_{\mathrm{D}0} = Ql$. In order to obtain the emitted field we need to evaluate either the potential Φ or the vector potential $\boldsymbol{A}$ from the time-retarded charges or time-retarded currents as outlined in B.2. Since the vector potential by itself is enough to obtain the B field, it is more convenient to calculate the current distribution for the Hertzian dipole and use its time-retarded value in (B.7) rather than to calculate Φ directly from the retarded charge distribution. $\boldsymbol{A}$ is obtained from

$$\boldsymbol{A}(\boldsymbol{r}, t) = \frac{\mu \mu_0}{4\pi} \int \frac{\boldsymbol{j}(r', t - |\boldsymbol{r} - \boldsymbol{r}'|/c)}{|\boldsymbol{r} - \boldsymbol{r}'|} \mathrm{d}^3 x' \, , \tag{2.10}$$

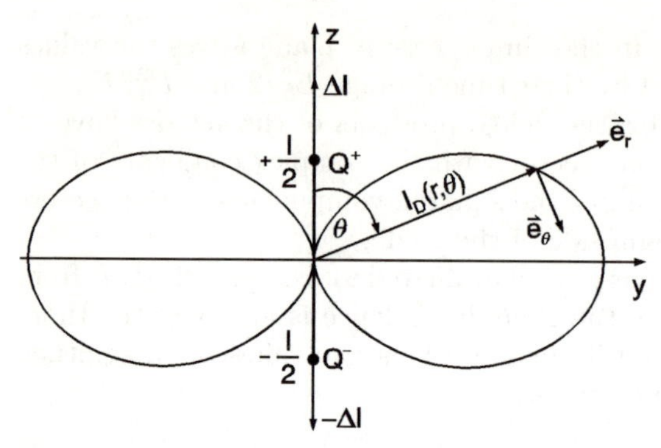

Fig. 2.1. Emission of electromagnetic radiation from a dipole. $I_{\mathrm{D}}(r, \theta)$ is the intensity observed at distance r under the angle θ. It has rotational symmetry around the dipole axis. $\boldsymbol{e}_\phi$ looks into the plane of the paper.

where $\boldsymbol{r}'$ represent the coordinates of the current distribution and $\boldsymbol{r}$ the coordinates of the field distribution.[2]

The time-dependent current distribution for the Hertzian dipole may be derived from the time-dependent charge distribution. If the dipole is very short and oriented in the z direction as in Fig. 2.1 we may consider the time-dependent charge $Q(t) = Q \cos \omega t$ as a source for a current in the z direction. This current then has the form

$$\boldsymbol{i}(t) = \frac{\mathrm{d}Q}{\mathrm{d}t} \boldsymbol{e}_z = -\omega Q \sin \omega t \, \boldsymbol{e}_z \, . \tag{2.11}$$

To evaluate $\boldsymbol{A}$ we need to integrate over the current density. Even though the latter nearly diverges locally because of the small size of the electron its integral is certainly finite. In a simplified form the integration along x' and y' yields the current $\boldsymbol{i}$ of (2.11), and the integration along z' yields the length of the dipole l. Here it is assumed that $\boldsymbol{r}' \ll \boldsymbol{r}$ for all points of interest so that we can use the expression $t - r/c$ for the retarded time and neglect $\boldsymbol{r}'$ in the denominator of (2.10). With these approximations the vector potential becomes

$$\boldsymbol{A}_{\mathrm{D}} = -\frac{\mu_0 Q l \omega \sin(\boldsymbol{k}\boldsymbol{r} - \omega t)}{4\pi r} \boldsymbol{e}_z \, , \tag{2.12}$$

or more generally

$$\boxed{\boldsymbol{A}_{\mathrm{D}} = -\frac{\mu_0 Q l \omega \sin(\boldsymbol{k}\boldsymbol{r} - \omega t)}{4\pi r} \boldsymbol{e}_{\mathrm{D}} \, ,} \tag{2.13}$$

where $\boldsymbol{e}_{\mathrm{D}}$ is the direction of the dipole moment and $\boldsymbol{k}$ is the wave vector $(\omega/c)\boldsymbol{e}_r$. The curl of $\boldsymbol{A}_{\mathrm{D}}$ yields the field $\boldsymbol{B}_{\mathrm{D}}$

[2] Note that here and for the remainder of the book limits for the integration extend from $-\infty$ to ∞ if not specified otherwise.

$$\boldsymbol{B}_\mathrm{D} = \mathrm{curl}\,\boldsymbol{A}_\mathrm{D} = -\frac{\mu\mu_0 Q l \omega^2 \sin\theta \cos(\boldsymbol{kr} - \omega t)}{4\pi c r}\boldsymbol{e}_\phi \;, \tag{2.14}$$

or more generally

$$\boxed{\boldsymbol{B}_\mathrm{D} = \mathrm{curl}\,\boldsymbol{A}_\mathrm{D} = -\frac{\mu\mu_0 Q l \omega^2 \cos(\boldsymbol{kr} - \omega t)}{4\pi c r}(\boldsymbol{e}_\mathrm{D} \times \boldsymbol{e}_r) \;,} \tag{2.15}$$

where in the first equation spherical polar coordinates $\boldsymbol{e}_r$, $\boldsymbol{e}_\theta$, and $\boldsymbol{e}_\phi$ have been used as shown in Fig. 2.1. The result for $\boldsymbol{B}_\mathrm{D}$ in (2.15) has been obtained by retaining only the term proportional to $1/r$. This term dominates for large distances as compared to terms $\propto 1/r^2$. This approximation is therefore only valid in the far field called the *wave* or *radiation zone*. For a more general solution (static or intermediate zone) see [2.1, 2.2]. Results for the wave zone are, in general, good enough for applications in spectroscopy.

The B field is independent of ϕ, perpendicular to $\boldsymbol{e}_r$ and $\boldsymbol{e}_\theta$, and decreases as $1/r$ with distance r from the radiating dipole. These proportions are evident from the second part of (2.15) where $\sin\theta\,\boldsymbol{e}_\phi$ was replaced by $(\boldsymbol{e}_\mathrm{D} \times \boldsymbol{e}_r)$.

The other characteristic quantities for the field follow immediately from the equations given in Sect. 2.1 and Appendix B.1. The electric field is calculated from Maxwell's equations

$$\boxed{\begin{aligned} \boldsymbol{E}_\mathrm{D} &= -\frac{c^2}{\omega}(\boldsymbol{k} \times \boldsymbol{B}_\mathrm{D}) = -\frac{\mu\mu_0 Q l \omega^2 \sin\theta \cos(\boldsymbol{kr} - \omega t)}{4\pi r}\boldsymbol{e}_\theta \\ &= -\frac{\mu\mu_0 Q l \omega^2 \cos(\boldsymbol{kr} - \omega t)}{4\pi r}(\boldsymbol{e}_r \times (\boldsymbol{e}_\mathrm{D} \times \boldsymbol{e}_r)) \;. \end{aligned}} \tag{2.16}$$

The power per unit area reaching point $\boldsymbol{r}$ (irradiance) is given by the Poynting vector evaluated from (2.7) and (2.8)

$$\boldsymbol{I}_\mathrm{D} = \frac{\mu\mu_0 (Ql)^2 \omega^4 \sin^2\theta \cos^2(\boldsymbol{kr} - \omega t)}{16\pi^2 c r^2}\boldsymbol{e}_r \;, \tag{2.17}$$

and the radiation per differential solid angle $\mathrm{d}\Omega$ is immediately obtained from this by

$$\mathrm{d}P = |\boldsymbol{I}_\mathrm{D}|\,\mathrm{d}F = |\boldsymbol{I}_\mathrm{D}| r^2 \mathrm{d}\Omega \qquad (\text{in W}) \;. \tag{2.18}$$

All quantities given above are time-dependent. To obtain time-average values for the radiation power the $\cos^2$ functions in (2.17) and (2.18) have to be replaced by their average value of $1/2$. Finally, the total average power emitted by the dipole is obtained from (2.18) by integration over the solid angle

$$P_\mathrm{tot} = \frac{P_\mathrm{D0}^2 \mu\mu_0 \omega^4}{12\pi c_0} = \frac{\mu\mu_0 (Ql)^2 \omega^4}{12\pi c} \qquad (\text{in W}) \;. \tag{2.19}$$

If more than two charges oscillate the radiation field becomes more complicated and is described by electric multipole radiation. For modern spectroscopic techniques quadrupole radiation is important. It still has a simple structure. The magnitude of the quadrupole moment for an arrangement of charges symmetric with respect to the z axis is[3]

$$P_Q = \sum_i Q_i(3z_i^2 - r_i^2)\,. \tag{2.20}$$

Charges arranged as shown in Fig. 2.2a therefore have a finite quadrupole moment (but no dipole moment). If the positive and negative charges oscillate 180° out of phase and with equal frequency, they establish a time-dependent quadrupole moment of the form $P_Q(t) = P_{Q0}\cos\omega t$. As a consequence they will emit a quadrupole radiation.

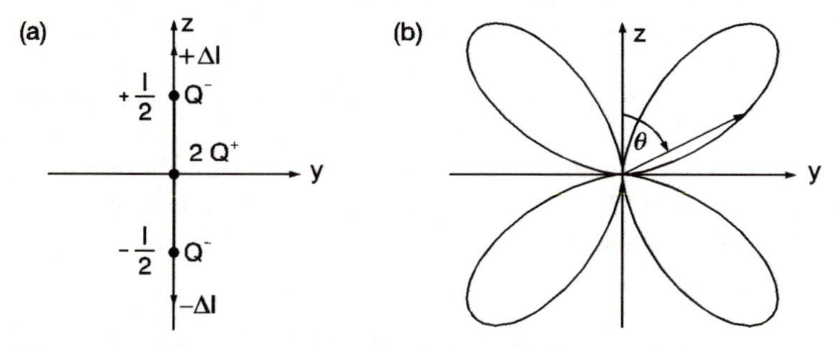

Fig. 2.2. Linear arrangement of charges, Q, with a finite quadrupole moment l^2Q (**a**), and the radiation pattern for a harmonic time dependence of Q (**b**). The pattern is rotationally symmetric about z.

The emitted electric field and magnetic induction are given for the arrangement in the figure as

$$\begin{aligned} \boldsymbol{E}_Q &= -\mathbf{e}_\theta \frac{\mu\mu_0\omega^3 P_{Q0}\sin\theta\cos\theta}{16\pi c r}\cos(\boldsymbol{kr}-\omega t) \\ \boldsymbol{B}_Q &= -\mathbf{e}_\phi \frac{\mu\mu_0\omega^3 P_{Q0}\sin\theta\cos\theta}{16\pi c^2 r}\cos(\boldsymbol{kr}-\omega t)\,. \end{aligned} \tag{2.21}$$

As a result of these relations the emitted power is proportional to $\sin^2\theta\cos^2\theta$ where θ is the direction between the z axis and the direction of emission. The radiation pattern for this geometry of the charges is depicted in Fig. 2.2b.

2.2.2 Emission from Arbitrarily Accelerated Charges

The Hertzian dipole described above may be considered as a special case of radiation from an accelerated point charge Q. The emission from an arbitrarily accelerated charge moving along $\boldsymbol{r}'(t)$ is also important. It is the

[3] For the definition of multipole moments, see Appendix B.3.

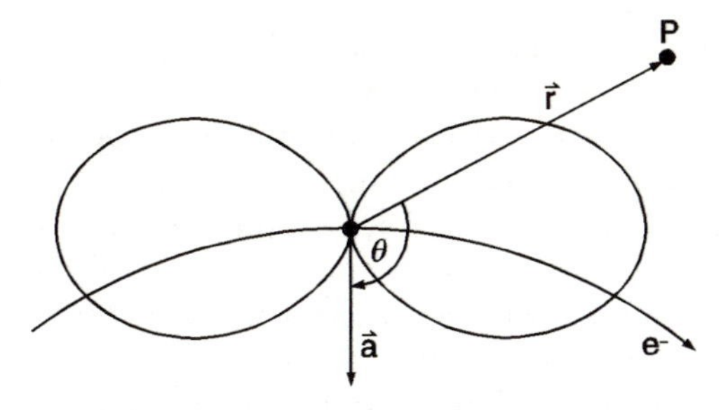

Fig. 2.3. Radiation characteristic for a particle moving with an arbitrary acceleration $\boldsymbol{a}$.

basis for the description of black-body radiation, x-ray *bremsstrahlung*, or synchrotron radiation. General expressions for the potential Φ and the vector potential $\boldsymbol{A}$ for this charge are known as Lienard–Wiechert potentials and given in Appendix B.5. From $\boldsymbol{A}$ and Φ the electric field $\boldsymbol{E}$ and the magnetic induction $\boldsymbol{B}$ for arbitrarily accelerated charges can be derived. If the distance between the field point P and the emitting charge is very large ($\boldsymbol{s} = \boldsymbol{r} - \boldsymbol{r}' \approx \boldsymbol{r}$ in Fig. B.1) and the particle is nonrelativistic $[(v/c_0)r = (\dot{r}'/c_0)r \ll r]$ the radiation intensity observed at distance r (irradiance) under the angle θ with respect to the direction of acceleration $\boldsymbol{e}_a$ is given, for Q equal to the elementary charge e, by

$$\boxed{I(r,\theta) = \frac{\mathrm{e}^2 \sin^2\theta}{16\pi^2\varepsilon_0 c_0^3 r^2}|\boldsymbol{a}|^2 \qquad (\text{in W/m}^2)\,,} \tag{2.22}$$

where $\boldsymbol{a} = \mathrm{d}^2\boldsymbol{r}'/\mathrm{d}t^2$ is the acceleration of the particle in Fig. B.1 or in Fig. 2.3. The radiation pattern for this particle is exhibited in the latter figure. The emission of radiation is strongest perpendicular to the direction of the acceleration, but it is independent of the direction of the particle velocity. The radiation pattern is, of course, directly related to the dipole radiation shown in Fig. 2.1.

The total power radiated from the particle is obtained by integrating over the full angular space. This yields

$$\boxed{P = \frac{\mathrm{e}^2}{6\pi\varepsilon_0 c_0^3}|\boldsymbol{a}|^2\,.} \tag{2.23}$$

This well-known formula for radiation emission from nonrelativistic, accelerated charged particles had already been derived by Larmor at the end of the last century.

2.3 Fourier Transforms

The radiation pattern described above for a plane wave and for an oscillator are still unrealistic as they assume a single value for the angular frequency ω

(or do not give any explicit frequency, as in the case of radiation from an arbitrarily moving point charge). Realistic fields always encompass a frequency spectrum with a given radiation energy per frequency interval. This means a realistic radiation field can be described either by a function of time such as $E(t)$ or by the distribution of the radiation energy over frequencies determined by the field $E(\omega)$. The relation between these two descriptions is given by the Fourier transform (FT) of one type of function into the other. Since FT is fundamental to modern spectroscopy we will review the basic concepts of this mathematical technique in this section. More details are given in Appendix B.6. Nearly all spectroscopic techniques described in the following chapters use this mathematical tool in one way or another. One of the main reasons for the importance of FT is the dramatic development in computer technology. Even personal-size computers can perform a Fourier analysis of very large numbers of data points in a very short time.

2.3.1 Fourier Theorem

In the usual terminology the Fourier theorem consists of two parts.

a) Any function $h(u)$ periodic in u as $h(u) = h(u+2\pi/v)$ with the period $2\pi/v$ can be represented by a sum of harmonic functions of the form $\sin(nuv)$ and $\cos(nuv)$ or $\exp(inuv)$ with the fundamental angular frequency v and overtone number n. The coefficients of the harmonic functions in the sum can be evaluated in an unique way from the original function $h(u)$.
b) Statement (a) also holds for non periodic functions in the sense that the sum is converted into an integral and the frequency range extends now from $-\infty$ to ∞. In other words, in this case, FT gives a rule for a well defined transform converting a function $h(u)$ in u-space to a corresponding function $g(v)$ in v-space. $h(u)$ and $g(v)$ are called Fourier pairs.

Since statement (b) actually includes periodic functions, it can be expected to cover statement (a). Indeed, it is not very difficult to show that this is true.

Fourier transforms are utilized in various disciplines of science such as optics, spectroscopy, communications, stochastic processes, solid-state science, etc. In spectroscopy u usually represents the time t and v the frequency f or ω. However, the use of space coordinate x for u and the wave numbers ν for v is also common. In solids FT is usually three-dimensional with space coordinate $\boldsymbol{r}$ for u and the unit vector of the reciprocal lattice $\boldsymbol{G}_0$ for v.

The mathematical expressions for the Fourier theorem, as they will be used in this text are:

a) A function $E(t)$ periodic in time with period T which means $E(t+T) = E(t)$, where $1/T = f$, and $2\pi f = \omega$ can be expressed by

$$E(t) = \sum_{n=-\infty}^{\infty} c_n e^{i2\pi nft} = C_0 + \sum_{n=1}^{\infty} [A_n \cos(2\pi nft) + B_n \sin(2\pi nft)] \tag{2.24}$$

with

$$c_n = \frac{1}{T} \int_{t_0}^{t_0+T} E(t) e^{-i2\pi nft} \, dt \, .$$

The coefficients c_n are complex with $c_{-n} = c_n^*$, and t_0 is arbitrary. The relation between the complex representation and the real representation given in (2.24) is obtained from a straight forward calculation and is left to the reader as an exercise.

b) For non-periodic functions $E(t)$

$$E(t) = \int E(f) e^{i2\pi ft} df \quad \text{with}$$

$$E(f) = \int E(t) e^{-i2\pi ft} dt \, . \tag{2.25}$$

$E(t)$ and $E(f)$ are the Fourier pairs. When $E(t)$ is even or odd in t, the transform can be obtained using only sine or cosine functions, respectively.

2.3.2 Examples of Fourier Transforms

In general FTs used in spectroscopy are rather simple. Advanced programs such as *Mathematica* can quite easily calculate FTs analytically on personal computers. We will give here two examples which may serve as a guide for related problems, and summarize some general properties of the Fourier pairs in Appendix B.6.

Let $E(t)$ be of the form $A|\cos 2\pi ft|$ with $f = 1/T$. A graph of this function is shown in Fig. 2.4a. Since the period of this function is $T/2$ the Fourier coefficients can be calculated from

$$c_n = \frac{2A}{T} \int_{-T/4}^{T/4} \cos(2\pi t/T) e^{-i2\pi n2t/T} \, dt \, . \tag{2.26}$$

The integration is performed by replacing the exponential by sine and cosine functions, and using the appropriate trigonometric relationships. c_0 is always the time average of the periodic function. It is equal to $2/\pi$ in our case. Since $E(t)$ is even only the real parts of the coefficients c_n are non zero. Consequently, $c_{-n} = c_n^* = c_n$. The explicit value of c_n is

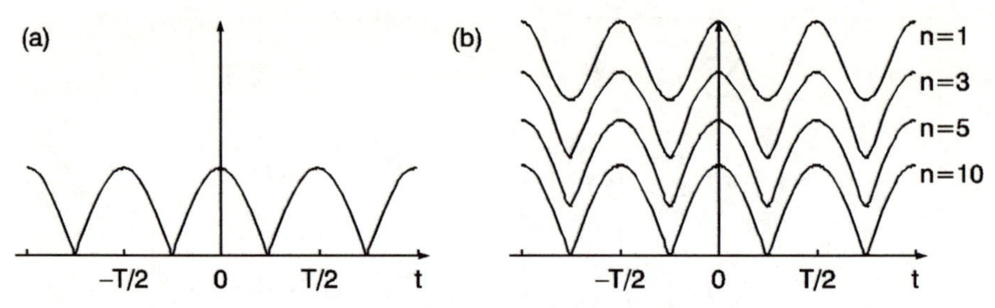

Fig. 2.4. Graph of the function $A|\cos 2\pi t/T|$ (**a**), and representation of the function by an increasing number of harmonic overtones (**b**). For clarity, curves for $n = 1$–5 are shifted in (b).

$$c_n = \frac{A\sin[\pi(2n+1)/2]}{\pi(2n+1)} + \frac{A\sin[\pi(2n-1)/2]}{\pi(2n-1)} . \tag{2.27}$$

Using these values, replacing $2\pi f$ by ω and recalling that the period of $|\cos 2\pi ft|$ is $T/2$ (2.24) yields

$$\begin{aligned} E(t) &= A|\cos 2\pi ft| = A|\cos\omega t| \\ &= \frac{2A}{\pi} + \sum_{n=1}^{\infty} \frac{4A}{\pi}\frac{(-1)^{n-1}}{4n^2-1}\cos 2n\omega t . \end{aligned} \tag{2.28}$$

Figure 2.4b displays $|\cos\omega t|$ represented by an increasing number of harmonic contributions. Obviously, the first ten contributions already represent the function $E(t)$ very well.

The two-sided exponential decay of a sine wave (wave packet) is a good example of a FT of a non-periodic function. The function is given by

$$E(t) = A\mathrm{e}^{-\gamma|t|}\sin\omega_0 t . \tag{2.29}$$

From this expression, $\tau = 1/\gamma$ is the lifetime of the oscillation. To obtain the FT we separate the Fourier integral into two parts

$$\begin{aligned} E(f) &= A\int_{-\infty}^{\infty} \mathrm{e}^{-\gamma|t|}\sin\omega_0 t\mathrm{e}^{-\mathrm{i}2\pi ft}\mathrm{d}t \qquad (2.30) \\ &= A\left(-\int_0^{-\infty} \mathrm{e}^{\gamma t}\sin\omega_0 t\mathrm{e}^{-\mathrm{i}2\pi ft}\mathrm{d}t + \int_0^{\infty} \mathrm{e}^{-\gamma t}\sin\omega_0 t\mathrm{e}^{-\mathrm{i}2\pi ft}\mathrm{d}t\right) . \end{aligned}$$

Substituting $-t$ for the integration variable t in the first integral we are left with two integrals which can be solved straightforwardly. We obtain for $E(f)$

$$E(f) = \frac{\mathrm{i}A\gamma}{\gamma^2 + 4\pi^2(f+f_0)^2} - \frac{\mathrm{i}A\gamma}{\gamma^2 + 4\pi^2(f-f_0)^2} . \tag{2.31}$$

The inverse transform according to (2.25) must give the original time function for the whole time space. The frequency spectrum (2.31) is imaginary and odd. In general, FTs of arbitrary functions $E(t)$ are complex. This is not really a problem. As we will see below, the distribution of the energy in the spectra is given by the square of the magnitude of $E(f)$ which is always real. Since the energy distribution for negative frequencies does not contain any new information products such as $E(f)E^*(f)$ must always be even. This is indeed the case in our example since the absolute square of any odd function is even.

The frequency spectra of harmonic functions are the δ functions. It is easy to show that the FT of $E(t) = A\cos\omega_0 t$ is $(A/2)[\delta(f-f_0)+\delta(f+f_0)]$. A similar, but odd and imaginary FT is obtained for $E(t) = A\sin\omega_0 t$. The FT of a time-independent function is $\delta(f)$.

More details about FTs, several examples, and useful general rules for the relations between Fourier pairs are given in Sect. 2.4 and Appendix B.6.

2.4 Radiation with a Finite-Frequency Spectrum

Let us return to radiation and apply a Fourier analysis to realistic radiation fields. A strictly monochromatic field, as described in Sects. 2.1 or 2.2, is only possible for waves which propagate fully undamped and extend in time from $-\infty$ to $+\infty$. This is unrealistic. In reality radiation is either damped (at least on an atomic scale) or switched on and off at certain points of time. The consequences of these experimental constraints will be discussed in this section for a damped harmonic oscillator and for a plane wave switched on and off at times $t = 0$ and T, respectively.

2.4.1 Damped Harmonic Oscillator

In a classical description, the time-dependence for the emission of an electric field E from a damped harmonic oscillator is given by

$$\ddot{E} + \gamma\dot{E} + \omega_0^2 E = 0\ . \tag{2.32}$$

The eigenvalues for this differential equation are

$$\alpha_{1,2} = -\frac{\gamma}{2} \pm \mathrm{i}\sqrt{\omega_0^2 - \gamma^2/4}\ , \tag{2.33}$$

so that the general complex solution is a linear combination of the terms $\exp\alpha_1 t$ and $\exp\alpha_2 t$. Since we are only interested in the real part of the field we take as the general solution

$$\mathrm{Re}\{E\} = C_1\mathrm{e}^{-\gamma t/2}\cos\omega t + C_2\mathrm{e}^{-\gamma t/2}\sin\omega t\ , \tag{2.34}$$

where ω is the detuned frequency $\sqrt{\omega_0^2 - \gamma^2/4}$. As long as the damping $\gamma/2$ remains smaller than ω_0, this is a damped harmonic oscillation, where the coefficients C_i serve to satisfy boundary conditions. If we want a maximum field and a zero derivative of the field at $t = 0$, the solution has the form

$$E(t) = E_0\mathrm{e}^{-\gamma t/2}[\cos\omega t + (\gamma/2\omega)\sin\omega t]. \tag{2.35}$$

We can also find a simpler special solution for (2.32):

$$E(t) = E_0 \mathrm{e}^{-\gamma t/2} \cos \omega t \quad \text{for} \quad t \geq 0$$
$$= 0 \quad \text{for} \quad t < 0. \tag{2.36}$$

In this case we have applied a more restrictive boundary condition which shuts the oscillator off for $t < 0$. A graph of this solution is displayed in Fig. 2.5.

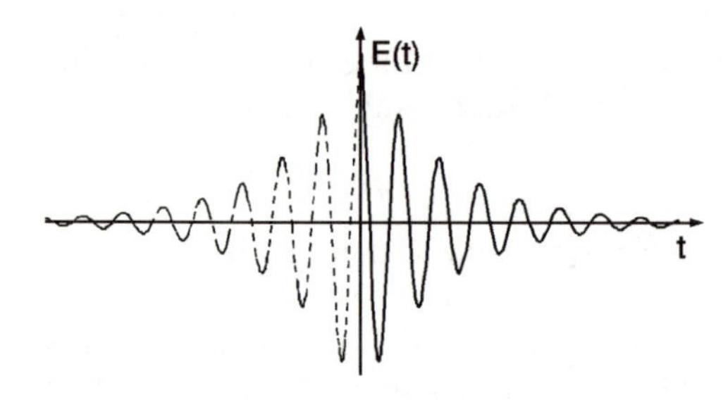

Fig. 2.5. Damped harmonic oscillation for $t > 0$ (—) and symmetrized form (– – –).

The function in (2.36) describing the damped wave field is often extended to negative values of time and, as we will see later, also to negative values of the frequencies. For the performance of the experiment time and frequency must, of course, always remain positive. The extension is often performed to a symmetric version with respect to $t = 0$ (replace t by $|t|$ for all values of t). Note that the new function is not a solution for (2.32). It is, however, convenient for mathematical treatments and therefore often used to describe the frequency spectrum of damped oscillations or wave packets.

The frequency spectrum for any time-dependent function is obtained from its FT as shown in the last subsection. Unfortunately the damped oscillation of (2.36) has a rather complicated frequency spectrum because of the abrupt change of the function at $t = 0$. The complex solution for the damped oscillator which may be written as

$$E = E_0 \mathrm{e}^{-\gamma t/2} \mathrm{e}^{\mathrm{i}\omega t} \quad \text{for} \quad t \geq 0$$
$$= 0 \quad \text{for} \quad t < 0 \tag{2.37}$$

gives a more simple but instructive frequency spectrum. Renaming the tuned frequency of the damped oscillator ω_0, for convenience, and applying FT yields for the frequency spectrum

$$E(f) = \int_0^\infty E_0 \mathrm{e}^{-\gamma t/2} \mathrm{e}^{\mathrm{i}\omega_0 t} \mathrm{e}^{-\mathrm{i}2\pi f t} \mathrm{d}t$$
$$= \frac{E_0}{\gamma/2 + \mathrm{i}2\pi(f - f_0)} \, . \tag{2.38}$$

As will be shown in detail in Sect. 2.4.3, the experimentally observed frequency distribution is obtained from the spectral intensity $S(f)$ given by the square of the magnitude of the frequency spectrum. Thus, returning to angular frequencies, we obtain the spectral intensity for the emission from the damped oscillator by

$$S(\omega) = E(\omega)E^*(\omega) = \frac{E_0^2}{(\gamma/2)^2 + (\omega - \omega_0)^2} \,. \tag{2.39}$$

This particular shape of the intensity spectrum is called a *Lorentzian line.* Its spectral width is obviously determined by the magnitude of γ. The full width at half maximum (FWHM) in ω space is exactly equal to γ in the present case.

Frequency and intensity spectra for the real solution of (2.32) are similar to (2.38) and (2.39).

Since γ is the spectral width (uncertainty in frequency) of the line as well as the reciprocal lifetime τ of the oscillation in intensity ($\gamma = 2\pi/\tau$) the relationship between these two quantities is an expression of the uncertainty relation in the following sense. The oscillator energy ϵ is only determined to the accuracy $\delta\epsilon = \hbar\delta\omega = \hbar\gamma$. Thus, the relationship between τ and $\delta\epsilon$ is

$$\tau\delta\epsilon = \tau\hbar\delta\omega = \tau\hbar\gamma = h \,. \tag{2.40}$$

The relation between lifetime, or pulse length, and bandwidth is quite general and very important. Pulses with shorter lifetimes have broader frequency spectra.

2.4.2 Frequency Spectrum for Electromagnetic Waves with a Finite Radiation Time

The frequency analysis of a plane wave oscillating only in a time interval from $t = 0$ to $t = T$ is a straightforward extension of the discussion above. Considering only the time-dependent part of the wave we have to study

$$\begin{aligned} E(t) &= E_0 \cos\omega_0 t \qquad \text{for} \qquad 0 \le t \le T \\ &= 0 \qquad\qquad\qquad \text{otherwise} \,, \end{aligned} \tag{2.41}$$

or its exponential analog. The latter is again more convenient. With FT, as demonstrated above, we obtain for the frequency spectrum

$$\begin{aligned} E(f) &= \int_0^T E_0 \mathrm{e}^{\mathrm{i}2\pi f_0 t} \mathrm{e}^{-\mathrm{i}2\pi f t} \mathrm{d}t \\ &= \frac{E_0 \sin[\pi T(f_0 - f)]}{\mathrm{i}\pi(f_0 - f)} \mathrm{e}^{\mathrm{i}\pi(f_0 - f)T} \,. \end{aligned} \tag{2.42}$$

The exponential term in this result is obviously a phase factor due to the fact that we did not start with a packet symmetric about $t = 0$. The intensity spectrum should, of course, not depend on this phase. This is indeed true as the product $E(\omega)E^*(\omega)$ yields for $S(\omega)$

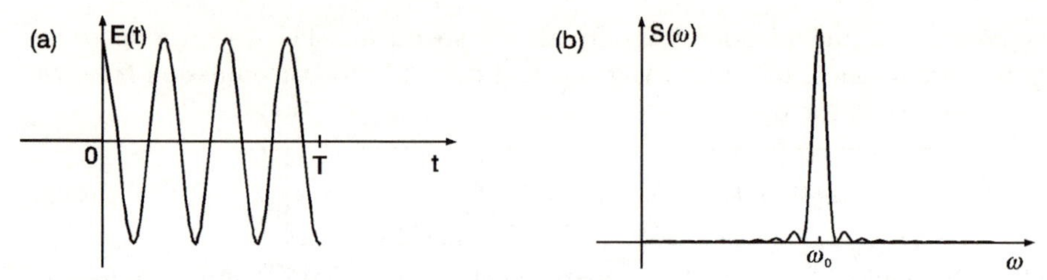

Fig. 2.6. Wave packet for a plane wave starting at time $t = 0$ and extending to $t = T$ (**a**) and intensity spectrum for its complex representation in time (**b**).

$$S(\omega) = \frac{4E_0^2 \sin^2(\omega_0 - \omega)T/2}{(\omega_0 - \omega)^2} \,. \tag{2.43}$$

The graph for the real part of the wave packet and the intensity spectrum (for its complex form) are shown in Fig. 2.6. From (2.43) the width of the spectral distribution is determined by the length T of the wave packet. For very large values of T it approaches a monochromatic structure which can be described by a delta function of the form $\delta(\omega_0 - \omega)$. This function is a very useful tool for the mathematical treatment of spectroscopic problems; Appendix B.7 lists some of its most important properties. The FWHM for the frequency spectrum of the wave packet is obtained from (2.43) by $\Delta\omega \approx 5.54/T$. As in the case of the damped waves shorter pulses have broader frequency spectra. A pulse of one femtosecond duration has an approximate bandwidth of 3 eV.

2.4.3 Frequency Spectrum and Power Spectrum

As mentioned above, we do not have to worry about complex frequency spectra. The physically meaningful quantity is the intensity or power spectrum given as

$$\boxed{S(f) = |E(f)|^2 \,.} \tag{2.44}$$

From the above definition it is not obvious that $S(f)$ is the relevant quantity to describe the spectral intensity. Alternatively one could have used the intensity $|E(t)|^2$ from any of the time functions given above and taken the FT. The result would not have been equal to $E(f)E^*(f)$. The experimentally observed quantity is indeed $S(f)$. The physical reason for this fact is that the electric field of the radiation interacts with the electrons of the detector, not its "power".

Note that the spectral intensity defined in (2.44) is not given in $\mathrm{W/m^2Hz}$. We need to prove that $E(f)E^*(f)$ is proportional to the intensity of the radiation, and the factors of proportionality must be evaluated. For a stationary field the intensity as measured with a detector over a period of time T is

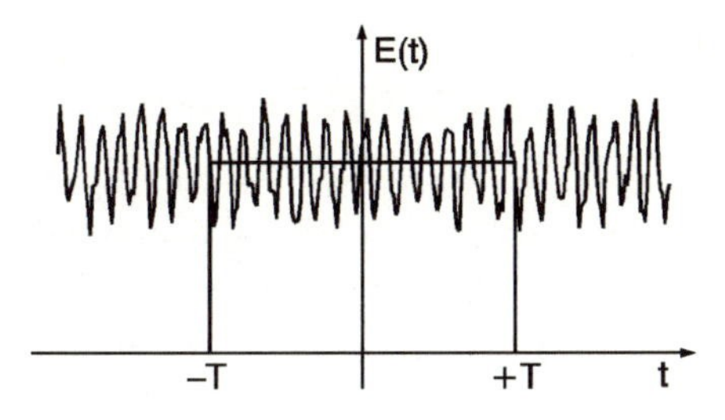

Fig. 2.7. Measurement of radiation intensity for a stochastic field.

$$I(T) = \varepsilon_0 c_0 \frac{1}{2T} \int_{-T}^{T} E(t) E^*(t) \mathrm{d}t \qquad (T \text{ arbitrary}) . \tag{2.45}$$

The rational for this measurement is illustrated in Fig. 2.7. The quantity actually measured is zero for $|t| \geq T$ and originates from a field with a Fourier transform $E(f, T)$. For T large enough $I(T)$ does not depend on T, and (2.45) can be written as

$$\begin{aligned} I &= \lim_{T \to \infty} \frac{\varepsilon_0 c_0}{2T} \int_{-T}^{T} E(t, T) \int E^*(f, T) \mathrm{e}^{-\mathrm{i}2\pi f t} \mathrm{d}f \mathrm{d}t \\ &= \lim_{T \to \infty} \frac{\varepsilon_0 c_0}{2T} \int \int_{-T}^{T} E(t, T) \mathrm{e}^{-\mathrm{i}2\pi f t} \mathrm{d}t E^*(f, T) \mathrm{d}f \\ &= \lim_{T \to \infty} \frac{\varepsilon_0 c_0}{2T} \int E(f, T) E^*(f, T) \mathrm{d}f = \frac{\varepsilon_0 c_0}{2T} \int S(f, T) \mathrm{d}f , \end{aligned} \tag{2.46}$$

where we have dropped the limes in the last line, but understand that T is large enough to have no influence on the measurement of the power. The second Fourier transformation in (2.46) is exact since we are considering a time function for the field E which is only finite between $-T$ and T. The derivation shows that $S(f, T)$ is indeed a spectral power and the intensity per unit frequency range is obtained by multiplying it with $\varepsilon_0 c_0 / 2T$. We may define a spectral intensity $I(f)$ explicitly as

$$I_\mathrm{f}(f) = \frac{\varepsilon_0 c_0}{2T} E(f, T) E^*(f, T) \qquad (\text{in W/m}^2\text{Hz}) . \tag{2.47}$$

For a pulse-like E field, the spectral intensity given in W/m^{2}Hz may not be very useful since, for example, in a damped oscillation, the intensity changes continuously with time. In this case a more meaningful description is obtained by considering the total energy W_T of the pulse

$$W_T = \varepsilon_0 c_0 \sigma \int E^2(t) \mathrm{d}t \qquad (\text{in Joule}) ,$$

where σ is the cross section of the pulse. The spectral energy density in J/Hz is then obtained from

$$W(f) = W_T \frac{|E(f)|^2}{\int |E(f)|^2 \, \mathrm{d}f} = \varepsilon_0 c_0 \sigma |E(f)|^2 , \tag{2.48}$$

where Parceval's theorem was used in the form

$$\int |E(t)|^2 \mathrm{d}t = \int |E(f)|^2 \mathrm{d}f \,. \tag{2.49}$$

The physical meaning of Parceval's theorem is that energy is conserved whether the total energy is expressed in time or in frequency space.

As we have seen, the intensity spectra are always real but we must still address the question of negative frequencies. They are understood as frequencies corresponding to negative times. This imposes a definite constraint on the intensity functions. If the field $E(t)$ is real, they must be even which is indeed always the case. (Note that the intensity spectrum in Fig. 2.6 is not even because the time function was complex.) Of course, the interpretation of negative frequencies does not mean that the reverse transform of $E(-f)$ gives the part of the time function with negative values of t. In fact, the part of $E(f)$ or $S(f)$ with negative f does not give any new information about the intensity spectrum and is, in this sense, useless.

Problems[4]

2.1 Show that the relation between the E Field and the B field for an electromagnetic plane wave is given as $\boldsymbol{B} = (1/\omega)(\boldsymbol{k} \times \boldsymbol{E}_0) \exp \mathrm{i}(\boldsymbol{kr} - \omega t)$.

(Purpose of exercise: use of Maxwell's equations.)

2.2 Show that the wave equation, as derived from the Maxwell's equations for a conducting system, is given by

$$\Delta \boldsymbol{E} = \frac{\varepsilon\mu}{c_0^2}\frac{\partial^2 \boldsymbol{E}}{\partial t^2} + \mu\mu_0\sigma\frac{\partial \boldsymbol{E}}{\partial t}$$

Discuss the equation for a good metal and study the behavior of $\boldsymbol{E}$ for a plane wave solution by performing an inverse transform.

(Purpose of exercise: use of Maxwell's formalism.)

2.3 Calculate the magnetic induction and the Poynting vector for a Hertzian dipole of the radiation zone from the vector potential.

Hint: Use spherical polar coordinates and neglect terms $\propto 1/r^2$.

(Purpose of exercise: use of spherical polar coordinates.)

2.4 Two positive and two negative charges are linearly arranged with the two negative charges coinciding at $z = 0$ and the two positive charges at $z = \pm l$. Calculate the electric dipole moment, the electric quadrupole moment, and the quadrupole radiation field if the charges vary harmonically as $Q = Q_0 \cos \omega t$.

Hint: For the evaluation of the field add the contributions from the two oppositely oriented dipole radiators.

(Purpose of exercise: multipole radiation; pick the right approximation.)

2.5 Show that $E = \exp(-\gamma t/2) \cos \omega t$ for $t > 0$ and $E = 0$ for $t < 0$ is a solution for the damped harmonic oscillator but $E = \exp(-\gamma |t|/2) \cos \omega t$ is not a solution.

(Subject of training: understanding damped oscillation.)

---

[4] A booklet with problem solutions which also includes some additional problems can be purchased from the author.

2.6* Show that the integral form of the Fourier theorem includes the representation of periodic functions as a sum of harmonic functions and evaluate the coefficients in the sum from the integral theorem.

Hint: Separate the integral from $-\infty$ to ∞ into a sum of integrals over the range equal to the period.

(Purpose of exercise: get a feel for the Fourier theorem.)

2.7* Evaluate the relationships between the complex Fourier coefficients c_n and the coefficients for the real representation A_n and B_n in (2.24). Show that

$$c_0 = \frac{1}{T}\int_{t_0}^{t_0+T} E(t)\,\mathrm{d}t, \quad A_n = \frac{2}{T}\int_{t_0}^{t_0+T} E(t)\cos n\omega t\,\mathrm{d}t,$$

$$B_n = \frac{2}{T}\int_{t_0}^{t_0+T} E(t)\sin n\omega t\,\mathrm{d}t. \tag{2.50}$$

(Purpose of exercise: equivalence of real and complex formalism of Fourier transforms.)

2.8 Calculate the difference in the Fourier transform of two functions generated from each other by an arbitrary shift in time.

(Purpose of exercise: gain an understanding of the spectroscopic meaning of a time shift.)

2.9 Show that the FT of the intensity $I(t) \propto |E(t)|^2$ for an asymmetric exponential decay $E(t) = A\mathrm{e}^{-\gamma t}$ is different from the power spectrum $S(\omega)$.

(Purpose of exercise: to prove the difference between the two quantities.)

2.10* Calculate the Fourier transform for the exponential decay $E(t) = A\mathrm{e}^{-\gamma t}$ for $t \geq 0$, $E(t) = 0$ otherwise and show that the inverse Fourier transform gives the correct time function.
Hint: use integration in the complex plane for the inverse transformation.

(Purpose of exercise: to prove that the full details of the time spectrum are retained when the inverse transform is taken.)

3. Light Sources with General Application

The light sources initially used in spectroscopy were tungsten lamps or more generally glowing solids and gas-discharge lamps. The emission of glowing solids is based on Planck's and Kirchhoff's law, whereas in gas-discharge lamps the radiation from characteristic transitions of valence electrons is used as well. Thus, in the latter a high density of radiation can be obtained in a narrow frequency range. Recently, the broad-band radiation emitted from synchrotron sources has been increasingly employed in various fields of spectroscopy. On the other hand, the rapid development of various types of lasers, particularly those relying on emission from semiconductor diodes, has opened up many new spectroscopic techniques such as laser ultraviolet or far-infrared applications, frequency tuning, or sub-picosecond resolution spectroscopy.

In this chapter we will discuss light sources with applications to different spectroscopic experiments. These sources may be classical, like black-body radiation and gas-discharge lamps, or more advanced, like synchrotron radiation and lasers. Radiation sources like microwave emission or x-ray and γ-ray sources with application to special spectroscopic methods will be discussed in the respective chapters.

3.1 Black-Body Radiation and Gas-Discharge Lamps

Black-body radiation is the oldest man-made light source and was used in spectroscopy from the beginning. Basically it originates from electrons which are statistically accelerated and decelerated by collisions and thus represent a system of emitting charges like the ones we discussed it in Sect. 2.2.2 (and in Appendix B.5). For the purpose of this text it is sufficient and more convenient to use a classical resonator model and some simple results from statistical thermodynamics to obtain a good description of black-body radiation. Within this approximation the radiation emitted is determined from the density of the transverse electromagnetic eigenmodes $n(\omega)$ of a black-body radiator.

We consider the momentum p of a photon and a space element $\mathrm{d}x$. From statistical physics the number $\mathrm{d}N$ of distinguishable states in the volume given by dp and dx is dpdx/h. This yields for three dimensions and for an isotropic distribution of the states

$$\mathrm{d}N = 2\frac{4\pi p^2 \mathrm{d}p\mathrm{d}^3x}{h^3} = \frac{p^2\mathrm{d}p\mathrm{d}^3x}{\pi^2\hbar^3} . \tag{3.1}$$

The factor 2 in front of the fraction takes care of the two possible states of polarization. Expressing the momentum of the photons by their frequency as $p = \hbar\omega/c_0$ the differential density of photon states per unit volume is derived from (3.1) as

$$\mathrm{d}n(\omega) = n_\omega(\omega)\mathrm{d}\omega = \frac{\omega^2\mathrm{d}\omega}{\pi^2 c_0^3} . \tag{3.2}$$

Multiplication of this quantity with the energy of the photons $\epsilon = \hbar\omega$ and the thermodynamic probability for the occupation of the state yields the well-known density of radiation of a black body. The occupation is given by the Bose–Einstein distribution $f_\mathrm{E}(\omega, T)$

$$f_\mathrm{E}(\omega, T) = \frac{1}{\exp(\hbar\omega/k_\mathrm{B}T) - 1} . \tag{3.3}$$

From this, the radiance (or brilliance) $L_\omega(\omega)\,\mathrm{d}\omega$ of radiation emitted per unit area to the outside of the black body, into a solid angle of 1 steradian in the frequency interval dω is given by Planck's radiation law

$$\begin{aligned} L_\omega(\omega, T)\,\mathrm{d}\omega &= \frac{c_0}{4}\frac{1}{2\pi} n_\omega(\omega)\hbar\omega f_E(\omega, T)d\omega \\ &= \frac{\omega^2}{8\pi^3 c_0^2}\frac{\hbar\omega\,\mathrm{d}\omega}{\exp(\hbar\omega/k_\mathrm{B}T) - 1} . \end{aligned} \tag{3.4}$$

$L_\omega(\omega, T)$ is expressed in Wm^{-2}ster^{-1}s. Note that the emission is into a total angle of 2π (not 4π)[1]. The total energy emitted per second, square meter, and steradian is obtained by integration over ω. The result is the Stefan–Boltzmann radiation law

$$L(T) = \frac{1}{2\pi}\left(\frac{\pi^2 k_\mathrm{B}^4}{60\hbar^3 c_0^2}\right)T^4 \qquad (\text{in W/m}^2\text{ster}) . \tag{3.5}$$

The expression in parentheses is the Stefan–Boltzmann radiation constant R with the value $R = 5.67 \times 10^{-8}\,\mathrm{W/m^2K^4}$.

Equation (3.4) is represented graphically in Fig. 3.1 where the temperature is the parameter for the curves, and the wavelength is used instead of the angular frequency as the abscissa. Note that the maximum in plotting (3.4) versus temperature for a certain frequency range does not appear for the same temperature as steradian is if the equation is plotted for a certain range in wavelengths.

[1] The factor 4 in the denominator comes from an integration over the solid angle outside the black body [3.1].

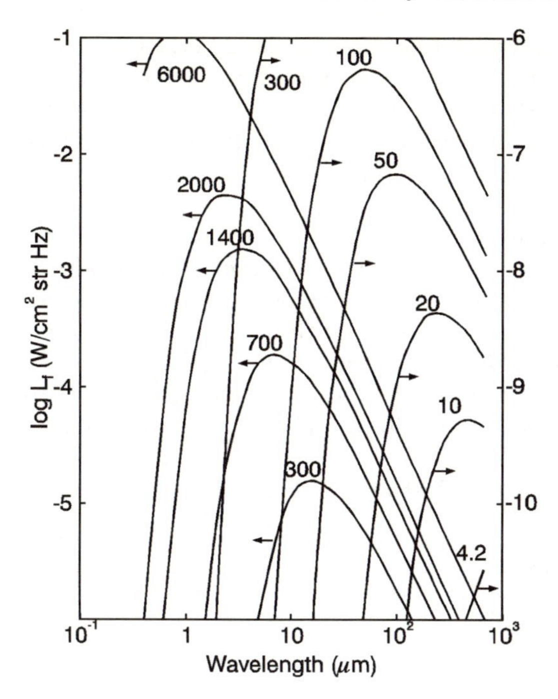

Fig. 3.1. Logarithm of black-body radiation power L_f per cm^2 steradian and 1 Hz bandwidth for various temperatures, as indicated in K.

The graphs in Fig. 3.1 are typical emission characteristics for glowing solids like tungsten filaments. Tungsten-filament lamps are particularly useful as light sources in the visible and near IR spectral range if a broad-band emission is required. To increase the lifetime of high-power lamps small amounts of halogen gases are often added to the low-pressure noble gas filling. Such lamps are known as tungsten-halogen filament lamps. The filament temperatures can be as high as 3300 K.

Gas discharges also emit a continuous spectrum similar to that shown in Fig. 3.1 because electrons are accelerated and decelerated by scattering processes in the gas plasma. The advantage of gas-discharge sources compared to solid sources is the possibility of reaching higher temperatures and using a smaller size for the emitting element. For example, the temperature in a mercury-arc lamp can be as high as 6000 K for a plasma pressure of 1–2 atm. This yields a very strong emission in the visible spectral region and a color temperature close to sunlight. The emitting arc in a Xe arc discharge lamp can be as small as (0.3×0.3) mm^2. If the pressure and temperature are very high broadening and overlapping of electronic states occurs in the excited atoms which results in a broad-band light emission. Whereas the efficiency given as the ratio between output light power in the visible spectral range to input electrical power does not exceed 10% for filament lamps, gas-discharge lamps can have efficiencies of up to 50% if their glass compartment is supplied with a phosphorescent layer. If gas-discharge lamps are used in the UV region, the covering glass must be quartz to allow transmission of the UV light.

Gas-discharge lamps filled with mercury, hydrogen or helium are employed for spectroscopy in the UV. In this case the radiation from characteristic transitions of the atoms is often used as it has a quite high intensity in a narrow spectral region.

High-power gas-discharge lamps (arc discharge) can operate at gas pressures as high as 100 bar (10 MPa) and need therefore an extra electrode for ignition. If the high gas pressure is only established as a consequence of the heat release after start up, ignition is possible with an extra electrode for glow discharge. If the pressure is high from the beginning ignition must be obtained with an extra electrode connected to a high-voltage pulse. Lamps up to 1000 W with a brightness up to 250 W/cm^2 ster are commercially available.

Models of gas-discharge lamps are illustrated in Fig. 3.2.

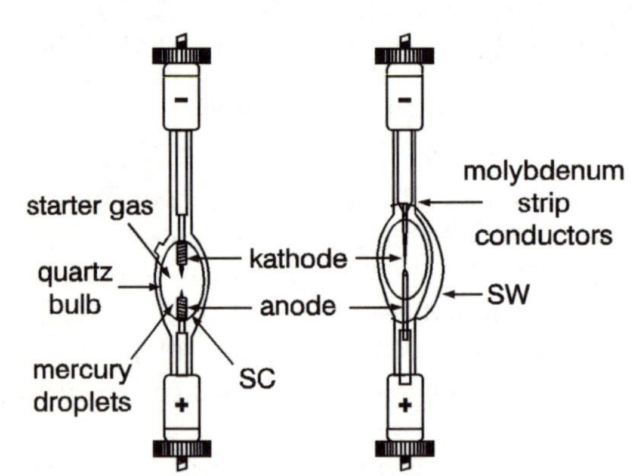

Fig. 3.2. Gas-discharge lamps for UV radiation; (SW: starter wire, SC: starting coil).

Note that ultra-violet radiation is very dangerous for the eyes and for the skin! When working with such sources absorbing goggles and skin protectors should always be worn. In addition, UV radiation below 250 nm generates toxic ozone from the oxygen in the air. This means that these types of high-power lamps must be well vented when used in closed rooms. Finally, lamps operated at very high pressure can be an explosion hazard.

Figure 3.3 compares the emission spectrum from a high-power filament lamp with spectra from a mercury and a xenon lamp. The irradiance I_λ in μW per cm^2 and bandwidth is plotted for a distance of 50 cm. Note the particularly strong emission for the gas-discharge lamps in the UV spectral region.

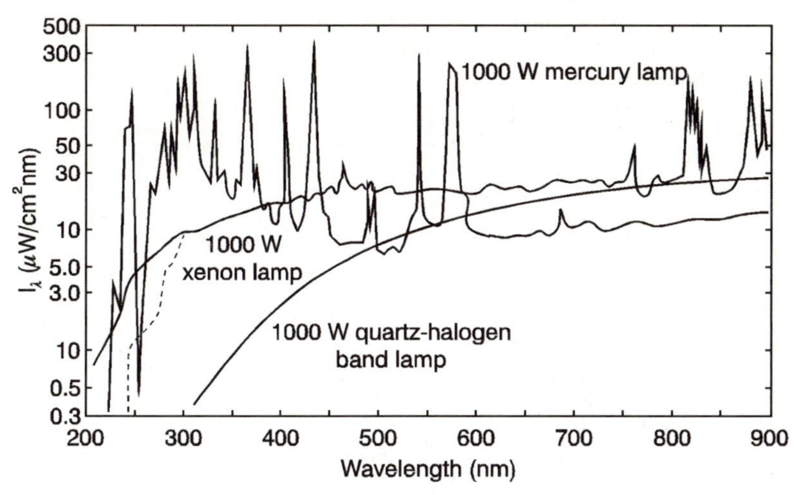

Fig. 3.3. Spectral distribution of irradiance I_λ for commercial high-power light sources. The dashed line labels the cut off in air due to ozone generation.

3.2 Spectral Lamps and Shape of Spectral Lines

Gas-discharge lamps can also be built to predominantly emit spectral lines. The most common line shapes for these emissions and spectral line shapes, in general, will be discussed in the following section.

3.2.1 Low-Pressure Spectral Lamps

To obtain spectral lines a low gas pressure is used and the lamps are operated in the glow-discharge region without ignition. Sources with slightly higher gas pressure are known as spectral lamps and need filament heating to provide enough electrons for start up. Spectral lamps are very important in spectroscopy for wavelength calibration. They are absolute standards. Typical linewidths are fractions of one Å. Usually, the lamps are filled with noble gases like Ne, Ar, Kr, and Xe or special metal vapors. The emission lines for these gases are well known and tabulated. A recommended source for spectral lines is *Landold-Björnstein* [3.2] and references therein. Some of the most important spectral lines are listed in Table 3.1.

3.2.2 Shape of Spectral Lines

Spectral lines are examples of a radiation field with a rather narrow frequency spectrum. The radiation originates from electrons returning to the ground state after having been excited into a higher orbital. The recombination to the ground state occurs after a certain lifetime and with a certain transition probability. Formally the spectral light may be considered as the emission

Table 3.1. Selected spectral lines in nm in the visible and near-visible spectral ranges. The numbers in parentheses are relative intensities. L indicates lines appropriate for stimulated emission.

Neon	Argon	Argon$^+$	Krypton	Krypton$^+$	Xenon	Mercury
	..					404.66
	415.86		427.40	417.18		434.75
	420.07	434.80(50)	431.96			435.84
	433.36	454.50(25L)	437.61		462.43	
		457.93(25L)	445.39	468.04(L)	473.42	
		465.79(25L)		469.44		
533.08		487.98(30L)		482.52(L)	480.70	
540.06		506.20(30)	557.03	484.56		546.07
585.25		514.53(25L)		501.65		576.96
621.73		617.23(40)		520.83(L)		
633.44	696.54	664.37(100)		521.79		
				647.09(L)		
717.39	727.29	663.82(50)	758.74	752.55		
753.58	763.51	668.43(50)	819.01	799.32	823.16	

from a damped (elementary) oscillator. The intensity spectrum of this process was described in Sect.2.4 and had the form of a Loretzian line. Since this line is fundamental for emission or absorption it is also called the *natural* line. Normalized to unit area in ω space, it has the form

$$I_{\mathrm{L}}(\omega-\omega_0)=\frac{\gamma/2\pi}{(\omega-\omega_0)^2+(\gamma/2)^2}\,. \tag{3.6}$$

γ is the FWHM of the line. A graph of a Lorentzian line is depicted in Fig. 3.4a. The dots correspond to the emission from a krypton laser.

The Lorentzian line is not the only line shape observed in experiments. If, for example, all emitting oscillators do not have exactly the same energy but rather are statistically distributed in energy, the emission line is statistically broadened. A typical example is a Gaussian distribution of emitters which may be due to the thermal motion of atoms or molecules. In this case the spectral line has the form

$$\begin{aligned} I_{\mathrm{G}}&(\omega-\omega_0)\\ &=\frac{2\sqrt{\ln 2}}{\gamma\sqrt{\pi}}\exp\frac{-\ln 2(\omega-\omega_0)^2}{(\gamma/2)^2}\\ &=\frac{1}{\sigma\sqrt{2\pi}}\exp\frac{-(\omega-\omega_0)^2}{2\sigma^2}\qquad\text{with}\qquad \sigma^2=\frac{\gamma^2}{8\ln 2}\,, \end{aligned} \tag{3.7}$$

where γ is again the FWHM of the line and (3.7) is normalized to one. σ^2 is the second moment (with respect to the mean value ω_0) or the variance

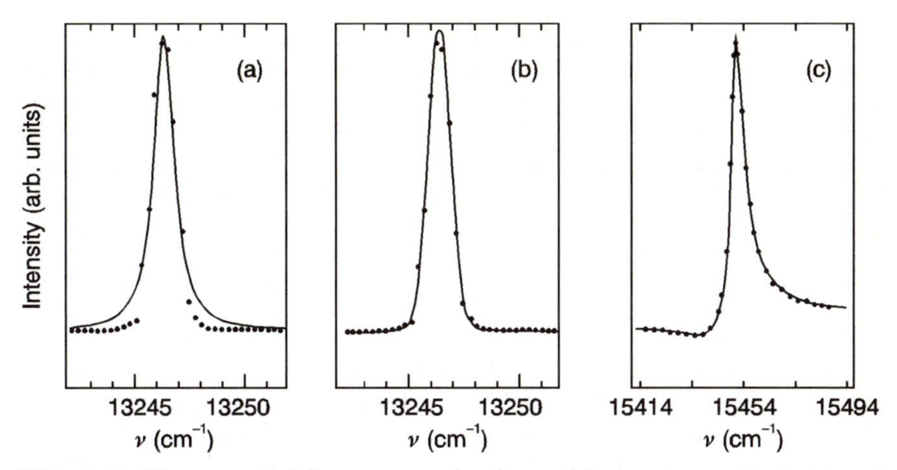

Fig. 3.4. Three spectral lines commonly observed in spectroscopic experiments. Lorentzian line (**a**), Gaussian line (**b**), Fano line (**c**). The dots in (**a**) and (**b**) are experimental results representing the emission from a laser. The dots in (**c**) are results from light-scattering experiments on GaAs with $\nu_0 = 15\,451$ cm^{-1} [3.3].

of the Gaussian distribution. Since the velocities of the ions in an emitting gas plasma have a Maxwellian distribution the spectral lines are very often Gaussian due to the Doppler effect. Even though in solids the atoms cannot move, Gaussian profiles are quite usual. They are due to perturbations of the energy levels of the electrons by lattice defects. A normalized Gaussian line is shown in Fig. 3.4b. For equal FWHM the Gaussian lines converge more rapidly to zero with increasing distance from the center as compared to the Lorentzian lines. The experimental results of the krypton laser undoubtedly match a Gaussian profile better than a Lorentzian profile.

Besides Lorentzian and Gaussian lines other line shapes are known and often used. An example is the Voigtian profile which represents a combination of a Gaussian and a Lorentzian line (Appendix C.2). Observed lineshapes and linewidths may originate from experimental conditions, and may not necessarily represent the shape and width of the line from the original source. For example, the instrumental response $G(\omega)$ (i.e., the spectrum obtained from the instrument for a strictly monochromatic input spectrum) or, in a scattering experiment, the width of the line used for excitation may distort or broaden the original line. In this case a deconvolution process is necessary to obtain the intrinsic line shape. In the simplest case of a combination of two Lorentzian lines or two Gaussian lines of widths γ_1 and γ_2 the resulting lines are again Lorentzian or Gaussian. More details about the convolution process are given in Appendix C.2.

For systems which consist of one or several discrete oscillators close to (or on top of) a continuum transition, interference effects may occur and lead to special emission characteristics known as *Fano lines*. These lines have an asymmetric shape, as shown in Fig. 3.4c. The experimental dots are from a light-scattering experiment. Fano lines can be observed for light emission, light absorption, and light scattering. A necessary condition for their occurrence is a coupling between the states responsible for the discrete and continuous transitions. A formal description of the simplest case of a Fano line (one discrete and undamped oscillator, energy independent coupling, and energy independent continuum transition strength) is obtained (for $\Gamma, Q \neq 0$) from

$$I_{\mathrm{F}}(\omega - \omega_0) = C \frac{(Q + \bar{\epsilon})^2}{1 + \bar{\epsilon}^2} \qquad \text{with} \qquad \bar{\epsilon} = \frac{\omega - \omega_0 - \Delta}{\Gamma/2} . \tag{3.8}$$

Δ describes the shift of the line position for the discrete transition from its position without interaction, and Γ is the damping (reciprocal lifetime) due to the interaction U and given as $\Gamma = 2\pi U^2$. Q is inversely proportional to U normalized by the relative strength of the (modified) discrete oscillator compared to the strength of the continuum transition. In Fig. 3.4c $\omega_0 + \Delta$ is used instead of ω_0, for simplicity, and the abscissa is scaled in wave numbers. Depending on the sign of Q the minimum of the intensity can appear to the left or to the right of the peak. More details on the Fano effect are given in Appendix C.3.

3.3 Synchrotron Radiation

Synchrotron radiation is a special type of radiation which has recently become readily available and has found many useful applications. Spectroscopically it is very similar to black-body emission. With respect to the geometry of the beam it is very different from the latter. Synchrotron radiation is widely applied today as a powerful UV and x-ray source for structural analysis, photoelectron spectroscopy, or photo lithography. Special sources have even been designed for IR spectroscopy.

3.3.1 Synchrotron Light Sources

Synchrotron radiation is the electric field emitted from charged particles in circular accelerators like synchrotrons or storage rings. The acceleration characteristics determine the spectral nature of the radiation. In synchrotrons the energy of the particles changes continuously, which is not very convenient for spectroscopy. It is more appropriate to store a bunch of charges in an extra ring and let them circulate with a constant velocity. In these storage rings the particles are accelerated by purely radial forces. This guarantees a very stable light source.

Synchrotrons and storage rings were originally developed for experiments in high-energy physics. At that time synchrotron radiation was considered a waste product, as it was responsible for a continuous and non negligible energy loss of the particles. To keep the particles circulating this energy must be provided by an accelerating ac electric field. Today many rings are specially built as a light source and are *dedicated* to spectroscopic experiments. Practically all these rings use electrons as the charge carrying particles and a storage ring as the region where particles circulate. Only such systems will be considered here.

The advantage of synchrotron radiation for spectroscopy originates from the following characteristic properties:

- very high emission density from a small spot,
- very small beam divergence, such as for lasers,
- very large bandwidth with the central position tunable by the particle energy,
- highly polarized radiation,
- very short light pulses.

A synchrotron light source has the following three main components: a particle source with a linear accelerator, a synchrotron for the circular acceleration of the electrons to relativistic velocities and a storage ring where the electrons circulate as a guided beam with several 100 MeV or several GeV. Figure 3.5 is a schematic of a synchrotron light source. On the left is the synchrotron where the electrons from the source are first linearly accelerated to several MeV and then pumped in the circular region to the final energy up to several GeV. Data given in the figure are typical for a medium-size light source.

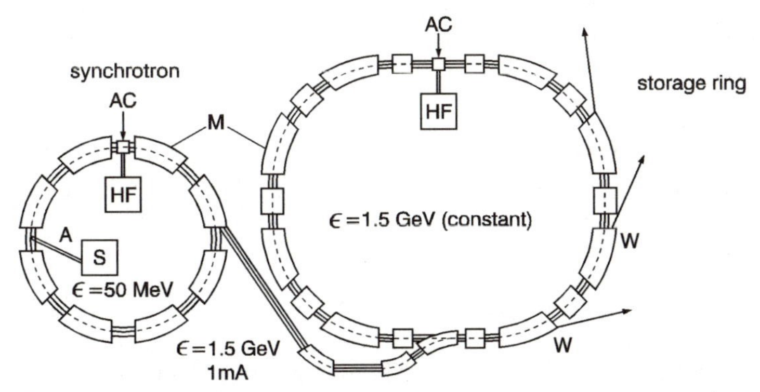

Fig. 3.5. Schematic view of a synchrotron with storage ring. (S,A: electron source with linear accelerator, HF: high frequency power supply, AC: acceleration facilities, M: magnets, W: windows).

The storage ring consists of a sequence of curved and straight magnetic lenses. It is evident from this constructive detail that the radius of curvature and the circumference of the ring are not directly related. Synchrotron radiation emitted in the curved parts is allowed to exit from the tube trough a window. Special components like wigglers or undulators may be accommodated in the straight sections. Large synchrotron sources, such as the *European Synchrotron Radiation Facility* (ESRF) in Grenoble or the *National Synchrotron Light Source* (NSLS) in Brookhaven, can have more than 30 windows and several special facilities. It is also possible for a single synchrotron to support more than one storage ring as is the case in Brookhaven.

Since the circulating electrons must be pumped continuously to balance the radiation losses they have to arrive with a well defined phase at the acceleration electrodes. Thus, they can only circulate in bunches. These bunches can be accumulated to provide a current of several 100 mA which may circulate for many hours in the storage ring. Eventually, as more and more electrons get lost, the electron beam dies out and the storage ring must be refilled. Beam lifetimes from 10 to 50 hours are not uncommon. Typical widths of the bunches are 100 to several 100 ps and the circulation time is of the order of milliseconds. Since the velocity of the particles is always close to c_0 the circulation frequency f_0 is c_0/l, l being the circumference of the electron track. Note that several bunches can circulate simultaneously.

3.3.2 Generation and Properties of Synchrotron Radiation

The emission of radiation by an arbitrarily accelerated charge was discussed in Sect. 2.2 and in Appendix B.5. The formulas derived there are, however, limited to non-relativistic particles. The extension to a relativistic description is not straightforward. Without extensive calculations only a relationship for the total emitted power can be derived from Larmor's formula in Sect. 2.2 by introducing relativistic expressions for time, momentum, mass, and energy. With

$$\beta = v/c_0 \,, \qquad \gamma = \frac{1}{\sqrt{1-\beta^2}} \,, \qquad \epsilon = \gamma m_0 c_0^2 \,,$$

where ϵ is the total relativistic energy, the total emitted power for highly relativistic particles turns out to be

$$\boxed{P = \frac{e^2 c_0 (\beta\gamma)^4}{6\pi R^2 \varepsilon_0} = \frac{e^2 (\beta\epsilon)^4}{6\pi m_0^4 c_0^7 R^2 \varepsilon_0} \qquad \text{(in W)}\,,} \tag{3.9}$$

where R is the radius of the bending magnet. Thus, for $\beta \approx 1$ the total energy emitted increases with the 4-th power of the particle energy ϵ.

From (3.9) the input power can be estimated which is needed to balance the emission. For a 2-GeV synchrotron with a radius of $R = 5.5\,\mathrm{m}$ the emission is 3.5×10^{-7} W/electron. If the storage ring is operated with 1 A the number N of circulating electrons is $N = 2\pi R/1.6 \times 10^{-19} c_0$ and hence the total emitted power

$$P_{\mathrm{tot}} = \frac{10^{19} \times 3.14 \times 11}{1.6 \times 3 \times 10^8} 3.5 \times 10^{-7} = 0.25\,[\mathrm{MW}]\,.$$

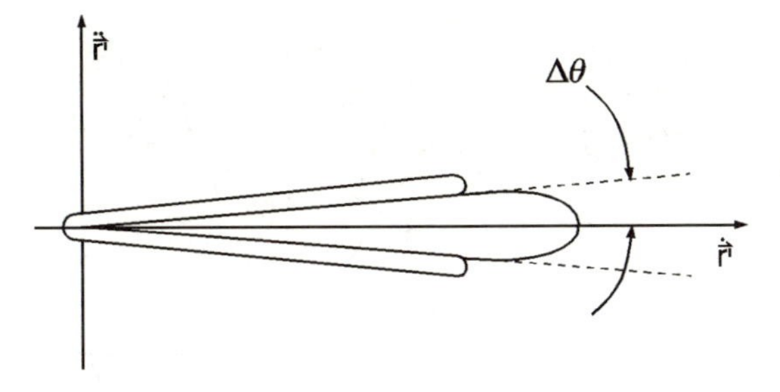

Fig. 3.6. Emission characteristic of a radially accelerated charge with relativistic speed.

The emission pattern differs radically from that for a non-relativistic particle shown in Fig. 2.3. Due to the relativistic Doppler effect, it is highly directional with a strong peak in the forward direction and the polarization is strictly confined to the plane defined by $\boldsymbol{a} = \mathrm{d}^2\boldsymbol{r}'/\mathrm{d}t^2$ and $\boldsymbol{v} = \mathrm{d}\boldsymbol{r}'/\mathrm{d}t$. Figure 3.6 exhibits the characteristic emission pattern of a radially accelerated relativistic particle. As in the case of non-relativistic propagation in Fig. 2.3 polar coordinates have been used. The direction of particle propagation is the polar axis. The figure demonstrates the characteristic difference for the two types of radiation. For a relativistic particle nearly all the radiation is emitted forward whereas for a non-relativistic particle forward and backward emission are almost the same. Note that the diagram does not have rotational symmetry any more. Symmetry is only retained with respect to the plane of the storage ring and the sagittal plane normal to it.

The angular width (half width half maximum) of the emitted radiation can be estimated from

$$\boxed{\Delta\theta \approx \frac{1}{\gamma} = \frac{m_0 c_0^2}{\epsilon}\ .} \tag{3.10}$$

For a 1 GeV electron the angular spread is only 0.5 mrad which corresponds to a linear aperture of 10^{-3}. The intrinsic angular spread is, of course, only observable perpendicular to the circulation plane because emission is smeared out for observation in the plane.

The tangential emission discussed above and its consequences for the detection of light are depicted in Fig. 3.7. The left part of the figure shows the angular spread of the emitted light for directions normal and parallel to the electron orbit. The right part demonstrates the limited time interval during which the emission can be detected by an observer.

One of the big advantages of synchrotron light is its extremely small emission area. This area is determined by the intrinsic width of the electron beam given by h_h and h_v, and by diffraction limits. The area may be evaluated from the dimensions H_h (horizontal width) and H_v (vertical width) both given by

$$H = \sqrt{h^2 + \left(\frac{\lambda}{\theta}\right)^2 + \left(\frac{R\theta^2}{8}\right)^2}\ , \tag{3.11}$$

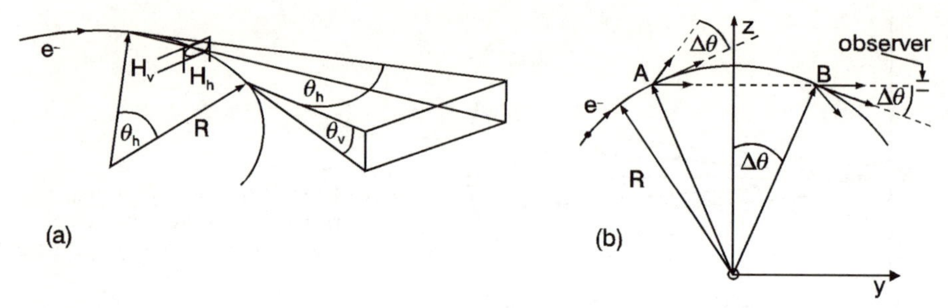

Fig. 3.7. Tangential light emission for electrons circulating with relativistic velocities (**a**) and resulting observation geometry for the relativistically chirped light pulses (**b**).

where R is the radius of the electron orbit and θ is the full angle of emission either in plane (h) or perpendicular to the plane (v) of the orbit. The *brightness* or *brilliance* L of the source, defined as emitted light in Joule (or number of photons) per second, area, and solid angle depends on the area $H_{\mathrm{h}} \times H_{\mathrm{v}}$. Typical values for $h_{\mathrm{v}} \times h_{\mathrm{h}}$ are are $0.15 \times 0.03\,\mathrm{mm}^2$ if the source is dedicated.

As already mentioned the emitted light is strongly polarized with the direction of polarization in the plane of electron motion. This holds only for emission precisely in the plane of the ring. With increasing angle Θ_{v} the emission becomes elliptically and finally circularly polarized.

The strongly forward peaked emission pattern also determines the time of observation of the radiation. Figure 3.7b indicates that the emission is only observed for the short time during which the electron bunch passes between the limiting positions A and B on their way around the storage ring. The locations of A and B are given by the direction of observation. The observation time is further reduced by a relativistic chirping effect. From the geometry in Fig. 3.7b the time spread Δt is given by the angular width of the emission. It can be estimated from

$$\Delta t = t_{\mathrm{e}} - t_{c_0} \approx \frac{2R}{\gamma v} - \frac{2R\sin(1/\gamma)}{c_0} = \frac{4R}{3c_0\gamma^3}\,, \tag{3.12}$$

where t_{e} and t_{c_0} are the time intervals which the electrons and the light need to travel from A to B. This interval corresponds to a typical frequency

$$\boxed{f_{\mathrm{typ}} = \frac{1}{\Delta t} = \frac{3c_0\gamma^3}{4R} = \frac{3\epsilon^3}{4Rm_0^3c_0^5}\,.} \tag{3.13}$$

For an electron with energy 1 GeV and a ring radius of 3 m this yields a frequency of $\approx 6 \times 10^{17}$ Hz or a radiation energy of 2 KeV.

The corollary of the short light pulses is a broad frequency spectrum. The first detailed calculation of this spectrum was reported by *Schwinger* [3.4]. According to [3.5] it can be described for one emitting electron by

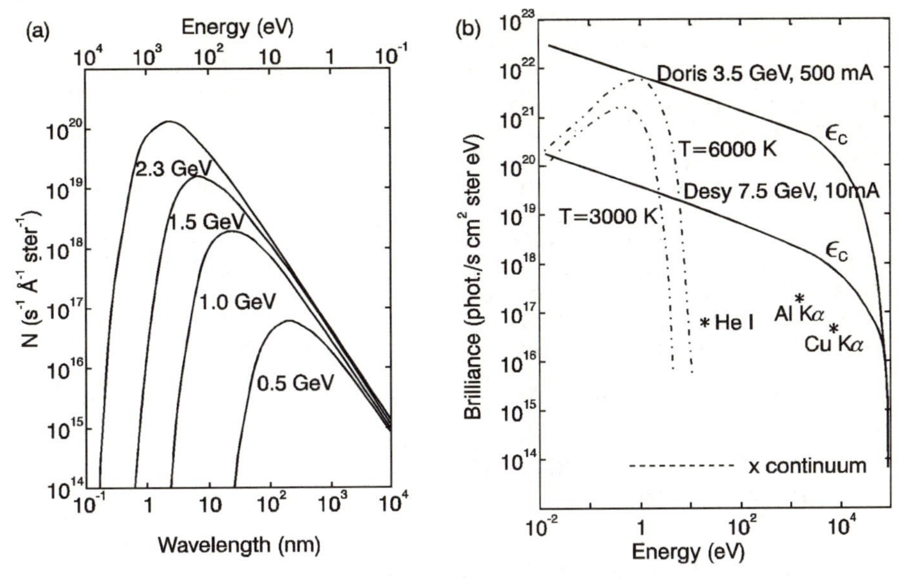

Fig. 3.8. Spectral energy distribution of synchrotron radiation for various particle energies (**a**) and comparison of brightness with conventional light sources (**b**); (—··) black body, * x-ray and VUV lines, (— — —) x-ray continuum, (—) storage rings; after [3.6].

$$P_\lambda(\lambda) = \frac{3^{5/3} e^2 c_0 \gamma^7 \lambda_c^3}{64\pi^3 R^3 \varepsilon_0 \lambda^3 10^6} \int_{\lambda_c/\lambda}^{\infty} \mathrm{K}_{5/3}(x)\mathrm{d}x \qquad (\text{in W/nm}) , \tag{3.14}$$

where $\mathrm{K}_{5/3}$ is a modified Bessel function, and

$$\lambda_c = \frac{h c_0}{\epsilon_c} = \pi \frac{4R}{3\gamma^3} \tag{3.15}$$

is a characteristic wavelength which determines the high-energy cut off of the spectrum. A plot of the emission normalized to 1 steradian is displayed in Fig. 3.8a for 1-Å bandwidth. The real advantage of synchrotron radiation is the high power in the VUV and x-ray spectral ranges and the small size of the emitting area. Therefore, the brightness should be used to compare it with conventional broad-band or x-ray sources. In Fig. 3.8b the spectral brightness (brightness per unit bandwidth) of a medium-size synchrotron source is compared with black-body radiation and with several narrow-band and continuous x-ray sources. The advantage of the synchrotron light is obvious, in spite of the rather large bandwidth of 1 eV used in the comparison. (The width of characteristic x rays is typically between 0.5 and 0.8 eV). Table 3.2 lists several well known sources used for synchrotron radiation experiments. Synchrotrons of the first generation were not dedicated light sources. Second- and third-generation synchrotrons are dedicated. Third-generation synchrotrons have special facilities like wigglers and undulators. Also, the

Table 3.2. Sources of synchrotron radiation.

Name	Location	Gen./Year	ϵ [GeV]	I [mA]	R [m]	ϵ_c [KeV]
PETRA	Hamburg	1/1978	6–13	90	200	75
PHOTON FACTORY	Tskuba	2/1981	2.5	500	8	4.3
ESRF	Grenoble	3/1992	6	175	135	200
DAPS	Daresbury	2/1980	2.0	1000	5.56	3.2
NSLS I	Brookhaven	2/1987	0.75	1000	8 (2)	0.4
NSLS II	Brookhaven	2/1987	2.5	1000	27(8)	4.3
BESSY II	Berlin	3/u.const.	1.5–2			
ELETTRA	Trieste	3/1993	1.5–2	400	41(5.5)	3.2
APS(ANL)[a]	Argonne	3/1993	7			

[a] The synchrotron at the Argon National Laboratory uses positrons as charge carriers.

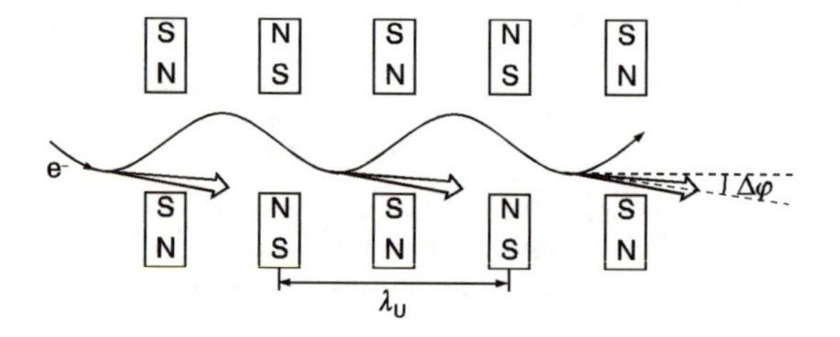

Fig. 3.9. Schematic arrangement of a wiggler.

storage-ring radius given is an average value which is no longer related to the particle bending path by the magnets because of the numerous linear segments along the track. (Bending radii are given in parentheses in column R[m]). The cut off energy is the cut off at the special facility (wiggler, undulator) with the highest energy.

3.3.3 Special Synchrotron Facilities

Considerable enhancement in the intensity of synchrotron light can be obtained by special arrangements for the extraction of the light beam. One possibility is presented in Fig. 3.9. The electron beam passes several alternating magnets on a straight section of the storage ring. As a result radiation is emitted at several positions and can be accumulated to form a beam with N-fold intensity for N consecutive alternating magnets. This arrangement is called a *wiggler*. If N is large the emission from the individual sections of the wiggler can be coherent, and the setup operates as an *undulator*. In this case the peak intensity is proportional to N^2. Values of N higher than 30 are possible. The length of the facilities are 1 to 2 meters. Since the undulator operates like an optical grating, interference occurs in certain directions correlated with particular wavelengths. Thus, radiation energy is accumulated at a particular wavelength with a very narrow bandwidth. As such it is similar to light emitted from laser oscillators, as will be discussed in Sect. 3.4.

Constructive interference is obtained for light of wavelength λ_n and direction $\Delta\varphi$ from the electron track if

$$\lambda_n = \frac{\lambda_U}{2\gamma^2 n}\left(1 + \frac{K^2}{2}\gamma^2\Delta\varphi^2\right) . \tag{3.16}$$

λ_U is the undulator period, n the overtone number, K the undulator parameter given as $e\lambda_U B_U/m_0 c_0$, and B_U the magnetic induction of the undulator magnets.

Figure 3.10 compares the emission from two magnets with the emission obtained with a 10-pole wiggler and several undulators at the ESRF.

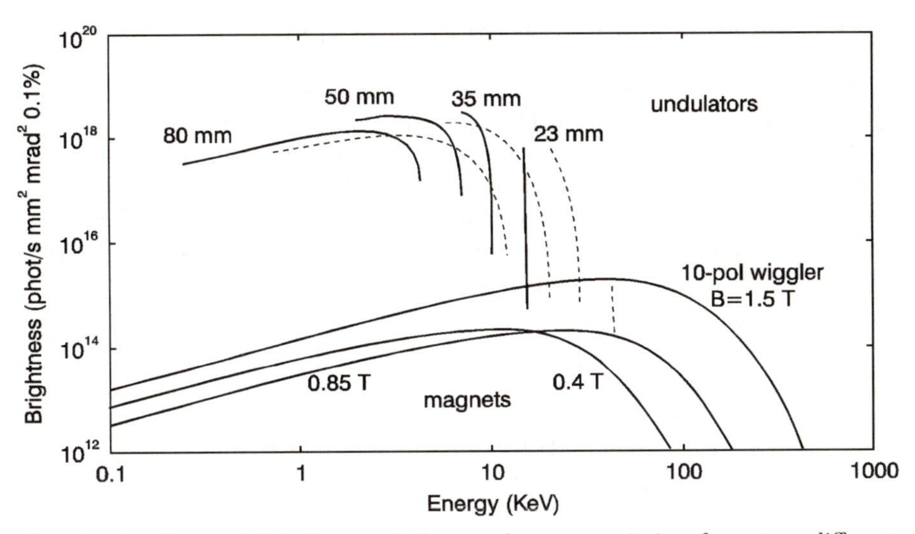

Fig. 3.10. Spectral brightness of the synchrotron emission from two different magnets compared to the emission from a wiggler and several undulators at the ESRF. The dashed lines represent third harmonic emission from the undulators. The period of the undulators is given in mm; after [3.7].

3.4 Lasers as Radiation Sources

Lasers are alternative light sources to tungsten-filament and to gas-discharge lamps. The name laser is an acronym of *light amplification by stimulated emission of radiation*. The principle for coherent amplification of electromagnetic waves was first discovered theoretically by C.H. Townes, N.G. Basov, and M. Prokhorov who received the Nobel prize in 1964 for their work. In 1958 A.I. Shawlow showed how to extend this principle to the optical spectral region and also received a Nobel prize in 1981, together with N. Bloembergen. T. Maiman demonstrated the first successful operation of an optical laser in

1960. Since then, applications in spectroscopy have been increasing continuously where lasers are used as the preferred light source. Initially this was only possible in the visible spectral range for a few selected lines and for some lines in the IR. Now applications have expanded far beyond this range, a very large number of lines is available and significant parts of the spectrum are covered by tunable lasers. In addition, spectral linewidths have been reduced to several hundred Hertz and pulsed lasers are available with pulse lengths in the femtosecond range. The most important properties of these lasers with particular emphasis to applications in solid-state spectroscopy will be reviewed in the following sections.

3.4.1 Generation and Properties of Laser Radiation

A laser consists basically of an "optical resonator", a "pump source", and an "optically active medium". The optically active medium is established by an "inversion" of population of electronic states. This means an electronic state 2 with energy ϵ_2 is populated with a higher number of electrons than an electronic state 1 with a lower energy ϵ_1. In some sense optical activity is already obtained if the ratio of the two population numbers N_2 and N_1 is larger than its value in thermal equilibrium.

$$\frac{N_2}{N_1} > \left(\frac{N_2}{N_1}\right)_{\mathrm{equ}} = \exp\left(-\frac{\epsilon_2 - \epsilon_1}{K_\mathrm{B}T}\right) . \tag{3.17}$$

To obtain amplification $N_2/N_1 > 1$ must be satisfied. The active material can either be a gas plasma, a crystal with optically active color centers, or the optically active region of a p-n junction in a semiconductor. In all cases the inversion is established by a pump. Figure 3.11 illustrates the three classical arrangements for lasers. In the case of a gas laser, an electrical discharge is responsible for the population inversion. The resonator consists of plane-parallel or confocal highly reflecting mirrors. The reflectivity of one of the mirrors is slightly reduced to 98–99% which allows the laser beam to exit from the cavity. The discharge tube is usually sealed with Brewster windows to minimize

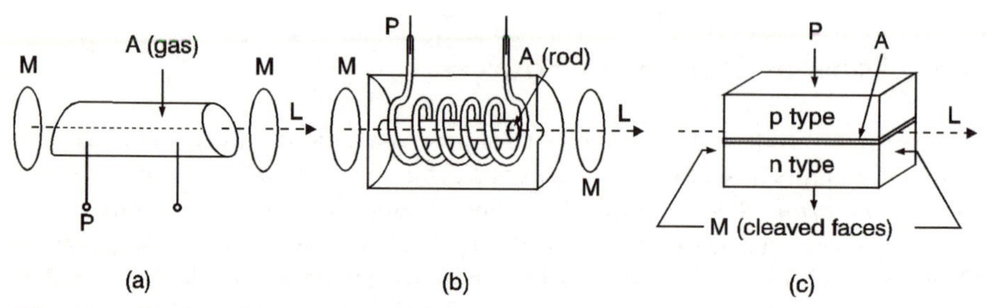

Fig. 3.11. Schematic representation of three classical laser systems; gas-discharge laser (**a**), optically pumped solid-state laser (**b**), and semiconductor laser (**c**); (M: mirrors, A: active material, P: pump).

surface reflection losses and to select one particular light polarization for the laser operation. Optically pumped solid-state lasers operate similarly but the inversion is established by flash lamps. For semiconductor lasers the cleaved faces of the p-n junction crystal are usually directly used as the reflectors of the cavity. The pump is the forward biased current and the active material extends over the volume of the junction where non-equilibrium carriers are generated by injection.

Laser light has the following characteristic properties:

a) *Very small beam divergence.* Typical values for a gas laser are 10^{-4} rad.
b) *Very narrow linewidths.* Special arrangements allow linewidths for gas lasers to be as low as 10 KHz and even less. Solid-state lasers can have linewidths as small as 150 Hz [3.8]. A narrow line is equivalent to a high coherence of the light. Here coherence is a measure of the length over which a well defined phase relation for the wave field exists. A more general and more exact definition of coherence will be given in the next section. Since the linewidth Δf is approximately given by the inverse of the coherence time τ the coherence length Δl for a 10-KHz bandwidth radiation can be estimated from

$$\Delta l = c_0 \tau \approx c_0 / \Delta f \approx 30\,\mathrm{Km}\ .$$

c) *Very high light intensity per bandwidth.* A reasonably good gas-discharge laser can emit 7 W/line with a typical linewidth of 75 MHz in the visible. This represents an intensity of 0.1 W/MHz, as compared to 0.9×10^{-12} W/MHz for a black-body radiator at 3300 K for the same angular width of 10^{-4} rad and for the same cross section of 2 mm^2.
d) If the active medium is terminated by Brewster windows the laser beam is *highly polarized in the sagittal plane*, perpendicular to the plane of the Brewster window.

These characteristic properties of the laser light are due to two important facts of the experimental setup.

a) The stimulated emission of radiation from the activated medium occurs in a very narrow frequency region around a center frequency f_0 determined by a gain coefficient $\alpha(f - f_0)$. (The gain coefficient $\alpha = \mathrm{d}n/\mathrm{d}x$ is defined as the increase in photon concentration n per incident photon and per unit of length.) The spectral distribution of α is given by the line-shape function $g(f - f_0)$. For gas lasers the width of this function is determined by the Doppler broadening of the electronic levels in the gas molecules. For α larger than a certain threshold laser oscillation is possible. This threshold is obtained from

$$2L\alpha - \gamma = 0\ ,$$

where L is the length of the active medium, and γ is a loss coefficient describing all sorts of losses during a round trip of the light. Details of the condition necessary for an effective stimulated emission process are given in Appendix C.4.

b) The resonator system leads to mode selection. This selection refers, in particular, to non-axial modes. Due to multiple reflections in the resonator these modes leave the cavity before they reach their full intensity. For this reason the divergence of the radiation is extremely narrow and, in most cases, diffraction limited.

Since the active medium is a cavity, only standing waves can develop as longitudinal resonator modes. In this case $2L$ must be equal to an integer number of wavelengths, and the distance between two neighboring longitudinal modes with wavelengths λ_1 and λ_2 is easily evaluated from

$$2L = n\lambda_1 = (n+1)\lambda_2$$

to be

$$\boxed{|\delta f| = \frac{c}{\lambda^2}\delta\lambda = \frac{c}{2L}\,.} \tag{3.18}$$

For a medium-size He-Ne laser with a typical cavity length of 75 cm the distance between longitudinal modes is only 200 MHz or $\approx 10^{-2}$ cm^{-1}.

The Doppler broadening of the energy levels is a consequence of the velocity distribution of the atoms in the plasma. The line-shape function and thus also the gain coefficient α may be much broader than the mode distance. Under these conditions several longitudinal modes can oscillate simultaneously. The situation is schematically demonstrated in Fig. 3.12a. Since the Doppler width for a He-Ne laser is typically 1500 MHz a considerable number of longitudinal modes can oscillate simultaneously. These oscillations are fluctuating as the light intensity statistically jumps between the various possible modes. The total spectral width of the laser line is given by the width of the gain curve and not by the width of the individual longitudinal modes.

In order to obtain a laser with a strongly reduced bandwidth, the fluctuations must be suppressed and the oscillations must be restricted to a single

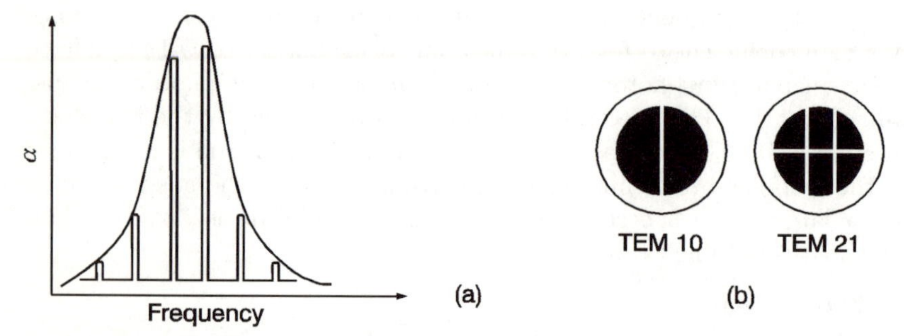

Fig. 3.12. Gaussian gain coefficient α and longitudinal modes (**a**) and field distribution of TEM-modes (**b**) in a laser resonator.

mode. This can be established by inserting an *etalon* filter into the cavity and tuning it to the frequency of one of the longitudinal modes. In this case the inverted population will be released in a single mode. The etalon or Fabry–Perot interference filter consists of a plane plate of high-quality quartz or dynasil 1000 glass. The filter is tuned by tilting it perpendicular to the beam axis. Due to multiple-beam interference in the plate only a very narrow frequency band is viable for each tilt angle. Thus only one longitudinal mode can propagate in the cavity. Etalon filters must be extremely flat (flatness of the order of 0.02 sec of arc) and require a temperature stability of the order of $\pm 0.01°$C. The theory of the Fabry–Perot etalon will be discussed in Sect. 4.3.

The width of the single modes is given by construction details which determine the temperature stability of the resonator size, the vibrational stability of the mirrors, the plasma tube oscillations, etc. Gas-cooled lasers have much narrower lines than water cooled lasers because of the turbulence-induced vibrations in the latter.

In addition to longitudinal modes, a laser also has certain transverse modes of oscillation. These modes are represented by a particular distribution of the electric field over the cross section of the beam and characterized with the symbols TEM_{ik} (transverse electromagnetic modes). Examples of TEM modes are exhibited in Fig. 3.12b. Dark areas characterize parts of the beam with high electric field. The indices i and k count the zeros along the x and y direction, respectively. In general, it is desirable to run the laser in a TEM_{00} mode since the possibilities for focusing are optimum in this case. A TEM_{00} mode can be obtained by a proper construction of the discharge tube or by inserting diaphragms to reduce the beam diameter.

The beam divergence θ is determined by diffraction at the exit mirror. Thus it depends on the beam diameter D and the wavelength λ as

$$\theta = \frac{2\lambda}{\pi D} . \tag{3.19}$$

θ is the half angle of the cone over which the beam intensity has dropped to $1/e$ of its maximum. A consequence of relation (3.19) is the possibility to decrease the divergence θ by increasing the beam diameter with a telescopic system.

An important advantage of lasers is their excellent focusing characteristics which is due to the small beam divergence. For a lens with a given focal length F the beam can be focused to a spot with diameter d and extent l (Fig. 3.13) according to

$$\boxed{d = \frac{4\lambda F}{\pi D} \quad \text{and} \quad l = \frac{16\lambda F^2}{\pi D^2} .} \tag{3.20}$$

For $D = 0.15$ cm, $F = 10$ cm, and $\lambda = 500$ nm, d is 4×10^{-3} cm, which results in an increase of the light intensity by a factor 10^3.

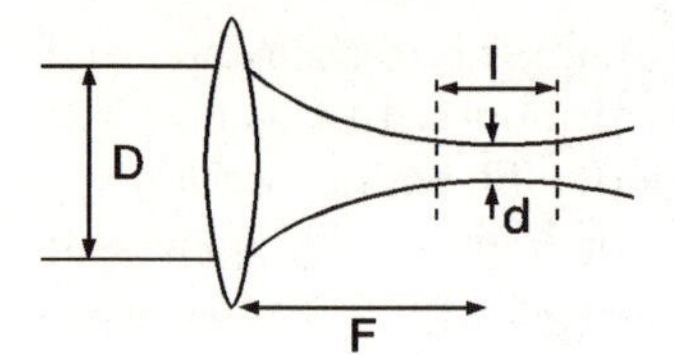

Fig. 3.13. Focusing of a laser beam by a collection optic; (D: beam diameter, F: focal length, d, l: diameter and length of focal area).

The good focusing characteristics of lasers are particularly useful for applications in micro-mechanic machining or surgery. Conversely, they result in serious eye hazards. Intensities smaller than 1 mW are enough to damage the retina because of the very good focusing characteristic of the eye lens.

3.4.2 Continuous-Wave Lasers

The He-Ne laser is a good example to study the principle of population inversion and stimulated emission. The electronic levels involved for the He and the Ne atoms are shown in Fig. 3.14. The active medium of the He-Ne laser consists of a mixture of 100 Pa He and 12 Pa Ne. A large number of He atoms are excited by a gas discharge into the metastable 2^3S and 2^1S states. By collisions with Ne atoms the latter are stimulated and $3s_2$ and $2s_2$ levels become occupied in a resonant energy transfer process. This results in a population inversion with respect to the non-excited Ne levels $2p_4$ and $3p_4$. The transition between the s levels and the p levels of the Ne atoms is dominated by stimulated emission. As indicated in Fig. 3.14, the transitions correspond to the wavelengths 3391 nm, 1152 nm, and 632.8 nm, respectively. Since the occupations of the Ne p_4 levels are short-lived and rapidly decay to the Ne $1s_2$ levels by spontaneous transition, the inversion is maintained at least as long as the gas discharge is on, and laser radiation can be emitted if the amplification factor α becomes large enough. A selection of the eventually

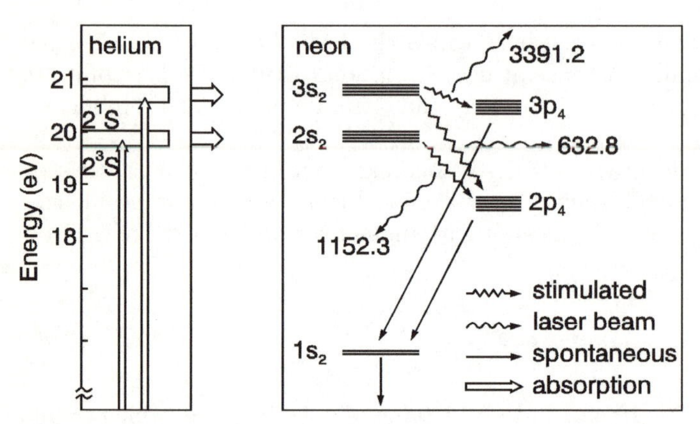

Fig. 3.14. Energy-level scheme and emission processes for the He-Ne laser. The lower case symbols in the scheme for Ne assign excited atomic levels.

Table 3.3. Emission lines for two krypton-ion gas lasers with different power.

Single line			Multiple lines		
Wavelength [nm]	Power [W] Weak	Strong	Wavelength [nm]	Power [W] Weak	Strong
799.3	0.03	1.2	752.5–799.3	0.25	3.8
752.5	0.1	3.0			
676.4	0.12	2.5	647.1–676.4	0.6	10.0
647.1	0.5	7.0			
568.2	0.2	2.5	520.8–568.2		7.5
530.9	0.2	3.0			
482.5	0.05	1.0	468.0–530.9		6.0
476.2	0.05	1.0			
468.0		0.8	406.7–422.6		3.0
413.1		2.4	333.9–356.4		2.5

excited laser line can be made by using wavelength selective mirrors or a dispersive prism in the cavity.

The He-Ne gas mixture was one of the first systems to be utilized for stimulated emission of radiation in a laser cavity. Today a very large number of other gas mixtures are available. Some of them can provide much higher laser powers and oscillate on a large number of lines for a given gas mixture. Emission lines for two krypton-ion lasers with different output power are compiled in Table 3.3. Single-line or multi-line operation can be selected by choosing the proper mirrors or by inserting a prisma for dispersion. More than 15 W of power on one line can be obtained with these lasers. However, to obtain 7 W for the red line at 647.1 nm 76 kW of input power for the gas discharge are needed, as shown in Table 3.4. Thus, the absolute efficiency for light generation is very low. However, the total power is not as important for spectroscopy as the stability, bandwidth, beam diameter, etc. Table 3.4 lists

Table 3.4. Technical data for two typical krypton-ion lasers.

Property	Weak	Strong
Light noise level (2 MHz bandwidth)	0.2%	0.5%
Long time stability (30 minutes)	±0.5%	±0.5%
Frequency stability	60.0 MHz/K	330 MHz/K
Beam diameter	1.23 mm	1.6 mm
Beam divergence	0.78 mRad	0.6 mRad
Cavity length	1m	3.44m
Longitudinal mode extent	150 MHz (0.05 cm^{-1})	43.5 MHz (0.014 cm^{-1})
Weight	105 kg	411 kg
Electrical input power	13.1 KW	76 KW
Cooling requirements	8.4 l/min	26.5 l/min

Table 3.5. Gas lasers and solid-state lasers used in solid-state spectroscopy. (YAG stands for YAl-Garnet ($Y_3Al_5O_{12}$))

Active species	Matrix	Most important lines
Cd	Helium	325,441.6 nm
Ar^+	Argon	351.1, 488.0, 514.5 nm
Kr^+	Krypton	799.3, 647.1, 413.1 nm
Ne	Helium	632.8, 1153, 3390 nm
N_2	Nitrogen	337 nm
CO_2	Helium, nitrogen	10.2 μm(several lines)
HCN	Gas with C, H, N	311, 337 μm
Cr^{3+}	Al_2O_3 (ruby)	694.3 nm
Sa^{3+}	CaF_2	708.3 nm
Nd^{3+}	Glass or YAG	1064 nm
Ti^{3+}	Al_2O_3(sapphire)	700 to 1100 nm
Cr^{3+}	$BeAl_2O_4$ (alexandrite)	720 to 800 nm

some technical data for the type of krypton lasers characterized in Table 3.3. Data for argon-ion lasers are similar. In both cases it is the positively charged rare gas ion which provides the energy level system for the lasing process. Krypton and argon lasers are the most commonly used for continuous-wave (CW) spectroscopy. For UV, IR, or pulsed laser spectroscopy many other systems are available. Table 3.5 lists selected gas and solid-state lasers used in solid-state spectroscopy, together with their most important lines. The visible and the near-IR spectral range are well covered by the various laser lines. In the mid- and far-IR various semiconductor lasers are used, particularly those based on III-V compounds and PbTe. They will be discussed in the next subsection and in Sect. 10.1. The highest possible intensities in CW operation can be obtained with CO_2 lasers; powers of 27×10^{12} W can be reached. However, such powerful lasers are not used in spectroscopy. They were developed for nuclear-fusion experiments.

Strong intentions exist to build lasers in the vacuum UV. This spectral range can be reached either by repeated frequency doubling or by the use of excimer lasers. Frequency doubling relies on the nonlinear optical properties of crystals described by the second-order susceptibility χ_{ikl}. This and the higher-order susceptibilities are defined by the generalized nonlinear response for the polarization P.

$$P_i = \chi_{ik}^{(1)} \varepsilon_0 E_k + \chi_{ikl}^{(2)} \varepsilon_0 E_k E_l + \dots \,. \tag{3.21}$$

Since the response from $E_k E_l$ has a frequency of 2ω, crystals with high values of $\chi^{(2)}$ are often used for frequency doubling. Up to the 13th harmonic with a wavelength of 80 nm has been observed for Nd:YAG radiation. However, a high degree of frequency multiplication can only be obtained for very high power and pulsed systems.

Excimer lasers use rare-gas halogen mixtures. Wavelengths down to 120 nm have been obtained. Besides spectroscopy such lasers are frequently used in material technology such as laser sputtering. Operating such a laser under ambient conditions is dangerous since the high-energy light quanta generate the hazardous gas ozone with high efficiency.

3.4.3 Semiconductor Lasers

The development of semiconductor lasers has proceeded rapidly in the last few years. They benefit from the continuous progress in semiconductor technology, from their easy integrability into electronic circuits, and from their small size. Semiconductor lasers can be operated as pulsed systems but also in CW, in spite of the very large currents needed for an efficient operation. For example, GaAs lasers oscillating at 840 nm need currents of 3×10^8 A/m^2. Efficient cooling of the laser diode is required and is achieved by mounting the diode on a good heat sink. Semiconductor lasers can be operated with an efficiency of 10% for electrical to light power conversion. The quantum efficiency for the emission process can be as high as 15%.

The advantages of the semiconductor lasers are unfortunately accompanied by several drawbacks. The lifetime of these lasers is rather short unless they are operated at low temperatures and in a pulsed mode. The optical transitions in the semiconductor are rather broad, of the order of 2 nm, yielding a line shape function with a width of the order of

$$\delta f = \frac{c_0 \Delta\lambda}{\lambda^2} \approx 900\,\mathrm{GHz} = 30\ \mathrm{cm}^{-1}$$

for $\lambda = 840$ nm. Since the lasing crystals are usually rather short, of the order of 0.1 cm, the longitudinal mode separation is of the order of 150 GHz. Finally, since the active zone at the junction is rather small, of the order of 20 μm, the beam divergence evaluated from (3.19) is as large as 2.5×10^{-2} rad.

Recent developments in the field have led to laser diodes arranged in arrays with a considerable number of individual emitting units. Standard numbers of coupled units are 19 or 24. Coupling is either direct or by guiding the light from the individual units with a glass fiber to a bundle. In both cases a strong gain in emitted power can be obtained. Table 3.6 lists data for a selection of commercially available semiconductor lasers operating in the visible and near-IR spectral range.

Semiconductor lasers with an external cavity (external cavity lasers, ECL) can be tunable by about ± 10 nm for emission in the deep red. For specially constructed lasers (vertical cavity surface emitting lasers, VCSEL) light emission can be parallel to the current flow through the junction.

3.4.4 Pulsed Lasers

In order to obtain very high powers either pulsed excitation systems or systems where the stimulated emission occurs only for a very short time are

Table 3.6. Semiconductor lasers for the visible and near-IR spectral range.

Type	Junction	Spectral range [nm]	Power [W]	Mode	Manufacturer
	ZnMgSSe/ZnSSe	460 nm	Few mW	RT,CW	[3.9]
T3	$Al_xGa_{1-x}As$	650–670	Up to 0.4	RT, CW	Laser 2000
OPC	$Al_xGa_{1-x}As$	790–980	Up to 60	Peltier, CW array,24 units	Spectra Physics
OPC	$In_xGa_{1-x}As$	980 and higher	Up to 60	Peltier, CW array, 24 units	Spectra Physics
ECL		670 ±8	0.01	RT, CW, tunable	Laser 2000

utilized. In the early days of pulsed lasers a static mirror and a rapidly rotating 180° reflecting prism were used for the termination of the laser resonator. In this case laser oscillation is only possible for the very short time in which the reflexion from the mirror and from the prism are collinear and the whole inversion accumulated from a flash lamp can be used for light generation. Lasers operating in such a mode are called Q-switched (Q describes the quality of a resonator cavity). If these lasers had a power of 1 W for quasi-CW operation (1 second flash) 100-ns pulses would already yield a power of 10 MW, provided the full inversion of the flash can be converted in the short time. Ruby lasers, Nd : YAG lasers, and nitrogen lasers are particularly useful for this mode of operation. Ruby lasers can produce 250 J of light energy within 5 ns which corresponds to a power of 50 GW. Very large intensities can be obtained by focusing the light. For a 50-kW neodymium laser peak intensities of 10^{12} W/cm^2 are quite common. This corresponds to a field strength of 2×10^9 V/m. Such lasers are used to study nonlinear optical properties of materials. For neodymium lasers intensities up to 10^{20} W/cm^2 were produced recently for nuclear-fusion experiments.

By improved Q-switches and other techniques such as mode coupling, pulse lengths of the order of 10^{-15} seconds can be reached with corresponding high light power. This has given rise to a new field in spectroscopy known as *picosecond spectroscopy.*

As we have already mentioned in Sect. 3.4.1, the intensities of the various longitudinal laser modes fluctuate statistically unless they are reduced to a single oscillator by an etalon filter. Another option to suppress the fluctuations and to obtain some order in the mode structure is mode locking. In contrast to the insertion of a frequency selective absorber, the relative phases of the modes in the cavity are coupled. In this way the phase fluctuations between modes are eliminated. To understand how this works we must recall that any radiation field can be represented in its time domain by $E(t)$ and $I(t)$ or in its frequency domain by $E(f)$ and $I(f)$. Both E and I occur with fluctuating phase ϕ. (See for this description also Sect. 3.5.1.) Figure 3.15 shows the situation for the intensity fluctuations and for the phase fluctu-

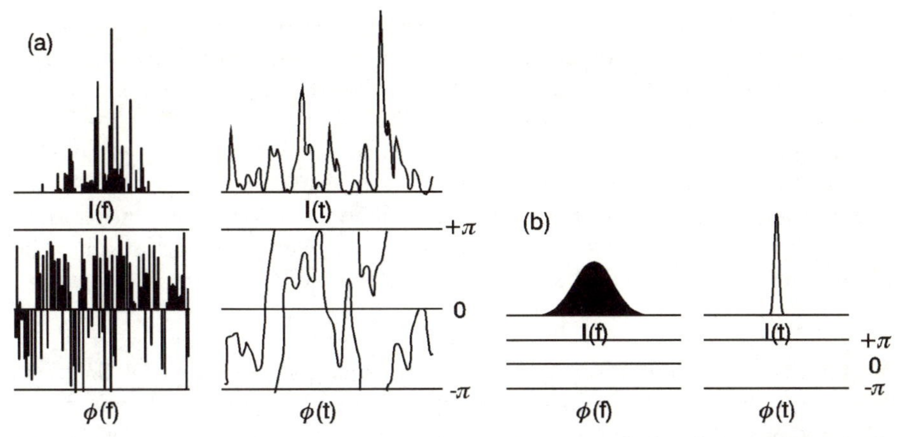

Fig. 3.15. Schematic representation of the intensity and the phase of laser oscillations in the frequency domain and in the time domain for a free laser (**a**) and for a mode-locked laser (**b**).

ations in the case of a free oscillating laser and a laser with mode locking, respectively. As long as there is no phase correlation between the oscillator modes their intensity in the frequency domain fluctuates under the gain curve and the radiation fluctuates in the time domain like thermal noise about a mean value (Fig. 3.15a). A similar situation holds for the phases. They oscillate between $\pm\pi$ in the frequency as well as in the time domain. In both cases there is a general relation between the time scale of the fluctuations and the bandwidth of the spectrum as will be discussed in the next section in detail.

If phase coupling exists, each mode is amplified according to its position under the gain curve and the intensity does not fluctuate (Fig. 3.15b). Note the completely different time pattern for the intensity in the two cases. In Fig. 3.15a the intensity fluctuations extend over all time whereas in Fig. 3.15b a spike is formed.

In practice, phase locking is performed by an appropriate modulation of the gain curve. As a consequence a train of extremely short spikes occurs with a period (repetition time) T equal to twice the transition time of the light in the resonator

$$T = \frac{2L}{c} \,. \tag{3.22}$$

Active or passive mode-coupling techniques are used to establish the phase locking. In either case a modulation of the gain curve is induced with a frequency $c/2L$ corresponding to the reciprocal turn-around time of the light in the resonator or, likewise, the distance of the longitudinal modes. For passive mode coupling a saturable absorber is inserted in the cavity. In this case absorption is minimum (or gain is maximum) for peak light intensities. This means that light pulses are shaped with a repetition rate equal to the round-trip transition time in the cavity. Active mode coupling is obtained by inserting a piezoelectric crystal into the resonator which operates as an amplitude modulator. Side bands develop for each longitudinal mode ω_s. The side bands are tuned to coincide with the closest neighboring longitudinal mode of frequency $\omega_{s\pm1}$. Thus, they contribute to the stimulated emission of the neighboring mode and all modes under the gain curve will be coupled.

The mathematical formulation for the establishment of picosecond-pulse trains results in both cases from an amplitude modulation of the longitudinal modes. It is most easily discussed for the case of the active mode locking. The modulation of mode s can be described in this case by a harmonic wave

$$E_s = T_\mathrm{M} E_{s0} \cos \omega_s t \qquad \text{with} \qquad T_\mathrm{M} = 1 - \delta(1 - \cos \Omega t) \leq 1. \tag{3.23}$$

δ is the depth of modulation, and T_M the transmission coefficient for the piezo modulator. Simple trigonometric manipulations yield the side bands $E^{(\pm s)}$ for the mode number s at $\omega^{(\pm s)} = \omega_s \pm \Omega$ by

$$E^{(\pm s)} = \frac{E_{s0}\delta}{2} \cos(\omega_s \pm \Omega)t \; . \tag{3.24}$$

When Ω is tuned to $2\pi c/2L$, the side bands of ω_s coincide with the modes $\omega_{s\pm1}$. The number m of coupled modes is given by the ratio between the width of the gain curve and the longitudinal mode distance. For simplicity, we assume a rectangular gain curve which means all E_{s0} in (3.23) are equal and the contribution of the side bands is negligible compared to the mode amplitude. Counting all modes with frequencies $\omega_s = \omega_0 + s\Omega = \omega_0 + s\pi c_0/L$ from a central frequency ω_0 yields the total field as a superposition of all modes:

$$E = \sum_{s=-m}^{m} E_0 \cos(\omega_0 + s\Omega)t \; . \tag{3.25}$$

Considering $I(t) \propto E(t)^2$ yields from this

$$\boxed{I(t) \propto \frac{E_0^2 \sin^2[(2m+1)\Omega t/2]}{\sin^2(\Omega t/2)} \cos^2 \omega_0 t \; ,} \tag{3.26}$$

which is, indeed, in the limit of large m a pulse train with the repetition period $T = 2L/c$ and the pulse width δT

$$\boxed{\delta T = \frac{2L}{mc} \; .} \tag{3.27}$$

m is the number of coupled modes equivalent to the number of modes under the gain curve and L is the resonator length. A plot of (3.26) for $m = 10$ is presented in Fig. 3.16a.

The more modes are coupled the shorter are the resulting laser pulses. For generation of very short pulses a large bandwidth or equivalently a broad gain curve of the laser is needed. Bandwidth γ and pulse length τ are related by the usual formula $\gamma\tau \approx 1$ (Sects. 2.4 and 3.2). A light pulse for which the duration is only determined by the bandwidth is called *bandwidth-limited* with a *natural linewidth.* The bandwidth of an Ar^+ laser is, e.g., 7 GHz which would give a natural pulse width of 140 ps. In contrast, the bandwidth of a dye laser is 3 THz for which a natural pulse width of 0.3 ps can be obtained. If the length of the cavity of the laser is 1 m 20000 modes are coupled.

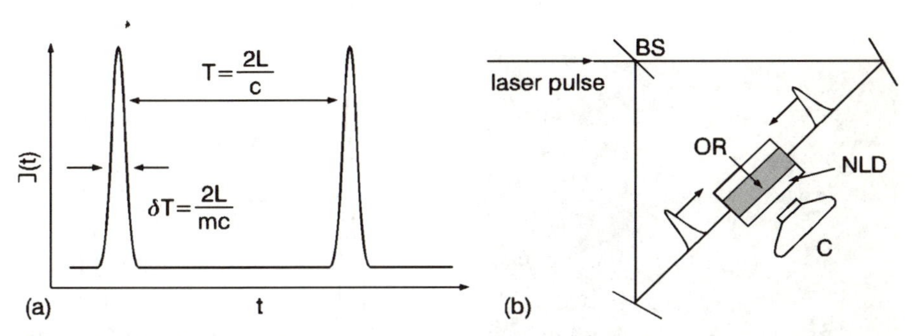

Fig. 3.16. Light pulses from a mode-coupled laser (**a**), and experimental setup for measuring ps laser pulses by two-photon luminescence (**b**); (BS: beam splitter, OR: overlap region, NLD: nonlinear dye, C: camera).

The detection of laser pulses in the ps time range is not trivial. Even though the photoelectric effect has a time constant of only 10^{-14} s this time can not be resolved since oscilloscopes are limited to 5 GHz allowing for a maximum resolution of 70 ps. Photodiodes also have time constants of at least 100 ps which excludes a direct measurement of pulse lengths in the picosecond or sub-picosecond region. A realistic possibility for the detection of such pulses consists in the nonlinear excitation of luminescence by a two-photon fluorescence (TPF) process according to Fig. 3.16b. In this experiment the light pulse is divided into two parts by a beam splitter. Subsequently both parts are again superposed colinearily but in opposite directions in a nonlinear medium. A strong luminescence which can be observed with a camera is emitted only during the time at which both pulses are at the same position in the crystal.

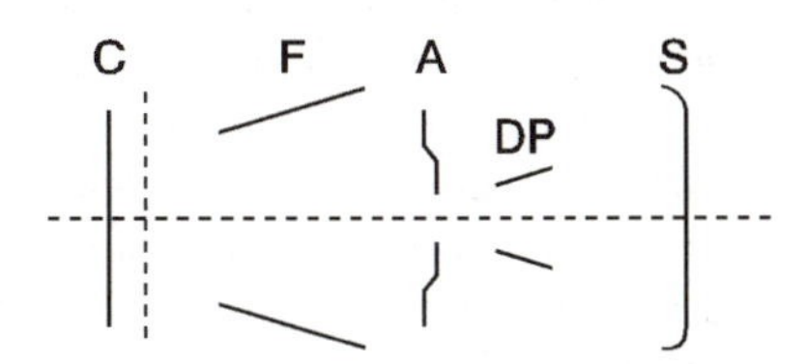

Fig. 3.17. Schematic view of a streak camera; (C: cathode, F: focusing plates, A: anode, DP: deflection plates with time ramp, S: screen).

A *streak camera* (Fig. 3.17) can also be used to detect picosecond pulses. Electrons are freed from the photocathode by the very fast external photoelectric effect. They are observed after a strong acceleration towards the anode and a subsequent deflection by a time-ramped electric field. It is immediately evident that the resolution in time depends only on the velocity distribution of the freed electrons. Figure 3.18 shows a streak photograph and the corresponding densitometer readings for two picosecond pulses of width 1.5 ps and 60 ps apart.

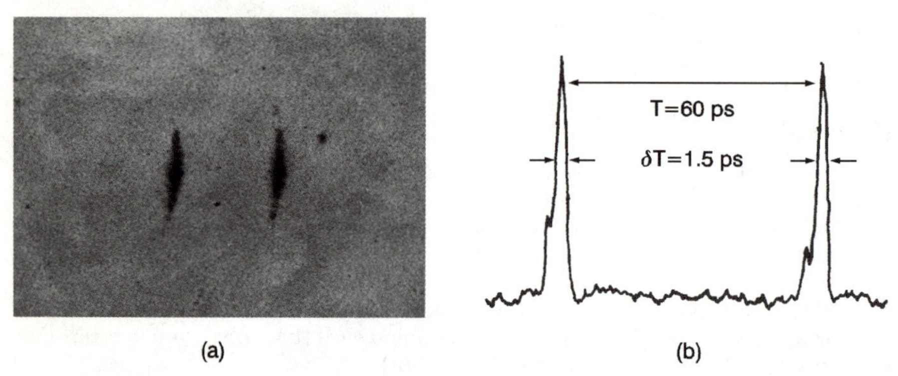

Fig. 3.18. Streak-photo (**a**) and densitometer reading (**b**) for two picosecond pulses; after [3.10].

3.4.5 Tunable Lasers

Initially the disadvantage of laser sources was their restriction to the emission of a few single and narrow lines. Early in the development process, and in addition as the result of recent work, lasers were devised which avoid this problem and allow wavelength tuning at least in a limited spectral range. These systems are known as *dye lasers*. They use a gas laser or a pulsed solid-state laser to pump an active medium in an extra cavity. Originally the active media were dyes like rhodamine, coumarine, or stilbene, etc., with broad optical transitions. The broad transitions originate from splitting of electronic levels by rotational states which are themselves broadened by intermolecular interactions. In this way a quasi-continuum of electronic transitions is established, at least within a limited spectral range. Only the lowest rotational levels are populated in thermal equilibrium so that a population inversion can be obtained by excitation into higher rotational states. This may lead to stimulated emission and eventually to a nonlinear coherent amplification. Figure 3.19 shows the luminescence of rhodamine 6G after excitation from the S_0 ground state to an S_1 excited state (solid line). This luminescence is more than 10 nm wide.

In many cases the excited state relaxes by intersystem crossing into a long-living triplet state T_1. This state introduces an additional destructive absorption process into the system. It overlaps partly the luminescence and

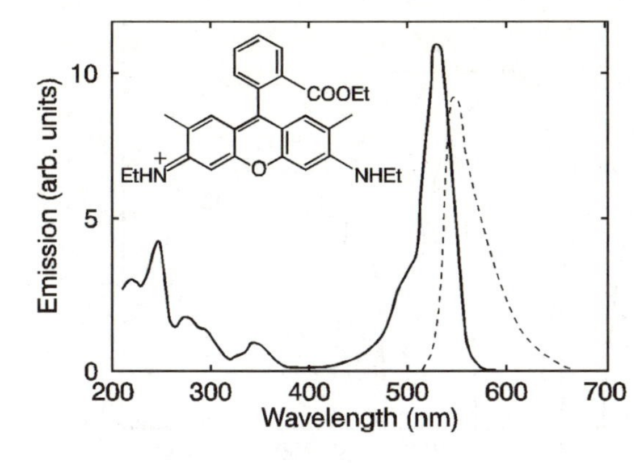

Fig. 3.19. Luminescence (—) and triplet absorption (– – –) for rhodamine 6G after excitation from the S_0 ground state to a S_1 excited state. The insert shows the rhodamine 6G molecule; after [3.11].

therefore decreases the overall gain for the nonlinear amplification. The problem can be solved by adding a *triplet quencher* to the dye which transfers the triplet states back to the S_1 state by collisions. Alternatively, the dye can be excited in a rapidly flowing jet stream where the molecules leave the active volume before the triplet state is generated.

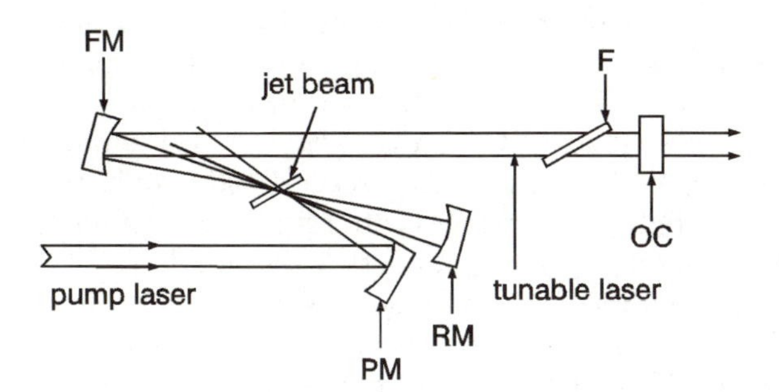

Fig. 3.20. Schematic arrangement of a dye laser; (FM: folding mirror, PM: pump mirror, RM: reflector mirror, OC: output coupler, F: tuning filter).

The schematic construction of a dye laser is shown in Fig. 3.20. The pump light is introduced into the resonator cavity by the pump mirror and excites the dye which is injected by a jet with a speed of about 10 m/s. The cavity is folded and consists of a reflector, a folding mirror, and an output coupler. The wavelength for resonance is tuned by a birefringent Lyot filter. Figure 3.21 displays efficiencies for various commercial dyes. From this figure it is evident that a large number of different dyes are needed to cover even the visible spectral range.

Recent developments in the field of dye lasers refer to tunable ring lasers and solid state lasers. In a ring laser the cavity is not terminated by a system of reflectors but consists of a closed optical ring. Instead of the build up of standing waves as in normal cavities propagating waves are generated. This results in a more efficient spatial geometry for the release of the inversion and

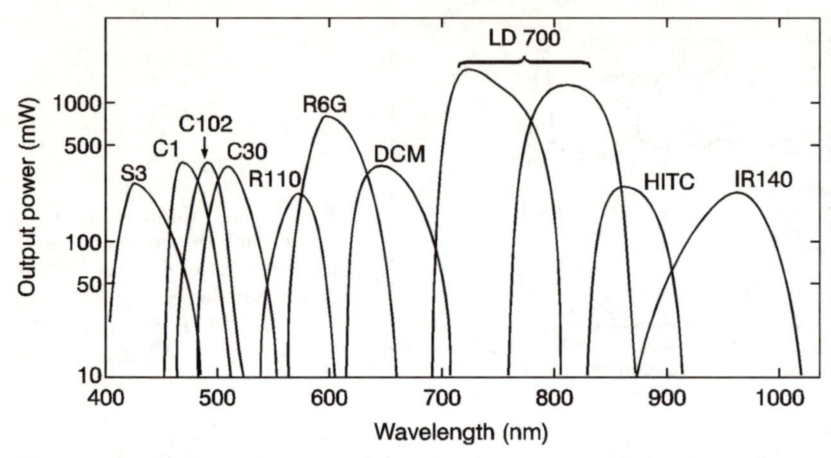

Fig. 3.21. Emission of dye lasers in various spectral ranges for pumping with a 4-W krypton laser in the blue-green range; after [3.12].

thus gives a higher gain. Ring lasers are more powerful than conventional dye lasers but require a better alignment and stabilization.

Solid-state tunable lasers are frequently used because they have a high efficiency and are simpler to operate. A widely used system is a Ti:sapphire cavity pumped by an argon laser. The optical active ion in the sapphire is Ti^{3+} which is embedded in a crystalline Al_2O_3 matrix (Table 3.5). Pumping with an all-line argon laser of 5 W yields emission of several 100 mW in the spectral range of 700–1000 nm. High-power Nd:YAG after frequency doubling can be utilized as an alternative pump system. (For details see also Sect. 3.4.7 below). Nd:YAG lasers excited with a lamp can reach 100 W in CW. Thus, even after frequency doubling to 532 nm with a KTP crystal considerable pump power is available for the Ti:sapphire system. Figure 3.22 presents the output power of a Ti:sapphire laser for three different resonator mirrors (a) and line intensities after frequency doubling with a 15-W pump. Frequency

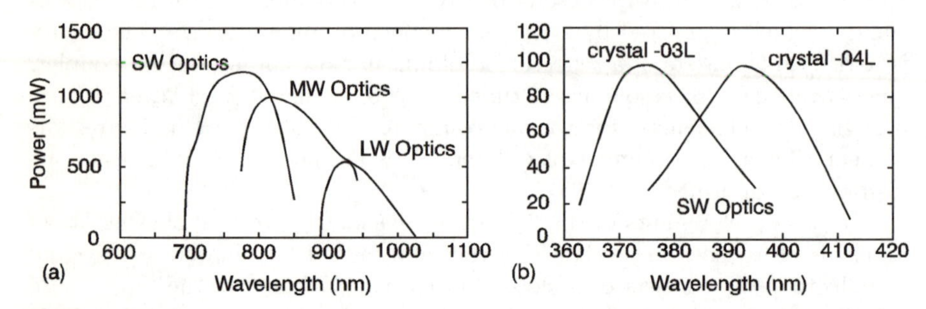

Fig. 3.22. Output of a Ti:sapphire laser versus wavelength for a 5-W all-line argon pump (**a**) and output after frequency doubling for a 15-W pump (**b**); SW, MW, and LW refer to short, medium long, and long wavelengths, respectively.

doubling was performed in this case with a BBO (β-barium borate, BaB_2O_4) crystal.

3.4.6 Free-Electron Lasers

The free-electron laser is an alternative to conventional tunable laser systems and operates in vacuum. It is based on the stimulated emission of synchrotron radiation from electrons in an undulator, and has very good tuning properties. In principle, it could be used from the IR to the UV spectral range. The active medium is the electron beam of the synchrotron. It is guided through the laser cavity which contains the undulator. Figure 3.23 sketches a setup. The radiation from the wiggles adds up like the partial beams of a grating. If the timing is such that after one round trip in the cavity, a photon pulse, emitted from an electron bunch in the undulator, passes the undulator again at exactly the same time as the next bunch arrives, stimulated emission of synchrotron radiation occurs. This is possible because the electrons acquire a weak transverse velocity component from the undulator, which is sufficient to initiate the stimulated emission process. Since the laser operates in vacuum no inversion in a classical sense is generated. If the gain from the stimulated emission is larger than the total loss during one round trip laser oscillations turn on.

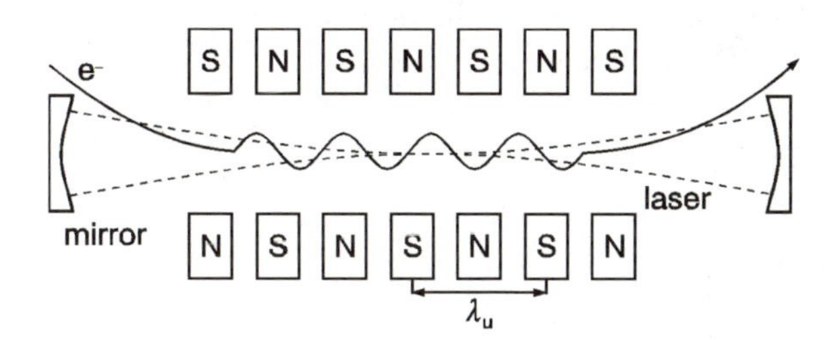

Fig. 3.23. Schematic set-up for a free-electron laser; (N and S: magnets of the undulator, (—): electron beam from the storage ring, (– – –): laser light).

The problem with the free-electron laser is the rapid decrease of the gain coefficient with increasing electron energy ϵ. α is obtained from

$$\alpha = C \frac{\lambda_U^3 N^3 K^2 F}{\epsilon^3} , \tag{3.28}$$

where C is a constant, N the number of periods in the undulator, and F a filling factor between the volume of light propagation and electron propagation. λ_U and K have the same meaning as in (3.16). Since, from the same equation the wavelength for constructive interference decreases as $1/\epsilon^2$, free-electron laser oscillation for short wavelengths is difficult. So far only oscillations in a spectral range of 570 to 650 nm have been reported. The expected spectral range should, however, cover wavelengths down to 130 nm [3.13].

3.4.7 New Developments

Recent developments in laser technology have concentrated more and more on solid-state lasers. In particular the Nd:YAG laser attracts the interest of laser engineers because of its very high efficiency. This system operates at 1064 nm but is easily frequency doubled to 532 nm with, e.g., a $MgO:LiNbO_3$ crystal. Lamp pumping and diode pumping are still in competition for the excitation of the Nd:YAG crystal. Krypton-arc lamps certainly have a high power and can produce multi-mode CW or mode-locked laser radiation from Nd:YAG with a power of more than 100 W [3.14]. On the other hand, diode-pumped Nd:YAG lasers operate with very high efficiency since the emission of GaAs diodes can be tuned exactly to the main absorption line in the Nd:YAG system at 808 nm, as shown in Fig. 3.24. The emission line matches exactly the transition energy between the ground state and the ^4F states in Nd^{3+}. The pump may be either a single diode, a diode laser, or even a diode-laser array. Since the diodes of the III-V compounds operate very efficiently in this spectral range, an all over laser efficiency of the order of 25% can be expected. In contrast, for broad-band excitation with a lamp all energy not matching this transition is wasted. In addition, the stability of the diode-pumped systems is an order of magnitude better than for lamp-pumped systems. Commercially available systems have powers of 10 W CW. Line widths can be less than 5 KHz, even after frequency doubling. For 532 nm radiation powers of 500 mW are available. All these systems work with wall-plug powers and without water cooling.

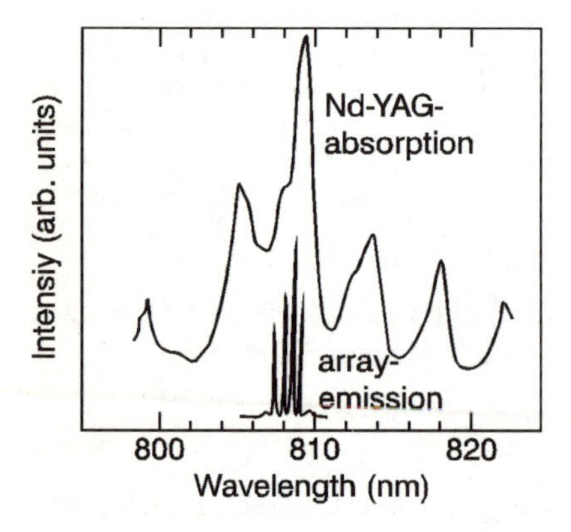

Fig. 3.24. Energy matching between Nd^{3+} excitation in Nd:YAG lasers and diode emission; after [3.8].

The rapid development of the diode-pumped Nd:YAG lasers suggests an all solid-state laser system as the coming source in laser spectroscopy, even for the visible spectral range. Such systems consist of a diode array as a pump and a Nd:YAG laser with frequency doubling which provides a suitable pump for the tunable Ti:sapphire laser (Fig. 3.25). The cavity of the latter contains a Lyot filter for frequency tuning. The advantage of the all solid systems is

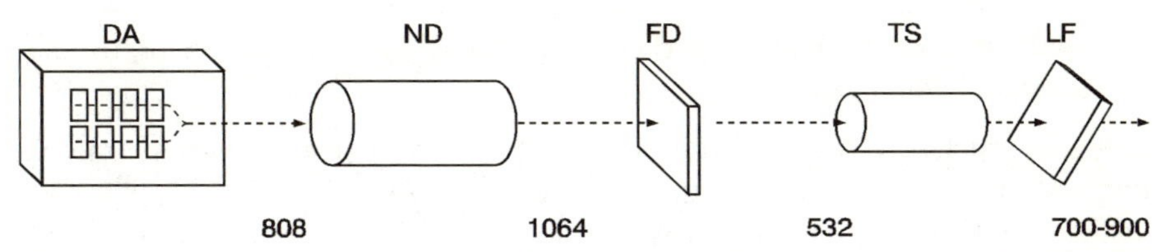

Fig. 3.25. All solid-state tunable laser system; (DA: diode array, ND: Nd:YAG pump laser, FD: frequency doubler, TS: Ti:sapphire laser, LF: Lyot filter). The numbers are the instantaneous laser wavelengths in nm.

their compact construction and their stability. At the moment these lasers are unfortunately still rather expensive.

3.5 Coherence and Correlation

In the preceding section we have already used the terminology of coherence and phase relation of waves. For a proper description of the EM fields a more precise definition is required. This section will summarize the concepts of coherence and correlation in order to provide a basic knowledge of these quantities. It will be sufficient for understanding the following chapters. Specialists in laser spectroscopy or correlation spectroscopy will need to study these subjects in greater depth.

3.5.1 Periodic and Non-Periodic Electromagnetic Fields

In Sect. 2.1 we discussed EM fields with periodic oscillations of infinite duration. This discussion was extended in Sect. 2.4 to non-periodic functions, most of which vanish for $t \to \infty$. Real radiation fields are different as the phases and amplitudes of the wave trains have a certain statistical or stochastic character. This character exists for laser radiation as well as for black-body emission even though it is much less prominent in the former. For convenience, we will distinguish four different types of EM fields:

a) fully periodic fields, to be described by harmonic functions,
b) quasi-periodic non-stationary fields which vanish for $t \to \infty$, to be described by damped oscillator functions, as discussed in Sect. 2.4,
c) quasi-stochastic fields where amplitude and phase vary statistically with time but the variations are weak. Such fields may be described by the real part of

$$E(t) = A(t)\mathrm{e}^{\mathrm{i}\alpha(t)} \qquad \text{with} \qquad \alpha(t) = 2\pi f_0 t + \phi(t) \, , \tag{3.29}$$

where the usual time-independent amplitude E_0 and phase ϕ have been replaced by more or less rapidly varying functions in time $A(t)$ and $\phi(t)$,

d) highly stochastic fields like those from black-body radiation or from a stochastic generator.

A highly stochastic field may be visualized as a statistically emitted train of damped oscillations as shown in Fig. 3.26. The coherence time for this light is given by the lifetime of the wave.

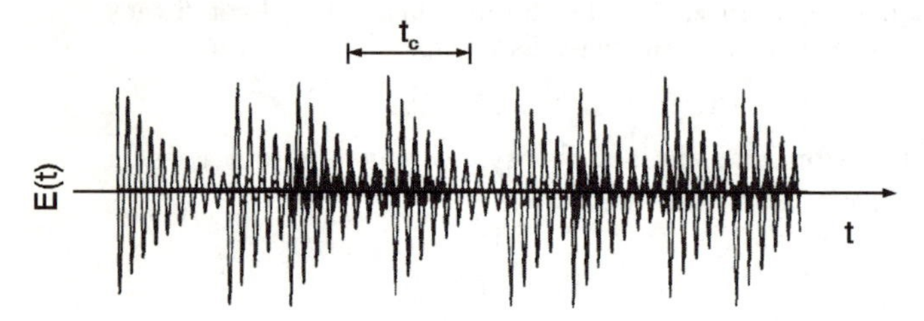

Fig. 3.26. Stochastic light consisting of randomly generated trains of oscillations with a lifetime t_c.

3.5.2 Coherent and Non-Coherent Superposition

The coherence of components in a wave field is of fundamental importance for their superposition. This superposition may be coherent or incoherent. The results are different and need careful consideration. The intensity of the radiation field is proportional to E^2 and is given by

$$\langle E^2 \rangle = \frac{1}{2T} \int_{-T}^{T} E^2(t)\mathrm{d}t \tag{3.30}$$

for an arbitrary function of time. $2T$ is the duration of the measurement, as discussed in (2.45). For two superposed harmonic waves E_1 and E_2 with amplitudes, frequencies and phases given by $E_{01,2}$, $f_{1,2}$ and $\phi_{1,2}$, respectively, the intensity is given by

$$\begin{aligned} \langle E^2 \rangle &= \langle (E_1 + E_2)^2 \rangle = \langle E_1^2 \rangle + \langle E_2^2 \rangle + 2\langle E_1 E_2 \rangle \\ &= \frac{E_{01}^2}{2} + \frac{E_{02}^2}{2} + 2E_{01}E_{02}\langle \cos\alpha_1 \cos\alpha_2 \rangle , \end{aligned} \tag{3.31}$$

where $\alpha_{1,2} = 2\pi f_{1,2}t + \phi_{1,2}$. The first two terms in the equation represent the intensities of the individual waves and the third term describes the interference.

For $\alpha_1 \neq \alpha_2$ the interference term vanishes from the time average, as it can be expressed by cosine functions of the phase difference $\alpha_1 - \alpha_2$ and sum $\alpha_1 + \alpha_2$. Then, the intensities of the two individual fields simply add together.

For fields with equal amplitude and α the situation is different. Since the amplitudes add to $2E_0$ the intensity increases from $2 \times (E_0^2/2)$ to $2E_0^2$. This result is also a direct consequence of (3.31), since the time average in the interference term becomes 1/2. This surprising increase in energy seems to violate energy conservation. In practice, energy is, of course, conserved even in an interference experiment of this type. Coherent and collinear superposition of two equal waves is indeed only possible for beams oriented perpendicular to each other and split into two equal parts each by inserting a beam splitter of 45°. In this case each of the original waves propagates, after leaving the splitter, in two mutually perpendicular partial beams with field amplitude $E_0/\sqrt{2}$. The constructive interference can then only occur for one of the partial beams, and superposition leads to an intensity of $\langle(2E_1/\sqrt{2})^2\rangle = E_0^2$ at maximum. Since in this case the intensity in the other superposed beam will be zero due to destructive interference, energy is conserved.

3.5.3 Temporary Coherence and Correlation

A discussion of coherence needs a more precise definition of this conception. Comparing one field with a second one which is shifted in space or time, is probably a good example. Coherence as we understand it from this is a phase correlation in time or space. In other words, we ask how well is $E(r_2, t_2)$ known for a general EM field $E(r, t)$ if we know $E(r_1, t_1)$. For small Δr and Δt the question can often be answered but this will become harder as the two functions move further apart in space and time. A good indicator to describe the problem could be the *mutual coherence* or *mutual correlation function* defined as

$$G(r_1, r_2, \tau) = \langle E^*(r_1, t) E(r_2, t + \tau)\rangle \, , \tag{3.32}$$

where $\langle \quad \rangle$ refers to a time average. Spatial coherence is relevant for interference experiments with light from extended sources. Since this will not be discussed in the frame of this book, we will restrict ourselves to temporal coherence. In this case coherence at the same point r in space is considered and the argument r for the coordinates can be dropped in (3.32). Then we obtain from (3.32) the *temporal coherence function* or *autocorrelation function* explicitly

$$\boxed{G(\tau) = \lim_{T\to\infty} \frac{1}{2T} \int_{-T}^{T} E^*(t) E(t + \tau) \mathrm{d}t \, .} \tag{3.33}$$

The autocorrelation function is maximum for $\tau = 0$ and may go to zero for $\tau \to \infty$. The autocorrelation function for a partially or highly stochastic light field vanishes for $\tau \to \infty$ since eventually $E(t)$ and $E(t + \tau)$ become random in phase and all contributions cancel due to the time averaging. The autocorrelation functions do not have to become zero for $\tau \to \infty$. For example the autocorrelation function for a constant field $E(t) = E_0$ is $E_0^2 \neq 0$ for all values of τ. The autocorrelation function for a sine or cosine wave is given by a cosine wave, and remains oscillating for all values of τ.

Note that the definition of the autocorrelation function in (3.33) cannot be used for non stationary fields like wave packets, etc. $G(\tau)$ would be identically zero. In this case we have to define it as

$$G(\tau) = \lim_{T \to \infty} \int_{-T}^{T} E^*(t)E(t+\tau)\mathrm{d}t \,. \tag{3.34}$$

For the case of a wave packet this function is indeed maximum for $\tau = 0$ and approaches 0 for $\tau \to \infty$. Since the only interesting behavior of $G(\tau)$ is its dependence on the time shift τ, a normalized autocorrelation function of the form

$$g(\tau) = G(\tau)/G(0) \tag{3.35}$$

is often used. In this case discrimination between stationary and non-stationary fields is not required.

The decay time of the autocorrelation function gives the coherence time $\Delta\tau$ and the coherence length $\Delta l = c_0\Delta\tau$. There are several possible definitions for the decay time of $G(\tau)$. The FWHM is one. Another straightforward definition is related to the variance (mean square deviation) of the autocorrelation function or to the second moment of the absolute square of this function. In the latter case the coherence time is defined by

$$(\Delta\tau_{\mathrm{v}})^2 = \frac{\int \tau^2 |G(\tau)|^2 \mathrm{d}\tau}{\int |G(\tau)|^2 \mathrm{d}\tau} \,. \tag{3.36}$$

This definition may be paralleled by a definition for the spectral bandwidth of the function $E(t)$ as the variance of the absolute square of the power spectrum.

$$(\Delta f_{\mathrm{v}})^2 = \frac{\int (f - f_0)^2 |S(f)|^2 \mathrm{d}f}{\int |S(f)|^2 \,\mathrm{d}f} \,. \tag{3.37}$$

The expressions for $\Delta\tau$ and Δf allow us to study the product of the two quantities. As might be expected, there is a fundamental connection between them.

The connection between the bandwidth of a fluctuating field $E(t)$ and its coherence time can be investigated in a very general way through an analysis of a field with a quasi-stochastic fluctuation given by (3.29). Splitting the complex exponent $\alpha(t)$ into a harmonic part oscillating with the frequency f_0 and a fluctuating phase $\phi(t)$ we have

$$E(t) = \int E(f)\mathrm{e}^{\mathrm{i}2\pi f t}\mathrm{d}f = A(t)\mathrm{e}^{\mathrm{i}\phi(t)}\mathrm{e}^{\mathrm{i}2\pi f_0 t} \,,$$

where $E(f)$ is the Fourier transform of $E(t)$. This expression can be used to study explicitly the relationship between the fluctuation of $E(t)$ in time and its bandwidth. Multiplying both sides with $\exp(-\mathrm{i}2\pi f_0 t)$ yields

$$A(t)\mathrm{e}^{\mathrm{i}\phi(t)} = \int E(f)\mathrm{e}^{\mathrm{i}2\pi(f-f_0)t}\mathrm{d}f \,. \tag{3.38}$$

A very small bandwidth of $E(t)$ allows only substantial values for $E(f)$ if f is very close to f_0, which means $(f-f_0)$ very small or very slow fluctuations with a long coherence. On the other hand, if $E(f)$ gives substantial contributions to the integral even for large values of $(f-f_0)$ rapid fluctuations with a short coherence exist. Thus, bandwidth Δf and coherence Δt are inversely related. The fundamental relationship

$$\Delta t \Delta f = \text{constant} \approx 1 \tag{3.39}$$

is quite general. The constant is determined by the particular type of the field as well as by the actual definition used for the bandwidth and for the coherence. There are numerous examples for this relationship in the literature. We will give only one and leave others as problems.

Let the field be the damped, one-sided complex exponential wave $E(t) = \exp(-\gamma t/2)\exp(\mathrm{i}2\pi f_0 t)$ where we assumed $E_0 = 1$, for convenience. We evaluate the autocorrelation function $G(\tau)$ from (3.34) and use (3.36) and (3.37) for the definition of $\Delta\tau_\mathrm{v}$ and Δf_v, respectively. This yields

$$\begin{aligned} G(\tau) &= \int E^*(t)E(t+\tau)\,\mathrm{d}t = \int_0^\infty \mathrm{e}^{-\gamma t/2}\mathrm{e}^{-\gamma(t+\tau)/2}\mathrm{e}^{\mathrm{i}2\pi f_0\tau}\,\mathrm{d}t \\ &= \mathrm{e}^{-\gamma\tau/2}\mathrm{e}^{\mathrm{i}2\pi f_0\tau}\int_0^\infty \mathrm{e}^{-\gamma t}\,\mathrm{d}t \\ &= \frac{1}{\gamma}\mathrm{e}^{-\gamma|\tau|/2}\mathrm{e}^{\mathrm{i}2\pi f_0\tau} \,. \end{aligned} \tag{3.40}$$

The power spectrum $S(f)$ for the field under consideration was already evaluated in (2.39) so that we can immediately calculate the correlation time $\Delta\tau_\mathrm{v}$ and the band width Δf_v from

$$\begin{aligned} (\Delta\tau_\mathrm{v})^2 &= \frac{\int \tau^2|G|^2\mathrm{d}\tau}{\int |G|^2\mathrm{d}\tau} = \frac{(1/\gamma^2)\int_0^\infty \tau^2 e^{-\gamma\tau}\mathrm{d}\tau}{(1/\gamma^2)\int_0^\infty e^{-\gamma\tau}\mathrm{d}\tau} \\ &= \frac{4/\gamma^5}{2/\gamma^3} = \frac{2}{\gamma^2} \end{aligned} \tag{3.41}$$

and

$$\begin{aligned} (\Delta f_\mathrm{v})^2 &= \frac{\int (f-f_0)^2|S|^2\,\mathrm{d}f}{\int |S|^2\,\mathrm{d}f} \\ &= \int \frac{(f-f_0)^2\,\mathrm{d}f}{[(\gamma/2)^2+(2\pi)^2(f-f_0)^2]^2} \Big/ \int \frac{\mathrm{d}f}{[(\gamma/2)^2+(2\pi)^2(f-f_0)^2]^2} \\ &= \frac{1}{(2\pi)^2 2\gamma}\frac{\gamma^3}{2} \,. \end{aligned} \tag{3.42}$$

The square root of the product of the two quantities yields the well known result

$$\Delta\tau_{\mathrm{v}}\Delta f_{\mathrm{v}} = \frac{\sqrt{2}}{4\pi} \,. \tag{3.43}$$

Note that the relationship we had in (2.40) between the lifetime of an oscillation and its bandwidth is very similar to (3.43). Physically these relationships are indeed based on the same fundamental principle of uncertainty but conceptually they are very different. The result from (3.43) can be compared with the simple product of the decay time $\tau = 2/\gamma$ and the FWHM $\Delta f = \gamma/2\pi$ of the Lorentz line which yields

$$\tau\Delta f = 1/\pi \,. \tag{3.44}$$

3.5.4 The Wiener–Khintchin Theorem

In many applications both functions, the Fourier transform $E(f)$ and the correlation function $G(\tau)$, of a fluctuating field $E(t)$ are needed. An example was just given. It turns out that the two functions are not independent but are correlated in a rather simple way known as the *Wiener–Khintchin* theorem. This theorem states that the Fourier transform of the correlation function $G(\tau)$ equals the power spectrum $S(f)$ of a field $E(t)$, and vice versa. In mathematical terms this means that

$$\boxed{\begin{aligned} \int G(\tau)\mathrm{e}^{-\mathrm{i}2\pi f\tau}\mathrm{d}\tau &= S(f) \quad \text{and} \\ \int S(f)\mathrm{e}^{\mathrm{i}2\pi f\tau}\mathrm{d}f &= G(\tau) \,. \end{aligned}} \tag{3.45}$$

The proof of the theorem is rather simple for non stationary fields. Using the definition for the Fourier transform of $G(\tau)$ and writing it as

$$\begin{aligned} \int G(\tau)\mathrm{e}^{-\mathrm{i}2\pi f\tau}\mathrm{d}\tau &= \int E^*(t)E(t+\tau)\mathrm{e}^{-\mathrm{i}2\pi f\tau}(\mathrm{e}^{-\mathrm{i}2\pi ft}e^{\mathrm{i}2\pi ft})\mathrm{d}t\mathrm{d}\tau \\ &= \int E^*(t)\mathrm{e}^{\mathrm{i}2\pi ft}E(t+\tau)\mathrm{e}^{-\mathrm{i}2\pi f(t+\tau)}\mathrm{d}t\mathrm{d}(t+\tau) \\ &= E^*(f)E(f) \,, \end{aligned} \tag{3.46}$$

which is exactly the power spectrum $S(f)$. The proof for stationary fields is similar but more laborious. The Wiener–Khintchin theorem is fundamental in coherent signal processing.

Problems

3.1 Use the Stefan–Boltzmann law to estimate the solar constant for the temperature of the sun ($T \approx 6000\,\text{K}$), the distance between sun and earth $\approx 150 \times 10^6$ Km, and the radius of the sun about 6.9×10^5 Km.

(Purpose of exercise: emission of radiation.)

3.2 Calculate the total emitted power per unit area for a black-body radiator from Planck's law.

(Purpose of exercise: difference between spectral power and total power.)

3.3 Show that the FWHM for a Gaussian line equals $\sqrt{\sigma^2 8 \ln 2}$.

(Purpose of exercise: spectral line shapes.)

3.4 Calculate the convolution of two Lorentzian lines and show that the resulting linewidth is the sum of the individual widths.

(Purpose of exercise: performing convolutions.)

3.5 Discuss the maximum and the minimum position for a Fano line. What happens for weak coupling?

(Subject of training: nature of Fano line.)

3.6* The velocities of atoms in a light-emitting gas are distributed according to a Maxwellian distribution. Show that the Doppler effect for the frequency shift leads to a Gaussian linewidth for the emitted radiation. How large is the width of the lines in an argon plasma of 3000 K?

Hint: neglect the intrinsic linewidths of the atoms.

(Purpose of exercise: study the origin of spectral line shapes.)

3.7 A laser operating at a wavelength of 633 nm has a beam diameter of 2 mm, a spectral width of 10 KHz, and a power of 5 W. How much brighter is the laser compared to the sun for equal bandwidth?

(Purpose of exercise: characterization of laser light.)

3.8* Evaluate the superposition of the side bands with the longitudinal resonator modes according to (3.25) and discuss the resulting pulse train on a personal computer.

(Purpose of exercise: summation of field components, presentation of calculated results on a personal computer.)

3.9 Calculate the autocorrelation function for a sine wave and for a cosine wave.

(Purpose of exercise: use correlation functions.)

3.10 Calculate the product of the coherence time $\Delta\tau$ and the bandwidth Δf for a Gaussian line with the definitions of (3.36) and (3.37).

(Purpose of exercise: verify the relation between coherence length and bandwidth.)

4. Spectral Analysis of Light

Information from a light beam after its interaction with a solid is obtained from an analysis of the change in its intensity spectrum. Usually monochromators or interferometers are applied to perform this analysis. These instruments will be discussed in the current chapter. It is often useful to "preprocess" the light on its way from the light source to the analyzer or from the analyzer to the detector. Optical elements like reflectors, lenses, filters, polarizers, etc., are appropriate for this process. Light pipes or fiber optics are convenient means to guide the light. Thus, before discussing in detail spectrometers and interferometers we will review some useful optical elements.

4.1 Optical Elements

Only optical elements not described in standard textbooks or elements of particular use in spectroscopy such as filters and polarizers will be discussed. Since much progress has been made recently in the field of fiber optics, this topic will also be addressed here.

4.1.1 Optical Filters

Optical filters are transmission elements which are only transparent in a well defined spectral range. They are available for the whole visible and near-visible spectral range. Filters are often used to remove unwanted spectral components from the light beam. They can be categorized either according to their construction (glass filters, or interference filters) or according to their spectral transmission range (narrow-band filters, line filters, broad-band filters, edge filters, heat filters, and neutral-density filters). Broad-band, edge, and neutral density filters are usually glass filters containing special color centers. In contrast, line or narrow-band filters are usually interference filters. They can be either all dielectric or metal dielectric. To obtain sharp cut offs an interference type of construction is used even for edge filters. Indeed, interference can be used as a basis for any filter. The drawback of interference filters compared to conventional ones is their high price and lower resistance to irradiation load.

Filters are characterized by a *filter function* defined as the transmission $T(\lambda)$

$$\boxed{T(\lambda) = \frac{I_T(\lambda)}{I(\lambda)}\,,} \tag{4.1}$$

where $I(\lambda)$ and $I_T(\lambda)$ are the incident and the transmitted light intensity. The filter function defines the wavelength of maximum transmission, the maximum transmission itself, the half width of the transmission band, the wavelength of cut on and cut off, and the blocking range. Another important characteristic of a filter is the maximum allowable irradiation load.

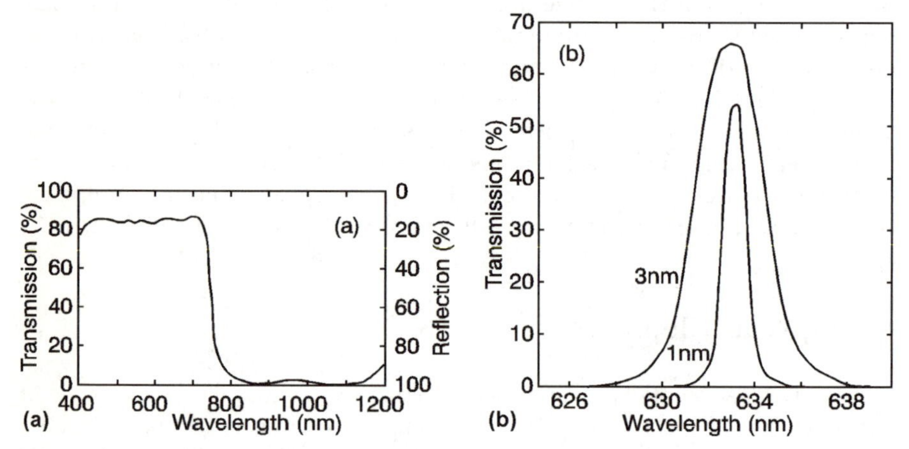

Fig. 4.1. Filter functions for a heat filter (**a**) and for a narrow-band filter (**b**). The full width half maximum is indicated for the latter.

Interference filters consist of several thin layers with varying index of refraction. The extinction of light is based on the interference from layers with thickness $\lambda/4$. This superposition of layers can either lead to a transmission of a narrow spectral range or to a suppression of the transmission in a well defined range. Narrow-band interference filters can be built with a half width as low as 1 nm. In this case the maximum transmission is reduced by 50%. Figure 4.1 shows filter functions for a heat filter and for two different narrow-band filters. Narrow-band filters can be slightly tuned by rotation of the filter out of the plane of normal incidence. The tuning range is about 10 nm for a filter with a FWHM of 10 nm. Tuning by rotation is only possible towards wavelengths longer than that of the center line for normal incidence since the optical wavelengths can only be increased. Narrow-band filters can be made for any specific wavelength. Standard filters exist for the most important laser lines. Narrow-band interference filters are particularly important in spectroscopy with gas lasers since they can suppress the plasma lines originating from the gas discharge.

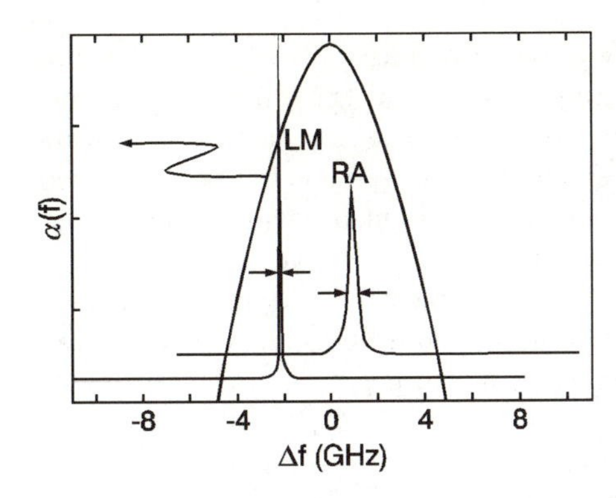

Fig. 4.2. Laser gain curve and spectral lines for the iodine filter; (LM: longitudinal laser mode, RA: rotational absorption of iodine).

In addition to the filters mentioned above, filters using materials in the vapor phase play an important role in spectroscopy. A good example is the iodine filter. Iodine has a rotational absorption spectrum with one line at 514.537 nm. This absorption line nearly coincides with the green line of the argon laser at 514.532 nm. If the cavity of the laser is tuned with an etalon filter to the longitudinal mode which coincides with the rotational absorption band of the vapor an extremely narrow-band filter for the laser line is obtained. The situation is shown in Fig. 4.2. With such a filter, elastically scattered stray light can be effectively suppressed in a light scattering experiment (Raman or Brillouin scattering) while the inelastically scattered light remains unattenuated.

4.1.2 Polarizers and Phase Plates

Polarizers are valuable optical elements because information on the structure of solids can be inferred from the change in the state of polarization of the light after interaction with the solid. Polarizers consist of two prisms of an uniaxial optical material glued together by a liquid with an appropriate index of refraction. The unwanted polarization component is deflected out of the beam by total reflection at the interface between the two prisms. Details about the optical path of the light can be found in standard textbooks. The prisms are known as *Nicol* or *Glan–Thomsen* prisms and are commercially available from companies selling optical accessories. With such prisms polarization ratios

$$R_{\mathrm{p}} = \frac{I_{\perp}}{I_{\|}}$$

of the order of 10^{-5} and a very good throughput for the light can be obtained. The disadvantage of the Nicol prisms is their high price, in particular if they are needed with a large cross section. Sheet polarizers are another option for polarization elements. They are constructed from sheets of highly oriented organic polymers which only allow transmission for light with polarization perpendicular to the orientation of the aligned polymer chains. Polarization ratios for such polarizers are as good as for crystal polarizers but their throughput is lower. Also, since the suppression of the unwanted component of light is by absorption, resistance to light intensity is low. They are not appropriate for laser powers beyond about one watt.

Very often, instead of polarized light, a completely random polarization is desirable. In this case a polarization *scrambler* can be used. A polarization scrambler consists of a birefringent material which generates arbitrary elliptically polarized light. Scramblers only work properly for linearly polarized incident light.

Finally, transformation of linearly polarized light to circularly or elliptically polarized light, or vice versa, may be required. In this case again plates of birefringent material can be used. Linearly polarized light incident with its direction of polarization under 45° to the material's optical axis becomes elliptically or circularly polarized by traversing a plate cut parallel to the optical axis. This is immediately evident from the phase difference generated between the two polarization components oriented parallel to the directions with the refractive indices n_e and n_o, respectively, as shown in Fig. 4.3a. The phase difference $\Delta\phi$ is

$$\boxed{\Delta\phi = \frac{2\pi(n_\mathrm{e} - n_\mathrm{o})d}{\lambda}} \,. \tag{4.2}$$

For $\Delta\phi = \pi/2$ circular polarized light is obtained.

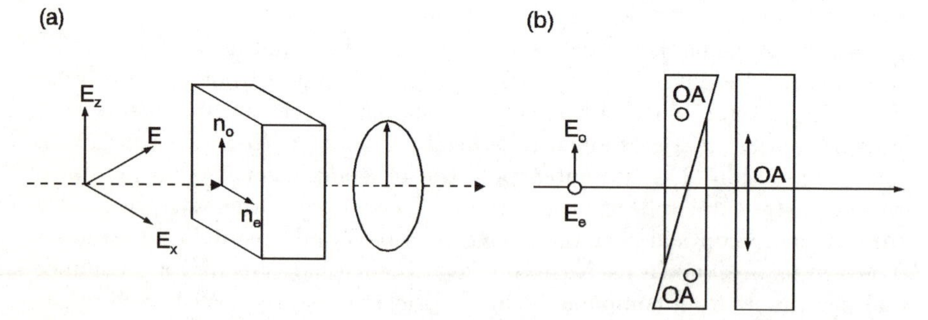

Fig. 4.3. Geometric arrangement for generating circular polarized light from linearly polarized light by a $\lambda/4$ plate (**a**) and compensator arrangement consisting of two perpendicularly oriented phase plates (**b**); (OA: optical axis).

For the arrangement of Fig. 4.3b an arbitrary and adjustable amount of phase difference between the ordinary beam (E_o) and the extraordinary beam (E_e) can be obtained. This means that the phase shift can be adjusted to give linear polarized light for any elliptically polarized light.

4.1.3 Glass Fibers and Light Pipes

The technology of glass fibers or light pipes for light transmission has improved dramatically over the last few years and gained considerable importance in spectroscopy. Instead of moving the spectrometer and the light source to the object for spectroscopy the fibers allow the light to be guided easily to the latter.

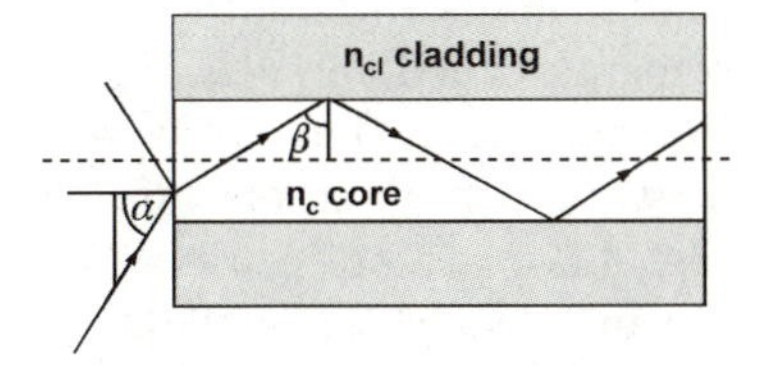

Fig. 4.4. Structure and optical path for the innermost part of an optical fiber.

Optical fibers consist of a highly transparent glass or quartz core surrounded by a "cladding" with a lower index of refraction. Then, up to a certain angle of incidence the incoming light is totally reflected at the interface between core and cladding, and thus guided through the fiber. The geometrical constraints for a single fiber are shown in Fig. 4.4. The minimum internal angle β for which total reflection is possible is related to the ratio between the index of refraction of the cladding and the core by $\sin\beta = n_{\mathrm{cl}}/n_{\mathrm{c}}$. From this the angle of the external acceptance cone is immediately obtained from the geometry of the fiber:

$$\boxed{\sin\alpha = \frac{1}{n_0}\sqrt{n_{\mathrm{c}}^2 - n_{\mathrm{cl}}^2}\,,} \tag{4.3}$$

where n_0 is the index of the medium outside the fiber. The value given by (4.3) is the *numerical aperture* of the fiber. For reasonable differences between the refractive indices of the two components of the fiber, numerical apertures close to 1 can be obtained. Standard commercial fibers usually have apertures of 0.5 or less. For single-mode fibers the apertures in use are even smaller. Instead of the numerical aperture the F/number is often quoted. It is related to the aperture by

$$F/\text{number} = \frac{1}{2\tan\alpha}\,.$$

The fibers must be thin to remain flexible. As a consequence light transfer from extended sources requires several thousand fibers which may be packed into a bundle. Under these circumstances the following types of fibers can be considered:

– single fibers (actually single-mode fibers),
– randomly oriented fiber bundles,

– coherent bundles for imaging.

Coherent fiber bundles are used in imaging systems. Typically 10 000 light points per mm^2 can be transferred coherently in one bundle. Imaging light detectors which will be discussed in the next chapter use such systems.

Randomly oriented fiber bundles are standard and low price light transmitters. Typical transmission curves for a 1-m glass fiber and a 50-cm quartz fiber are shown in Fig. 4.5a. The transmission losses are due mainly to the (non-transparent) interfaces between the individual fibers. Standard fibers can be connected easily by optical coupling elements called *ferruels*. Standardized input and output couplers with appropriate collection optics are available. Bifurcated or trifurcated cables with two or three outlets on one end and a common outlet on the other end (Fig. 4.5b) are often used. For example, for an input light coming through the one branch stray light or reflected light can be detected by the other branch. Coupling a circular to a rectangular cable is possible. This change of cross section may be useful for optimum adaption to the slit of a spectrometer.

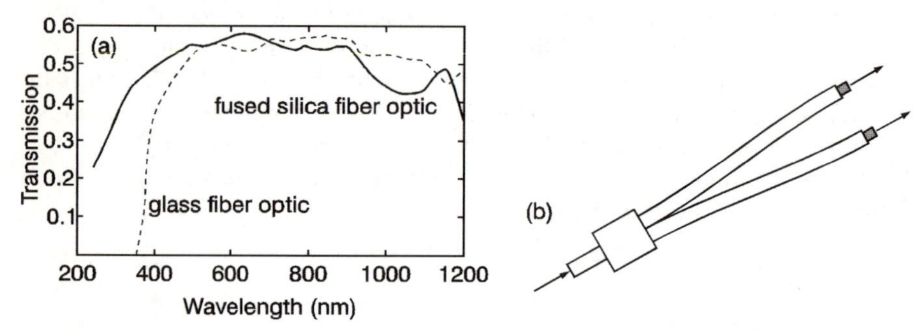

Fig. 4.5. Transmission of a 1-m standard glass fiber optic with 3.2 mm diameter (– – –), and of a fused silica fiber optic, 0.5 m long (—) (**a**), and a bifurcated cable (**b**).

Single-mode fibers are real optical waveguides. They can be used to transfer single-mode laser light without destroying its typical characteristics. Even the polarization direction can be preserved. Like in any waveguide the core diameter is the principal constraint. For a single-mode fiber conserving polarization and operating in the blue/green spectral range, the fiber diameter must not exceed 3 μm, otherwise coupling to other modes and thus multimode transfer will occur. The *normalized frequency number*, also called the *V-number*, is a useful specification for a fiber.

$$V = \frac{2\pi a \sin\alpha}{\lambda}, \tag{4.4}$$

where $\sin\alpha$ and a are the numerical aperture and the radius of the core, respectively. V is 2 for typical single-mode fibers. It determines the maximum number of modes which can propagate in the fiber or the maximum size of the laser focus required for optimum coupling into the fiber. The maximum number of modes is $V^2/2$ and the focus must not exceed the core diameter by more than 30%. The numerical apertures of single-mode fibers are usually rather low, of the order of 0.1 which means a difference in the indices between core and cladding of only 0.02. Since single-mode fibers are used for telecommunications they are manufactured with very low loss. The standard attenuation is 0.1 dB/Km (for the minimum absorption in the spectrum, which is at 1.3 μm). The very small diameter of the fibers requires a very careful alignment of the laser focus for optimum light input.

4.2 Monochromators and Spectrometers

Light can be analyzed with monochromators or spectrometers. The terminology spectrometer is used for the whole setup whereas the monochromator is the optical element itself. The monochromator can be *dispersive* or *non-dispersive*. In dispersive systems like prisms or gratings, a spatial separation is obtained for the spectral components of the light beam. An example of a non-dispersive element is the interferometer used in Fourier spectroscopy. Fourier spectroscopy has been applied for a long time in the far-IR since there are neither strong light sources nor sensitive detectors for this spectral region. Even though it is used today from the far-IR to the near-UV, this technique will be discussed only together with IR spectroscopy in Chap. 10.

4.2.1 Characteristics of Monochromators

As shown schematically in Fig. 4.6, a monochromator consists of an entrance slit, a dispersive element, and an exit slit. The entrance slit is imaged onto the exit slit by a set of mirrors or lenses with focal length F. For convenience, the height H and the width W are assumed equal for the two slits. The most important properties of the monochromator are its *brightness* and its *resolution*. The brightness is given by the ratio A/F and is equal to the *etendue* E

$$E = \frac{WHA^2}{F^2} \qquad (\text{in m}^2) \tag{4.5}$$

of the instrument. The power P (in Watts) transmitted through the spectrometer is obtained from the incident light intensity I_0 and E by

$$\boxed{P = T(\lambda) I_0 E\,,} \tag{4.6}$$

where $T(\lambda)$ is the transmission coefficient of the instrument. If A^2/F^2 in (4.5) is expressed in steradian E is called the *optical conductance* $G = WH\Omega = WH\pi A^2/F^2$.

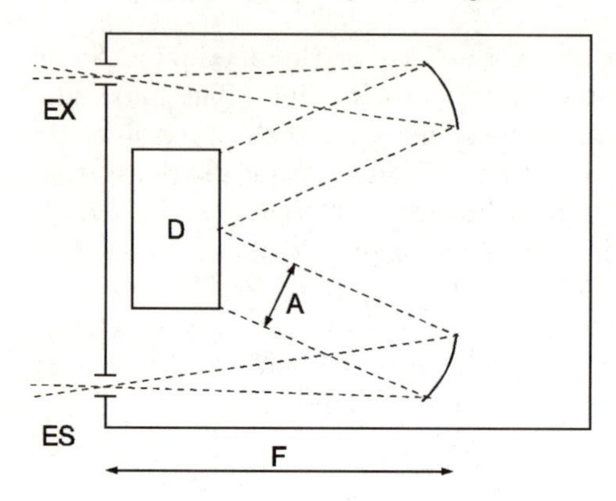

Fig. 4.6. Schematic arrangement of a monochromator; (ES: entrance slit, A: aperture, D: dispersive element, EX: exit slit, F: focal length).

The resolution is determined by diffraction effects or by the nature of the multiple-beam interference. The *Rayleigh criterion* is the basis for defining the resolution. Two beams with wavelengths λ and $\lambda + \delta\lambda$ are resolved if the maximum of the diffraction pattern (or the constructive interference) for one beam coincides with the minimum of the diffraction pattern (or the destructive interference) for the other beam. With this definition of $\delta\lambda$ the resolution is

$$\boxed{R_0 = \frac{\lambda}{\delta\lambda} = \frac{f}{\delta f} = \frac{\nu}{\delta\nu} .} \tag{4.7}$$

Its value is the same whether the spectral distribution is given in wavelengths λ, frequencies f, or wave numbers ν.

4.2.2 The Prism Monochromator

Prism monochromators rely on the dispersion of light propagating in solids. This dispersion is determined by the oscillations of bound charges within the material. As will be shown in Chap. 6, the dispersion as well as the absorption of light strongly increases close to such oscillators.

The brightness and resolution of a prism monochromator depend only on the size and dispersion of the prism. The resolution is given by the Frauenhofer diffraction

$$I(\theta, \delta\theta) = I(\theta, 0)\frac{\sin^2 \pi\nu A\delta\theta}{(\pi\nu A\delta\theta)^2} . \tag{4.8}$$

Here $I(\theta, \delta\theta)$ means the distribution of the light intensity as shown in Fig. 4.7, ν is the light frequency in wave numbers, A the aperture, and $I(\theta, 0)$ the intensity at the maximum. The angular dispersion depends on the dispersion of the prism $\mathrm{d}n/\mathrm{d}\nu$ and is given by[1]

[1] For a derivation of this relation see [4.1].

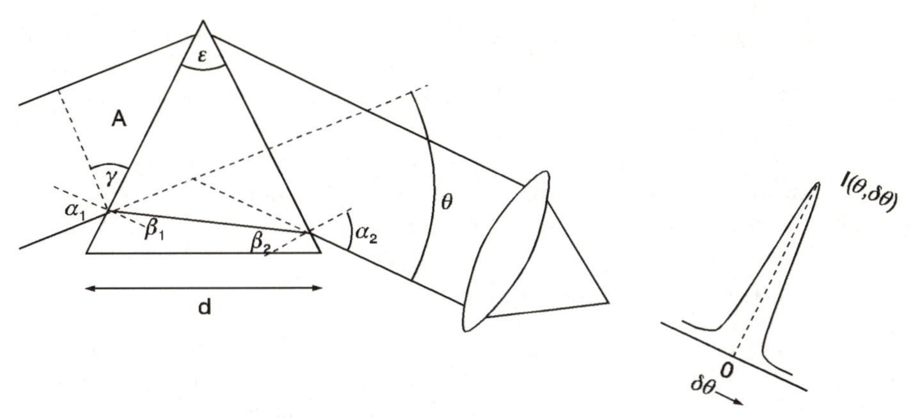

Fig. 4.7. Optical path through a prism monochromator.

$$\frac{\mathrm{d}\theta}{\mathrm{d}\nu} = \frac{\mathrm{d}\theta}{\mathrm{d}n}\frac{\mathrm{d}n}{\mathrm{d}\nu} = \frac{\sin\varepsilon}{\cos\alpha_2\cos\beta_1}\left|\frac{\mathrm{d}n}{\mathrm{d}\nu}\right| . \tag{4.9}$$

For a symmetric transition with a total angle of deflection θ we have $\alpha_1 = \alpha_2 = \alpha = (\theta+\varepsilon)/2$, $\beta_1 = \beta_2 = \beta = \varepsilon/2$, and $\sin\alpha/\sin\beta = n$. Then

$$\boxed{\frac{\mathrm{d}\theta}{\mathrm{d}\nu} = \frac{2\sin(\varepsilon/2)}{[1-n^2\sin^2(\varepsilon/2)]^{1/2}}\frac{\mathrm{d}n}{\mathrm{d}\nu} = \frac{d}{A}\frac{\lambda^2\mathrm{d}n}{\mathrm{d}\lambda} .} \tag{4.10}$$

From the Frauenhofer diffraction pattern the distance between the maximum and minimum diffraction intensity is

$$\delta\theta = \frac{1}{\nu A} . \tag{4.11}$$

With this and (4.10), the Rayleigh criterion yields for the resolution

$$R_0 = \frac{\nu}{\delta\nu} = A\nu^2\frac{2\sin(\varepsilon/2)}{\cos((\theta+\varepsilon)/2)}\left|\frac{\mathrm{d}n}{\mathrm{d}\nu}\right| = \nu^2 d\frac{\mathrm{d}n}{\mathrm{d}\nu} \tag{4.12}$$

or

$$\boxed{R_0 = \lambda/\delta\lambda = d\frac{\mathrm{d}n}{\mathrm{d}\lambda} ,} \tag{4.13}$$

where d is the base of the prism. Consequently, for a high resolution, a prism must have a wide base and be made of a material with a high dispersion. Thus, the resolution increases as the wavelength of the light approaches the absorption lines for phonons. However, with this the absorption increases so that finally no more light is transmitted. This limits the use of the prism. Note that the brightness of a prism increases with height. Thus, large prisms are necessary for good monochromators.

Prism monochromators are still used for the near-UV, visible, and the near-IR spectral range in simple instruments. Compared to grating spectrometers their advantage is the dispersion of the light into only one spectral order. This means that all the light of a particular wavelength is deflected by the same angle. In contrast, in the case of a grating spectrometer, only a certain fraction of the light is diffracted and this fraction is distributed to several orders of diffraction.

A prism spectrometer's usefulness is limited to long wavelengths by phonon absorption. In reality prisms can be applied down to wave numbers of 280 cm^{-1}, or in special cases to wave numbers as low as 180 cm^{-1}.

4.2.3 The Grating Monochromator

In general, grating monochromators are superior to prism monochromators. It is obviously more difficult to make large prisms than to make large gratings. Moreover, the spectral resolution depends on the number of lines rather than on the size of the grating. To calculate the spectral resolution we have to start from the equation for the difference in optical wavelength for the partial beams of the grating, which can be read directly from Fig. 4.8

$$\Delta l_{\mathrm{opt}} = d(\sin\theta_{\mathrm{i}} \pm \sin\theta_{\mathrm{d}}) \, . \tag{4.14}$$

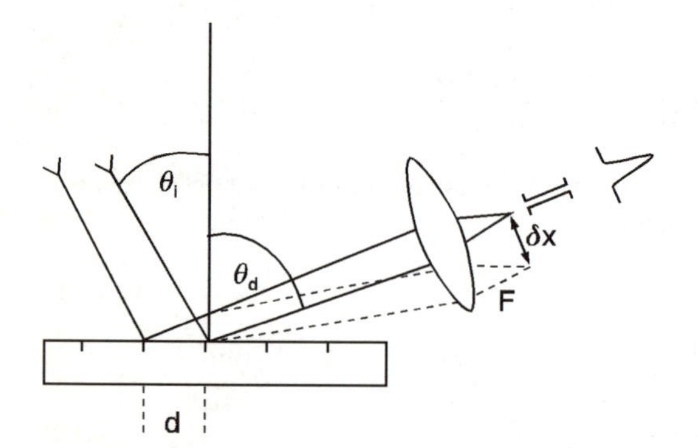

Fig. 4.8. Diffraction geometry for a grating and spectral resolution; (F: focal length, d: grating constant).

θ_{i}, θ_{d}, and d are the angle of incidence, the angle of diffraction, and the grating constant, respectively. The positive or the negative sign is used depending on whether the diffracted beam is on the same or on the opposite side of the incident beam with respect to the grating normal. Constructive interference is obtained for $\Delta l_{\mathrm{opt}} = m\lambda$.

To sum the partial beams their relative phase difference ϕ must be considered, i.e.,

$$\phi = 2\pi\frac{\Delta l_{\mathrm{opt}}}{\lambda} = \frac{2\pi d}{\lambda}(\sin\theta_{\mathrm{i}} - \sin\theta_{\mathrm{d}}) \, . \tag{4.15}$$

With the reflection coefficient R, the total field in the diffracted beam is

$$E_{\mathrm{r}} = \sqrt{R}\sum_{s=1}^{N} E_0 e^{\mathrm{i}s\phi} = \sqrt{R}E_0\frac{1 - e^{\mathrm{i}N\phi}}{1 - e^{\mathrm{i}\phi}} \tag{4.16}$$

for a grating with N lines. Since the average intensity in the beam is EE^*, the interference pattern has the form

$$I_\mathrm{r} = RI_0 \frac{\sin^2(N\phi/2)}{\sin^2(\phi/2)} \,. \tag{4.17}$$

In the directions for which $\phi/2 = 0$ or $m\pi$ I_r is maximum which means constructive interference. Figure 4.9 exhibits interference fringes as a function of ϕ for $N = 20$ and $N = 5$. The spectral resolution is determined by the width of the constructive interference line. If $\delta\lambda$ is the minimum distance between two lines which can be resolved, the spectral resolution for the m-th order spectrum is

$$R_0 = \frac{\lambda}{\delta\lambda} = Nm \,. \tag{4.18}$$

For good gratings N is of the order of 2×10^5 which, in the first order, corresponds to a resolution, expressed as a band pass $\delta\nu$, of 0.1 cm^{-1} for visible light. In Fig. 4.9 all orders have the same peak intensity. In reality this is not so. As a consequence of diffraction of the individual partial beams by the entrance slit higher order spectra are suppressed as $1/m^2$.

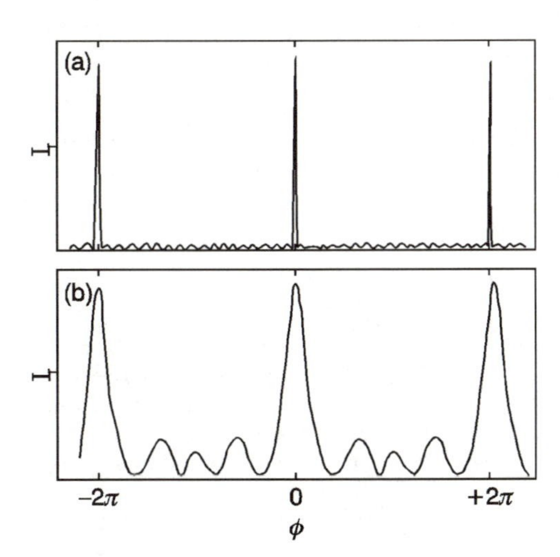

Fig. 4.9. Multiple-beam interference for a grating with $N = 20$ lines (**a**) and with $N = 5$ lines (**b**) from (4.17).

In practice, the spectrum is recorded by rotating the grating and measuring the light passing through a small slit on the exit side of the spectrometer. This step by step method of recording is called single channel detection. In contrast, for multichannel recording wide parts of the spectrum are recorded simultaneously, as discussed below and in Chap. 5.

The spectral resolution of a grating spectrometer is not only determined by the resolution of the grating itself but also by the width of the slits. This is illustrated by the dashed lines in Fig. 4.8. The reflected light is imaged on the exit slit by an optical element of focal length F. Hence, the distance δx of two images generated by a beam with wavelengths λ and $\lambda + \delta\lambda$ is $F\delta\theta$. To resolve the two images they must be separated by more than the slit width. This means, at most, $\delta x = F\delta\theta$ must be equal to the width W of the slit. Hence[2]

$$W = \delta x = \frac{\delta x}{\delta\lambda}\delta\lambda = F\left|\frac{\mathrm{d}\theta}{\mathrm{d}\lambda}\right|\delta\lambda\,. \tag{4.19}$$

Using (4.14) for $\Delta l_{\mathrm{opt}} = m\lambda$ the spectral resolution is obtained for very small angles of incidence as

$$\boxed{R_0 = \frac{\lambda}{\delta\lambda} = \frac{\lambda F m}{W d \cos\theta}\,.} \tag{4.20}$$

For the numerical values $d = 5\times10^{-5}$ cm, $W = 100\,\mu\mathrm{m}$, $F = 100$ cm, $\lambda = 0.5\,\mu\mathrm{m}$, $\theta = 0°$, and $m = 1$ the resolution, expressed as a spectral bandpass, is $\delta\nu = 2\ \mathrm{cm}^{-1}$.

The resolution allows to express the etendue by the parameters of the spectrometer. From (4.5) and (4.20) we have for $\lambda \approx d$

$$E^{\mathrm{gr}} = \frac{WHA^2}{F^2} = \frac{A^2H}{F}\frac{\delta\lambda}{\lambda} = \frac{AR_0H\delta\lambda}{F}\,, \tag{4.21}$$

where we have used $R_0 = \lambda/\delta\lambda \approx A/d$.

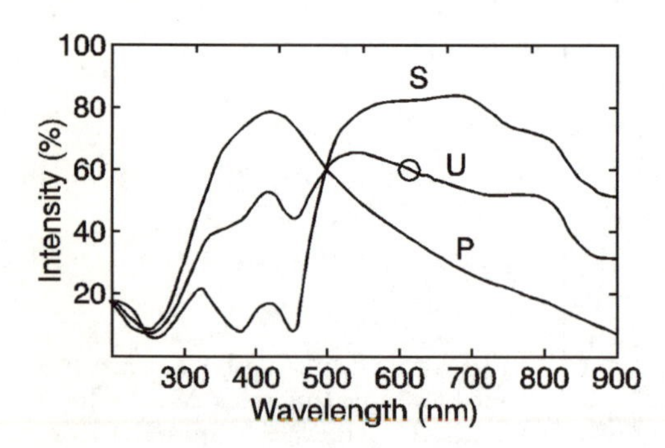

Fig. 4.10. Efficiency of holographic gratings with 180 000 lines for unpolarized (U), parallel polarized (P) and perpendicularly polarized (S) light.

The amount of light transmitted through the spectrometer is determined by the brightness and the transmission coefficient. The latter depends strongly on the direction of polarization of the incident light with respect to the orientation of the ruling on the grating. Figure 4.10 plots the relative intensities observed through a spectrometer with holographic gratings versus light wavelength for a given incident intensity. Obviously light at 647.1 nm is much easier to observe if it is polarized perpendicularly to the ruling

[2] This equation holds as well for the spectral resolution in a prism spectrometer.

compared to light with a polarization parallel to the ruling. For blue light it is just opposite. This characteristic of the response must definitely be taken into account if intensities for the two directions of polarization are compared. In addition any partial polarization of the light in either direction may result in incorrect relative intensities along a spectrum. To avoid this problem it is possible to use a scrambler, as described in Sect. 4.1. Unfortunately a scrambler only works properly for highly linearly polarized light.

The distribution of the diffracted light over the different orders is a disadvantage of grating monochromators since there is less energy in any single order. The distribution can be avoided by using special gratings called *echellet* gratings. Since such gratings are mainly employed in the IR they will be discussed in Chap. 10.

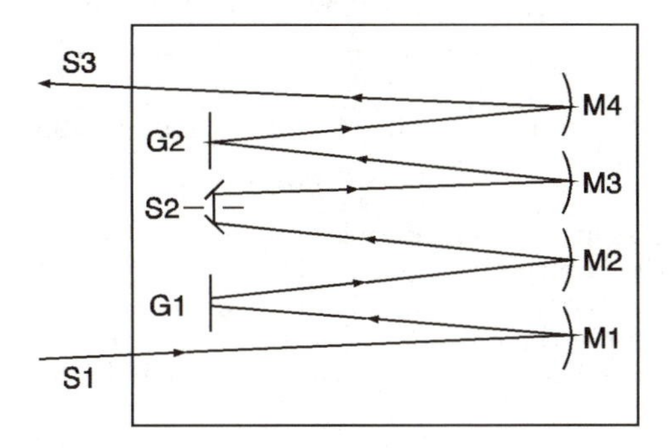

Fig. 4.11. Schematic setup of a double monochromator; (M: mirrors, G: gratings, S: slits).

In modern spectroscopy usually multiple monochromators are employed. They can either be operated in an additive or in a subtractive mode. Figure 4.11 sketches the setup for a double monochromator in an additive mode for a Cherny-Turner arrangement. In this case two gratings, three slits, and four concave mirrors are needed. The incoming light is two times additively dispersed by the gratings G1 and G2. Scanning occurs by synchronous rotation of both gratings. The resolution and suppression of stray light is greatly improved compared to a single monochromator.

For further improvement a third monochromator can be arranged behind the exit slit resulting in a triple monochromator system. The third monochromator must be tuned to be synchronous with the two other gratings. Such systems are often used for spectroscopic analyses in the immediate vicinity of a very strong line such as a laser line. In this case the third monochromator operates with a wide slit and a sharp cut off towards the laser line.

For a subtractive mode the first two monochromators have oppositely directed spatial dispersions. This means that after passing the first two monochromators the light which satisfies the bandpass condition is refocused on the exit slit which serves as the entrance slit of the third monochromator. Only the latter provides the spatial dispersion. The path of the light beam is sketched in Fig. 4.12a for the additive and for the subtractive mode. The

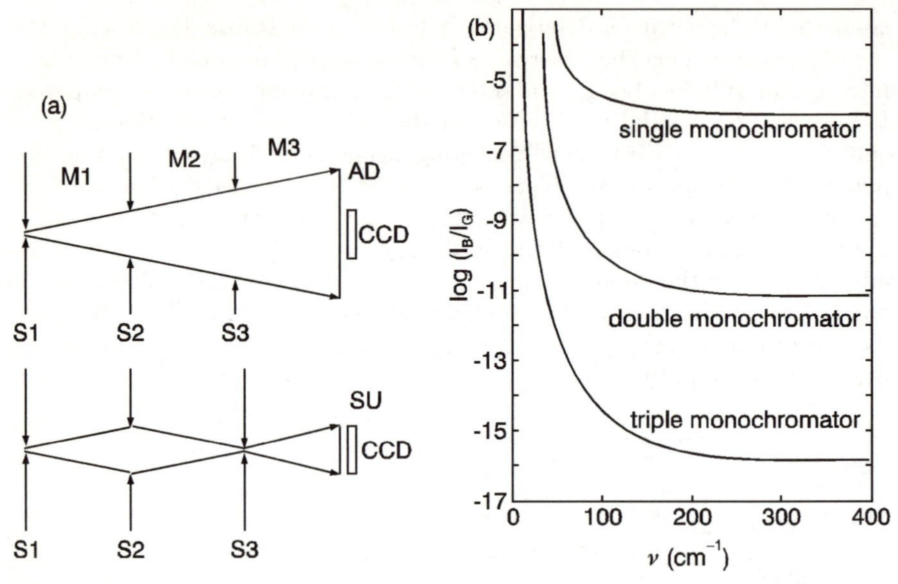

Fig. 4.12. Schematic drawing of the light path for a triple monochromator operating in the additive (AD) and in the subtractive (SU) mode (**a**), and ratio of background scattering I_{B} to grating scattering I_{G} versus distance ν from a laser line for various multiple monochromators in the additive mode and single-channel detection; after [4.2] (**b**); Mi and Si refer to the three monochromators and to the three slits in use.

design is for use with an optical multichannel analyzer described in the next chapter. For the additive mode the slit S3 is wide, the linear dispersion and the resolution are about a factor three higher than for the subtractive mode but the detector can only record a small part of the spectrum. For the subtractive mode the slit S3 is narrow, the linear dispersion and the resolution are low but the detector can record a much larger part of the spectrum. Stray-light rejection is also much better for the subtractive option. When using an optical multichannel analyzer the gratings are not rotated.

The suppression of stray light for a single, double, and triple monochromator is compared in Fig. 4.12b. As, for example, in Raman-scattering experiments a stray-light suppression of 10^{-10} is needed with a double monochromator the laser line can be approached to about 100 cm^{-1} and with a triple monochromator to about $20\,\mathrm{cm}^{-1}$.

4.3 Interferometers

If a very high spectral resolution, albeit on a rather narrow spectral range, is required interferometers must be applied. High quality *Fabry–Perot* interferometers which are commercially available are the most frequently used

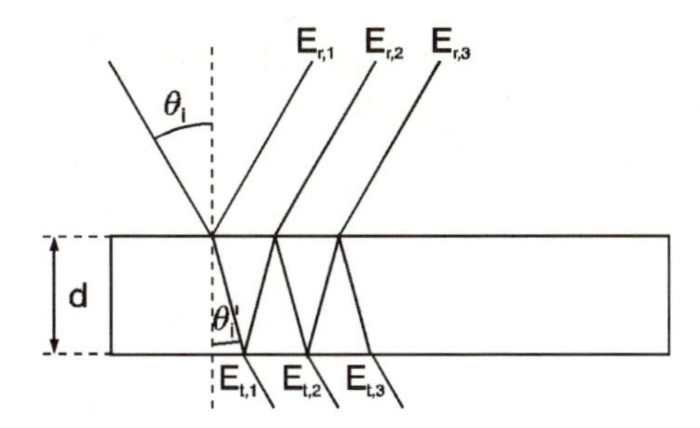

Fig. 4.13. Reflection and transmission on a plane-parallel plate of thickness d; (E: partial beam amplitudes).

instruments for this purpose. We will therefore restrict the discussion to this type and start with a study of plane and parallel plates which are its basic constructive elements.

4.3.1 Multiple-Beam Interference for a Parallel Plate

The Fabry–Perot interferometer relies on the interference of light multiply reflected and refracted by a plane parallel plate. As shown in Fig. 4.13, the EM wave is split on the front and back side of the plate into partial beams with the amplitudes $E_{\mathrm{r,i}}$ and $E_{\mathrm{t,i}}$, respectively. The phase difference ϕ between two consecutive partial beams is

$$\boxed{\phi = \frac{4\pi n d}{\lambda} \cos\theta_{\mathrm{i}}' \,,} \tag{4.22}$$

where d and θ_{i}' are the thickness of the plate and the internal diffraction angle, respectively, and n is the refractive index of the plate. The partial beams are subjected to interference resulting in an enhancement or quenching of the reflected and transmitted light. Performing a phase coherent summation of the field amplitudes and then taking the absolute square of the resulting field the intensities of the two beams are

$$\frac{I_{\mathrm{r}}}{I_0} = \frac{4R\sin^2(\phi/2)}{(1-R)^2 + 4R\sin^2(\phi/2)} \,, \tag{4.23}$$

$$\frac{I_{\mathrm{t}}}{I_0} = \frac{(1-R)^2}{(1-R)^2 + 4R\sin^2(\phi/2)} \,. \tag{4.24}$$

R is the reflection coefficient at the interface from outside to inside the plate.

With increasing phase angle ϕ both equations yield a periodic structure (with sharp peaks and broad minima in the transmission) representing the increasing order of the interference.

The relations (4.23) and (4.24) are known as the *Airy formulae.* Any absorption in the plate is neglected. The derivation of the formulae is given in Appendix D.1

It is common practice to introduce the symbol F for the expression $4R/(1-R)^2$. Then (4.23) and (4.24) have the simpler forms

$$\frac{I_\mathrm{r}}{I_0} = \frac{F\sin^2(\phi/2)}{1+F\sin^2(\phi/2)} , \tag{4.25}$$

$$\frac{I_\mathrm{t}}{I_0} = \frac{1}{1+F\sin^2(\phi/2)} . \tag{4.26}$$

The quantity F^* which is directly related to F by

$$F^* = \frac{\pi}{2}\sqrt{F} = \frac{\pi\sqrt{R}}{1-R} \tag{4.27}$$

is called the *finesse* of the interference. Since for good mirrors $R \geq 0.99$, F^* can be of the order of 300.

In general, $d \gg \lambda$ so that phase differences between the partial beams are large. Then, for components of the spectrum located at ν_1 and ν_2 the maximum for the m-th order of interference for the first frequency may coincide with the maximum for the $m+1$-th order for the second frequency. In this case we can obviously not discriminate between these two frequencies. The minimum distance $\nu_1 - \nu_2 = \Delta\nu$ where this happens is obtained from (4.22) and the Airy formulae. For perpendicular incidence it occurs where $\Delta\phi = 2\pi = 4\pi n d \Delta\nu$ or

$$\Delta\nu = \frac{1}{2nd} . \tag{4.28}$$

Since for this and any larger value of the difference between frequencies the individual frequencies can not be identified, $\Delta\nu$ is called the *free spectral range* of the interferometer plate. On the other hand (4.25) and (4.26) define also the full width at half maximum $\phi_H = \phi_1 - \phi_2 = \phi(\nu_1) - \phi(\nu_2)$ for the interference pattern by

$$I_\mathrm{t}(\phi_1) = I_\mathrm{t}(\phi_2) = I_0/2 .$$

Expressed in phase angles this yields for reasonably large values of R (F or $F^* \gg 1$)

$$\phi_\mathrm{H} = \frac{4}{\sqrt{F}} . \tag{4.29}$$

Since the free spectral range expressed in phase angles is 2π we obtain for the ratio between the free spectral range and the half width

$$\boxed{\frac{\Delta\nu}{(\delta\nu)_\mathrm{H}} = \frac{2\pi}{\phi_\mathrm{H}} = \frac{\pi}{2}\sqrt{F} = F^* ,} \tag{4.30}$$

with $(\delta\nu)_\mathrm{H} = \nu_1 - \nu_2$. Thus, the finesse has a very intuitive meaning. It is the ratio of the distance of two consecutive interference fringes to their width. Hence, it determines also the spectral resolution $\nu/\delta\nu$ of the interference. If we define two frequencies as being resolved if their separation $\delta\nu$ is $\geq$ the half width of their interference fringes $(\delta\nu)_{\mathrm{H}}$, $\delta\nu$ equals $1/\pi n d\sqrt{F}$. From this we obtain

$$\boxed{\frac{\nu}{\delta\nu} = \frac{\nu F^*}{\Delta\nu} = \nu F^* 2nd\,.} \tag{4.31}$$

4.3.2 The Fabry–Perot Interferometer

The Fabry–Perot interferometer relies on a multiple-beam interference from a parallel plate of air contained between two highly reflecting mirrors. Figure 4.14a is a schematic of the interferometer. The light is multiply reflected between the mirrors M_1 and M_2, and the transmitted fraction is imaged on a screen. For a cylindrical geometry and monochromatic radiation circular fringes are generated as a result of the constructive and destructive interferences for off-axis partial beams. The off-axis beams are a consequence of the finite extension of the light source. The distance between the rings determines the free spectral range of the interferometer and the width of the rings the resolution $\delta\nu$. Since $n = 1$ in the present case the two quantities are given by

$$\Delta\nu = \frac{1}{2d} \tag{4.32}$$

and

$$\delta\nu = \frac{\Delta\nu}{F^*} = \frac{1}{2dF^*}\,. \tag{4.33}$$

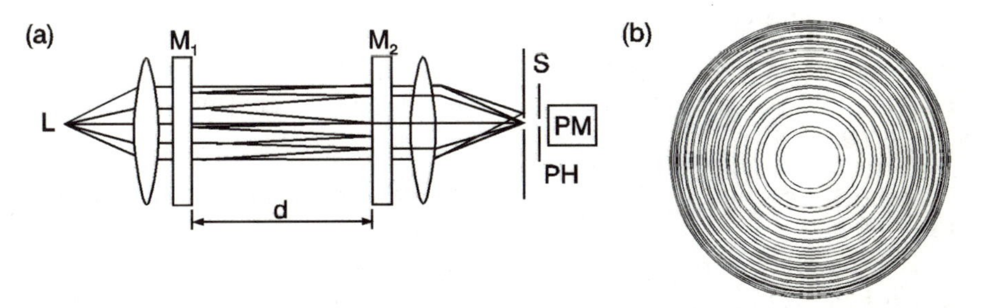

Fig. 4.14. Schematic drawing of a single-pass Fabry–Perot interferometer (**a**) and interference fringes for the yellow line of Na vapor (**b**); (L: light source, M_1, M_2: mirrors, S: screen, PH: pinhole, PM: photomultiplier).

The finesse F^* of the interferometer is again given by the ratio between the free spectral range and the resolution $\Delta\nu/\delta\nu$. Figure 4.14b shows the fringes for a spectral line with a doublet structure. The small distance between the concentric rings in one fringe gives the amount of splitting.

Equation (4.24) can be used to evaluate the transmission of the interferometer, but the absorption A of the mirrors must be taken into account.

Because of the large value of R (R is very close to 1) a large number of reflections occurs and even small losses at each reflection cannot be neglected. The absorption is given by $A = 1 - R - T$ where T is the transmission coefficient. The absorption is taken into account by multiplying (4.24) with T^2/T_0^2 where $T_0^2 = (1-R)^2$ is the square of the transmission without absorption. The correction factor contains the square of the transmission ratio because there are two reflections per round trip. With this correction the transmission is obtained from

$$\frac{I_\mathrm{t}}{I_\mathrm{i}} = \frac{1 - 2A/(1-R)}{1 + (4F^{*2}/\pi^2)\sin^2(2\pi d/\lambda_0)} . \tag{4.34}$$

The spectral bandpass $\delta\nu$ is directly related to the inverse of the finesse, as can be seen from (4.33). Thus, the resolution improves as the finesse or equivalently the reflectivity of the mirrors increases.

Very often as, for instance, in inelastic light scattering experiments, we wish to study a very weak line very close to a very strong line. In these cases it is crucial for the light extinction ratio between the interference fringes to be very high. This is described by the *contrast* C of the interferometer, obtained from (4.34) by

$$\boxed{C = \frac{I_0(\max)}{I_0(\min)} = 1 + \frac{4F^{*2}}{\pi^2} .} \tag{4.35}$$

This means that a high finesse gives also a good contrast. However, it is important to recall that with increasing R the maximum for the transmission is reduced by $1 - 2A/(1-R)$. Typical values for a good Fabry–Perot interferometer are:

$R = 0.985$	$F^* = 207$	$\Delta\nu = 0.5\,\mathrm{cm}^{-1}$
$d = 1\,\mathrm{cm}$	$C = 5 \times 10^4$	$\delta\nu = 1/2dF^* = (1/414)\,\mathrm{cm}^{-1}$
$A = 0.002.$		

Thus, the resolution of these instruments is of the order of $0.002\ \mathrm{cm}^{-1}$ or 3×10^{-4} meV.

For the practice of spectroscopy it is convenient to have one of the mirrors in Fig. 4.14 on a piezoelectric translation stage. Then, the length of the cavity can be tuned and the interference rings move from or to the center of the screen. The fringes and the space between them can be tuned very accurately by replacing the screen in Fig. 4.14 with a pinhole and watching the transmission through it with a photomultiplier. The latter records a signal, as shown in Fig. 4.15, where the light was assumed to consist of a main line and a small satellite close by. This is a very convenient way to record the spectrum electronically. The physical meaning of the finesse as the ratio between $\Delta\nu$ and $\delta\nu$ is particularly evident in this picture.

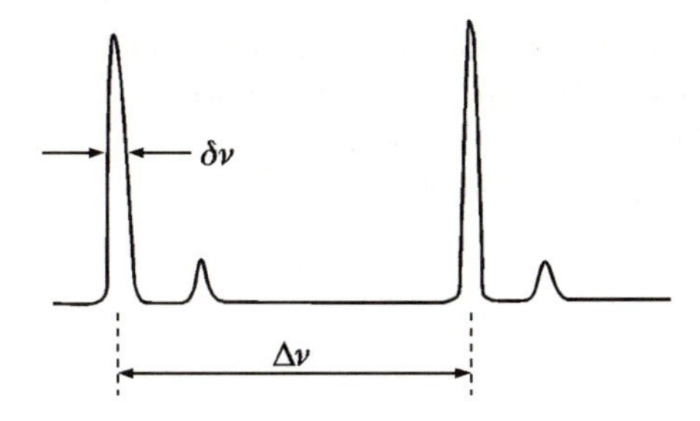

Fig. 4.15. Output of a Fabry–Perot interferometer recorded with a photomultiplier.

4.3.3 The Multipass Fabry–Perot Interferometer

Since the contrast in the interferometers described above is only 10^4, several attempts were made to improve it by arranging two or more interferometers in series. The most successful instrument along this line is the *multipass* Fabry–Perot interferometer where the interference process is repeated several times by inserted corner cube reflectors. A possible set up of a five-pass interferometer is illustrated in Fig. 4.16. The diaphragms D were inserted to reduce stray light. As can be expected, the overall contrast is increased as

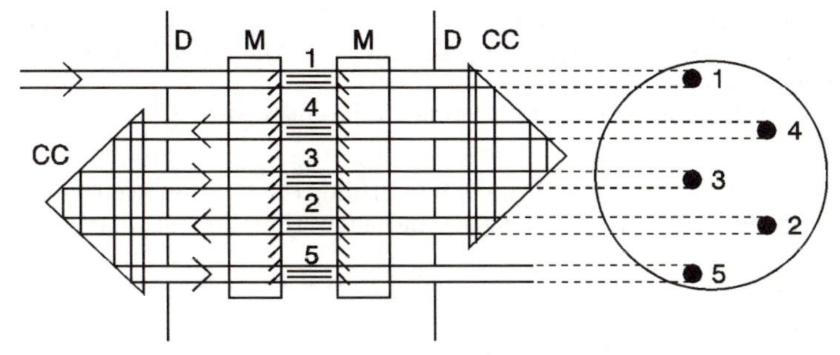

Fig. 4.16. Schematic drawing of a five-pass interferometer; (M: mirrors, CC: corner cube reflectors, D: diaphragm with pinholes); after [4.3].

$$C = C_1^p \,, \tag{4.36}$$

where p is the number of passes, and C_1 is the contrast for a single pass. The finesse increases as

$$F^* = \frac{F_1^*}{\sqrt{2^{1/p} - 1}} \,. \tag{4.37}$$

However, the transmission decreases as

$$\frac{T}{T_0} = \left(1 - \frac{A_1}{1 - R}\right)^{2p} \,. \tag{4.38}$$

For multipass operation of the interferometers a contrast of the order of 10^{10} is easily obtained. This is already very close to stray-light suppression in double monochromators. As a result multipass Fabry–Perot interferometers are widely used for inelastic light-scattering spectroscopy from acoustic phonons, for example.

Problems

4.1 A doublet structure of a spectral line in the green spectral range has a peak-to-peak separation of 3 cm^{-1}. Which slit width is needed to resolve it in the first order with a $F = 1$ m grating monochromator for a 10 cm size grating with 180 000 lines

(Purpose of exercise: use of an important formula for practical work.)

4.2[a] Show that for the interference on a plane parallel plate the ratio of the amplitudes from two consecutive partial beams $E_{\text{r},m+1}/E_{\text{r},m}$ equals the reflection coefficient $R = I_\text{r}/I_0$ for $m \geq 2$.

(Purpose of exercise: get familiar with reflection and transmission geometry for a plane parallel plate.)

4.3 A thin film with index of refraction $n = 1.4$ shows interference patterns for excitation with red light close to the line of a HeNe-laser. The next neighbor distance of the fringes is 50 cm^{-1}. How thick is the film if the light incidence is normal to the film surface? Note: In a wave number linear representation, the distance between the fringes is independent of ν.

(Purpose of exercise: application of interference relationships to thin film optics.)

4.4 Express the free spectral range and the half width for the interference fringes of a Fabry–Perot interferometer in wavelengths and show that the ratio is again equal to the finesse.

(Purpose of exercise: change from phase angles to wave numbers and wavelengths.)

4.5 Show that the contrast for a Fabry–Perot interferometer is given by $1 + 4F^{*2}/\pi^2$.

(Purpose of exercise: use of Airy formulae.)

4.6* Show that the finesse for the multipass Fabry–Perot interferometer is given by $F^* = F_1^*/\sqrt{2^{1/p} - 1}$ where p is the number of passes.

(Purpose of exercise: get familiar with multipass problems.)

5. Detection of Electromagnetic Radiation

The radiation used in spectroscopic experiments needs to be measured. Due to the wide field of applications, radiation detectors must satisfy a large number of different constraints. Requirements can be high sensitivity, large bandwidth, high speed, high stability, etc. Hence, a large number of different concepts and technical realizations exist for the detectors. Their characteristic and basic properties are:

- the quantum efficiency,
- the signal response,
- the detection limit, and
- the speed.

As in the previous section we will first discuss general forms of radiation detectors applicable to various spectroscopic techniques. Detectors dedicated to special spectroscopic methods will be discussed in the corresponding chapters.

5.1 Signal and Noise

Since the radiation intensities in spectroscopic experiments are usually very low, noise and statistical errors play an important role. The *signal-to-noise ratio* is the quantity which characterizes this problem. The question is very general and applies to photographic films as well as to photon counting or photoelectric detection. Figure 5.1 gives an example from a scattering experiment. The figure shows a signal intensity of about 26 counts per second, sitting on a background signal from the detector of about 23 counts. The noise of the background is about 4 counts. This gives a signal to noise ratio of 6.5.

The detection of light is determined by the quantum nature of the radiation. The photon is either absorbed in a process relevant to the detection scheme or not. These are the only two possibilities. The probabilities of such processes are thus similar to the probabilities of tossing a coin. For coins either head or tail may occur and these are exclusive events. The only difference with the photon absorption is that the probabilities of the two events p

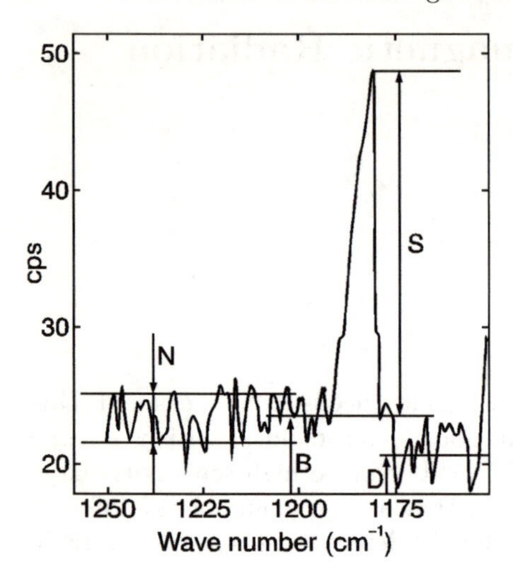

Fig. 5.1. Light signal from a scattering experiment; (S: signal intensity, B: background signal, D: dark signal, N: noise).

and $q = 1 - p$ are equal in the coin tossing experiment. Each of them is 1/2. For the photon absorption the probability p for contributing to the detection process may be much smaller than 1. Considering the problem of light detection in a form appropriate for a description by probability theory we call "signal" the number of events k of photon absorption processes out of a number of $n \geq k$ incident photons. Each single absorption process has a probability p and $q = 1 - p$ is the probability that no absorption occurs or that absorption does not contribute to the signal. This approach is valid for a wide range of different radiation detectors. The probability for the "constructive" absorption of k photons is given by the distribution

$$P(k, n, p, q) = \binom{n}{k} p^k q^{n-k} \,. \tag{5.1}$$

This distribution peaks for $k = k_{\max} \approx p(n + 1)$. The total signal scales, of course, with n. For small values of p the binomial distribution can be well approximated by the Poisson distribution

$$\boxed{P(k, n) = \frac{y^k}{k!} e^{-y}, \qquad \text{with} \qquad y = np \approx \langle k \rangle \,.} \tag{5.2}$$

For n large enough y is the mean value $\langle k \rangle$ of the distribution or the average magnitude of the signal. Since P is a probability other values of k may be observed as well. This means the mean signal $\langle k \rangle$ is dressed with noise. The square root of the variance σ^2 of the distribution P should be a good definition for the noise. A rather simple calculation yields $\langle k \rangle$ for the variance of the Poisson distribution. This means the variance is equal to the mean value. Then, according to the definition above the noise is

$$\sigma = \sqrt{\langle (k - \langle k \rangle)^2 \rangle} = \sqrt{\langle k^2 \rangle - \langle k \rangle^2} = \sqrt{\langle k \rangle} \,. \tag{5.3}$$

Since the signal $\langle k \rangle$ increases linearly with the length of the measuring time T, the noise increases as the square root of T. The important quantity defining the accuracy of the measurement is the signal-to-noise ratio which accordingly increases as $\sqrt{T}$.

5.2 Photographic Films

Photographic films are common light detectors used in the spectral range from the near-IR to x rays. A large variety of emulsions has been developed which vary in spectral sensitivity, granularity, speed, resolution, etc., to satisfy specific requirements. The basic process of the photographic recording is a reduction of a silver-halide salt during light exposure according to a reaction of the type

$$\mathrm{AgBr} + \hbar\omega \rightarrow \mathrm{Ag} + \mathrm{Br}.$$

The latent image created in this way is amplified by the developing process of the film in which additional silver-halide is reduced, particularly at the sites of the latent image. The process is terminated by a stopping reaction. In a final process the unreacted salt is removed from the emulsion and a stable image is obtained.

The most important signature of an emulsion is its quantum efficiency η. It determines which fraction of the incoming light quanta participates in the reduction process. Typical values for η are between 0.01% and 1%. The quantum efficiency depends strongly on the light wavelength and on the grain size of the emulsion.

From a practical view point the *sensitivity* equal to the signal response, the *spectral sensitivity*, and the *resolving power* are the most important characteristics of the emulsion. The sensitivity is defined as the optical density of the film after development for a given light exposure. The light exposure is measured in incident light power times irradiation time, technically described in units of (candela meter seconds). Figure 5.2 shows the relation between

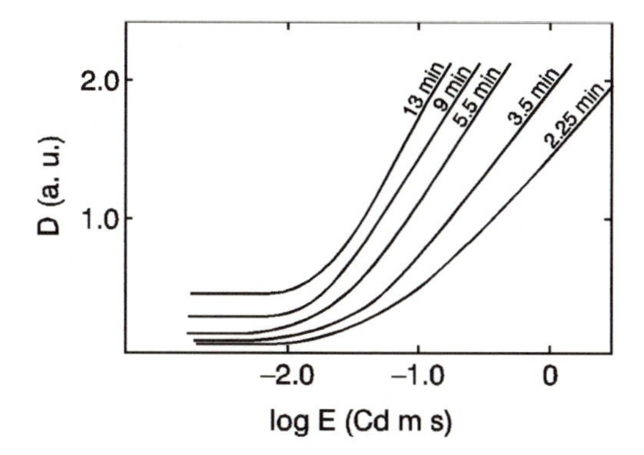

Fig. 5.2. Optical density D versus logarithm of exposure E for a Kodak 103a-0 emulsion. The parameter of the curves is the developing time.

the optical density and the logarithm of the exposure for a commercial emulsion (Kodak 103a-0). The working range of the emulsion is the part of the curves where optical density and logarithm of exposure are linearly related. The gradient (steepness) of the curves can be slightly varied by changing the developing time. The spectral sensitivity is the change in quantum efficiency with the light wavelength. For the emulsion 103a-0 the quantum efficiency starts to decrease rapidly for wavelengths longer than 500 nm.

The resolution is determined by the grain size: the smaller the grain the higher the resolution. The sensitivity decreases, on the other hand, with decreasing grain size so that the grains must not be too small. Resolutions of 50–100 lines/mm are standard.

The advantage of the film detectors is the simultaneous recording of all spectral components of a light beam, provided these components have been "dispersed" in space by a dispersive spectrometer. Photographic recording is still widely used in x-ray diffraction experiments. Its disadvantages are mainly the laborious processing of the films, and the cumbersome and inaccurate evaluation procedures of the recorded data. Nowadays direct electrical recording with photoelectric detectors (photoconductors, photodiodes or diode arrays) or photomultipliers presents a significant advantage over the photographic techniques.

5.3 Photomultipliers

In the visible and near-visible spectral range the photomultiplier is the detector of choice in spite of its complicated construction and the related high manufacturing costs. The simple reason for this is its extremely high sensitivity. A schematic drawing of a photomultiplier tube is illustrated in Fig. 5.3. The main parts of the tube are:

- the photocathode,
- the dynodes, and
- the collector (anode).

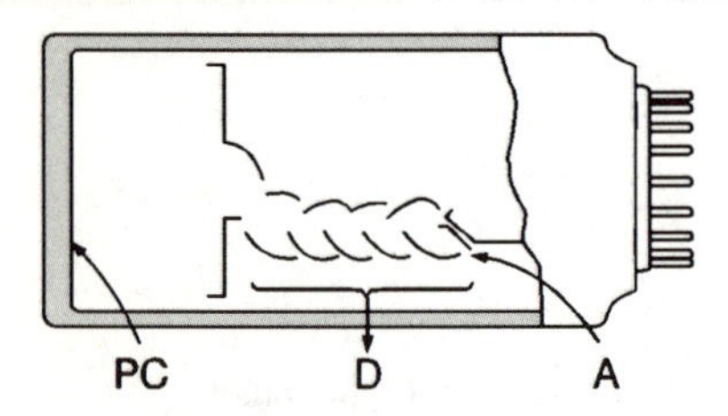

Fig. 5.3. Schematic drawing of a photomultiplier tube; (PC: photocathode, D: dynodes, A: anode).

Electrons are released at the photocathode by an external photoelectric effect. Hence, the photocathode must consist of a material with a very low work function. Multi-alkali compounds are widely used. Semiconductors like GaAs or InGaAs are employed in more recent models. With such electrodes a much higher sensitivity is obtained, particularly in the deep-red or near-IR spectral region. Photocathodes are classified according to their chemical composition like bialkali, trialkali or semiconducting and according to their spectral sensitivity. The quantum efficiency η is the fraction of released electrons per incident photon. It can be as high as 35%. In the near IR the quantum efficiency drops rapidly to zero. The maximum wavelength where a photomultiplier can be used is 1.2 μm. At this wavelength only a photocathode of S1 type is appropriate with a quantum efficiency of only 0.004%.

The *radiation sensitivity* or *absolute sensitivity* $E(\lambda)$ represents the signal response. It is characteristic of the photomultiplier and can be derived immediately from the quantum efficiency

$$E(\lambda) = \frac{I_c}{P} = \frac{e\eta(\lambda)}{\hbar\omega} \quad \text{(in A/W)}\,, \tag{5.4}$$

where I_c is the current at the photocathode, and P the incident light power (in Watts). A *cathode sensitivity* in (amperes/lumen) is defined in a similar way. Typical magnitudes for these quantities are 80 mA/W or 90 A/lm. Figure 5.4 displays the sensitivity versus wavelength for various photocathodes. The main disadvantage of photomultipliers is the rapid drop of the sensitivity in the near IR. This drop results from the drop in the quantum efficiency. If the photomultiplier is used for UV radiation, the entrance window must be made of quartz.

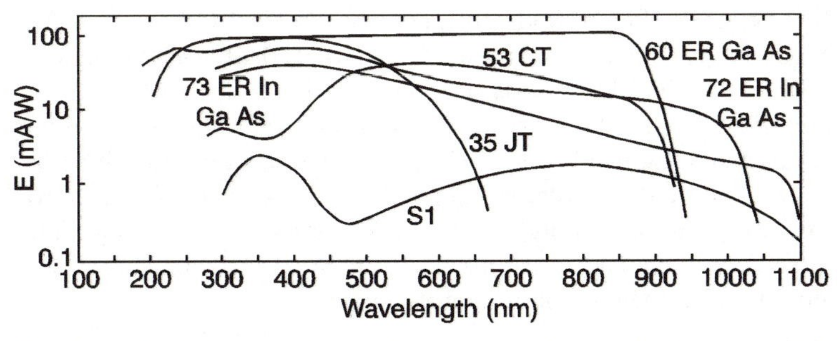

Fig. 5.4. Radiation sensitivity E of various photocathodes versus light wavelength.

The second most important characteristic of the photocathode is the dark current due to the thermal emission of electrons (Richardson emission). It is determined by the work function Φ_A, the temperature, and the size of the cathode. Thus, small dark currents can be obtained by cooling the photocathode and by reducing its size. In general, cooling to -40 °C is good enough to lower the dark current to 1–10 electrons per second and cm^2. Cooling in this range is easily performed with Peltier elements. By reducing the area of the photocathode to the order of one mm^2, a dark current of the order of 0.1

to 1 electron/s can be obtained. Then signals of the order of 1 photon/s or less can be measured. This is the detection limit. Dark current characteristics for several photocathodes are shown in Fig. 5.5.

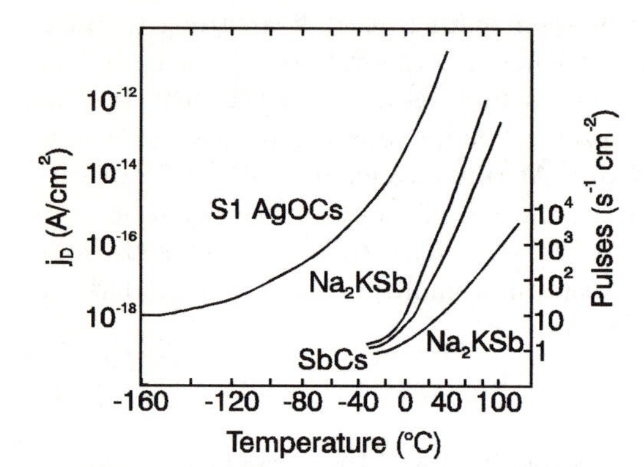

Fig. 5.5. Photomultiplier dark-current density j_D versus temperature for various cathodes.

Photomultipliers are extremely light-sensitive. After irradiation with daylight their dark current may be enhanced by several orders of magnitude. It is therefore quite common to keep photomultipliers cool and in the dark for several months. Windows made of quartz glass usually give a lower dark current than windows of commercial glass because of radioactive inclusions in the latter.

Electrons emitted from the photocathode are multiplied by secondary emission at the dynodes. The dynode surfaces are made of CsSb which has a high coefficient for secondary emission. The potential at each dynode is pinned by a voltage divider. The construction of the latter determines speed, sensitivity, linearity, etc., of the system. The cross current through it must be at least an order of magnitude larger than the current along the dynodes to guarantee linearity of the detector. Typical operating values for photomultipliers are 1500 V at the cathode and 1 mA current at the anode. The multiplication factor G for the electrons is given by the coefficient of secondary emission δ and by the number n of dynodes

$$G = \delta^n \, . \tag{5.5}$$

A typical value of δ is 5 which, for 10 dynodes gives a gain of $5^{10} \approx 10^7$. δ depends, of course, strongly on the voltage applied.

At the anode the arriving electrons are measured either as a current or, for very low incident light intensities, as a charge pulse per incoming photon. The *total sensitivity* of the detector is defined by A/lumen, where A is the current at the anode and lumen the incident light flux. Maximum acceptable power dissipation at the anode is usually 1 W for a resistor R_A of 50 Ω.

The speed of the detector is determined by R_A and the capacitance C_A of the anode. These stray capacitances can be as low as 10 pF which yields a time constant of 0.5 ns. This is, however, a lower limit. An average time resolution for a good photomultiplier is of the order of 2 ns.

For very low light intensities a *photon counting* system is appropriate. In this case the charge pulses arriving at the anode are counted. Discrimination of the pulse height with respect to a certain threshold enables background contributions to be suppressed to a large degree.

Photomultipliers are not only used for the detection of visible light but also for the detection of electrons, x rays or γ rays. In the last two applications the high energy of the light quanta must be transformed first to lower values by a scintillation process. A discussion of such system will be given in Chap. 12.

5.4 Photoelectric Detectors

The continuous progress in semiconductor technology has led to an ever increasing use of photoelectric detectors. This has been the case for the near-IR, IR, and far-IR spectral range but photoconductors and photodiodes are also becoming serious competitors to the photomultipliers even in the visible spectral range. Diode arrays and charge-coupled devices are particularly attractive.

In this section we will first discuss some fundamental properties of the photoelectric detectors and then study specifically photoconductors, photodiodes, and diode arrays.

5.4.1 Fundamentals of Photoelectric Detectors

For photoelectric detectors the signal response and the detection limit are known as *responsivity* and *detectivity* . The responsivity is defined as

$$R = \frac{\Delta V_{\mathrm{S}}}{\Delta P_{\mathrm{I}}} \quad (\text{in V/W}) , \tag{5.6}$$

where ΔV_{S} is the change in the detector output signal for a change ΔP_{I} in the incoming light power. To characterize the detectivity we need the *noise equivalent power* (NEP) defined as the light power (in Watts) which generates a signal (in volts) equal to the signal produced by the noise. The inverse of the NEP is the detectivity D. Obviously this quantity does not only depend on the magnitude of the generated signal but also on the intensity of the noise. There are many origins for noise in photoelectric detectors. The noise power is very often proportional to the square root of the detection bandwidth Δf_{R} and to the square root of the detector area A. It is therefore common practice to characterize a detector material by its *specific detectivity* D^* where

$$\boxed{D^* = \frac{\sqrt{\Delta f_{\mathrm{R}}}}{NEP}\sqrt{A} \quad (\text{in m/s}^{1/2}\text{W}) .} \tag{5.7}$$

D^* is a standard symbol for this quantity. Since the electric noise is, like any noise, of statistical origin, the noise signal increases with the square root of the duration of the measurement. Since the signal increases linearly in time, the signal-to-noise ratio increases with the square root of the duration of the measurement, as discussed before.

For the detectors to be discussed in this section, as well as for those to be treated in Chap. 10, the following mechanisms are the most important sources of noise.

a) *Johnson noise (Nyquist noise)*: The Johnson noise is due to the thermal motion of the carriers in the resistor. The mean square noise voltage $\langle V_\mathrm{R}^2 \rangle$ across a resistor R_D is determined by the temperature from

$$\langle V_\mathrm{R}^2 \rangle = 4k_\mathrm{B} T R_\mathrm{D} \Delta f \, . \tag{5.8}$$

Δf is the bandwidth of the detection system. The noise voltage is the square root of this quantity. It is always present, even without radiation incident on the resistor. For example, the noise voltage is 5 nV for a 1-KF resistor at room temperature and a 1-Hz bandwidth detection.

b) *Generation-recombination noise*: This noise, similar to the *shot noise* in diodes, is generated by thermal generation and recombination of carriers. Generation-recombination noise is only observed if current flows through the detector.

c) *1/f noise*: $1/f$ noise is determined by the detector surface. It is obviously dominant at low frequencies.

d) *Amplifier noise*: Each signal amplifier also amplifies the noise at the input. The *noise figure* of the amplifier measures the amplified input noise compared to the noise on the output side. The noise figure should be close to 1.

e) *Background noise*: Background light incident on the detector is another source of noise. This noise is particularly important for the far-IR since in this spectral range the room temperature radiation already leads to a considerable noise. The background noise can be reduced by using a cold filter. If the square of the background noise voltage is larger than the sum of the squares of the other contributions to the noise the detector is called *ideal* or *background limited* with the acronym BLIP (background limited photodetector).

5.4.2 Photoconduction Detectors

Photoconduction (PC) detectors play a dominating role, particularly in IR spectroscopy. For intrinsic photoconduction electrons are excited from the valence band to the conduction band by the absorbed photon as shown in Fig. 5.6a. The conductivity increases with the increased number of carriers in the conduction band and in the valence band. The process of excitation is possible provided the quantum energy of the radiation is larger than the energy gap ϵ_g of the semiconductor. Obviously, like in photomultipliers, there is also a hard quantum limit below which the PC detectors can not be used. Fortunately, there are numerous semiconductors with rather small bandwidths so that the condition $\hbar\omega \leq \epsilon_g$ is *de facto* not a real limiting relation. In addition, photoexcitation of electrons or holes from impurities can also lead to photoconductivity and is therefore appropriate for radiation detection. The real limitation for PC detectors comes from the thermal excitation of the carriers across the gap or from the impurity levels. If there is a large dark current the sensitivity of the photodetector becomes low.

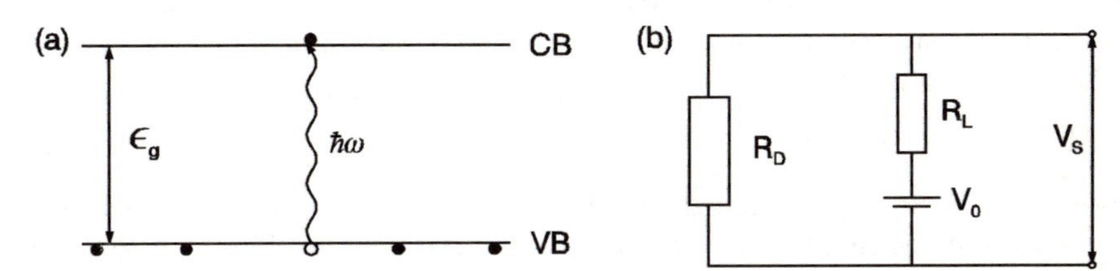

Fig. 5.6. Intrinsic photo excitation between the valence band (VB) and the conduction band (CB) of a semiconductor (**a**) and circuit for a photoconduction detector (**b**); (R_L: load resistor, R_D: intrinsic detector resistor, V_0: applied voltage, V_S: signal voltage).

An electric circuit representing PC detectors is shown in Fig. 5.6b. From this the signal can be obtained by the change of the conductance ΔG

$$\Delta G = \frac{\Delta\sigma A}{l} \tag{5.9}$$

of the resistor R_D induced by the incident light. $\Delta\sigma$ is the change of the conductivity of the detector due to irradiation, and A and l are its cross section and length, respectively. The signal ΔV_S is finally obtained from Fig. 5.6b by

$$\boxed{\Delta V_S = \frac{V_0 R_L R_D^2 \Delta G}{(R_D + R_L)^2} = \frac{V_0 R_L R_D^2 A}{(R_D + R_L)^2 l}\Delta\sigma\ .} \tag{5.10}$$

At least for the range of linear photoconductivity ΔG is proportional to the incident light intensity I_0. Then, from (5.10) the signal is maximum for $R_D = R_L$.

The change of the conductivity $\Delta\sigma$ is evaluated from the generation rate

$$g = \frac{\alpha\eta I_0}{\hbar\omega} \qquad (\text{in } 1/\text{m}^3\,\text{s}) \tag{5.11}$$

and the equation of continuity for the equilibrium state

$$\Delta n = g\tau_n, \qquad \Delta p = g\tau_p \tag{5.12}$$

as

$$\Delta\sigma = e\mu_{\mathrm{n}}\Delta n + e\mu_{\mathrm{p}}\Delta p \propto g \propto I_0 , \tag{5.13}$$

where α, η, τ_{n}, and τ_{p} are the absorption coefficient, the quantum efficiency, and the lifetimes for the generated carriers. Δn and Δp are the number of carriers generated by the light per unit volume. Obviously, the lifetimes play a crucial role in determining the sensitivity of the detector. As τ increases so does the sensitivity of the detector. We can define an amplification factor for the photoconductor as the ratio between the number of photo carriers passing the electrode and the number of generated photo carriers, both per unit time. Amplification factors for good photoconductors are as high as 10^4.

The lifetime also controls the speed of the detector since it determines the time scale for the change of the carrier concentration with changing light for non-steady-state conditions. In this case the equation of continuity yields for $\Delta n(t)$

$$\Delta n(t) = \Delta n(0)\mathrm{e}^{-t/\tau_{\mathrm{n}}} , \tag{5.14}$$

where $\Delta n(0)$ is the light-induced carrier concentration at $t = 0$. If τ_{n} is large, only slow changes in the light intensity can be detected. This is a general rule. The speed of the detection and its sensitivity are inversely related.

In (5.11) to (5.13) it was assumed that the distribution of the photo carriers is homogeneous which implies a weak light absorption. For strong absorption the carrier distribution will be inhomogeneous and carrier diffusion becomes important. The basic properties of the detector such as linearity, sensitivity and speed are, however, retained. The same holds for photodetectors on the basis of an extrinsic photoconduction.

The best-known photoconduction detectors are CdS for the visible spectral range and PbS for the IR. For specific applications, particularly in the IR and far-IR a large number of more sophisticated detectors are available. Each of them covers only a limited spectral range. Characteristics for selected photoconduction detectors are presented in Fig. 5.7, together with some photodiodes. From the figure it is evident that several detectors are needed to cover the spectral range from the visible to the far-IR. For long wavelengths cooling of the detectors to liquid-He temperatures is required.

5.4.3 Photodiodes

Photodiodes are becoming more and more important. They are widely used in the IR spectral region but are also useful for high-energy radiation. Diode arrays or imagers also make extensive use of photodiodes. Photodiodes are usually operated in a reverse-biased mode where the change of the very small reverse current due to light-induced carrier generation is measured. An alternative but rarely used possibility is to measure the light-induced photovoltaic effect for an unbiased diode. With respect to light absorption, lifetime, and quantum limit the same relationships apply as for photoconductors. The active volume is the depletion layer of the pn junction. Unfortunately this

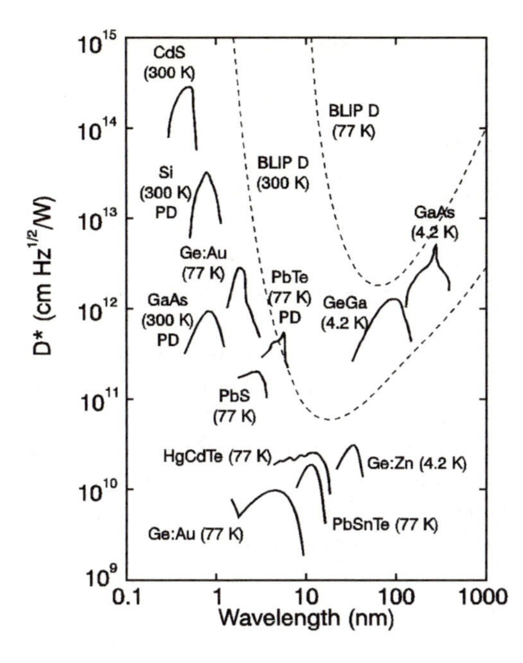

Fig. 5.7. Specific detectivity D^* versus wavelength for various photoelectric detectors. Photodiodes are labeled PD. The dashed lines are ideal values for 77 K and 300 K, respectively; after [5.1].

volume is usually very small so that the probability of absorbing a photon is quite low. The situation can be improved by inserting a small intrinsic conducting region between the p-type semiconductor and the n-type semiconductor. Such devices are called p-i-n diodes. The acceptable dimension of the intrinsic layer is given by the lifetime and diffusion or drift length of the carriers. If the carriers recombine before they have left the active zone they will not contribute to the photocurrent.

Alternatively the sensitivity of the diodes can be improved by using a solid-state type photomultiplier. This is possible since free carriers can accept energy from an applied field and generate secondary carriers by impact ionization across the energy gap or from impurity centers. If the applied voltage is high enough, avalanche-type multiplication occurs. Such devices are called *avalanche photodiodes*. Unfortunately the amplification process applies also to the dark current which means that the noise increases as well. As a consequence the detectivity decreases for too high an avalanche multiplication. This effect could be suppressed by the discriminator in the case of the photomultipliers. Since suppression is not possible in avalanche photodiodes, the latter are not as sensitive as the photomultipliers but they have a faster response. One can reach a high amplification without loss in bandwidth. Response times down to 10^{-10} seconds and amplification factors up to 10^4 are common. The *amplification-bandwidth* product for good avalanche photodiodes is 100 GHz. With improved p-i-n structures of the form pipn the speed can be further increased and amplification-bandwidth products of 250 GHz can be obtained. These are the RAPD diodes (reach through avalanche photodiodes). For very fast detection Schottky diodes (metal-semiconductor junctions) are more appropriate.

5.4.4 Detector Arrays and Imagers

Arrays of very narrowly spaced diodes can be prepared with modern semiconductor technology. The arrangements of the diodes can be one-dimensional, quasi-one-dimensional or two-dimensional. In a one-dimensional arrangement the diodes are lined up exactly in one row (diode array). In quasi-one-dimensional detectors several arrays are used shifted in the direction perpendicular to the array. In two-dimensional detectors the diodes are arranged in a true two-dimensional matrix. This latter arrangement allows a complete imaging of a two-dimensional signal. The big advantage of the array detectors is the simultaneous recording of a whole spectrum if the light has been dispersed in the spectrometer. This obviously saves measuring time or equivalently results in a gain in signal-to-noise ratio. Detectors of this type are called *optical multichannel analyzers* with the acronym OMA.

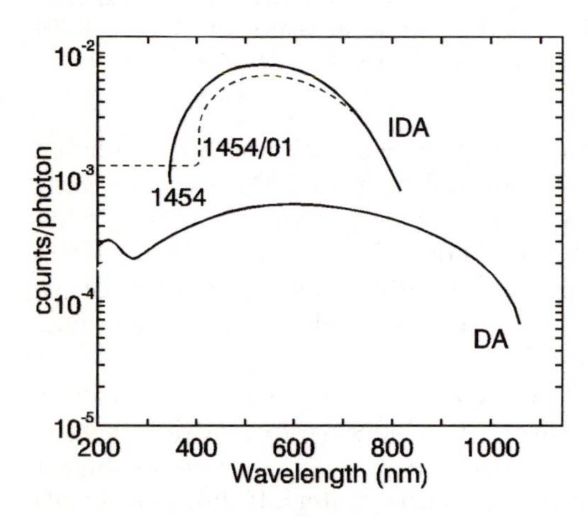

Fig. 5.8. Quantum yield for a standard diode array (DA) and for two intensified diode arrays (IDA).

Figure 5.8 presents the effective quantum yield (counts per photon, not photoelectrons per photons) for one standard and two intensified diode arrays of silicon. Even though the quantum yield for the single Si diode is between 70 % and 80 % the effective quantum yield is less than the one for photomultipliers even for the intensified array. However, if the array consists of 512 diodes a factor of about 500 is gained in measuring time. Another advantage of such detectors is the much better response in the deep red and near-IR as compared to photomultipliers. This is evident from the figure. Si diode-array detectors cover a spectral range from 185 to 1100 nm. Typical dimensions for the diodes (pixels) are 25 μm×2.5 mm with a pixel distance of 25 μm. With fast scanning diode arrays scan times down to 13 μs/diode can be obtained which means 70 spectra per second with a 1024 pixel array.

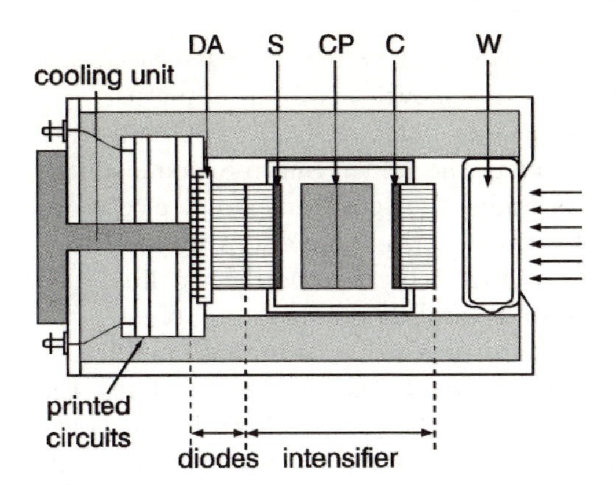

Fig. 5.9. Construction details of a diode array detector with image intensifier; (W: window, C: cathode, CP: channel plate, S: screen, DA: diode array). The horizontally hatched areas represent fiber optics.

The low effective quantum efficiency of the diodes can be raised dramatically by image intensifiers as demonstrated in Fig. 5.8. Intensified diode arrays are hybrid constructions between photomultipliers and array detectors. The basic concept is illustrated in Fig. 5.9. The light enters the system through a window and gets absorbed by an external photo effect at the photocathode. The released photoelectrons are accelerated by about 250 V towards the micro-channel plate. The photocurrent gets amplified in the micro channels across which 700 V are applied. The micro channels act like a photomultiplier with continuous dynodes. The important point of the imaging plates is the conservation of the spatial distribution of the light pattern during all processes. From the channel plate the electrons are accelerated to a luminescence screen by 6 KeV to yield an intensified image. This image is finally detected with the diode array. The photonic gain can be as high as 5×10^3. Coupling from the screen to the diode array can be performed with a coherent fiber optic.

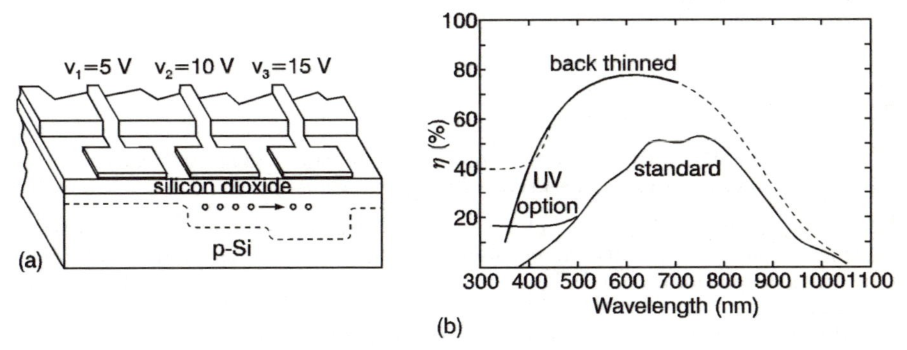

Fig. 5.10. Schematic drawing of a CCD-system (**a**) and quantum yield η for a standard and a back thinned chip (**b**).

Quasi-one-dimensional detectors are very often constructed as *charge coupled devices* (CCD detectors). Figure 5.10a shows a schematic picture of a CCD system. Closely spaced MOS diodes are arranged on a SiO substrate and biased to establish a depletion layer in the underlying p-Si semiconductor. Light generates minority carriers under the semitransparent electrodes which can be read out eventually by transporting them step by step to a collector electrode. This is indicated in Fig. 5.10a by the dashed line. The signal is the response from either one or several diode lines.

CCD detectors were introduced in light measuring techniques in 1970. A typical pixel size is $5 \times 25\,\mu\text{m}^2$ with 512 or 1024 pixels per array. Thus, the total width of the device is 5 mm for about 1000 pixels. CCD detectors are characterized by very high sensitivity and an extremely low dark current. For a liquid nitrogen-cooled CCD system the dark current is as low as 1 electron per pixel per hour. The read out process is accompanied by a noise of 5 to 6 counts. The quantum efficiency of a CCD camera is about 40 % in the red, and the acceptable spectral range is 420 to 1000 nm. In contrast to diode arrays, with a good electronic an efficiency of 1 count per photoelectron can be obtained. Quantum efficiency and spectral range can be increased by *back-thinning* the CCD chip. In this case the chip is thinned from the back by chemical etching until even UV light can reach the active zone of the MOS units. Illumination occurs in these systems from the back. The quantum efficiency can be raised in this way to 70 % in the whole visible spectral range. With an UV option the chip can be used down to 200 nm. Quantum efficiencies for a standard and for a back-thinned CCD chip are shown in Fig. 5.10b. Since for high gain 1 count per photoelectron is obtained the total effective quantum efficiency can be 0.7 in contrast to 5×10^{-4} for the diode array. The spectral resolution of the detector is limited. As for all multichannel systems it is determined by the spectral range incident on one pixel. This depends, of course, on the amount of spatial dispersion generated by the spectrometer used.

Problems

5.1 Show that the variance for the Poisson distribution equals its mean.
(Purpose of exercise: handle probability distributions.)

5.2 The probability distribution for a signal is a Poisson distribution, the measuring time available is 10 hours. An optimum signal-to-noise ratio should be obtained. Is it better to accumulate one signal for 10 hours or to accumulate ten 1-h signals and take the average?
(Purpose of exercise: a very practical measuring problem.)

5.3 Calculate the signal ΔV_{S} for a photoconduction detector with internal resistance R_{D} and working resistance R_{L} and show that the highest signal is obtained for $R_{\mathrm{D}} = R_{\mathrm{L}}$.
(Purpose of exercise: importance of circuit parameters.)

5.4* In a strongly absorbing semiconductor carriers are generated close to the surface. Calculate the steady state distribution of carriers as a function of depth and the resulting total current for a given applied voltage, a lifetime τ, and a mobility μ of the carriers.

(Purpose of exercise: recall inhomogeneous carrier distribution.)

6. The Dielectric Function

The dielectric function and the functions directly related to it are fundamental in solid-state spectroscopy. Their derivation is based on a very general description of the reaction of a system to an external force. As long as the reaction is linear, the response is obtained according to a *linear response model* and the relations describing the reactions are called the *linear response functions*. Thus, the linear response functions are properties of the solid-state system itself and are independent of the driving force. This concept is applicable to the whole spectral range from radio waves to γ rays as well as to spectroscopy with particles. The linear response is formulated in time and space. Since the response is, in general, frequency- and wave vector-dependent and since it is convenient to operate with harmonic functions, a discussion in Fourier space, both with respect to time and coordinates, is more appropriate. Thus, rather than studying the response function directly the linear relation between the Fourier transform of the driving force and the Fourier transform of the system response are considered. One of the most fundamental response functions is the electric susceptibility which describes the polarization $P(q,\omega)$ generated by an incident electric field $E(q,\omega)$.

$$P(q,\omega) = \chi(q,\omega)\varepsilon_0 E(q,\omega)\,, \tag{6.1}$$

which is simply a generalization of the definitions given in Appendix B.1 in connection with the Maxwell theory.

Another important example for a response function is the relationship between the displacement D and an applied field E

$$D(q,\omega) = \varepsilon(q,\omega)\varepsilon_0 E(q,\omega)\,. \tag{6.2}$$

$\varepsilon(q,\omega)$ is known as the dielectric function (DF) or relative permittivity of the solid. Note that we have used in both equations a scalar terminology for convenience even though field and polarization are vectors and susceptibility and DF are second-rank tensors. Also, for the purpose of this chapter we will simplify the problem and use the response functions $\chi(\omega)$ and $\varepsilon(\omega)$ which depend only on the frequency. An introduction to a more general description of the linear response is given in Chap. 14 and in Appendix L.

We start this chapter by recalling some general relationships between the optical constants including the Kramers–Kronig relation and continue with a description of the response of the solids by several simple DF models. For an extended discussion of DFs of solids in the context of many-body theory special textbooks like [6.1] must be studied.

6.1 Optical Constants and Kramers–Kronig Relations

6.1.1 Optical Constants

From the Maxwell equations some important relationships can be derived for optical constants and for the propagation of EM waves in homogeneous media. Using the wave equation from Appendix B.1 for a non-conducting and non magnetic medium in one dimension

$$\frac{\partial^2 E(x,t)}{\partial x^2} = \frac{\varepsilon}{c_0^2}\frac{\partial^2 E}{\partial t^2}\,, \tag{6.3}$$

together with the complex representation of a plane wave from (2.2) we obtain the dispersion relation for the propagation of the EM wave from

$$k^2 = \frac{\omega^2}{c_0^2}\varepsilon(\omega)\,. \tag{6.4}$$

Obviously, $\varepsilon(\omega)$ determines the dispersion of the wave. Note that in Appendix B.1 we have used a relationship in the time domain for the definition of ε. However, this definition does not change if we go to the frequency domain since, so far, ε was only used for harmonic waves. Expressing the propagation constant k by the index of refraction, as it was done in (2.3), we obtain

$$k_{\mathrm{c}} = \frac{N\omega}{c_0} \qquad \text{with} \qquad N = \sqrt{\varepsilon(\omega)}\,. \tag{6.5}$$

We have used the symbols k_{c} and N for the propagation constant and for the index of refraction since we now allow also complex values for both, and hence also for the DF. The two functions may be represented as

$$\boxed{N(\omega) = n(\omega) + \mathrm{i}\kappa(\omega)\,,} \quad \text{and} \quad \boxed{\varepsilon(\omega) = \varepsilon_{\mathrm{r}}(\omega) + \mathrm{i}\varepsilon_{\mathrm{i}}(\omega)\,.} \tag{6.6}$$

The sign for the imaginary part may be chosen either positive or negative. We have selected the positive sign, for convenience, in the formulae used later on. The complex notation for the index of refraction leads immediately to EM waves damped in space of the form

$$\begin{aligned} E(x,t) &= E_0 \mathrm{e}^{\mathrm{i}[(N\omega/c_0)x-\omega t]} \\ &= E_0 \mathrm{e}^{-(2\pi\kappa/\lambda)x}\mathrm{e}^{\mathrm{i}(kx-\omega t)}\,. \end{aligned} \tag{6.7}$$

The expression $2\pi\kappa/\lambda$ describes the attenuation of the field. Since the absorption coefficient is usually defined by the attenuation in intensity written in the form of Lambert's law

$$I(x) = EE^* = I_0 \mathrm{e}^{-\alpha x} \tag{6.8}$$

we obtain for the relationship between κ and α

$$\boxed{\alpha = \frac{4\pi\kappa}{\lambda}\,.} \tag{6.9}$$

From Maxwell's relationship $N = \sqrt{\varepsilon}$ the real and the imaginary parts of the two functions $N(\omega)$ and $\varepsilon(\omega)$ are correlated via

$$\varepsilon_{\mathrm{r}} = n^2 - \kappa^2\,, \qquad \varepsilon_{\mathrm{i}} = 2n\kappa \tag{6.10}$$

or

$$\boxed{n = \frac{\sqrt{\varepsilon_{\mathrm{r}} + |\varepsilon|}}{\sqrt{2}}\,, \qquad \kappa = \frac{\sqrt{-\varepsilon_{\mathrm{r}} + |\varepsilon|}}{\sqrt{2}}\,.} \tag{6.11}$$

With these relationships the absorption coefficient $\alpha(\omega)$ can be expressed by the imaginary part of the DF as

$$\boxed{\alpha(\omega) = \frac{\omega\varepsilon_{\mathrm{i}}(\omega)}{c_0 n(\omega)}\,.} \tag{6.12}$$

According to its definition the electric susceptibility $\chi(\omega) = \varepsilon(\omega) - 1$ is a complex function as well.

Finally, even the conductivity defined from $j(\omega) = \sigma(\omega)E(\omega)$ may be considered as a linear response function in a conducting medium. To be general enough we may, in this case, express the displacement current $\partial D/\partial t$ in the first Maxwell equation by the polarization in the form

$$\frac{\partial D}{\partial t} = \frac{\partial(\varepsilon_0 E + P)}{\partial t} = \frac{\partial(\varepsilon_0 E)}{\partial t} + j\,,$$

where j represents the current density from the polarization of the bound carriers. Since there is no real need to discriminate between bound and free carriers as bonding may be arbitrarily weak or strong, we can write the total current density as the time-derivative of the polarization

$$j = j_{\mathrm{free}} + j_{\mathrm{bound}} + \ldots = \frac{\partial P}{\partial t}\,. \tag{6.13}$$

P consists of a contribution $P_{\mathrm{free}} = -j_{\mathrm{free}}(0)/\mathrm{i}\omega$ from the truly free carriers and another contribution P_{bound} from the more or less bonded carriers. Likewise, both the susceptibility from $P = \chi\varepsilon_0 E$ and the conductivity from $j = \sigma E$ will include both components. With this terminology and assuming a time dependence for the field of the form $\exp(-\mathrm{i}\omega t)$ we can write

$$j = \frac{\partial P}{\partial t} = \frac{\chi\varepsilon_0 \partial E}{\partial t} = -\chi\varepsilon_0 \mathrm{i}\omega E = \sigma E\,.$$

This yields

$$\boxed{\sigma(\omega) = -\mathrm{i}\omega\varepsilon_0\chi(\omega)} \tag{6.14}$$

and

$$\boxed{\varepsilon(\omega) = \chi(\omega) + 1 = 1 + \frac{\mathrm{i}\sigma(\omega)}{\omega\varepsilon_0} .} \tag{6.15}$$

The frequency-dependent conductivity is now also a complex function with the components σ_r and σ_i. Finally, it is very useful to express the absorption from (6.12) as a function of the conductivity. Inserting from (6.15) yields

$$\boxed{\alpha(\omega) = \frac{\sigma_\mathrm{r}(\omega)}{n(\omega)\varepsilon_0 c_0} .} \tag{6.16}$$

6.1.2 Reflection and Transmission

Since information about solid materials is very often obtained from reflection experiments, the propagation of EM waves across planar interfaces between materials with different optical properties must be studied. This can be done by considering the continuous transitions of the tangential component of the E and H fields, and the conservation of energy. For the special case of incidence perpendicular to the boundary between two media with relative index of refraction $N(\omega)$ these conditions are

$$E_\mathrm{i} - E_\mathrm{r} = E_\mathrm{t} \qquad \text{and} \qquad H_\mathrm{i} + H_\mathrm{r} = H_\mathrm{t} , \tag{6.17}$$

where i, r, and t refer to the components of the incident, reflected, and transmitted fields, respectively. The minus sign in the relationship for the electric field follows from the assumption that the reflection occurs at the boundary to a medium with higher optical density. If the reflection was at a boundary to a medium with lower optical density, the minus sign would be for H_r. Using the relation between E and H from (2.4) this yields for the complex field reflection coefficient r_c

$$r_\mathrm{c} = \frac{E_\mathrm{r}}{E_\mathrm{i}} = \frac{N-1}{N+1} . \tag{6.18}$$

The reflectivity is the reflection coefficient for the intensities and immediately obtained from (6.18) as

$$\boxed{R = r_\mathrm{c} r_\mathrm{c}^* = \frac{(n-1)^2 + \kappa^2}{(n+1)^2 + \kappa^2} .} \tag{6.19}$$

For a non-perpendicular light incidence a more general treatment is required. The reflection coefficients for the field are then given by the *Fresnel formulae*. Appendix E.1 discusses these formulae in more detail.

The situation becomes much more complicated if the radiation is transmitted through a parallel plate. In this case transitions at two boundaries (from medium 1 to medium 2 and from medium 2 to medium 3) and multiple-beam interference has to be considered. The geometrical situation is similar to the case of the Fabry–Perot interferometer discussed in Sect. 4.3, but it is more complicated because absorption now plays an essential role. Even for perpendicular incidence the equations become rather complicated. However, experiments of this type are very important and are often applied to measure absorption coefficients of solids. Generalized relations for transmission and reflection are discussed in Appendix E.2.

For the special case where medium 1 and 3 are the same and the interference fringes are not resolved, simplified formulae can be given for the averaged transmission $\langle T \rangle$ and reflection $\langle R \rangle$. In the case of a plate with thickness d we have

$$\langle T \rangle = \frac{(1-R)^2(1+\kappa^2/n^2)\mathrm{e}^{-\alpha d}}{1-R^2\mathrm{e}^{-2\alpha d}} \tag{6.20}$$

and

$$\langle R \rangle = R(1 + \langle T \rangle \mathrm{e}^{-\alpha d}) \,. \tag{6.21}$$

Here averaging means that the transmission or reflection coefficient is averaged over the phase angle for the partial beams. This averaging may occur automatically in the experiment if the surfaces are not smooth enough or the absorption is too strong to allow for the development of interference fringes. A more general discussion of the optical response from optical multi-layers is given in [6.2].

6.1.3 Kramers–Kronig Dispersion Relations

The Kramers–Kronig dispersion relations are integral relationships between real and imaginary parts of a function $f(\omega)$ defined in the complex frequency plane. The linear response functions such as $\varepsilon(\omega)$, $\chi(\omega)$, $N(\omega)$, etc., are examples. A generalized description including the conditions for which Kramers–Kronig relations hold is given in Appendix L.1. One of the requirements for the relationships to be valid is that the response function vanishes for $\omega \to \infty$. This can be expected for the susceptibility since the DF approaches 1 for very high frequencies. The Kramers–Kronig dispersion relations for the susceptibility can be written in the form

$$\chi_{\mathrm{r}}(\omega_0) = \frac{2}{\pi}\mathrm{P}\int_0^\infty \frac{\omega\chi_{\mathrm{i}}(\omega)}{\omega^2-\omega_0^2}\,\mathrm{d}\omega \tag{6.22}$$

and

$$\chi_{\mathrm{i}}(\omega_0) = -\frac{2\omega_0}{\pi}\mathrm{P}\int_0^\infty \frac{\chi_{\mathrm{r}}(\omega)}{\omega^2-\omega_0^2}\,\mathrm{d}\omega \,. \tag{6.23}$$

P represents the principal value of the integral. Similar relationships are available for the DF and for $N(\omega)$. With these relationships one component of the response function can be calculated step by step if the other component is known for the whole spectral range. This is extremely important since often only one component can be determined easily from an experiment. However, this component must be known for the whole frequency range. Since the full frequency range cannot be covered experimentally proper extrapolations are required.

To see how good the Kramers–Kronig relationships work even for a finite range of integration we give an example in Fig. 6.1 for the case of $\varepsilon(\omega)$. It is assumed to have the form

$$\varepsilon(\omega) = 1 + \frac{A}{\omega_{\mathrm{T}}^2 - \omega^2 - \mathrm{i}\gamma\omega} \tag{6.24}$$

with $A = 2250$, $\gamma = 20$, and $\omega_{\mathrm{T}} = 50$. We can of course calculate the real part and the imaginary part and plot them as shown in Fig. 6.1. Alternatively we may insert the real part into a Kramers–Kronig relation for the DF and calculate ε_{i} numerically. To mimic experimental errors the function was only integrated up to the frequency 600. Despite this approximation the reproduction of the imaginary part is very good.

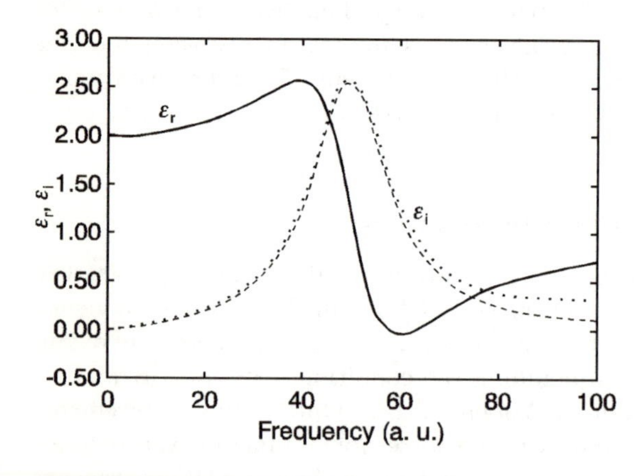

Fig. 6.1. Dielectric function for a damped harmonic oscillator; (—) ε_{r}, (– – –) ε_{i} as calculated from the function (6.24), and (...) ε_{i} as calculated from the Kramers–Kronig relation.

A technique used very often to find the DF of a solid is to measure the reflectivity $R(\omega)$ over a wide frequency range. The square root of this value equals the amplitude of the complex field reflectivity r_{c}. Thus, with $r_{\mathrm{c}} = |r_{\mathrm{c}}|\mathrm{e}^{\mathrm{i}\phi}$ and (6.19) we find

$$\ln r_{\mathrm{c}}(\omega) = \ln \sqrt{R(\omega)} + \mathrm{i}\phi(\omega) \,, \tag{6.25}$$

where $r_{\mathrm{c}}(\omega)$ is given by (6.18). $\phi(\omega)$ is the phase of $r_{\mathrm{c}}(\omega)$. The real part and the imaginary part of (6.25) are Kramers–Kronig-related in the form

$$\phi(\omega_0) = -\frac{\omega_0}{\pi} \int_0^\infty \frac{\ln\sqrt{R(\omega)} - \ln\sqrt{R(\omega_0)}}{\omega^2 - \omega_0^2} \, \mathrm{d}\omega \,. \tag{6.26}$$

This means, from a measurement of $R^{1/2}$ we can determine the phase $\phi(\omega)$, and, in turn, $n(\omega)$ and $\kappa(\omega)$ or the DF $\varepsilon(\omega)$.

Another important response function is the *energy loss function* $\mathrm{Im}\{-1/\varepsilon(\omega)\}$. It is related to $\varepsilon(\omega)$ by

$$\mathrm{Im}\left\{-\frac{1}{\varepsilon(\omega)}\right\} = \frac{\varepsilon_\mathrm{i}(\omega)}{\varepsilon_\mathrm{r}^2(\omega) + \varepsilon_\mathrm{i}^2(\omega)} \tag{6.27}$$

and describes the loss of energy of particles or quasi-particles on their way through the solid. We will return to this function and to more details about DFs in Chaps. 14 and 15.

6.2 Physical Origin of Contributions to the Dielectric Function

The DF is determined by the possible excitations of the solid. These excitations can be of very different nature depending on the frequency range considered. Qualitatively the following processes, listed by increasing spectral energy, can be expected to dominate the DF:

- dielectric relaxation processes,
- lattice vibrations (optical phonons and librons),
- free carrier absorption,
- excitons and absorption across the energy gap,
- valence electron polarization and interband transitions,
- valence band plasmon absorption (only in second order for optical absorption),
- transitions into higher bands,
- transitions from core levels, etc.

In principle, even magnetic excitations on the low-energy side and nuclear excitations on the high-energy side could be added to the list. In some ranges of the spectrum and for some applications it is enough to characterize the DF by its imaginary part and thus to study only the optical absorption versus excitation energy. In these cases we can schematically plot the absorption coefficient α versus excitation energy (Fig. 6.2).

All transitions require the conservation of energy and momentum. Since, at least for the visible spectral range and below, the wave vector of the photons is very small only excitations with wave vectors $q \approx 0$ can contribute to the DF. This condition is, of course, not required for excitations with electrons, neutrons, or γ quanta. The DF for the latter cases will be discussed in Chaps. 14 and 17.

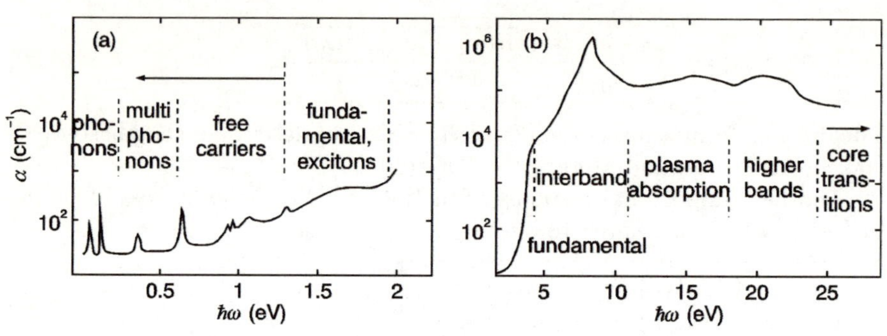

Fig. 6.2. Schematic representation of the absorption coefficient in solids for the various excitation processes; low-energy range (**a**), high-energy range (**b**).

6.3 Model Dielectric Functions

Since in general, it is very difficult to calculate the DF directly from the electronic structure model DFs are very useful.

6.3.1 Dielectric Function for Harmonic Oscillators

A simple but very useful DF can be derived for a set of damped harmonic oscillators. A harmonic electric field $E(t)$ excites a harmonic oscillator with mass m, charge e, damping γ and eigenfrequency ω_{T}. In a one-dimensional picture the equation of motion reads

$$m\ddot{x} + m\gamma\dot{x} + m\omega_{\mathrm{T}}^2 x = eE_1\mathrm{e}^{-\mathrm{i}\omega t} \,. \tag{6.28}$$

The subscript T for ω_{T} indicates a transverse oscillator frequency since only for this mode a coupling to the transverse electric field is possible. The representation (6.28) is very general. It is useful for the description of lattice modes as well as for electronic transitions and in the limit of $\omega_{\mathrm{T}} \to 0$ even for free carriers. A particular solution of (6.28) is $x = x_0 \exp(-\mathrm{i}\omega t)$. Inserting into (6.28) yields

$$x = \frac{e}{m}\frac{E_1 \exp(-\mathrm{i}\omega t)}{\omega_{\mathrm{T}}^2 - \omega^2 - \mathrm{i}\omega\gamma} \,. \tag{6.29}$$

If there are n oscillators per unit volume we obtain for the polarization P

$$P = nex \tag{6.30}$$

and from the relationship between electric field and polarization

$$P = \chi\varepsilon_0 E = (\varepsilon - 1)\varepsilon_0 E \tag{6.31}$$

the susceptibility for the oscillators is

$$\chi_{\mathrm{osc}} = \frac{ne^2/m\varepsilon_0}{\omega_{\mathrm{T}}^2 - \omega^2 - \mathrm{i}\omega\gamma} \,. \tag{6.32}$$

The corresponding DF is $\chi_{\mathrm{osc}} + 1$.

To be general enough for a realistic system a contribution from the deformation of the ion cores by the electric field must be added to the polarization of (6.30). This contribution is described by an optical susceptibility χ_{opt} (or χ_∞). While the contributions of the oscillators dominate the susceptibility in a frequency range close to ω_{T}, the polarization from the ion cores dominates at high frequencies. The corresponding DF given as $\chi_\infty + 1$ is therefore often assigned as ε_∞. The DF from the ion cores is determined by the relation of Clausius and Mosotti

$$\varepsilon_\infty = \chi_\infty + 1 = \frac{\sum_j n_j \alpha_j}{\varepsilon_0 + \sum_j (n_j \alpha_j / 3)} + 1 , \tag{6.33}$$

where α_j (in $\mathrm{A\,s\,m^2/V}$) and n_j (in $\mathrm{m^{-3}}$) are the atomic polarizabilities and the density of the atoms of type j in the crystal, respectively. This relationship holds at least as long as the generated local field is given by the Lorentz equation [6.3]. For very high frequencies, that is in the spectral range of x rays, α becomes 0 and ε_∞ becomes 1.

Considering the additivity of the polarization we can write a very useful model DF as

$$\varepsilon = \chi_\infty + \chi_{\mathrm{osc}} + 1 = \chi_{\mathrm{osc}} + \varepsilon_\infty ,$$

or explicitly

$$\boxed{\varepsilon = \varepsilon_\infty + \frac{ne^2}{\varepsilon_0 m} \frac{1}{(\omega_{\mathrm{T}}^2 - \omega^2 - \mathrm{i}\omega\gamma)} .} \tag{6.34}$$

This relation is often called the *Kramers–Heisenberg* dielectric function. Separating the real and imaginary parts and introducing the plasma frequency

$$\boxed{\omega_{\mathrm{p}} = \sqrt{\frac{ne^2}{\varepsilon_0 m}} ,} \tag{6.35}$$

we obtain

$$\begin{aligned} \varepsilon_{\mathrm{r}} &= \varepsilon_\infty + \omega_{\mathrm{p}}^2 \frac{\omega_{\mathrm{T}}^2 - \omega^2}{(\omega_{\mathrm{T}}^2 - \omega^2)^2 + \omega^2\gamma^2} , \\ \varepsilon_{\mathrm{i}} &= \omega_{\mathrm{p}}^2 \frac{\omega\gamma}{(\omega_{\mathrm{T}}^2 - \omega^2)^2 + \omega^2\gamma^2} . \end{aligned} \tag{6.36}$$

In the case of a mechanical oscillator, like a phonon, ω_{p} is the ion plasma frequency, and e and m are the effective charge and the reduced mass of the oscillator.

The Kramers–Heisenberg DF can easily be generalized to an arbitrary number of different oscillators, as for example, for a set of polar lattice modes. In this case we write

$$\varepsilon = \varepsilon_\infty + \sum_j \left(\frac{\omega_{jp}^2(\omega_{jT}^2 - \omega^2)}{(\omega_{jT}^2 - \omega^2)^2 + \omega^2\gamma_j^2} + \mathrm{i}\frac{\omega\omega_{jp}^2\gamma_j}{(\omega_{jT}^2 - \omega^2)^2 + \omega^2\gamma_j^2} \right) . \tag{6.37}$$

An important further generalization of the Kramers–Heisenberg DF is obtained by introducing a generalized oscillator strength. From (6.34) and (6.37) $\omega_\mathrm{p}^2 = ne^2/\varepsilon_0 m$ obviously determines the strength of the response of the system to the electric field. Thus, instead of just counting the number of oscillators we may replace ω_p^2 by a generalized function S which describes the strength of the response. It turns out to be convenient to introduce S as the dimensionless (reduced) *oscillator strength*

$$S_j = \omega_{jp}^2/\omega_{jT}^2 . \tag{6.38}$$

The Kramers–Heisenberg DF has then the generalized form

$$\varepsilon = \varepsilon_\infty + \sum_j \left(\frac{S_j\omega_{jT}^2(\omega_{jT}^2 - \omega^2)}{(\omega_{jT}^2 - \omega^2)^2 + \omega^2\gamma_j^2} + \mathrm{i}\frac{S_j\omega\omega_{jT}^2\gamma_j}{(\omega_{jT}^2 - \omega^2)^2 + \omega^2\gamma_j^2} \right) . \tag{6.39}$$

In this form the oscillator strength S_j is used as a parameter or may be calculated eventually from a quantum mechanical model.

Returning to the simple form of the DF given in (6.34) or (6.36) we note that an imaginary part in the DF is only obtained for $\gamma \neq 0$. In contrast, the imaginary part κ of the index of refraction can be nonzero even for $\gamma = 0$. In this case $\varepsilon_\mathrm{i} = 0$ but κ may be $\neq 0$ if $\varepsilon_\mathrm{r} = -\kappa^2$ is negative. For these conditions n will be zero from (6.11) and the reflectivity will be 1 from (6.19). Following (6.36) this situation is indeed possible for $\omega^2 > \omega_\mathrm{T}^2$. With ω increasing further, however, the contribution of ω^4 in the denominator makes ε_r again positive. The frequency for which ε_r becomes 0 is labeled ω_L and called the *longitudinal* component of the oscillator. From (6.36) ω_L is obtained for $\gamma = 0$ as

$$\boxed{\omega_\mathrm{L}^2 = \omega_\mathrm{T}^2 + \omega_\mathrm{p}^2/\varepsilon_\infty .} \tag{6.40}$$

In the spectral range where ε_r is negative, i.e., between ω_T and ω_L, reflection is very strong. Since multiple reflections from an interface selectively emphasizes this spectral range, it is called *Reststrahlenbande.* If ω_T lies in the far-IR as it is the case for optical phonons, the multiple reflections can be used to select a narrow band of far-IR radiation.

Inserting ω_L for ω_p in (6.36) we obtain for $\gamma = 0$

$$\varepsilon_\mathrm{r} = \varepsilon_\infty \frac{\omega_\mathrm{L}^2 - \omega_\mathrm{T}^2}{\omega_\mathrm{T}^2 - \omega^2} + \varepsilon_\infty = \varepsilon_\infty \frac{\omega_\mathrm{L}^2 - \omega^2}{\omega_\mathrm{T}^2 - \omega^2} . \tag{6.41}$$

In this form a discussion of the DF is particularly instructive. A qualitative plot of the equation is shown in Fig. 6.3. ε_r becomes negative for for $\omega_\mathrm{T} \leq \omega \leq \omega_\mathrm{L}$. For negative ε_r and vanishing ε_i n will be zero from (6.11) and thus the reflectivity will be 1 from (6.19). Between TO and LO frequency the light is totally reflected.

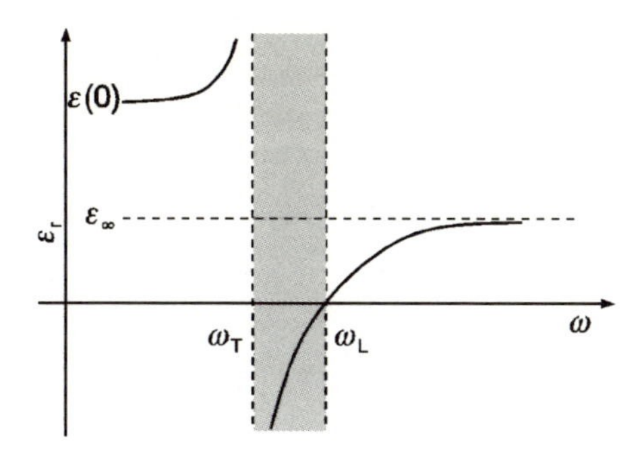

Fig. 6.3. Real part of the Kramers–Heisenberg dielectric function for an undamped oscillator.

The famous Lyddane–Sachs–Teller relation follows immediately from (6.41).

$$\boxed{\frac{\varepsilon(0)}{\varepsilon_\infty} = \frac{\omega_\mathrm{L}^2}{\omega_\mathrm{T}^2}\ .} \tag{6.42}$$

Since $\varepsilon(0)$ is always larger than ε_∞, $\omega_\mathrm{L} > \omega_\mathrm{T}$ follows.

When discussing the dispersion of the prism we already mentioned the influence of the lattice oscillator on the propagation constant $n(\omega)$. Note also that the function used in Sect. 6.1.3 to demonstrate the Kramers–Kronig relation obviously had the form of a Kramers–Heisenberg DF. Finally, it should be stressed again that it is enough to know the frequency dependence for one of the functions ε_r, ε_i, χ_r, χ_i, n, κ, α, or R to obtain the full DF. Figure 6.4 shows the experimentally observed reflectivity for a CdS single crystal together with two components of the index of refraction as evaluated from (6.26) and (6.18).

6.3.2 The Dielectric Function for Free Carriers

The dielectric response function for free carriers can be derived immediately from (6.28) by taking $\omega_\mathrm{T} = 0$. Again, as for the Kramers–Heisenberg DF, this response function is strongly simplified as it does not take into account any dependence of the excitations in the electron gas on the wave vector of the electrons. The differential equation (6.28) can be reduced to first order by introducing the particle velocity $\dot{x} = v$. Expressing γ by an inverse collision time as $\gamma = 1/\tau$ it can be written in the form

$$m^*\dot{v} + m^*v/\tau = eE_1\mathrm{e}^{-\mathrm{i}\omega t}\ . \tag{6.43}$$

For $v = v_0 \exp(-\mathrm{i}\omega t)$ the amplitude j of the current density becomes for n carriers per cm^3

$$j = nev = \frac{ne^2}{m^*}\frac{\tau E_1}{1 - \mathrm{i}\omega\tau} = \sigma E_1\ . \tag{6.44}$$

With $\sigma_0 = ne^2\tau/m^*$, the conductivity is

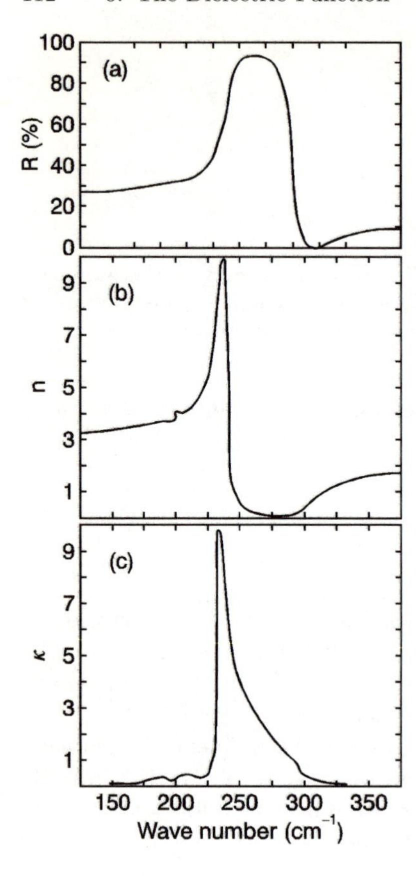

Fig. 6.4. Measured reflectivity R of a single crystal CdS and calculated components n and κ of the index of refraction; after [6.4].

$$\sigma = \frac{\sigma_0}{1 - \mathrm{i}\omega\tau} = \frac{\sigma_0}{1 + \omega^2\tau^2} + \mathrm{i}\frac{\sigma_0\omega\tau}{1 + \omega^2\tau^2} . \tag{6.45}$$

Using the relationship (6.15) between DF and conductivity from Sect. 6.1.1, the DF for the free carriers is

$$\varepsilon_{\mathrm{D}}(\omega) = \varepsilon_\infty + \frac{\mathrm{i}\sigma_0}{\varepsilon_0\omega(1 - \mathrm{i}\omega\tau)} , \tag{6.46}$$

or

$$\boxed{\varepsilon_{\mathrm{D}}(\omega) = \varepsilon_\infty + \frac{\mathrm{i}\omega_{\mathrm{p}}^2\tau/\omega}{1 - \mathrm{i}\omega\tau} = \varepsilon_\infty - \frac{\omega_{\mathrm{p}}^2}{\omega^2 + i\omega/\tau} ,} \tag{6.47}$$

where for the purpose of generality the contribution of the core electrons was added. This DF corresponds to the Drude model for free carriers and is therefore often called the *Drude dielectric function*. Separating the real and imaginary parts yields

$$\varepsilon_{\mathrm{D}} = \varepsilon_{\mathrm{Dr}} + \varepsilon_{\mathrm{Di}} = \varepsilon_\infty - \frac{\omega_{\mathrm{p}}^2\tau^2}{1 + \omega^2\tau^2} + \mathrm{i}\frac{\omega_{\mathrm{p}}^2\tau/\omega}{1 + \omega^2\tau^2} . \tag{6.48}$$

An important special case of (6.48) is obtained for a loss-free plasma which means $\tau = \infty$. In this case $\varepsilon_{\mathrm{D}}(\omega)$ is real and has the value

$$\boxed{\varepsilon_{\mathrm{D}}(\omega) = \varepsilon_{\infty} - \frac{\omega_{\mathrm{p}}^2}{\omega^2}\,.} \tag{6.49}$$

The zeros of this DF correspond to the longitudinal plasma oscillation with the frequency ω_{pl} at $\omega_{\mathrm{p}}/\sqrt{\varepsilon_{\infty}}$. This oscillation is a possible excitation of the plasma in the form of a quantized quasi-particle called the plasmon. In first order the plasmon cannot be excited by an EM wave because of its longitudinal character. Thus, it does not immediately contribute to the response for EM radiation. More details about the response and the dispersion of plasmons will be given in Sect. 15.1.

Equation (6.49), and in a related way also (6.48), define a frequency range for which $\varepsilon = \varepsilon_{\mathrm{r}} \leq 0$. In this range the index of refraction will be purely imaginary and thus R will be 1. The corresponding range is limited by the longitudinal plasma oscillation frequency ω_{pl}. This behavior is called a *plasma reflection* and is very well known in semiconductor and metal physics. For the undamped plasma it starts approximately at the plasma oscillation frequency and extends to zero frequency. Since τ was assumed ∞ ($\gamma = 0$) plasma reflection takes place without energy loss. It is due to a powerless current of the electrons.

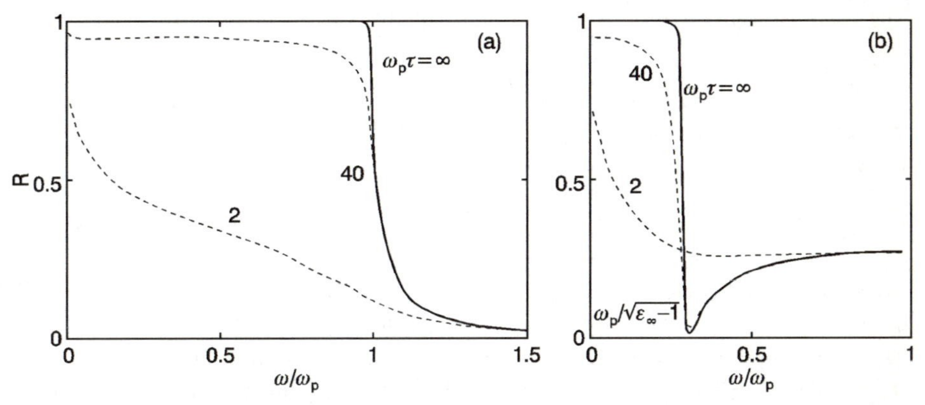

Fig. 6.5. Reflection for systems with free carriers; $n = 10^{22}\,\mathrm{cm}^{-3}$, $\varepsilon_{\infty} = 1$ (**a**) and $n = 10^{17}\,\mathrm{cm}^{-3}$, $\varepsilon_{\infty} = 11$ (**b**). The dashed line is for finite damping. The numbers indicate $\omega_{\mathrm{p}}\tau$.

Interestingly the shape of the plasma reflection depends strongly on the magnitude of ε_{∞}, as shown in Fig. 6.5 for various values of $\omega_{\mathrm{p}}\tau$. Therefore it has different character for semiconductors and metals. For the latter $\varepsilon_{\infty} \approx 1$ and the reflection drops rapidly from 1 at ω_{p} but approaches 0 only for very large frequencies. In contrast, for semiconductors where ε_{∞} is large, ω_{p} is quite different from ω_{pl}. The reflectivity drops sharply to zero close to ω_{pl} (for $\tau = \infty$) and increases back to the value given by ε_∞. Figure 6.6 exhibits experimental results for the plasma reflection of aluminum and heavily doped InSb. The large value for the plasma edge for Al is due to its high carrier concentration. The plasma reflection passes through zero in the case of InSb (Fig. 6.6b).

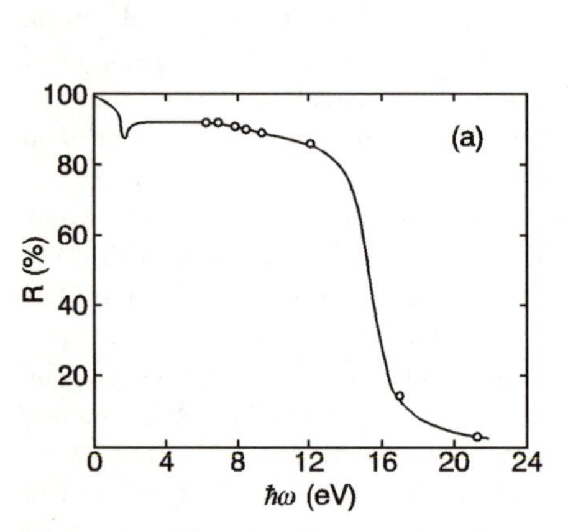

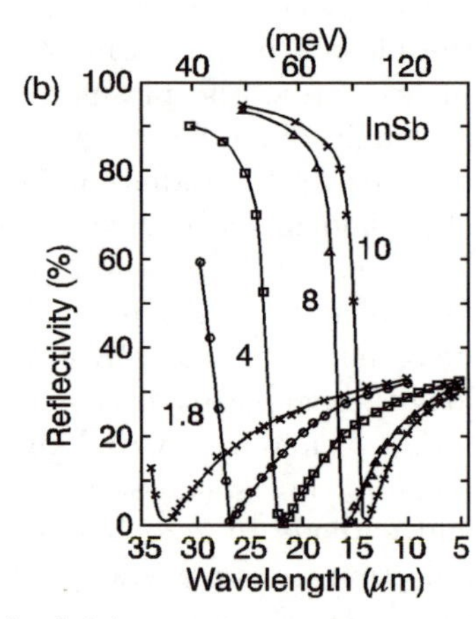

Fig. 6.6. Plasma reflection for aluminum, after [6.5] (**a**), and for heavily doped InSb, adapted from [6.6] (**b**). The carrier concentration for Al is $n = 18 \times 10^{22}\,\text{cm}^{-3}$, the concentration for InSb is given in units of $10^{18}\,\text{cm}^{-3}$ in the figure.

For another limit which is often observed in semiconductors the absorption coefficient α as given by (6.12) can be approximated for $\omega\tau \gg 1$ (collision-free plasma) and $\omega_{\text{p}} \ll \omega$. In this case $n(\omega) \approx \sqrt{\varepsilon_\infty}$ and

$$\alpha = \frac{\omega_{\text{p}}^2 \lambda_0^2}{4\pi^2 \sqrt{\varepsilon_\infty} \tau c_0^3} \propto \omega_{\text{p}}^2 \lambda_0^2 \,. \tag{6.50}$$

An example for a free carrier absorption is presented in Fig. 6.7 for InAs. The linear relationship between λ and α in the double logarithmic plot confirms the power law of (6.50). The slope for the lines is about 2.7. For $n = 2.5 \times 10^{17}\,\text{cm}^{-3}, \lambda = 15\,\mu\text{m}$ and $\alpha = 30\,\text{cm}^{-1}$ (as taken from Fig. 6.7) τ becomes $10^{-10}\,\text{s}$ for a typical value of 10 for ε_∞ and an effective mass for InAs of $3 \times 10^{-2} m_0$. With these values the limiting conditions $\omega\tau = 10 \gg 1$ and $\omega_{\text{p}} = 10^{12}\,\text{s}^{-1} \ll \omega = 10^{13}\,\text{s}^{-1}$ are well satisfied.

The quantity ω_{p} is determined by the density of oscillators which, in the present case, is the density of free carriers in the conduction band. More generally the value of ω_{p} depends on the energy range considered or, more precisely, on the energy range which can be excited in the solid. As an example, for low-energy excitation, only the electrons and holes from the conduction band and from the valence band contribute to ω_{p}. For higher excitation

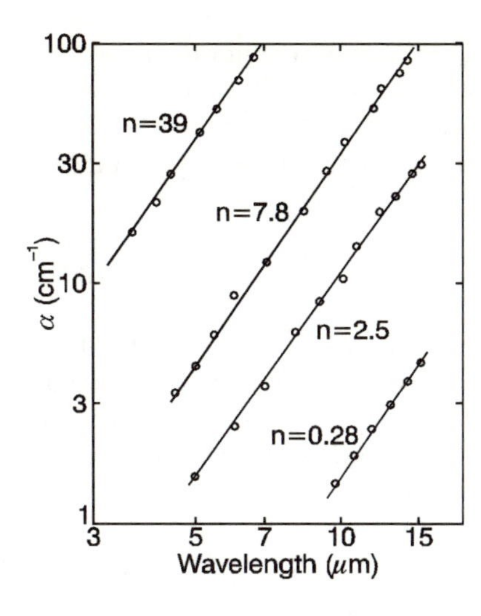

Fig. 6.7. Free carrier absorption in InAs for different carrier concentrations given in units of $10^{17}\,\mathrm{cm}^{-3}$; after [6.7].

energies the relevant plasma is established by all valence electrons. If the solid is excited in the energy range of 20-30 eV even electrons from the lower orbitals (for instance, the *d* electrons in Ge) contribute to the plasma frequency. This means that several plasma oscillations can be excited in a solid. However, a high plasma reflection which extends to zero frequency can only originate from free carriers since it implies a vanishing oscillator energy for the single-particle excitation.

6.3.3 Oscillator Strength and Sum Rules

In Sect. 6.3.1 a constant S was introduced to provide a generalized description for the strength of the oscillators. The same could have been done in Sect. 6.3.2 for the transitions of the free carriers. Physically S is related to the probability f that an oscillator is actually excited. Hence, another quantity f is often defined instead of S by

$$S = \frac{\omega_\mathrm{p}^2 f}{\omega_\mathrm{T}^2} \; .$$

The total activity of the oscillators for the whole frequency range is described by the sum of all oscillators and can be expressed by a *sum rule* (or *f-sum rule*) for the oscillator strengths. Several sum rules exist. A commonly used rule which holds for the first moment of the imaginary part of the DF has the form

$$\boxed{\int_0^\infty \omega\varepsilon_\mathrm{i}(\omega)\,\mathrm{d}\omega = \sum_j \frac{\pi\omega_{\mathrm{jT}}^2 S_\mathrm{j}}{2} \; .} \tag{6.51}$$

This relation is independent of the form of ε_i. The integral on the left-hand side of the equation can be determined from a numerical evaluation of an experiment. The result can be used to calibrate the experiment on an absolute scale. If only one oscillator is considered the right-hand side of (6.51) yields $\pi\omega_{\mathrm{p}}^2/2$ for S inserted from (6.38). For free carriers in a Drude model ω_{p} is the plasma frequency given by (6.35). For phonons the corresponding ion plasma frequency ω_{ip} with

$$\omega_{\mathrm{ip}}^2 = \frac{n e_{\mathrm{T}}^{*2}}{\varepsilon_0 M_{\mathrm{R}}} \tag{6.52}$$

must be used, where n is the concentration of the oscillators, M_{R} their reduced mass, and e_{T}^* the effective transverse charge. Phonons which have an effective transverse charge $\neq 0$ are called *polar*. Only polar modes can interact (directly) with the light. The sum rule for ε_{i} is easily rewritten for the real part of the conductivity σ as

$$\boxed{\int_0^\infty \sigma_{\mathrm{r}}(\omega)\,\mathrm{d}\omega = \frac{\pi\varepsilon_0\omega_{\mathrm{p}}^2}{2}\,.} \tag{6.53}$$

Another important sum rule is obtained from the Kramers–Kronig relation for the DF of lattice modes

$$\varepsilon_{\mathrm{r}} - \varepsilon_\infty = \frac{2}{\pi}\mathrm{P}\int_0^{\omega_{\mathrm{c}}} \frac{\omega\varepsilon_{\mathrm{i}}\mathrm{d}\omega}{\omega^2 - \omega_0^2}\,, \tag{6.54}$$

where ω_{c} is a frequency above the lattice modes but below optical transitions. For $\omega_0 = 0$

$$\boxed{\int_0^{\omega_{\mathrm{c}}} \frac{\varepsilon_{\mathrm{i}}(\omega)}{\omega}\,\mathrm{d}\omega = \frac{\pi}{2}[\varepsilon(0) - \varepsilon_\infty]} \tag{6.55}$$

follows. This shows that the difference between the static and the optical dielectric constant comes from the integrated activity of the oscillators at finite frequency.

6.4 Experimental Determination of Dielectric Functions (Ellipsometry)

As we have learned above, it is enough to measure one of the components of the DF and use the Kramers–Kronig transformation to obtain the other component. This process may still be cumbersome since the one component must be measured over the whole frequency range. In alternative methods both components of the DF are determined simultaneously over a limited frequency range. In this case a Kramers–Kronig analysis is not needed. A standard procedure for this approach consists in the analysis of elliptically polarized light obtained after reflection from a plane surface of the material under investigation. Accordingly, this technique is called *ellipsometry*.

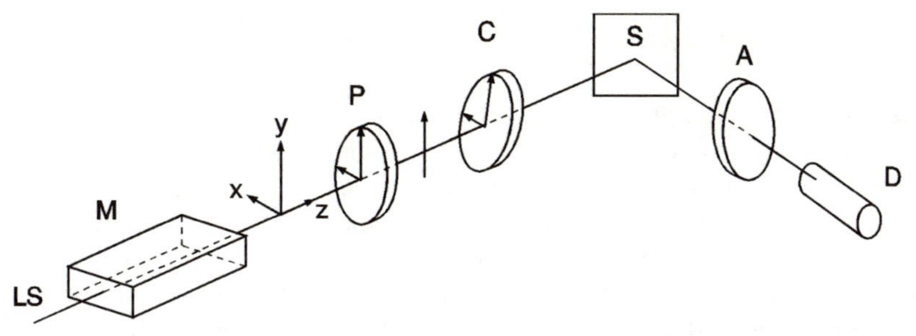

Fig. 6.8. Schematic set up for a PCSA ellipsometer; (LS: light source, M: monochromator, P: polarizer, C: compensator, S: sample surface, A: analyzer, D: detector); PCSA is the acronym for polarizer, compensator, surface, and analyzer.

Figure 6.8 shows schematically a setup for an ellipsometer. The basic elements are the polarizer which gives linearly polarized light, the compensator which provides a well defined ellipticity, the sample surface to be investigated and the analyzer to check the ellipticity of the light after reflection. These instruments are called PCSA ellipsometers according to the main elements of the system. In addition to these basic elements, a broad band light source, a monochromator, and a detector are needed for spectroscopy.

To fully characterize elliptically polarized light of frequency ω seven parameters are needed. Three parameters define the orientation of the wave in space and the remaining four describe the intrinsic polarization properties. In the following discussion propagation in the z direction is assumed. Then, the remaining four parameters determine the orientation of the ellipse described by the electric-field vector in the xy plane, the shape of the ellipse, and the magnitude and absolute phase angle (as seen from the origin) of the electric field. These parameters are used to express the two components of the field in the x and y direction as a plane wave

$$\begin{aligned} E_x &= E_{0x}\mathrm{e}^{\mathrm{i}(kz-\omega t-\delta_x)} \\ E_y &= E_{0y}\mathrm{e}^{\mathrm{i}(kz-\omega t-\delta_y)} \,. \end{aligned} \tag{6.56}$$

To describe the state of polarization of the wave only the two amplitudes E_{0x} and E_{0y} and the two phases δ_x and δ_y are necessary. Thus, any state of polarization can be characterized by a two-dimensional vector of the form

$$\boldsymbol{E} = (E_{0x}\mathrm{e}^{\mathrm{i}\delta_x}, E_{0y}\mathrm{e}^{\mathrm{i}\delta_y}) \,, \tag{6.57}$$

where the two components are complex numbers. These vectors are called *Jones vectors* and are utilized in ellipsometry to describe elliptically polarized light. The intensity of the light is given by the product $\boldsymbol{E}\boldsymbol{E}^*$. The Jones vectors for light linearly polarized parallel to x or y and left-circularly polarized are

$$\begin{aligned} \boldsymbol{E}(\|x) &= E_0(1,0), \\ \boldsymbol{E}(\|y) &= E_0(0,1), \end{aligned}$$

$$\boldsymbol{E}(\mathrm{lcp}) = \frac{E_0}{\sqrt{2}}(1, -\mathrm{i}) \,. \tag{6.58}$$

Note that the relative magnitude $|E_y/E_x|$ and the phase difference $\exp[\mathrm{i}(\delta_y - \delta_x)]$ are enough to characterize the ellipticity of the light.

If light is transmitted or reflected by an optical element which is active with respect to the state of polarization, the Jones vector will be changed to another Jones vector. Typical examples of such elements are polarizers, phase plates or any type of reflectors, including the reflecting surface of the sample to be investigated. Since the transformation of one vector into another is performed by a matrix, each optical element is represented by a matrix $\boldsymbol{T}$ which is called the *Jones matrix*. As an example the Jones matrix for a phase plate of thickness d is given in diagonalized form by

$$\boldsymbol{T} = \begin{pmatrix} \exp(-\mathrm{i}\delta_1) & 0 \\ 0 & \exp(-\mathrm{i}\delta_2) \end{pmatrix} ,$$

with $\delta_j = 2\pi n_j d/\lambda$. $j = 1, 2$ stands for the extraordinary and for the ordinary beam, respectively. The matrix is diagonal for x and y oriented parallel and perpendicular to the optical axis of the phase plate.

For each optical element there exists a state of polarization which remains unchanged upon transmission. The corresponding Jones vector is called an *eigenvector* of the Jones matrix for the element. For example, the Jones vectors for polarization directions parallel and perpendicular to the plane of incidence are eigenvectors for any reflection element. Each Jones matrix has *eigenvalues* which are calculated from the diagonalization procedure in the usual way. For a reflection element the eigenvalues are the complex reflection coefficients for the amplitudes of waves polarized parallel and perpendicular to the plane of incidence, defined as

$$r_{\mathrm{p}} = \frac{E_{\mathrm{r,p}}}{E_{\mathrm{i,p}}} = |r_{\mathrm{p}}|\mathrm{e}^{\mathrm{i}\delta_{\mathrm{p}}} \qquad r_{\mathrm{n}} = \frac{E_{\mathrm{r,n}}}{E_{\mathrm{i,n}}} = |r_{\mathrm{n}}|\mathrm{e}^{\mathrm{i}\delta_{\mathrm{n}}} \,. \tag{6.59}$$

If the light traverses several optical elements the product of the corresponding Jones matrices describes the total change of polarization. The change in the state of polarization in the setup of Fig. 6.8 can be calculated as the product of the matrices for the elements P, C, S, and A from which the Jones matrix for the surface of the sample can be determined. The ratio $r_{\mathrm{s}} = r_{\mathrm{p}}/r_{\mathrm{n}}$ of the complex eigenvalues of the resulting matrix written in the form

$$\boxed{r_{\mathrm{s}} = \frac{r_{\mathrm{p}}}{r_{\mathrm{n}}} = \tan\psi \mathrm{e}^{\mathrm{i}\Delta}} \tag{6.60}$$

with

$$\tan\psi = \left|\frac{r_{\mathrm{p}}}{r_{\mathrm{n}}}\right| \qquad \text{and} \qquad \Delta = \delta_{\mathrm{p}} - \delta_{\mathrm{n}}$$

is used to determine the DF $\varepsilon(\omega)$. Equation (6.60) is called the *ellipsometer equation* and ψ and Δ are the *ellipsometric angles*. Physically $\tan\psi$ is the ratio of the field attenuation upon reflection for light polarized parallel and perpendicular to the plane of incidence, and Δ gives the corresponding phase difference. From the ratio of the eigenvalues of the Jones matrix the DF is obtained from

$$\boxed{\varepsilon(\omega) = \sin^2\phi + \sin^2\phi \tan^2\phi \left(\frac{1-r_s}{1+r_s}\right)^2 ,} \tag{6.61}$$

where ϕ is the angle of incidence of the light with respect to the sample surface and the reflection is versus vacuum. (Note: $\phi = 90°$ is not allowed since it does not define a plane of incidence and therefore r_s can not be evaluated.)

For the explicit experimental determination of r_s either a zero method (zero ellipsometer) or a rotating analyzer (rotating analyzer ellipsometer) is used. More details of such instruments are given in Appendix E.3. An extensive description of ellipsometry can be found in [6.8].

Like all reflectivity measurements ellipsometry is very sensitive to contamination of the surface of the sample. Hence this technique is also very useful for the analysis of thin films on a surface.

Problems

6.1* Determine the reflection coefficient R at an interface for perpendicular incidence using energy conservation and the continuity of the electric field at the transition.

(Purpose of exercise: derivation of a very important quantity in spectroscopy.)

6.2 The frequency dependence of the absorption coefficient is often directly related to the frequency dependence of ε_i. Discuss to what extent this is justified by evaluating $\alpha(\omega)$ and $\varepsilon_i(\omega)$ explicitly for a Kramers–Heisenberg DF with strong absorption.

(Purpose of exercise: prove an often used approximation.)

6.3* Specify the equations for the reflection from a thin film on a substrate for the case of a plane-parallel plate in vacuum and thus provide proof of the Airy formulae.

(Purpose of exercise: training in thin film optics.)

6.4 Show how the DF can be obtained from the magnitude and the phase angle for field reflection.

(Purpose of exercise: gain experience with the Kramers–Kronig transformation.)

6.5 Show explicitly that the Kramers–Kronig relation holds for a Kramers–Heisenberg DF.

(Purpose of exercise: gain experience with the Kramers–Kronig transformation.)

6.6 Show that the reflectivity from a Drude DF has a minimum for semiconductors but not for metals.

Hint: Study the simplified case for $\tau = \infty$ and show that the reflectivity becomes zero for $\omega^2 = \omega_p^2/(\varepsilon_\infty - 1)$ in this case.

(Purpose of exercise: use of the Drude DF.)

6.7 Give a proof of the sum rule for the real part of the conductivity for the Drude DF and for the Kramers–Heisenberg DF of a polar phonon.

(Purpose of exercise: gain experience with sum rules.)

6.8 Show that $\lambda/4$ platelets generate circularly polarized light from linearly polarized light by using the formalism of Jones vectors and Jones matrices.

(Purpose of exercise: gain experience with the Jones formalism.)

7. Spectroscopy in the Visible and Near-Visible Spectral Range

The spectral range where spectroscopy has been used for the longest time and where it is still most often applied is the visible and near-visible. In this range spectroscopy can be performed with the naked eye without any scientific instruments. However, even here, highly sophisticated experimental techniques have been applied and have revealed many details about electronic structures. In the present chapter we review first some details about the quantum-mechanical description of optical absorption and then apply this formalism to absorption in systems with extended states such as semiconductors and to systems with localized states like color centers, defect states and transitions in molecular crystals. An introduction to luminescence processes is presented in the last section. More details about the quantum-mechanical formalisms used can be found in Appendices F.1 to F.3.

7.1 Quantum-Mechanical Description of Optical Absorption

Optical absorption is dominated by the imaginary part of the DF. This is true for the absorption across the energy gap in a semiconductor as well as for the absorption by defects or by deep laying electronic levels. It is therefore convenient to discuss first a general quantum-mechanical description of the absorption process and then apply the results to various special configurations in a solid.

The strength of an absorption process is determined by the quantum-mechanical probability for the transition rate of a system, changing from an initial electronic state i to a final electronic state f. In general the initial state is the ground state and the final state is an excited state. The transition probability is proportional to the square of the magnitude of the matrix element $H'_{\rm fi}(0)$, where H' is the perturbation driving the transition. The transition energy is $\hbar\omega_{\rm fi}$. In the present chapter the perturbation is an EM radiation with vector potential $\boldsymbol{A}(x,t)$. For small EM fields the perturbation is explicitly given (in operator form) by

$$\boxed{\mathbf{H}' = -\frac{e}{m}\mathbf{pA}\,,} \tag{7.1}$$

where **p** is the momentum operator for the electrons. (For proof of this see Appendix F.2). The wave functions and the matrix elements of the perturbed system are usually obtained by first-order perturbation theory, as outlined in Appendices F.1 and F.2. With the matrix elements $H'_{\mathrm{fi}}(0)$ of the perturbation the *golden rule* of quantum mechanics gives the probability for a transition per unit of time

$$\boxed{P_{\mathrm{fi}} = \frac{2\pi |H'_{\mathrm{fi}}(0)|^2}{\hbar^2}\delta(\omega_{\mathrm{fi}} - \omega)\,.} \tag{7.2}$$

The matrix element $H'_{\mathrm{fi}}(0)$ is evaluated from the time-dependent matrix element of the perturbation

$$H'_{\mathrm{fi}} = \langle \mathrm{f}|\mathbf{H}'|\mathrm{i}\rangle = -\frac{e}{m}\langle \mathrm{f}|\mathbf{pA}|\mathrm{i}\rangle\,. \tag{7.3}$$

The time-dependent part of this matrix element yields the δ function in (7.2) (see Appendix F.2). It describes the density of allowed final states. The transition to these states is subjected to energy conservation. The time-independent part is evaluated by the multipole approximation. Within the dipole approximation $\boldsymbol{A}= \boldsymbol{A}_0\mathrm{e}^{\mathrm{i}kx}$ is replaced by its amplitude $\boldsymbol{A}_\mathbf{0}$ which yields

$$\boldsymbol{H}'_{\mathrm{fi}}(0) = -\frac{eA_0}{m}\langle \mathrm{f}|\mathbf{p}|\mathrm{i}\rangle = -\frac{eA_0}{m}\boldsymbol{p}_{\mathrm{fi}}\,. \tag{7.4}$$

In this case, the transition matrix element is given in the momentum representation by the momentum matrix elements

$$(p_j)_{\mathrm{fi}} = -\mathrm{i}\hbar \int \psi_{\mathrm{f}}^* \frac{\partial \psi_{\mathrm{i}}}{\partial x_j}\mathrm{d}^3x, \qquad j = 1,2,3\,. \tag{7.5}$$

Note that these matrix elements are the components of a vector. ψ_{f} and ψ_{i} are eigenfunctions of the unperturbed system. The transition matrix element H'_{fi} can be expressed in the dipole representation (Appendix F.3) by replacing the momentum matrix element with the dipole matrix element $(M_j)_{\mathrm{fi}}$. This matrix element is defined from

$$\boxed{(p_j)_{\mathrm{fi}} = \mathrm{i}\omega_{\mathrm{fi}}\frac{m}{e}(M_j)_{\mathrm{fi}} = \mathrm{i}\omega_{\mathrm{fi}}\frac{m}{e}\int \psi_{\mathrm{f}}^* e x_j \psi_{\mathrm{i}}\,\mathrm{d}^3x\,.} \tag{7.6}$$

Again, the dipole matrix elements are vectors with x, y and z components.

Representing the square of the vector potential of the radiation by the intensity $I(\omega) = nA_0^2\varepsilon_0 c_0\omega^2$ the absolute square of the matrix elements of the perturbation become

$$|H'_{\mathrm{fi}}|^2 = |\langle \mathrm{f}|\mathbf{H}'|\mathrm{i}\rangle|^2 = \frac{e^2 I(\omega)|p_{\mathrm{fi}}|^2}{m^2\varepsilon_0 c_0\omega^2 n}\,, \tag{7.7}$$

where n is the refractive index.

The δ function on the right-hand side of (7.2) selects a single transition and insures energy conservation. If there are several states in the immediate vicinity of the initial and the final state, a summation is required over all states with equal distance in energy. This is, for example, the case for transitions between two energy bands.

For states which are characterized by a k vector conservation of momentum requires, in addition, that $\sum \boldsymbol{k}_i = 0$ where the sum extends over all k vectors contributing to the transition process. Momentum conservation is automatically fulfilled when the transition matrix element is evaluated.

From (7.2) the absorption may be evaluated as the ratio between the rate at which energy is absorbed per volume $\mathcal{V}$ and the rate at which energy is incident per unit area

$$\alpha(\omega) = \frac{\hbar\omega P_{\mathrm{fi}}}{I(\omega)\mathcal{V}} . \tag{7.8}$$

This definition is equivalent to the formal definition given in (6.8). Using (7.2), (7.4), and (7.7) we finally find

$$\boxed{\alpha(\omega) = \frac{2\pi}{\mathcal{V}} \frac{e^2 |p_{\mathrm{fi}}|^2}{m_0^2 \varepsilon_0 c_0 n \omega} \delta(\hbar\omega_{\mathrm{fi}} - \hbar\omega) .} \tag{7.9}$$

With (7.9) and the relationship between α and the imaginary part of $\varepsilon(\omega)$ given by (6.12), the latter becomes

$$\varepsilon_{\mathrm{i}}(\omega) = \frac{2\pi}{\mathcal{V}} \frac{e^2 |p_{\mathrm{fi}}|^2}{m_0^2 \varepsilon_0 \omega^2} \delta(\hbar\omega_{\mathrm{fi}} - \hbar\omega) . \tag{7.10}$$

7.2 Absorption from Extended States in Semiconductors

Absorption by extended states plays a dominant role in semiconductors. The transition for electrons from the valence band to the conduction band starts abruptly for a quantum energy of the radiation $\hbar\omega$ which exceeds the energy gap ϵ_{g}. The corresponding increase of the absorption by several orders of magnitude is called the fundamental absorption or the absorption edge. The energies at which the absorption edges occur range from several meV to more than 10 eV. Table 7.1 lists some examples.

7.2.1 The Physical Background and the Shape of the Absorption in Semiconductors

Figure 7.1 exhibits the absorption coefficient for three different semiconductors. It starts at a well defined energy and increases immediately like an edge by several orders of magnitude. At higher energies the increase slows down and characteristic structures appear. Details at the edge depend on the band structure and the nature of the electronic transitions. In special cases, such as for GaAs, the initial slope on an α versus ω diagram is 1/2 when a double logarithmic plot is used. The corresponding transition is called *allowed* and

Table 7.1. Lowest energetic distance between valence band and conduction band for various solids.

Crystal	ϵ_g [eV]	Crystal	ϵ_g [eV]	Crystal	ϵ_g [eV]
αSn	0.08	GaAs	1.47	ZnO	3.3
PbTe	0.19	GaP	2.24	BaO	4.4
InSb	0.23	NiO	2.3	$LiGaO_2$	5.2
PbS	0.29	CdS	2.5	CaO	6.5
Bi_2Te_3	0.31	SiC	2.8	Quarz	6.7
Ge	0.67	ZnSe	2.8	KCl	8.69
Si	1.10	$SrTiO_3$	3.3	Al_2O_3	10
InP	1.37	$SrTiO_3$	3.3	KF	10.9

direct. An allowed transition corresponds to a finite dipole matrix element at the position of minimum distance between valence band and conduction band. Direct transitions are possible if the maximum of the valence band and the minimum of the conduction band occur for the same k vector, as shown schematically in Fig. 7.2a. A Structure on the band edge as, for example, in the case of Ge in Fig. 7.1, indicates phonon assistance for the electronic transition. This is always the case if the minimum of the conduction band and the maximum of the valence band do not occur for the same k vector. An example is sketched in Fig. 7.2b. The full arrow drawn from the valence band to the conduction band indicates the transition with minimum energy. This case is called an *indirect* transition. In the first step of the transition the electron is excited to a virtual intermediate state where it stays for a very short time. From there a phonon with wave vector q_{ph} is required to finally transfer it to the real state for which energy and momentum is conserved.

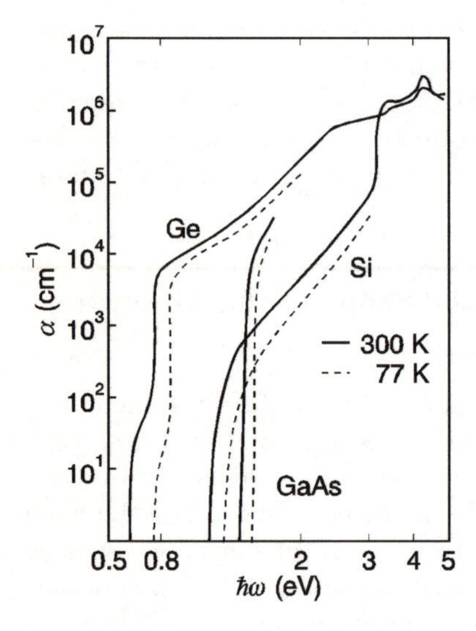

Fig. 7.1. Absorption coefficient α for the semiconductors Ge, Si, and GaAs close to the absorption edge; after [7.1].

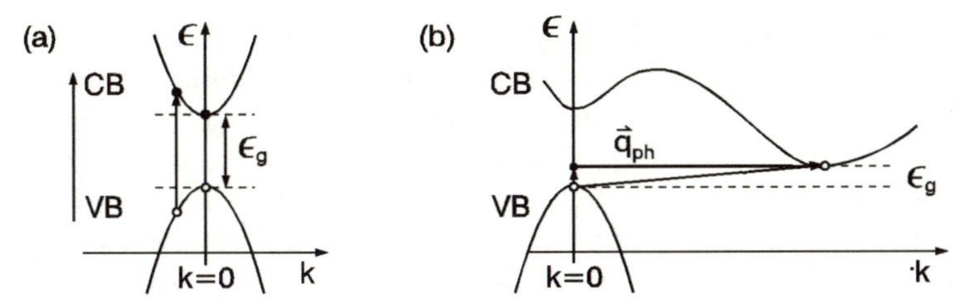

Fig. 7.2. Direct electronic transitions between two simple bands (**a**) and indirect transitions including the absorption or emission of a phonon (**b**); (VB: valence band, CB: conduction band).

The shape of the absorption curve beyond the edge is determined by transitions starting from deeper in the valence band or ending higher in the conduction band. Eventually also transitions into bands beyond the conduction band or from bands below the valence band contribute to the absorption. Analysis of structures in the absorption can give detailed information on the shape of the bands.

The fundamental absorption is of particular interest. In this case, i and f in (7.2) and (7.8) correspond to the valence band and the conduction band. In addition, the electronic states are characterized by their k vector in the band, and wave vector conservation is required for any transition. Since the wave vector of the light is small compared to the wave vector of the electrons, k conservation is only possible for direct transitions which means $\boldsymbol{k}_\mathrm{V} + \boldsymbol{k}_\mathrm{C} \approx 0$ for any transition. In other cases like in Fig. 7.2b phonons must assist to establish k-conservation. As shown in Fig. 7.2a the direct transitions need not start at $\boldsymbol{k} = 0$. Since for a given energy difference many transitions starting from various values for $\boldsymbol{k}$ are possible the density of states in the initial and in the final band are important.

7.2.2 Direct and Allowed Transitions at the Absorption Edge

In this section we will study direct and allowed transitions in more detail. Whether a transition is allowed or not depends on the symmetry of the initial and final states. (Further discussion of this problem will be found in Chap. 8). For example, transitions between s-waves or between s- and d-waves are dipole forbidden. Transitions between s- and p-waves are dipole allowed. For the evaluation of the matrix elements in (7.5) or in (7.6) the Bloch functions for the valence-band electrons and for the conduction-band electrons

$$\psi_\mathrm{i} = u_\mathrm{V}(r)\mathrm{e}^{\mathrm{i}\boldsymbol{k}_\mathrm{V}\boldsymbol{r}}, \qquad \psi_\mathrm{f} = u_\mathrm{C}(r)\mathrm{e}^{\mathrm{i}\boldsymbol{k}_\mathrm{C}\boldsymbol{r}} \tag{7.11}$$

are required. The explicit evaluation of the matrix elements is difficult since the exact band structure and the correct wave function must be known. Approximations such as free carrier wave functions instead of Bloch functions are common. Details of such calculations can be found in special books on semiconductor physics [7.2, 7.3].

Since the matrix elements depend only weakly on the energy, the shape of the absorption curve at the edge is mainly determined by the *joint* density of states $\sigma_{\mathrm{CV}}(\omega)$ between the valence band and the conduction band. The joint density of states $\sigma_{\mathrm{CV}}(\omega)$ is the density of states available for a given transition energy $\hbar\omega$. Thus, the value of α obtained from (7.9) must be integrated over all states in k space which satisfy the energy conservation. These states are defined by the δ function in (7.9). Since the δ function is the only strongly k-dependent part in the equation, the integration can be restricted to this function.

$$\begin{aligned}\sigma_{\mathrm{CV}}(\omega) &= \frac{2\mathcal{V}}{8\pi^3}\int \delta[\hbar\omega_{\mathrm{CV}}(k) - \hbar\omega]\,\mathrm{d}^3k \\ &= \frac{\mathcal{V}}{4\pi^3}\int \delta[\hbar\omega_{\mathrm{C}}(k) - \hbar\omega_{\mathrm{V}}(k) - \hbar\omega]\,\mathrm{d}^3k\,. \end{aligned} \tag{7.12}$$

With this the absorption coefficient α is given by

$$\alpha(\omega) = \frac{2\pi}{\mathcal{V}}\frac{e^2|p_{\mathrm{CV}}|^2}{m_0^2\varepsilon_0 c_0 n\omega}\sigma_{\mathrm{CV}}(\omega)\,. \tag{7.13}$$

In the case of absorption, $\hbar\omega$ is the quantum energy of the incident light. For a spherical band we can replace d^3k by $4\pi k^2\,\mathrm{d}k$ and integrate explicitly using the special property point 5 of the δ function given in Appendix B.7.2 with

$$g(k) = \hbar[\omega_{\mathrm{C}}(k) - \omega_{\mathrm{V}}(k) - \omega]\,.$$

The result is

$$\boxed{\sigma_{\mathrm{CV}}(\omega) = \mathcal{V}\frac{k^2}{\pi^2}\left(\frac{\mathrm{d}\epsilon_{\mathrm{C}}(k)}{\mathrm{d}k} - \frac{\mathrm{d}\epsilon_{\mathrm{V}}(k)}{\mathrm{d}k}\right)^{-1}\,.} \tag{7.14}$$

If the band is spherical and parabolical the relations between ϵ and k are[1]

$$\epsilon_{\mathrm{V}} = \frac{\hbar^2k^2}{2m_{\mathrm{V}}^*} \qquad \text{and} \qquad \epsilon_{\mathrm{C}} = \frac{\hbar^2k^2}{2m_{\mathrm{C}}^*} + \epsilon_{\mathrm{g}}\,.$$

From this the joint density of states for a crystal with the volume $\mathcal{V}$ results in

$$\boxed{\sigma_{\mathrm{CV}}(\omega) = \mathcal{V}\frac{(2m_r^*)^{3/2}}{2\pi^2\hbar^3}\sqrt{\hbar\omega - \epsilon_{\mathrm{g}}}\,.} \tag{7.15}$$

m_{r} is the reduced mass obtained from the effective mass of the holes and of the electrons in the bands.

This result agrees well with the observation of the square root dependence of the absorption constant on the light energy described above. (See Sect. 7.2.1 for the direct allowed transitions). Thus, for this case α has the form

---

[1] If the bands are not spherical and parabolical k must be considered as a vector and the derivatives with respect to k must be replaced by ∇_k.

$$\begin{aligned} \alpha(\omega) &= B\sqrt{\hbar\omega - \epsilon_\mathrm{g}} && \text{for} \quad \hbar\omega \geq \epsilon_\mathrm{g} \\ &= 0 && \text{for} \quad \hbar\omega \leq \epsilon_\mathrm{g}, \end{aligned} \tag{7.16}$$

where B is a constant given from equations (7.2) to (7.15) by

$$B = \frac{e^2(2m_\mathrm{r}^*)^{3/2}}{12\pi^5 m_0^2 n \varepsilon_0^2 c_0 \hbar^3} |p_\mathrm{CV}|^2 .$$

7.2.3 Forbidden Transitions and Phonon-Assisted Transitions

A transition is dipole forbidden if the dipole matrix element is zero at $k = 0$. It may be nonzero for other wave vectors. In this case an analysis similar to that presented above yields for the energy dependence of the optical absorption

$$\begin{aligned} \alpha_\mathrm{forb}(\omega) &= C(\hbar\omega - \epsilon_\mathrm{g})^{3/2} && \text{for} \quad \hbar\omega \geq \epsilon_\mathrm{g} \\ &= 0 && \text{for} \quad \hbar\omega \leq \epsilon_\mathrm{g} . \end{aligned} \tag{7.17}$$

For semiconductors with an indirect bandgap the energy dependence at the absorption edge is still different. A famous example of such a material is Ge. The valence band has its maximum at the center of the Brillouin zone (Γ point). The minimum of the conduction band is at the boundary of the zone (L point). The difference in energy is only 0.67 eV. Electronic transitions between the two points are only possible if a zone-boundary phonon is simultaneously absorbed or emitted to balance the momentum-conservation. If the phonon is absorbed, the net photon energy can be even smaller than the gap energy by the amount of the phonon energy $\hbar\Omega$. Therefore, in indirect semiconductors the optical absorption starts at a quantum energy $\hbar\Omega$ lower than the energy gap. For light with a quantum energy higher than $\hbar\Omega + \epsilon_\mathrm{g}$ phonon emission and absorption are possible. Accordingly, a similar calculation as above yields

$$\begin{aligned} \alpha_\mathrm{indir} &= 0 && \text{for } \hbar\omega < \epsilon_\mathrm{g} - \hbar\Omega \\ &= \frac{A(\hbar\omega - \epsilon_\mathrm{g} + \hbar\Omega)^2}{\exp(\hbar\Omega/k_\mathrm{B}T) - 1} && \text{for } \epsilon_\mathrm{g} + \hbar\Omega \geq \hbar\omega \geq \epsilon_\mathrm{g} - \hbar\Omega \\ &= \frac{A(\hbar\omega - \epsilon_\mathrm{g} + \hbar\Omega)^2}{\exp(\hbar\Omega/k_\mathrm{B}T) - 1} \\ &\quad + \frac{A(\hbar\omega - \epsilon_\mathrm{g} - \hbar\Omega)^2}{1 - \exp(-\hbar\Omega/k_\mathrm{B}T)} && \text{for } \hbar\omega > \epsilon_\mathrm{g} + \hbar\Omega . \end{aligned} \tag{7.18}$$

With the light energy further increasing even direct transitions across the gap at the Γ point may be possible. Quantitative details about the evaluation of the absorption for forbidden transitions and for indirect transitions can be obtained from [7.4].

Because of the selective behavior of the absorption information on the band structure can be obtained from an analysis of the band edge. Since the lower part of the edge is determined by phonon-assisted transitions even phonon energies can be determined. This is a possibility to analyze non-zone-center phonons. It works particularly well for indirect semiconductors at low temperatures. Figure 7.3 shows the low-energy edge of the indirect semiconductor GaP. Several acoustical and optical modes can contribute to the absorption. The structures are particularly prominent for the data at 77 K. The subscripts A and E identify the symmetry of the phonons to be discussed in Chap. 8.

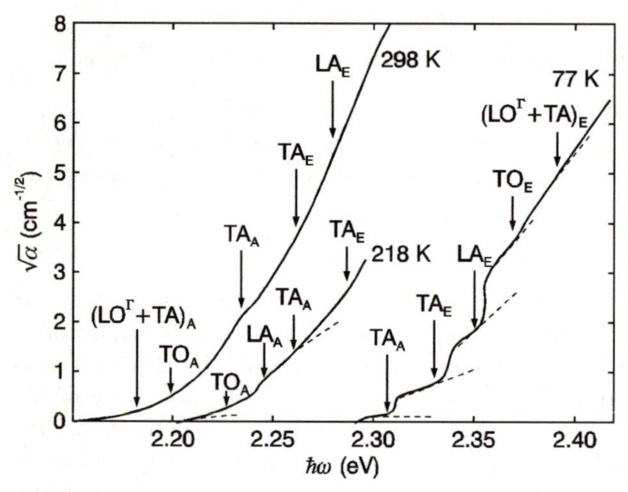

Fig. 7.3. Low energy part of the absorption edge in GaP for three different temperatures plotted as $\sqrt{\alpha}$ vs. light energy; (TO, LO: transverse and longitudinal optical modes, TA, LA: transverse and longitudinal acoustical modes); the dashed lines are guides for the eye; after [7.5].

7.2.4 Absorption from Higher Transitions

For excitation of the solid with energies higher than ϵ_g transitions into higher bands become important, as well as vertical transitions at positions in k space where the energy separation is larger than the gap. In this case structure in the absorption function is obtained for energies where the joint density of states (7.12) has singularities or critical points. These points are called *van Hove* singularities in analogy to the singularities in the phonon density of states. The gaps are assigned as M_0, M_1, M_2, or M_3 depending on their analytical structure in k space. M_0 gaps refer to local minima, M_1 and M_2 to saddle points, and M_3 to maxima in the band-to-band distance for a given k vector. Figure 7.4a shows the measured and the calculated imaginary parts of the dielectric function for Ge. The assigned transitions correspond to the critical points in the corresponding density of states. The Arabic and Greek letters identify points in the first Brillouin zone where the transitions occur. The discrepancy between the experimental result and calculation for the $\Lambda_3 \rightarrow \Lambda_1$ transition is artificial. The splitting of the band in the experiment is due to a spin–orbit interaction which was not taken into account in the calculation. The corresponding band structure with the symmetry points is displayed in Fig. 7.4b.

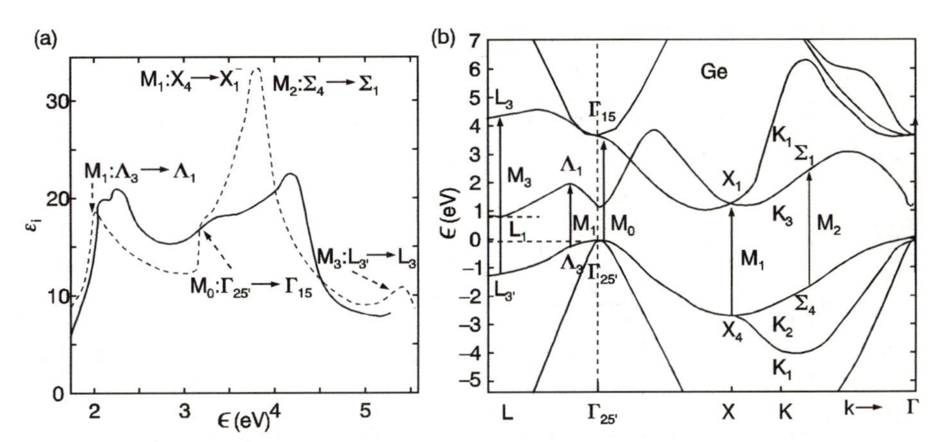

Fig. 7.4. Imaginary part of the DF for germanium, (—) as measured [7.6], and (- - -) as calculated [7.7] (**a**) and band structure of germanium; after [7.8] (**b**).

To emphasize the structures in the spectrum of ε_i the technique of *modulation spectroscopy* can be applied. In this case the first or second derivative of $R(\omega)$ with respect to the frequency is measured rather than the reflectivity itself. Alternatively any other external interaction with the crystal such as an electric field or hydrostatic pressure can be used. The iteration will be successful provided it has a reasonable influence on the electronic structure of the crystal. The critical points can be obtained with high accuracy with this technique, as demonstrated in Fig. 7.5. The top spectrum is the measured reflectivity. The calculated derivative of the reflectivity (part b) shows clearly the critical points in the band structure. The derivative was taken with respect to the photon energy. The observed derivative of the reflectivity (part c) agrees very well with the calculations.

More details about modulation spectroscopy can be found in [7.9].

7.3 Absorption from Localized States

In addition to band-to-band transitions, also transitions between localized states can be important even for solids. These transitions may be intrinsic to the solid such as excitons or transitions between molecular units in the crystal or they may originate from crystal point defects. As will be discussed below, the basic concepts are the same as those considered in the last section but experimental features and quantitative formulations are different.

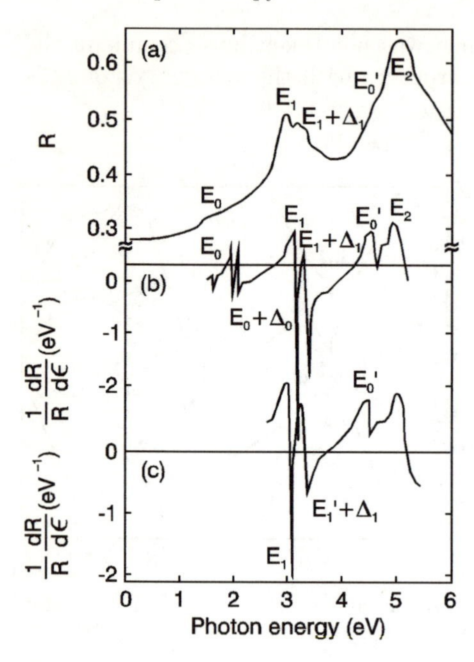

Fig. 7.5. Reflectivity of GaAs (**a**), calculated derivative of the reflectivity (**b**), and experimentally observed derivative of the reflectivity (**c**). E and Δ are symbols for the critical points; adapted from [7.3].

7.3.1 Absorption of Extended and Localized Excitons

In classical semiconductors exciton absorption occurs close to the fundamental absorption. Thus, excitonic features as well as phonon sidebands can cover the true shape of the fundamental absorption. Excitons are fundamental and intrinsic excitations of electrons (or holes) for which a certain amount of Coulomb interaction between the electron and the hole is retained. Different types of excitons are defined depending on how far apart the electron and the hole are. Weakly bonded excitons where the two particles are separated by many lattice constants are of the *Wannier–Mott* type. Strongly bonded excitons which correspond to highly localized pairs are of the *Frenkel* type. In classical semiconductors excitons are usually of the Wannier–Mott type because the high dielectric constant strongly shields the Coulomb interaction. Frenkel-type excitons are usually observed in wide-band ionic semiconductors, molecular crystals, or noble-gas crystals. The electronic states of Wannier-Mott excitons are schematically shown in Fig. 7.6a. They are well described by a hydrogen model using the reduced mass $m_\mathrm{r}^* = m_\mathrm{e}^* m_\mathrm{h}^* / (m_\mathrm{e}^* + m_\mathrm{h}^*)$ where m_e^* and m_h^* are the effective masses of electrons and holes, respectively. Counting the binding energy from the bottom of the conduction band downwards yields

$$\boxed{\delta\epsilon_{\mathrm{exc}} = -\frac{1}{n^2}\frac{m_\mathrm{r}^* e^4}{2\hbar^2(4\pi\varepsilon\varepsilon_0)^2} \qquad n = 1, 2, 3, \ldots\,.} \tag{7.19}$$

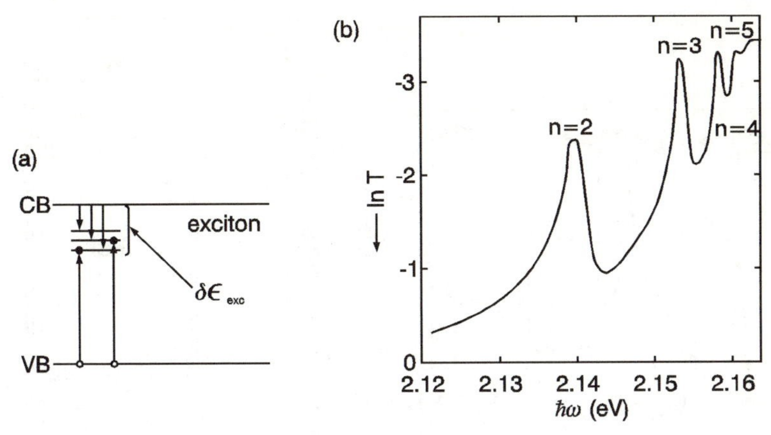

Fig. 7.6. Energy levels for exciton transitions in a semiconductor (**a**) and exciton absorption for CuO_2 at 77 K, plotted as the logarithm of the transmission T; after [7.10] (**b**). The down-arrows in (a) indicate the binding energy.

Similarly, the radius a_{exc} of the exciton is obtained from

$$\boxed{a_{\mathrm{exc}} = \frac{\varepsilon m_0}{m_{\mathrm{r}}^*} \frac{4\pi\hbar^2\varepsilon_0}{m_0 e^2}} \,. \tag{7.20}$$

Except for the factor $\varepsilon m_0/m_{\mathrm{r}}^*$ the right-hand side of (7.20) is equal to the Bohr radius of the hydrogen atom. The large radius of the excitons in semiconductors is due to their large values of the dielectric constant and the small reduced effective masses. Exciton binding energies are usually a few meV. Figure 7.6b shows the exciton absorption of CuO_2 measured at 77 K. The different absorption lines correspond to different values for n in (7.19).

Wannier–Mott excitons are extended, neutral, and highly mobile particles which have a momentum and kinetic excitations subjected to a dispersion relation. In contrast to electrons and holes excited to the bands, excitons do not contribute to photoconductivity. They are observed in absorption as well as in emission. Thus, they will be further discussed in Sect. 7.5.

For non-metallic solids with a low dielectric constant like noble gases or alkali halides the exciton energy can be more than one eV. The oscillator strengths are also larger, and the reflection or absorption spectra are dominated by exciton series. In Fig. 7.7a,b a typical absorption spectrum for KBr is shown together with the band structure. Obviously the Γ-excitons and the L-excitons are more strongly expressed than the band transitions at Γ ($\Gamma_{15}^{3/2} \to \Gamma_1$) and the L-band transitions ($L_{3'} \to L_2$ and $L_{3'} \to L_1$).

7.3.2 Absorption by Defects

Electronic transitions between an extended band state and a localized state in the gap or transitions between two localized states in the gap are very

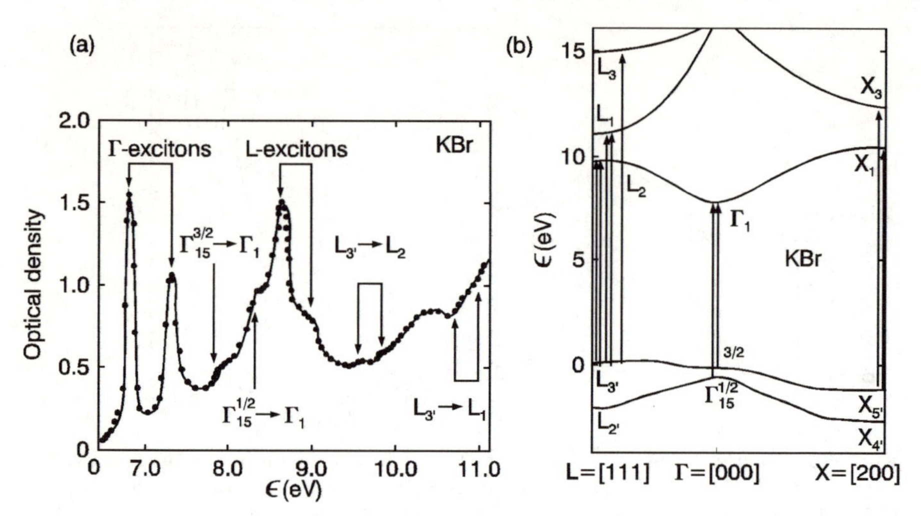

Fig. 7.7. Optical density for KBr measured at 80 K (**a**) and band structure for the corresponding lattice (**b**); after [7.11].

important too. Such states originate from defects like impurity atoms, vacancies, or interstitials. We already saw some examples of such localized states for the Cr ions in Al_2O_3 or for the various attributes to the filter glasses. In general in many of these cases luminescence turns out to be of more practical importance but absorption is more fundamental. The color centers played, for example, a fundamental role in the early days of solid-state spectroscopy. In particular, color centers in alkali halides have been studied in depth. They consist of lattice defects which develop special optical properties by capturing an electron. The most prominent example is the F-center in NaCl which has an absorption line at 480 nm. As shown in Fig. 7.8, it consists of a Cl-vacancy which has captured an electron. The dipole transitions of this electron determine the absorption. The same lattice defect occurs in other alkali halides with absorption at a slightly shifted position. The position for maximum absorption of the F-centers turns out to depend quadratically on the inverse of the lattice constant:

$$\hbar\omega_{\max} = Cd^{-2} . \tag{7.21}$$

This is known as the *Molwo relation.*

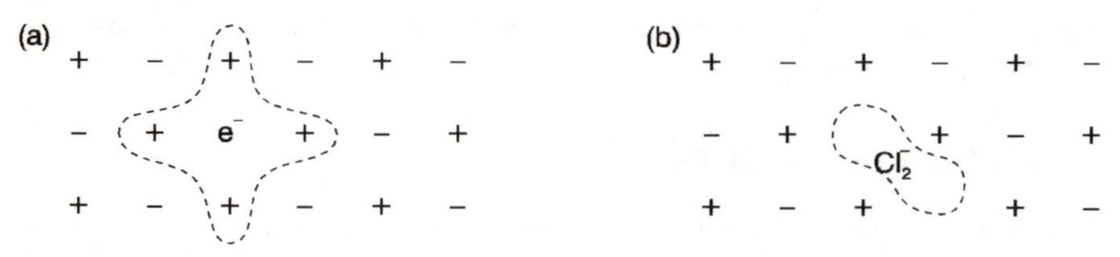

Fig. 7.8. Color centers in alkali halides; F-center (**a**), V_k-center (**b**).

The other defect shown in Fig. 7.8 is the famous V_k-center. It is represented by a Cl_2^- molecule or, in other words, by two Cl^- ions which have captured a hole. The absorption line for the V_k-center is in the UV spectral range.

Color centers can be generated by irradiation of crystals with UV light or with x rays. Historically they were the first systems in solids for which theoretical concepts could be compared with experiments. For a long time such investigations were purely academic. However, recently, light emission from color centers has found an interesting technical application in the development of lasers.

For a theoretical treatment of the absorption from color centers as well as for absorption from the other localized states (7.9) is an appropriate starting point as demonstrated in the next section.

7.4 Theoretical Description of Absorption by Localized States

The theoretical description of the absorption process by localized states is again based on the relations derived in Sect. 7.1. Using the dipole matrix elements instead of the momentum matrix elements we start with

$$\alpha_{\mathrm{fi}} = \frac{2\pi}{\mathcal{V}} \frac{\omega_{\mathrm{fi}}^2 |M_{\mathrm{fi}}|^2}{\varepsilon_0 c_0 \omega n} \delta(\hbar\omega_{\mathrm{fi}} - \hbar\omega) . \tag{7.22}$$

In this case the matrix element turns out to be the crucial term for the energy dependence. Since the localized state has no periodicity, Bloch functions cannot be used to evaluate α_{fi}. In contrast to the behavior of extended states, any change in electronic structure due to excitation leads to a change in the local configuration of the center. The center *relaxes* as a consequence of the excitation. This phenomenon is represented by a *configuration interaction* since in a calculation the relaxed state is represented by a superposition of different configurations. Identification of the relaxed state is crucial since in (7.3) i corresponds to the wave function for the ground state but f to the wave function for the relaxed excited state which is not simply equal to the wave function for the state f in the ground-state configuration. This situation implies that for the calculation of the matrix elements wave functions are required which describe both the electrons and the lattice.

In addition, changes in local geometry induced by phonons contribute to the energy of the localized state and to the matrix elements much more strongly than was the case for the extended states. These local lattice vibrations are called vibrons. In contrast to phonons, they do not have a wave vector and thus wave vector conservation is not required. Therefore, vibrational contributions to the electronic transitions are the rule rather than the exception.

Referring to the situation described above the wave functions must describe the vibronic and the electronic system simultaneously. Calculations of this generality cannot be performed, and one relies on good approximations. The most important simplification is the *adiabatic* or *Born–Oppenheimer* approximation where the total wave function $\psi(r,R)$ for electrons with the coordinates r and atoms with the coordinates R is factored into a wave function for the electrons $\varphi(r,R)$ and one for the atoms $\varrho(R)$.

$$\psi(r,R) = \varphi(r,R)\varrho(R) \, . \tag{7.23}$$

The wave function for the electrons may still contain the atomic coordinates R as a parameter. Then, the eigenvalues for these wave functions will also depend on R. The wave functions for the atoms are harmonic-oscillator wave functions. (For a summary see Appendix F.4). A further simplification is obtained from the *Condon* approximation. Here the matrix elements for the pure electronic transitions are assumed to be independent of the atomic coordinates R. In this case we can write the dipole matrix element (7.22) as

$$\begin{aligned}(M_j)_{\mathrm{fi}} &= e\int \varrho_{\mathrm{f}}^*(R)\varphi_{\mathrm{f}}^*(r,R)x_j\varphi_{\mathrm{i}}(r,R)\varrho_{\mathrm{i}}(R)\mathrm{d}^3x\mathrm{d}^3X \\ &= (M_j^{\mathrm{el}})_{\mathrm{fi}}\int \varrho_{\mathrm{f}}^*(R)\varrho_{\mathrm{i}}(R)\mathrm{d}^3X \, ,\end{aligned} \tag{7.24}$$

where $(M_j^{\mathrm{el}})_{\mathrm{fi}}$ is the pure electronic matrix element. Note that the integral on the right-hand side of the equation is not zero even though integration is performed over a product of two different harmonic-oscillator wave functions. The two oscillators are different as one is in the ground-state configuration and the other in the excited-state configuration. This situation is conveniently described by the adiabatic potentials expressed in a configuration coordinate Q. The configuration coordinate (given in meters) is similar but not identical to a normal coordinate (given in meters/$\sqrt{\mathrm{kg}}$). The adiabatic potentials for the excited state U_{f} can be obtained from the adiabatic potentials for the ground state U_{i} through a Taylor series expansion and by adding the transition energy ϵ_{fi} for the unrelaxed system.

$$\begin{aligned}U_{\mathrm{i}}(Q) &= M\Omega^2Q^2/2 \qquad \text{and} \\ U_{\mathrm{f}}(Q) &= U_{\mathrm{i}}(Q) + \epsilon_{\mathrm{fi}} + AQ = \epsilon_{\mathrm{fi}} + AQ + M\Omega^2Q^2/2 \, .\end{aligned} \tag{7.25}$$

A describes the linear coupling between the excited electronic state and the lattice. The oscillator frequency Ω is assumed to be the same for the ground and the excited states. It is convenient to introduce another, dimensionless coupling constant a by

$$a = \frac{A}{\hbar\Omega}\left(\frac{\hbar}{M\Omega}\right)^{1/2} \, . \tag{7.26}$$

a is called the *Franck–Condon* or *electron-vibration* coupling constant. The right-hand side of (7.25) is the potential for a harmonic oscillator shifted in

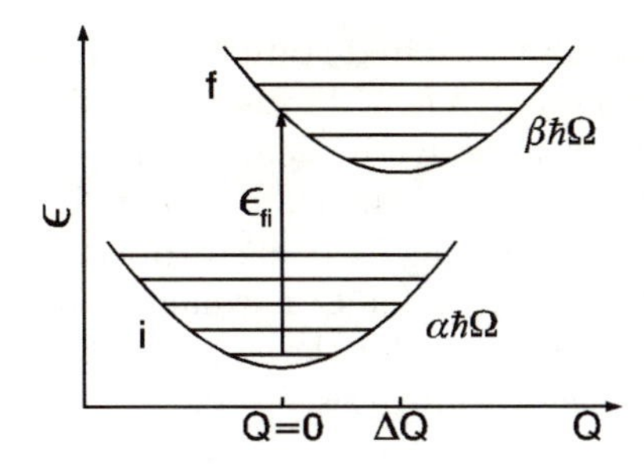

Fig. 7.9. Adiabatic potentials for the ground state i and for the excited state f as a function of the configuration coordinate.

Q and ϵ. It is convenient to draw the adiabatic potentials as in Fig. 7.9. The configuration coordinate is assumed to be zero for the bottom of the ground-state oscillator. The magnitude of the shift for the excited-state oscillator is easily evaluated from (7.25) and (7.26)

$$\boxed{\Delta Q = a\left(\frac{\hbar}{M\Omega}\right)^{1/2}} . \tag{7.27}$$

Thus, the Franck–Condon shift ΔQ is another signature of the strength of the electron–vibron interaction.

The final approximation for the calculation of the absorption utilizes the so called *semiclassical Franck–Condon* principle. According to the latter the electronic transition from i to f is so fast that the atoms do not have time to relax into the state f. As a result, with respect to the relaxed excited state, the electrons end up in a higher vibronic state.

The quantum-mechanical eigenvalues for the motion of atoms in the adiabatic potentials of Fig. 7.9 are

$$\begin{aligned} \epsilon_{\mathrm{i}\alpha} &= (\alpha + 1/2)\hbar\Omega , \\ \epsilon_{\mathrm{f}\beta} &= (\beta + 1/2)\hbar\Omega + \epsilon_{\mathrm{fi}} - a^2\hbar\Omega/2 . \end{aligned} \tag{7.28}$$

The last term in (7.28) describes the decrease in transition energy due to relaxation. It is easily evaluated from the excited-state oscillator by considering the loss in energy for a shift of ΔQ. The horizontal lines in Fig. 7.9 give the vibronic excitations for the ground state i and for the excited state f in units of $\hbar\Omega$. α and β are the vibronic quantum numbers.

Instead of a the quantity

$$S = \frac{a}{\sqrt{2}} \tag{7.29}$$

is sometimes used. S is called the *Huang–Rhys* factor.

The transition shown in Fig. 7.9 is not the only one which can occur. From each vibronic state $\mathrm{i}\alpha$ a transition to a vibronic state $\mathrm{f}\beta$ is possible, and the probability of this transition is given by the square of the corresponding matrix element (7.24). Thus, the δ function in (7.22) must be replaced by $\delta(\epsilon_{\mathrm{f}\beta} - \epsilon_{\mathrm{i}\alpha} - \hbar\omega)$. The matrix element is a product of a pure electronic part, and the overlap integral for the ground-state and excited-state harmonic-oscillator wave functions. The latter are usually given in terms of the dimensionless quantity $q = x\sqrt{M\Omega/\hbar}$ which has exactly the same structure as the Franck–Condon coupling constant a in terms of the shift ΔQ of the configuration coordinate. Thus, the wave functions in (7.24) are expressed by Hermite polynomials, as outlined in Appendix F.4 for Q normalized to $q = Q\sqrt{M\Omega/\hbar}$. The arguments for the ground-state and for the excited-state functions are q and $q + a$, respectively. The relevant integrals

$$\langle\beta|\alpha\rangle = F_{\beta\alpha} = \int \varrho^*_\beta(q+a)\varrho_\alpha(q)\mathrm{d}q \tag{7.30}$$

are called the *Franck–Condon* integrals. They can be evaluated analytically in terms of the associated Laguerre polynomials and are given in Appendix F.4. The Franck–Condon integral for $\alpha = 0$ is

$$F_{\beta 0} = (-1)^\beta \frac{a^\beta \exp(-a^2/4)}{2^{\beta/2}(\beta!)^{1/2}} . \tag{7.31}$$

To obtain the observed absorption all transitions from the state i with various values for α to the state f with various values for β must be considered with their individual matrix elements. For zero temperature only the state with $\alpha = 0$ is occupied in the ground state. Thus only contributions from $\alpha = 0$ to the various values of β have to be considered in the δ function of (7.22). Hence, the absorption has the form

$$\alpha_{\mathrm{fi}}(\omega, T=0) = K|M^{\mathrm{el}}_{\mathrm{fi}}|^2 \sum_\beta |F_\beta 0|^2 \delta(\epsilon_{\mathrm{f}\beta} - \epsilon_{\mathrm{i}0} - \hbar\omega) , \tag{7.32}$$

where K represents the constant factor in (7.22) and $M^{\mathrm{el}}_{\mathrm{fi}}$ is the appropriate component of the dipole matrix element, determined by the polarization of the light. Contributions to the absorption for the different transitions are indicated in Fig. 7.10 for various values of the Huang–Rhys factor. The figure shows clearly that the strongest contribution to absorption does not necessarily come from the 0-0 vibronic transition. In fact, it shifts to higher values of β with increasing electron–phonon coupling. This is due to the dependence of the maximum overlap between the wave function for $\alpha = 0$ and arbitrary values of β on the magnitude of the shift in the configuration diagram of Fig. 7.9. For a practical calculation the δ function in (7.32) may be replaced by an oscillator with a certain damping constant γ_e. This yields for the absorption

$$\alpha_{\mathrm{fi}}(\omega, 0, \gamma_e) = K|M^{\mathrm{el}}_{\mathrm{fi}}|^2 \sum_\beta \frac{|F_{\beta 0}|^2}{[\hbar\omega - (\epsilon_{\mathrm{fi}} - a^2\hbar\Omega/2 + \beta\hbar\Omega)]^2 + \gamma_e^2} . \tag{7.33}$$

For finite temperatures several vibronic levels in the ground state are thermodynamically occupied. Therefore the results of (7.32) and (7.33) have

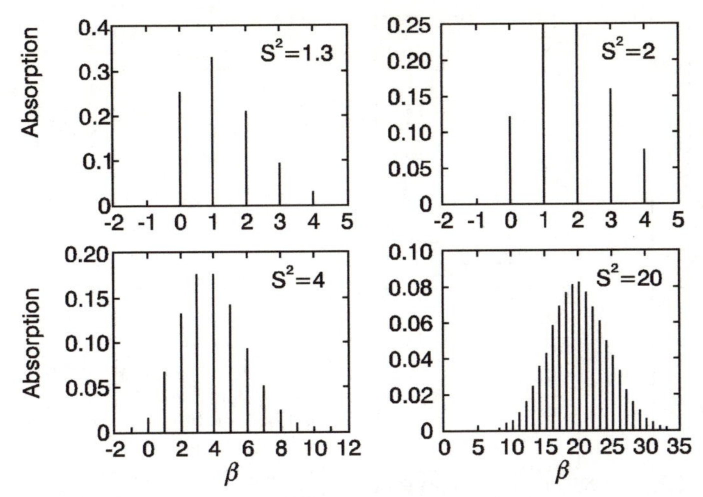

Fig. 7.10. Optical absorption of a localized state in relative units for $T = 0$ and various values of the electron–vibron coupling.

to be extended to finite values for α and a thermal averaging must be performed. Thermal averaging means that the contributions from the various vibronic states α must be included and weighted by their thermal occupation. A distribution of the form

$$W(\alpha, T) = \exp(-\alpha\hbar\Omega/k_B T)$$

is appropriate for the averaging. Thus, the temperature- and frequency-dependent absorption is finally obtained from

$$\boxed{\alpha_{\mathrm{fi}}(\omega, T) = K|M_{\mathrm{fi}}^{\mathrm{el}}|^2 \sum_{\beta,\alpha} W(\alpha, T)|F_{\alpha\beta}|^2 \delta(\epsilon_{\mathrm{f}\beta} - \epsilon_{\mathrm{i}\alpha} - \hbar\omega)\ .} \tag{7.34}$$

The equations derived above are only valid for the case of a linear coupling as assumed in (7.25). If A becomes zero, which may happen for symmetry reasons, quadratic coupling must be considered. In this case the difference between the harmonic oscillators in the ground state and in the excited state is important.

Experimental results for the optical absorption at 4.2 K are presented in Fig. 7.11 for the F-center in various alkali halides. At these low temperatures the vibronic sidebands to the electronic transitions are prominent. LiF has the lowest and NaCl the highest coupling constant. This type of absorption spectra are also observed for impurity states in semiconductors and for molecular crystals like anthracene or benzene. In the latter systems the excitations remain localized due to weak interactions between the molecules.

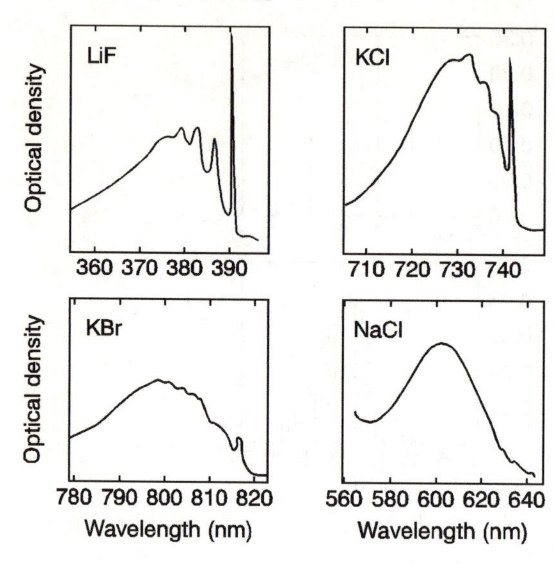

Fig. 7.11. Optical absorption for the F-center in various alkali halides at 4.2 K; after [7.12]. The strong lines at the long wavelength sides of the spectra are the 0-phonon transitions.

7.5 Luminescence

Luminescence is in some sense the inverse of absorption. Whereas in absorption a light quantum is destroyed by the excitation of an electron, luminescence is a consequence of the radiative recombination of the excited electrons. In competition with luminescence other, nonradiative recombination processes may occur. For luminescence to be efficient the radiative recombinations must dominate over the nonradiative recombinations.

Luminescence from solids is not only of scientific but also of considerable technological interest. There are two main reasons for the latter:

1) Luminescence is the basis for the construction of solid-state lasers. This is, in particular, true for luminescence from impurity states in semiconductors.
2) Luminescence from semiconductors and from other materials is important for displays in electronic equipment.

The process of luminescence requires a non-equilibrium carrier concentration in the electronic bands or in the electronic states of a defect structure. If the non-equilibrium is obtained by irradiation with light the radiative recombination is called *photo luminescence*, if it is obtained electronically, for instance, by forward biasing a p-n junction it is called *electro luminescence*. In Fig. 7.12 several possible processes for radiative and nonradiative recombinations are illustrated. The first three radiative processes and the last nonradiative process are intrinsic, whereas for the others at least one impurity is required. Reactions 1 and 2 are band–band recombinations which are usually not very efficient, reaction 3 is the exciton luminescence already mentioned. Reaction 6 gives rise to donor–acceptor pair spectra, as will be discussed below. The trapping centers labeled Tr in the figure can be very efficient for luminescence since they can first capture and localize an electron or hole which then strongly increases the probability of capturing the other partner for the radiative recombination. The last type of radiative recombination is particularly important for laser processes in the visible spectral range and for chromophores in organic crystals.

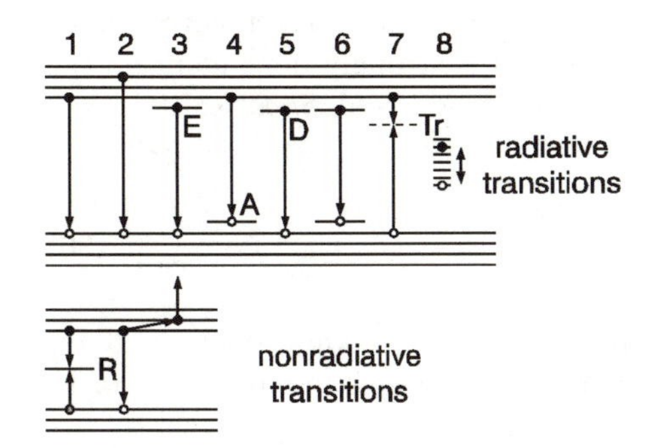

Fig. 7.12. Radiative and nonradiative recombination processes for excited electrons; (E: exciton, A: acceptor, D: Donors, Tr: trapping center, R: recombination center).

Nonradiative recombination occurs either via *recombination centers* or by *Auger* processes. Recombination centers are usually deep impurity levels close to the center of the gap. In Auger processes the energy released as a consequence of the recombination is transferred to another electron. This electron gets excited into a higher state in the band from where it can stepwise return to its ground state without radiation. Auger processes represent a very general phenomenon and occur in many different configurations. For example, the radiative recombinations from the band states to the impurities in Fig. 7.12 (*4* and *5*) have to compete with the corresponding Auger processes. The likelihood of Auger processes increases as the electrons become closer in space, that is as the electron density increases. In an n-type semiconductor the probability of an Auger process is proportional to $n^2p = nn_\mathrm{i}^2$. For carrier concentrations $n > 10^{18}\,\mathrm{cm}^{-3}$ Auger recombination is usually dominant. This is a serious problem for the construction of efficient electroluminescence devices. To obtain a large number of active species high concentrations of electrons and holes are required. The increase in the concentrations is unfortunately limited by the onset of Auger recombination.

The dominant process for the recombination is determined by its lifetime in the excited state. The process with the shortest lifetime wins. The quantum efficiency for a radiative transition is defined as

$$\eta = \frac{1/\tau_R}{1/\tau_R + 1/\tau_0} = \frac{\tau_0}{\tau_R + \tau_0}\,, \tag{7.35}$$

where τ_R is the lifetime of a radiative transition and τ_0 that of a nonradiative transition.

7.5.1 Luminescence from Semiconductors

Luminescence from semiconductors is particularly important. *Van Roosebroek* and *Shockley* provided the first reasonable useful theory in 1954 for the lifetime of excited quasi-particles [7.13]. Basically, the same relationships are valid for emission from a luminescence transition as for emission from a blackbody. The equilibrium between absorption and emission must be retained. For example, the band-to-band emission starts for an energy which corresponds to the bandgap. The lifetime for this transition is strongly dependent on the magnitude of the gap. According to *van Roosbroeck* and *Shockley* it is well described by

$$\boxed{\tau_{BB} = C \exp\left(\frac{\epsilon_{\mathrm{g}}}{k_{\mathrm{B}}T}\right) .} \tag{7.36}$$

This implies that the band-to-band luminescence becomes very small for transparent crystals. The line shape of the luminescence or, in other words, the luminescence spectrum is obtained from

$$\boxed{I(\omega) = C'\alpha(\omega) \exp\left(-\frac{\hbar\omega}{k_{\mathrm{B}}T}\right) ,} \tag{7.37}$$

where $\alpha(\omega)$ is the absorption constant. This equation clearly reveals the connection between absorption and emission. All relations for absorption obtained above can therefore be used for evaluating emission as well.

Basically, the lifetime for a radiative process is much shorter for a semiconductor with a direct gap compared to semiconductors with indirect gaps. This is true at least for band to band emission. Therefore, initially semiconductors with a direct bandgap were preferred as sources of electroluminescence. Typical examples of such sources are the III-V compounds such as GaAs for which a reasonably good technology has been developed. Most of these compounds are direct semiconductors but unfortunately their gap hardly reaches the red spectral range. Many of the III-V compounds are actually narrow gap semiconductors such as InSb. With decreasing size of the atoms the gap widens and can even extend into the green spectral range for GaP. Unfortunately, here the band structure has changed to an indirect configuration with a very low efficiency for luminescence.

In order to avoid the problem with indirect gaps localized states can be included into the luminescence process. The nitrogen impurity in GaP is a well known and important example. Since nitrogen is isoelectric to phosphorus it creates only a very shallow impurity state just below the conduction band. Luminescence from this level has nearly the same energy of 2.22 eV as the bandgap for pristine GaP ($\epsilon_{\mathrm{g}} = 2.24\,\mathrm{eV}$). Since the state of the N-center is localized, momentum selection rules are of no relevance and the luminescence efficiency can be high. Since nitrogen is isoelectrical to phosphorus it is

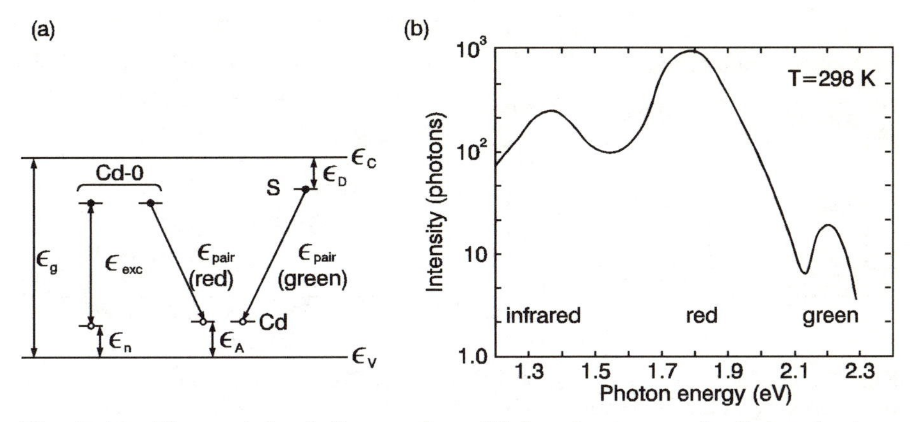

Fig. 7.13. Electronic level diagram for a Cd-doped p-type and a S-doped n-type material in a GaP p-n junction (**a**) and emission spectrum from the junction (**b**); ϵ_n is the level of the Cd-O complex which develops exciton states; adapted from [7.14, 7.15].

neither an acceptor nor a donor but rather a trapping center which enhances the cross section for radiative recombination even further.

A realistic example for light emission from impurity levels in a GaP diode is depicted in Fig. 7.13. The diode consists of a sulfur-doped n-type semiconductor and a cadmium-doped p-type semiconductor. The green-light emission at 2.2 eV (563.6 nm) results from the donor–acceptor pair recombination. Transitions between the exciton level of the Cd-O complex and the acceptor level give rise to the red light emission.

The donor–acceptor pair recombination can have an interesting structure if there remains a Coulomb interaction between the charged donors and the charged acceptors. Emission of this type is known as *donor–acceptor pair luminescence*. Well known examples are complexes of Zn and O or Si and Te in GaP. Figure 7.14 represents the electronic levels for the Zn and O doped crystal in (a) and the pair spectrum for Si/Te at 1.6 K in (b). Independent Zn and O impurity levels have an energetic distance of 1.27 eV and would give rise to luminescence in the IR. If the impurities are close enough in space they interact and the energy level separation rises to 1.8 eV. Thus, the luminescence is shifted to the visible and strongly enhanced due to the interaction. Interestingly, the exact difference in energy depends on the strength of interaction and is therefore determined by the spacial distance between the two impurity centers. The luminescence resonates in this case for the energies

$$\hbar\omega_i = \epsilon_\mathrm{g} - (\epsilon_A + \epsilon_D) + \frac{e^2}{4\pi\varepsilon\varepsilon_0 r_i} \,. \qquad (7.38)$$

The last term in the equation represents the Coulomb interaction between the donor and acceptor for a distance r_i. Since r_i are multiples of the lattice constant this term gives rise to a long series of sharp lines on the high-frequency

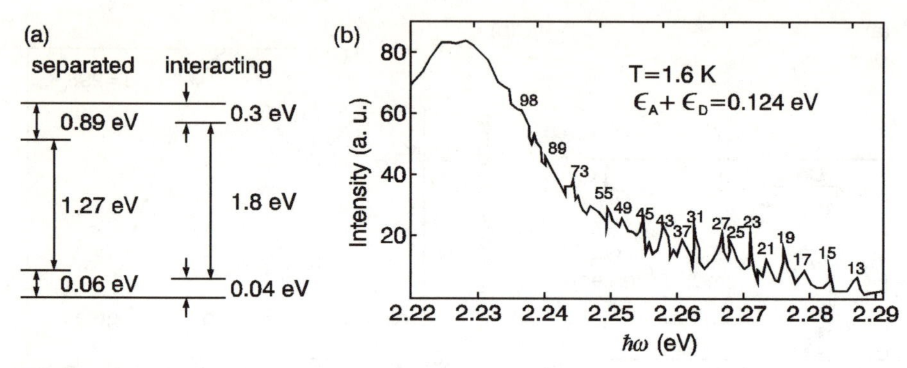

Fig. 7.14. Energy levels for Zn and O levels in GaP (**a**), and pair spectrum for GaP doped with Si/Te complexes (**b**). The numbers at the peaks are distances in units of the lattice constant; adapted from [7.16].

side of the luminescence spectrum. It can be observed experimentally for emission at very low temperatures as shown in Fig. 7.14b.

The interaction between impurity centers can give rise to other phenomena. At low temperatures a radiative transition can be effective between donor D and acceptor. At higher temperatures the electrons from the donor state can be drained to another donor state D′ from where they recombine non-radiatively (Fig. 7.15). In this case the luminescence will be quenched for higher temperatures.

Finally, a thermodynamically interesting phenomenon originates from an interaction between excitons in highly excited semiconductors. Excitons exhibit a luminescence close to the bandgap energy. At low temperatures and for high excitations the excitons interact strongly and can even finally condensate into a liquid state. Exciton droplets can grow and the exciton luminescence changes suddenly. This is due to the high carrier concentration in the droplets which shields the Coulomb interaction. As a result, the excitons are not stable any more and decay into an electron–hole micro plasma. This plasma has a characteristic recombination spectrum different from that of free excitons. An example is presented in Fig. 7.16 for highly excited Ge at three different temperatures. For the given excitation 2.78 K is above the condensation temperature. Thus only the line for the free excitons at 714.2 meV is observed. At 2.52 K a new strong line at 709.6 meV suddenly appears which

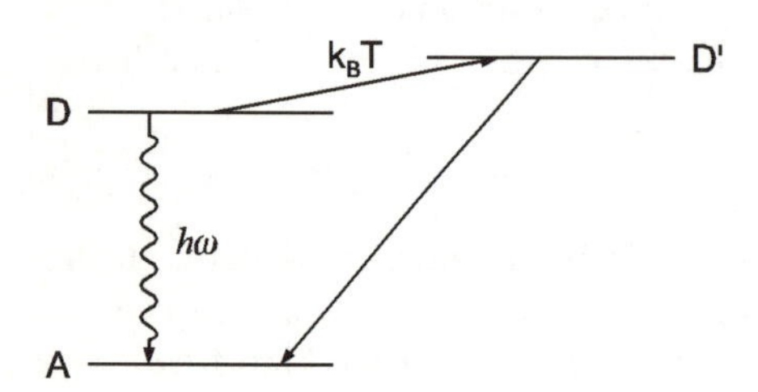

Fig. 7.15. Thermal quenching of a luminescence from a donor–acceptor recombination; (D,D′: donors, A: acceptor).

corresponds to excitons in the droplets. However, the emission from the free excitons is still there. For still lower temperatures like 2.32 K it has completely disappeared, and the droplets are the only radiating species. Exciton droplets have only been observed for indirect semiconductors since it is only in these systems that a high enough concentration of excitons can be obtained. Direct semiconductors form bound exciton states (exciton molecules) at high excitation.

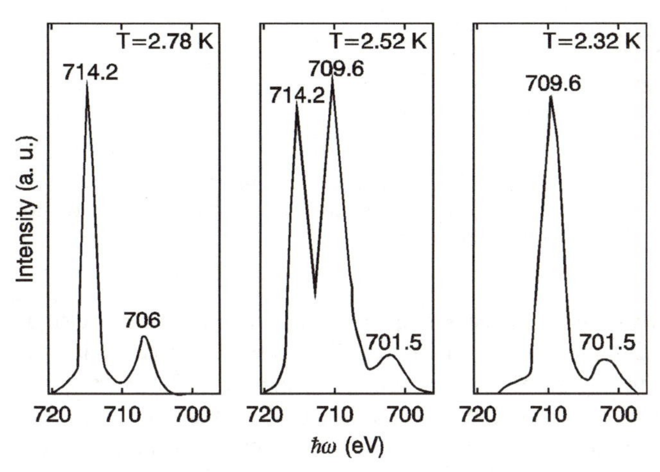

Fig. 7.16. Luminescence of Ge at low temperatures and high excitation; after [7.17]

Luminescence from semiconductor p-n junctions is an important subject in research and engineering for the development of semiconductor lasers. Whereas in the early days the research and development concentrated on IR and far-IR lasers, today strong efforts are directed towards the visible spectral range (see also Sect. 3.4). More details on these efforts are given in special works such as [7.18].

7.5.2 Luminescence from Point Defects in Insulators

The relation between absorption and emission is as valid for point defects as for extended states. This means, in particular, that we can apply the adiabatic approximation and the Franck–Condon principle to study the luminescence from point defects. The corresponding processes can be described with the same adiabatic potentials, as shown in Fig. 7.9 and redrawn in Fig. 7.17 for the case of luminescence. The origin of the transition processes is however different. Although absorption can start from any vibronic level α and can end at any vibronic level β, emission always starts from the lowest vibronic level $\beta = 0$ in the lowest excited state. This is independent of the level to which the electron was excited by the incident light, and is well established experimentally. It is known as *Kasa's* rule and due to the overlap of the wave functions between the excited state and the ground state. Note that the most likely transition occurs between two singlet states (S_0, S_1) which implies spin conservation. This is true for absorption as well as for emission from the S_1 state, as shown in the figure. The spin of the absorbed or emitted photon is converted to an angular momentum of the electron as $\Delta l = 1$ is required for this type of optical transition.

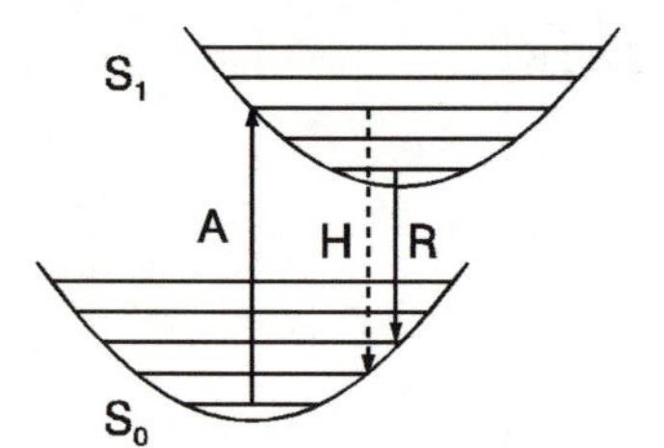

Fig. 7.17. Adiabatic potentials for the luminescence process; (A: absorption, R: emission, H: hot luminescence).

Radiative and nonradiative processes are in competition here likewise as for transitions from extended states. The situation is well described by the *Jablonski diagram* (Fig. 7.18). If no paramagnetic centers are present, the ground state is, in general, a singlet S_0. The excited states are again singlets S_1, S_2, S_3, etc., combined with any vibronic state β. From this state the excitation relaxes at first mainly by an interaction with the lattice into the corresponding pure electronic state with $\beta = 0$. Now, the next lower electronic state with a high value for β becomes occupied by a process of *internal conversion*, and relaxation into the corresponding state with $\beta = 0$ starts again. This process continues until the lowest excited electronic state is reached (state S_1 in Fig. 7.18). From there three processes are possible. Either the relaxation via the lattice continues and the electron returns to the ground state without radiation, or it returns to the ground state by emission of a photon or, alternatively, it undergoes a process of *intersystem crossing* with a spin flip to a triplet state and finite β. After relaxation to the triplet ground state it either recombines radiatively to the ground state in the form of a phosphorescence or nonradiatively via another internal conversion. The radiative recombination from the singlet state is very fast. It occurs spontaneously within picoseconds or nanoseconds and is called a luminescence process. The radiative recombination from the triplet state is considerably delayed since it involves a spin flip. It occurs only after microseconds or milliseconds and is called a phosphorescence process. According to the Kasa rule radiative recombination is very difficult before the lowest excited state is reached. If it happens it is called *hot luminescence*.

In contrast to the case of semiconductors where the excitation is usually from a forward biased p-n junction, color centers or chromophores are excited optically. Hence, two types of spectra can basically be measured: the total

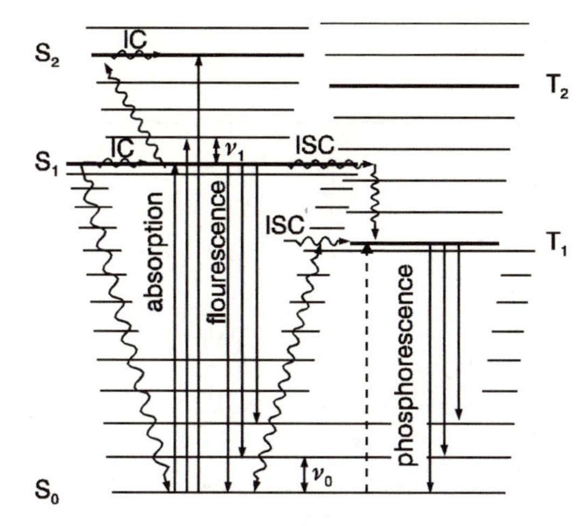

Fig. 7.18. Jablonski diagram for luminescence processes from localized states; (IC: internal conversion, ISC: intersystem crossing, S_i: singlet states, T_i: triplet states).

(or even energy selective) emission as a function of the spectral excitation, and the spectral distribution of the emitted light itself. A spectrometer for fluorescence consists therefore of two monochromators operating in parallel. One is used for the spectral selection of the exciting light, the other for the analysis of the emitted light. Figure 7.19a represents schematically the operational layout for these spectrometers. The light for the excitation of the crystal is spectrally selected by monochromator 1 and the emitted light is spectrally analyzed by monochromator 2.

The immediate correlation between optical absorption and emission can often be seen by comparing excitation and emission spectra. Figure 7.19b plots the phosphorescence for polystyrol at 77 K. The right part of the spectrum represents the spectral distribution of the emission from a T_1-S_0 transition. The peaks correspond to the individual vibronic Franck–Condon modes. The energetic distance of the maxima indicates that they originate from a C=O stretching vibration. Since there is no C=O bond in polystyrol the corresponding chromophore must be a defect. The left part of Fig. 7.19b is, except for the intensity, a mirror image of the right side and shows the excitation spectrum for the S_0-S_1 transition. The energetic difference between the first maximum of the phosphorescence and the first maximum of the excitation corresponds to the energy difference between the zero vibron states of S_1 and T_1. The strong structures in the excitation spectrum at 300 nm are due to excitations of the monomer.

The process of absorption from point defects is evidently always localized. However, if the density of the chromophores is high enough a migration of the excitation is possible. This occurs, in particular, if the absorption spectrum of one chromophore overlaps the emission spectrum of the other chromophore. Then the first center can act as a donor (with respect to excitation energy) and the second center as an acceptor. The energy transfer occurs by dipole–dipole interaction and not by emission and reabsorption. Therefore

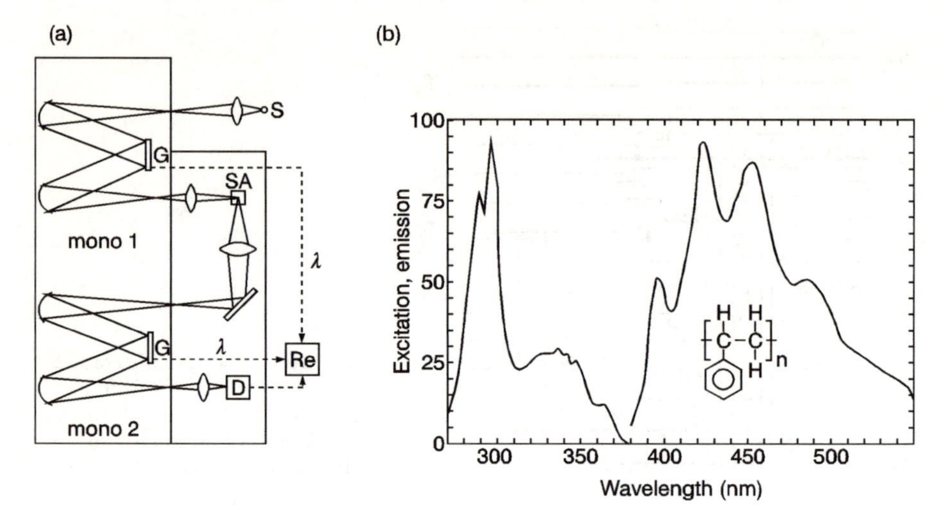

Fig. 7.19. Schematic setup for measuring the excitation and the emission spectrum of a luminescence; (S: source, G: grating, SA: sample, λ: drive, Re: recorder, D: detector) (**a**), and excitation spectrum (*left*) and emission spectrum (*right*) for polystyrol at 77 K (**b**); after [7.19].

this *energy migration* is only possible for centers located within a critical distance r_{F} which is called the *Förster radius*. The probability of energy transfer decreases with the 6th power of the distance. This energy transfer is particularly important for polymers since in this case singlet excitations can migrate step by step over wide distances. Special chromophores can be attached to the backbone and act as transmitters for the excitons. If the conformation of the chain changes with progression along the chain, the orientation of the dipole moment for the emission is continuously changing. This means that the conformation of the polymer can be studied by an investigation of the time resolved polarization of the luminescence.

Problems

7.1 Prove that the absorption given in (7.9) has its correct value in SI units.
(Purpose of exercise: convince yourself of the meaning of all symbols in the equation.)

7.2 Starting from (7.12) evaluate the joint density of states for spherical and parabolic bands.
Hint: Use the relation point 5 in Appendix B.7.2.
(Purpose of exercise: use of δ function, obtain final results for simple band structures.)

7.3 Show that the relation between the Franck–Condon coupling constant a and the shift in the configuration coordinate ΔQ is $a = \Delta Q\sqrt{M\Omega/\hbar}$. What is the value of the relaxation energy expressed in units of a?

(Purpose of exercise: get insight into the meaning of the adiabatic potentials.)

7.4 Study the optical absorption from a localized state as a function of the Huang–Rhys factor for zero temperature using a personal computer.

(Purpose of exercise: probe the relationship between electron–vibron coupling and maximum response for the absorption.)

7.5 Evaluate the real part of (7.10) from a Kramers–Kronig transformation.

(Purpose of exercise: a useful example of the application of the Kramers–Kronig relation.)

8. Symmetry and Selection Rules

Many laborious calculations in solid-sate physics can be simplified if the symmetry of the system under consideration is properly included. Furthermore, statements about possible transition processes can very often be made just from symmetry considerations, without any calculation. In these considerations group theory plays a fundamental role. This chapter summarizes the symmetry properties of molecules and crystals, and reviews some elements of group theory. Selection rules for electronic and vibronic transitions will be discussed. An extended summary of group theory, the character tables for the point groups, and more about transformation of coordinates are given in Appendix G.

8.1 Symmetry of Molecules and Crystals

Symmetry and symmetry operations of figures or objects can be defined formally. For extended work a mathematical description is recommended.

8.1.1 Formal Definition and Description of Symmetry

The symmetry of a figure or of an arrangement of points is the set of considered or mechanically performed operations which transform the object into a position where it cannot be discriminated from its initial position. An example is given in Fig. 8.1. It shows in part (a) a triangle stepwise rotated by 120°. If the corners of the triangle are identical, the three positions cannot be discriminated. The operation C_3 has transformed the triangle into positions undistinguishable from the initial position. The same would happen for a reflection with the mirror plane perpendicular to the plane of the triangle, and intersecting one corner and the midpoint of the opposite edge. We designate this operation as σ_v, where σ means the mirror plane and the index v indicates its vertical orientation to the rotation axis for the operation C_3. C_3 and σ_v are *symmetry operations* (SO) for the triangle in the above definition. If the corner 1 is different from the corners 2 and 3 the reflection with the mirror plane through 1 is still a SO but C_3 is not a SO any more. However, a mirror plane in the plane of the triangle is certainly a SO for both systems. We call it σ_h as it is oriented horizontally to the rotation axis.

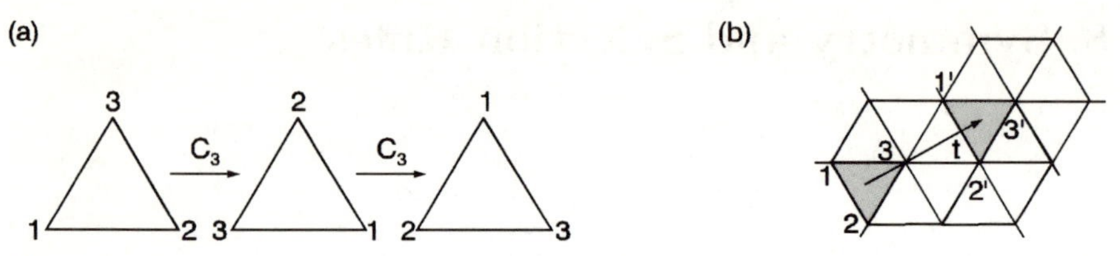

Fig. 8.1. Symmetry operations for a triangle: point symmetry (**a**), and translational symmetry (**b**).

The translation of the triangle to a different position in space is not a SO since in the new position it can be discriminated from the initial position. The situation is different if we consider an infinite arrangement of triangles as shown in Fig. 8.1b. If we shift the triangle 1, 2, 3 by a well defined translation vector $\boldsymbol{t}$ to the position $1', 2', 3'$, the new arrangement cannot be distinguished from the initial arrangement. Thus, $\boldsymbol{t}$ is a SO of the object in Fig. 8.1b. Since in the former type of SO at least one point of the object remains unchanged we have a *point symmetry*, whereas the latter type is called *translational symmetry*.

SOs are characterized by *symmetry elements* such as n-fold rotation axes, mirror planes, etc. The possible symmetry elements are listed in Table 8.1, together with their symbolic representation in the *Schönflies* and in the *international* notation. The first four symmetry elements refer to point symmetry,

Table 8.1. Symmetry elements in crystals

Symmetry element	Schönflies	International
Rotation axes	$C_n(U_n)$	$n = 1, 2, 3, 4, (5), 6, (\ldots)$
Mirror planes	$\sigma_h, \sigma_v, \sigma_d,$	m
Inversion	I	$\overline{1}$
Rotatory reflection axes	S_n	$\overline{n}$ (rotation inversion) $= \overline{1}, \overline{2}, \overline{3}, \overline{4}, \overline{6}$
Translations	t_n	t_n
Screw axes	C_n^k	n_k
Glide planes	σ^g	a, b, c, n, d

the last three elements are only applicable for infinitely extended objects. For symmetries which allow translations the counting of the rotational axes can only be 1, 2, 3, 4, and 6. The index k for the screw axes indicates how many translations are accomplished for one rotation by 360°. In other words, per SO k/n translations are performed. More details about symmetry elements and their properties and geometrical symbols can be found in corresponding reference data [8.1]. In the following pages we will restrict ourselves to the operations for point symmetry.

Symmetry operations can be carried out successively and the result is a new SO of the object. This means SOs can be combined or *multiplied*. We may write this formally as

$$\begin{aligned} \mathrm{SO}(1) \circ \mathrm{SO}(2) &= \mathrm{SO}(3) \qquad \text{or simply} \\ \mathrm{SO}(1)\mathrm{SO}(2) &= \mathrm{SO}(3)\,. \end{aligned} \tag{8.1}$$

For example, a reflection by a mirror plane through point 1 of the triangle in Fig. 8.1a and a successive reflection by a mirror plane through point 2 gives the same result as the rotation by 120°, or

$$\sigma_v(1)\sigma_v(2) = C_3\,.$$

For the combination of several SOs the associative law holds. This is a consequence of the validity of the associative law for the transformation of coordinates. As we will see in detail later SOs can be represented by such transformations. The SO which leaves the object unaltered can be taken as a unit element E. Since the object returns to its initial position after a finite number of successive SOs an inversion element exists within the set of the SOs for one object. The above properties of elements in a set define a group in a mathematical sense. Thus the symmetry elements of a finite object are elements of a group.

The total number of different groups including all subgroups which can be constructed from the point symmetry elements of Table 8.1 is 32. This anticipates the exclusion of symmetry elements like C_∞ (rotational symmetry) or molecular symmetries like C_5, C_7, etc. Since the elements of the groups are point symmetry operations the groups are called *point groups.* The point groups are well known and well tabulated. For the assignment of the point groups the Schönflies as well as the international symbols are used. Table 8.2 presents the 32 point groups in both notations as they are distributed over the seven crystallographic systems or the 14 Bravais lattices.

Table 8.2. Point groups for the seven crystal classes.

Tricline		Monocline mp,mc		Trigonal (Rhombohedral)		Tetragonal tp,tb		Hexagonal		Cubic cp,cf,cb	
C_1	1	C_2	2	C_3	3	C_4	4	C_6	6	T	23
C_i	$\bar{1}$	C_s	m	C_{3i}	$\bar{3}$	S_4	$\bar{4}$	C_{3h}	$\bar{6}$		
		C_{2h}	$2/m$			C_{4h}	$4/m$	C_{6h}	$6/m$	T_h	$m3$
		orthorhombic op,oc,oi,of		C_{3v}	$3m$	C_{4v}	$4mm$	C_{6v}	$6mm$		
		C_{2v}	$mm2$	D_{3d}	$\bar{3}m$	D_{2d}	$42m$	D_{3h}	$6m2$	T_d	$\bar{4}3m$
		D_2	222	D_3	32	D_4	422	D_6	622	O	432
		D_{2h}	mmm			D_{4h}	$4/mmm$	D_{6h}	$6/mmm$	O_h	$m3m$

In Appendix G.1 more extensive information on the groups is given in the form of character tables. These tables show the point groups, their elements, the irreducible representations, and their characters. Details about these quantities will be discussed below.

For the following we need a few more definitions and notations from group theory which are more or less selfexplanatory.

a) The *order* of the group is the number of elements in the group.
b) Most of the 32 point groups are not *commutative*. This means the group element obtained from the combination AB is not equal to the element from the combination BA if A and B are elements of group $\boldsymbol{G}$.
c) A group may consist of several *subgroups*.
d) Two elements $A, B \subset \boldsymbol{G}$ are *conjugated* if an element X exists in the group for which $XAX^{-1} = B$ holds. If X runs over all elements of $\boldsymbol{G}$ we obtain a set of elements conjugated to A. Since all these elements are also conjugated to each other the set is called a *class* of conjugated elements.
e) Each group consists of *classes* of *conjugated elements*. In other words, each element of $\boldsymbol{G}$ belongs to exactly one class. For example, the unit element E is only conjugated to itself. It always generates its own class. The symmetry elements for the point groups are given as distributed into individual classes Appendix G.1.

8.1.2 The Mathematical Description of Symmetry Operations

The mathematical description of SOs is obtained by the matrices for orthogonal transformation of coordinates in three-dimensional space. If ϕ describes the rotation around the z axis these matrices have the form

$$R(\phi) = \begin{pmatrix} \cos\phi & -\sin\phi & 0 \\ \sin\phi & \cos\phi & 0 \\ 0 & 0 & \pm 1 \end{pmatrix} . \tag{8.2}$$

Similar matrices are obtained for rotations about the x and the y axis. Allowing a minus sign for the R_{zz} component of the matrix includes improper rotations (reflections and inversion). A particular transformation of the type (8.2) is a SO of the object if it transforms the object onto itself. In this sense $R(\phi)$ represents the rotations C_n with $\phi_n = 360/n, n = 1, 2, 3, 4, 6$ including the unit element E for the positive sign of R_{zz}, and the reflection and the inversion with $\phi = 0$ and $180°$, respectively, for the negative sign of R_{zz}. If the rotation is about an arbitrary axis (8.2) must be subjected to an orthogonal transformation. It is most important to note that the trace d_R for the transformation matrix (8.2) is always

$$\boxed{d_R = 2\cos\phi \pm 1 .} \tag{8.3}$$

This is independent of any orthogonal transformation.

8.1.3 Transformation Behavior of Physical Properties

For the evaluation of the selection rules the transformation behavior of physical properties and of mathematical expressions with respect to SOs is important. They are explained and listed in the following.

Scalar quantities such as the density or the temperature are independent of transformations of coordinates. Polar vectors such as electric fields or dipole moments behave like coordinates and transform therefore according to (8.2). For axial vectors transformations of the proper rotations are performed with R from (8.2) but transformations for the improper rotations are transformed with $-R$. Tensors such as the dielectric constant ε_{ik} transform like products of coordinates.

For a transformation matrix C_{kl} the above statements can be formulated mathematically by the following expressions:

$$\begin{aligned} x'_m &= \sum_k C_{mk} x_k \;, \\ x'_m x'_n &= \sum_{k,l} C_{mk} C_{nl} x_k x_l \;, \\ \varepsilon_{mn} &= \sum_{k,l} C_{mk} C_{nl} \varepsilon_{kl} \;. \end{aligned} \tag{8.4}$$

As a simple example we study the transformation of the coordinates of the triangle in Fig. 8.1 by a rotation of 120° about z. The rotation matrix is

$$R(2\pi/3) = \begin{pmatrix} -1/2 & -\sqrt{3}/2 & 0 \\ \sqrt{3}/2 & -1/2 & 0 \\ 0 & 0 & +1 \end{pmatrix} .$$

The x, y coordinates are assumed to cross at the center of gravity, and one of them is assumed parallel to one side of the triangle. The sides of the triangle have the length 1. Using the first equation in (8.4) it is straightforward to show that the coordinates $(-1/2, -\sqrt{3}/6)$ for the corner 1 transform to $(1/2, -\sqrt{3}/6)$, which are exactly the initial coordinates of corner 2.

The second equation in (8.4) is general enough to construct a product for two (squared) matrices with an arbitrary number of lines and columns. The dimensions of the two matrices may even be different. The product $C_{mk}C_{nl}$ of the transformation matrices C_{kl}, or more generally the product $C_{mk}D_{nl}$ defines the *Kronecker* product of the two matrices. The Kronecker product can be written again as a square matrix. The arrangement of the new matrix elements is such that the product matrix consists of the submatrices $C_{mk}\boldsymbol{D}$ where each matrix element of $\boldsymbol{D}$ is multiplied by C_{mk}. If the dimensions of the two original matrices are M and N, respectively, the dimension of the product matrix is $M \times N$. For example, the Kronecker product for a matrix with 3 lines and 3 columns with a matrix with 2 lines and 2 columns yields a square matrix with 6 lines and 6 columns. The trace of the product matrix equals the product of the traces of the two original matrices.

8.2 Representation of Groups

A representation of a group is a mapping of the group $\boldsymbol{G}$ on a group of matrices $\boldsymbol{G}'$. The mapping between the elements of the two groups is defined in the following way. If for the elements $A, B, C \subset \boldsymbol{G}$ the relation

$$A \circ B = C \tag{8.5}$$

holds the corresponding relationship must also hold for the matrices $M(A)$, $M(B)$, and $M(C)$ from $\boldsymbol{G}'$

$$M(A) \circ M(B) = M(C) \tag{8.6}$$

but the correspondence between the elements does not have to be bijective. Each element form $\boldsymbol{G}$ corresponds exactly to one element of $\boldsymbol{G}'$ but each element of $\boldsymbol{G}'$ may correspond to several elements of $\boldsymbol{G}$. This type of mapping is known as a *homomorphism*.

The simplest, so-called trivial representation is a mapping of each element of $\boldsymbol{G}$ on the matrix (1). A slightly more complicated but still very simple mapping is obtained for a correlation of the symmetry elements to the matrices (1) or (−1). For example, using (8.5) and (8.6) together with the information on the point group $\boldsymbol{C_{3v}}$ from the tables in Appendix G.1 it is easy to show that the following mapping is correct:

$$E \to (1),\ C_3 \to (1),\ C_3^2 \to (1)\ , \tag{8.7}$$

$$\sigma_{v1} \to (-1),\ \sigma_{v2} \to (-1),\ \sigma_{v3} \to (-1)\ . \tag{8.8}$$

In other words, the one-dimensional matrices (1), (1), (1), (−1), (−1), (−1) are a representation of the point group $\boldsymbol{C_{3v}}$. A pyramid with a regular triangle as the basis is a geometrical representation of this group. Note that the sum over all matrices is zero! This is a special result of a very general sum rule which originates from the orthogonality of the representation matrices explicitly expressed by (G.4). In order to check the mapping defined in (8.8), it is convenient to set up two *multiplication tables.* In these tables all products of symmetry elements of the group and their representations are inserted, which allows the validity of (8.5) and (8.6) to be proven by comparison.

In principle, the result of a multiplication of symmetry elements can be obtained from a multiplication of the matrices for the orthogonal transformations given in (8.2). In a more simple procedure it can also be obtained from geometrical considerations by following the change of the positions of the coordinates as a consequence of the SO. In Fig. 8.2 two examples are shown: Application of the inversion to point 1 in part (a) of the figure transforms it to point 1′ and a subsequent application of a rotation by $\phi = \pi$ around z

yields the point 1″. Altogether the transformation is a reflection on the xy plane. This is similar in part (b) of the figure. Starting with an inversion a subsequent rotation by $\pi/2$ yields a $3\pi/2$ rotation-reflection.

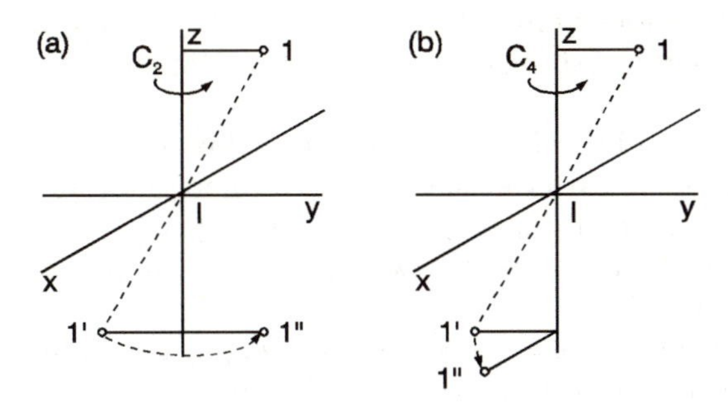

Fig. 8.2. Combination of symmetry elements. $I \circ C_2 = \sigma_h$ (**a**), $I \circ C_4 = S_4^3$ (**b**).

Some of the most important definitions and properties for the representation of groups are listed in Appendix G.2. Details can be found in [8.2, 8.3].

In particular, the following statements are valid, most of them can easily be proven by group theory:

a) The dimension of the representation is the dimension of the correlated matrix. For example, the matrices for the orthogonal transformations are a three-dimensional representation.

b) For each group an infinite number of representations exists. A d-dimensional representation for a group $\boldsymbol{G}$ of order g can be obtained from an equation for a transformation of d linear independent coordinates x_k of the form

$$\boxed{x_i' = \sum_{k=1}^{d} D_{ik} x_k} \tag{8.9}$$

if this transformation is subjected to the following conditions. From the infinite number of possible matrices D_{ik} we select g matrices in a way so that they represent a group and a homomorphic mapping of the group $\boldsymbol{G}$. Under these conditions the matrices

$$D_{ik} = D_{ik}(R)$$

are a representation of $\boldsymbol{G}$. The conditions required are fulfilled if the transformations $D_{ik}(R)$ map the set of coordinates x_k for all elements $R \subset \boldsymbol{G}$ onto themselves or onto a linear combination of themselves. In other words, the set of coordinates x_i' in (8.9) is the same set as the coordinates x_k or a linear combinations of the coordinates x_k. The coordinates x_k are called a *basis* for the d-dimensional representation $D(R)$ of group $\boldsymbol{G}$.

Let us consider as an example the representation for $\boldsymbol{C}_{2v}$. The matrices (1) and (−1) transform the x axis (or the y, z axis) either onto itself or onto minus itself. Thus, the x axis (or the y, z axis) is a basis for a one-dimensional representation. The z axis is a basis for the trivial representation.

Other representations can be obtained from the coordinates of geometrical figures by applying the symmetry operations of the figures, as explained in Appendix G.3.

Note: The result that the x axis is a basis for a representation does not mean that the x axis transforms, in general, with the matrices of this representation, for all symmetry elements R. This is, in fact, not true. A rotation by 120° about z (C_3) yields a new axis x' which needs a two-dimensional matrix in order to be generated from x and y!

c) Applying orthogonal transformations to the matrices $D_{ik}(R)$ we obtain other representations of $\boldsymbol{G}$. The traces of the matrices for the various elements R are the same for all these representations as the traces of matrices are invariant versus orthogonal transformations. The traces are therefore called the character $\chi(R)$ of the element R in the representation $D(R)$. A trivial but nevertheless important statement results: The character of the unit element equals the dimension of the representation.

d) Since the matrices for the representations of $\boldsymbol{G}$ can have an arbitrary dimension, an infinite number of representations exists. However, for any set of matrices there is an orthogonal transformation which groups the matrix elements in an optimum way around the diagonal elements. We call this new set a *fully reduced representation* of $\boldsymbol{G}$. In this case already smaller subsets of the coordinates in (8.9) are mapped onto themselves by matrices with lower dimensions. Since these matrices cannot be further reduced they are called *irreducible*. (For details of this problem see Appendix G.2.)

e) There is a well defined number of irreducible representations for each point group. This number turns out to be equal to the number of classes in the group.

f) The irreducible representations are denoted by the *Mullikan* symbols A, B, E, F, with subscripts $1, 2, g, u$ and superscripts $'$ and $''$. The symbols have the following meaning. A and B are one-dimensional representations. A labels representations symmetric with respect to the main rotations ($\chi(C_n) = 1$), B stands for representations antisymmetric with respect to the main rotations ($\chi(C_n^k) = (-1)^k$). E and F labels two- and three-dimensional representations. g and u refer to symmetry and antisymmetry with respect to the inversion and 1 and 2 denote symmetry and antisymmetry with respect to additional rotation or rotatory reflection axes and mirror planes. Finally, $'$ and $''$ indicate symmetry and antisymmetry to additional mirror planes.

The irreducible representations for the 32 point groups and for the icosahedral group are listed in Appendix G.1 together with the corresponding characters of the symmetry elements. Since the characters for elements of the same class are equal only the characters for the various classes are given.

g) There are several orthogonality relations. (For more details see Appendix G.2.) The most important relation holds for the characters $\chi^{(\alpha)}(R)$ and $\chi^{(\beta)}(R)$ of two irreducible representations $\Gamma^{(\alpha)}$ and $\Gamma^{(\beta)}$

$$\sum_R \chi^{*(\alpha)}(R)\chi^{(\beta)}(R) = g\delta_{\alpha\beta} \; . \tag{8.10}$$

The summation extends over all symmetry elements of $\boldsymbol{G}$.

h) Using the orthogonality relations for the characters one can immediately derive a formula which counts how often a particular irreducible representation $\Gamma^{(\alpha)}$ with characters $\chi^{(\alpha)}(R)$ occurs in an arbitrary d-dimensional reducible representation with characters $\chi(R)$. This number n_α is given by the famous *magic counting formula*

$$\boxed{n_\alpha = \frac{1}{g}\sum_R \chi^*(R)\chi^{(\alpha)}(R) \; .} \tag{8.11}$$

g is the order of the group.

i) The Kronecker product of the matrices for the two representations $\Gamma^{(n)}$ and $\Gamma^{(m)}$ are again a representation. If the set of coordinates x_n and x_m form a basis for the two representations the products $x_n x_m$ form a basis for the product representation. The characters of the product representation are the products of the characters of the two original representations:

$$\boxed{\chi^{(n\times m)}(R) = \chi^{(n)}(R)\chi^{(m)}(R) \; .} \tag{8.12}$$

j) Finally, the Kronecker product of two different irreducible representations never contains the trivial representation. The Kronecker product of two equal irreducible representation contains the trivial representation exactly once. Note in this connection: The Kronecker product of the representations $\Gamma^{(\alpha)}$ and $\Gamma^{(\beta)}$ may neither contain the representation $\Gamma^{(\alpha)}$ nor the representation $\Gamma^{(\beta)}$!

The three most important statements h) to j) follow immediately from the definition for a representation by (8.9), the orthogonality of the characters, and the definition of the Kronecker product.

We can now proceed to the first application of group theory in spectroscopy.

8.3 Classification of Vibrations

Lattice or molecular vibrations are represented by a total energy $H = T + U$ as

$$H = \frac{1}{2}\sum_{n,i} \dot{u}_i^2(n) M_n + \frac{1}{2}\sum_{i,j,m,n} a_{ij}^{mn} u_i(n) u_j(m) , \tag{8.13}$$

where i, j runs from 1 to 3 and m, n over all atoms N in the molecule or in the crystal. $u_i(n)$ are the displacement coordinates for atom n with mass M_n, and a_{ij}^{mn} are the second derivatives of the crystal potential U with respect to these displacements. Applying a proper orthogonal transformation and using mass-weighted coordinates for the displacements the total energy can be represented in normal coordinates Q_k for N mass points as

$$H = \frac{1}{2}\sum_{k=1}^{3N} (\dot{Q}_k^2 + \Omega_k^2 Q_k^2) , \tag{8.14}$$

where Ω_k are the normal oscillations of the ensemble. If several normal coordinates Q_1 to Q_{d_α} belong to the same Ω_α the normal mode is degenerate to the degree d_α. In this case (8.14) has the form

$$H = \frac{1}{2}\left(\sum_{k=1}^{3N} \dot{Q}_k^2 + \sum_\alpha \Omega_\alpha^2 \sum_{k_\alpha} Q_{k_\alpha}^2\right) . \tag{8.15}$$

Since the energy of an ensemble does not change if a SO of its point group is applied each normal coordinate Q_k in (8.14) must be mapped on itself or on $-Q_k$ and in (8.15) all Q_{k_α} of one normal mode must be mapped on themselves except for the sign. In both cases the normal coordinates are therefore a basis for a representation according to the definition of (8.9). Since the number of normal coordinates is $3N$ the representation is $3N$-dimensional and therefore, in general, reducible. (As written in (8.14) and (8.15) it is already presented in a fully reduced form). Using (8.11) it can be decomposed into the irreducible representations of the point group. This means all vibrations (including the translations and the rotations) can be classified according to the irreducible representations of the point groups. The usual way of expressing this fact is to assign the vibrations to a particular symmetry type or symmetry *species*. As a consequence the vibrations are denoted with the same Mullikan symbols as the irreducible representations. Figure 8.3 shows the possible in-plane vibrations for a square together with the assignment of their symmetry species. A_{1g} is totally symmetric whereas B_{1g} is antisymmetric with respect to the rotation. E_u is two-dimensional and antisymmetric with respect to the inversion.

For the classification of the vibrations of N mass points it is not necessary to know any of the infinite $3N$-dimensional representations and to reduce the matrix according to (G.1). It is enough to know the traces of the representation matrices for each symmetry element R. The traces or characters $\chi(R)$

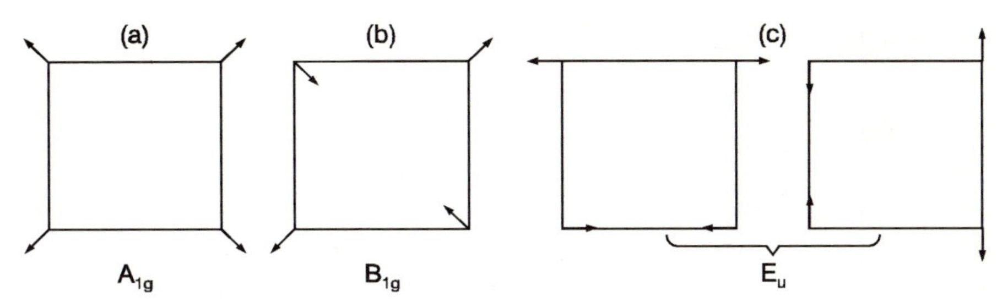

Fig. 8.3a–c. Vibrations of a square with symmetry $\boldsymbol{D}_{4h}$. (**a**) and (**b**) are described by one normal coordinate. (**c**) is a twofold degenerate oscillation and therefore needs two normal coordinates.

are the sum of the matrix elements on the main diagonal. A simple consideration which is explicitly demonstrated in Appendix G.3 shows that a mass point contributes only to the main diagonal of $M(R)$ if it is not moved by the application of the R. In this case the displacements of the mass point are transformed onto themselves and the matrix of (8.2) applies for the transformation. For these considerations it is convenient to return from the normal coordinates to Cartesian coordinates which is no problem since the character for the representations will remain the same. Also, since we need only the traces of the representation matrices it does not matter whether we regard rotation or reflection in z or x or any other direction. The traces remain the same and are always given by (8.3). If the number of atoms which remain unchanged for the application of the symmetry element R is $N_c(R)$ the characters of the $3N$-dimensional representation are

$$\chi^{(3N)}(R) = N_c(R)d_R = N_c(R)[\pm 1 + 2\cos\phi(R)] \, . \tag{8.16}$$

In this way the characters for all SOs of the ensemble can be evaluated. From these values for $\chi(R)$ (8.11) allows immediately to decompose the total representation into the vibrational symmetry species. The $3N$-dimensional representation $\Gamma^{(3N)}$ is alternatively called the *total* representation $\Gamma^{(\text{tot})}$. Since the normal coordinates for the pure translations and for the pure rotations correspond to particular transformations both can be selected from the total irreducible representations found from the application of (8.11). The transformations for the pure translations and for the pure rotations are those for the polar vectors (or coordinates) and for the axial vectors, respectively, as discussed in Sect. 8.1. In many character tables like in those of Appendix G.1 the corresponding irreducible representations are already identified with the letters x, y, z for the translations and with the letters X, Y, Z for the rotations.

To demonstrate the described procedure a simple example is appropriate. Let us consider the water molecule H_2O (Fig. 8.4). The molecule has $3N = 9$ motional degrees of freedom and the symmetry elements $E, C_2, \sigma_v, \sigma_{v'}$.

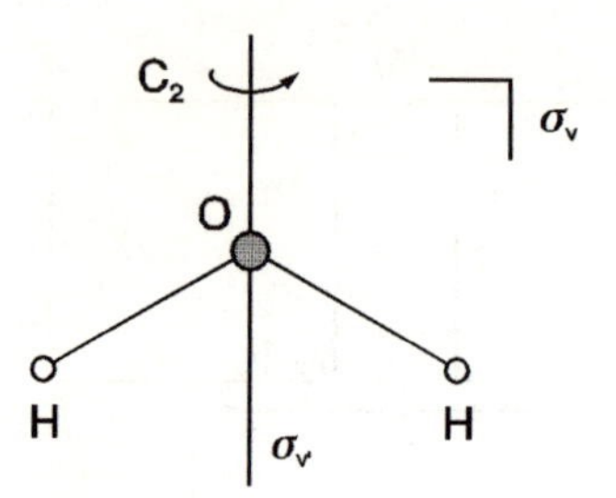

Fig. 8.4. Geometry and symmetry elements of the water molecule.

Table 8.3. Classification of motional degrees of freedom for H_2O.

C_{2v}	E	C_2	σ_v	$\sigma_{v'}$
d_R	3	−1	1	1
N_c	3	1	1	3
$\chi^{(3N)}$	9	−1	1	3
$\chi^{tr} = \pm 1 + 2\cos\phi$ (translations)	3	−1	1	1
$\chi^{rot} = 1 \pm 2\cos\phi$ (rotations)	3	−1	−1	−1
$[\chi]_2 = 2\cos\phi(\pm 1 + 2\cos\phi)$ (sym. tensors)	6	2	2	2

This means the molecule has $\boldsymbol{C}_{2v}$ point symmetry and the corresponding irreducible representations can be taken from Appendix G.1. It is convenient to set up a table like Table 8.3 where the important data and partial results from the calculation are inserted. The first line in the table lists the point group and the symmetry elements in the group separated into classes of conjugated elements. The second line contains the trace d_R of the symmetry elements, the third line the number of stationary atoms for the particular SO and the fourth line the characters for the total representations according to (8.16). The fifth and sixth lines give the characters for the translations and for the rotations. Finally, line 7 presents the characters for the representation according to which the symmetric tensors transform. The importance of this set of characters will be discussed later. To reduce the total representation we use the magic counting formula (8.11). We may ask, for example, how often is the irreducible representation A_1 of $\boldsymbol{C}_{2v}$ contained in the total representation $\Gamma^{(3N)}$. With (8.11) we find

$$n_{A_1} = (1/4)[(+9)\cdot(1) + (-1)\cdot(1) + (1)\cdot(1) + (3)\cdot(1)] = 3\,.$$

A_1 occurs three times in $\Gamma^{(3N)}$. Similarly we obtain for n_{B_2}, n_{A_2}, and n_{B_1} the values 3, 1, and 2. Hence the composition of $\Gamma^{(3N)}$ is

$$\Gamma^{(3N)} = 3A_1 + 3B_2 + A_2 + 2B_1\,. \tag{8.17}$$

To obtain the vibrations of the water molecule we must find out which of these representations are covered by the pure translations and by the pure rotations of the molecule. This means we have to check which irreducible representations are contained in the representations for the translations and for the rotations (lines 5 and 6 in Table 8.3). Alternatively we can look into

Table 8.4. Symmetry points in different notations.

Group	Symmetry point	Notation									
$\boldsymbol{O_h}$		A_{1g}	A_{1u}	A_{2g}	A_{2u}	E_g	E_u	F_{1g}	F_{1u}	F_{2g}	F_{2u}
	Γ, R, H	1	1′	2	2′	12	12′	15′	15	25′	25
$\boldsymbol{T_d}$		A_1	A_2	E	F_1	F_2					
	Γ	1	2	3	25	15					
$\boldsymbol{C_{4h}}$		A_{1g}	A_{1u}	A_{2g}	A_{2u}	B_{1g}	B_{1u}	B_{2g}	B_{2u}	E_g	E_u
	X, M	1	1′	4	4′	2	2′	3	3′	5	5′
$\boldsymbol{C_{4v}}$		A_1	A_2	B_1	B_2	E					
	T, Δ	1	1′	2	2′	5					
$\boldsymbol{D_{2d}}$		A_1	A_2	B_1	B_2	E					
	W	1	2	1′	2′	3					
$\boldsymbol{D_{3d}}$		A_{1g}	$A1u$	A_{2g}	A_{2u}	E_g	E_u				
	L	1	1′	2	2′	3	3′				
$\boldsymbol{C_{3v}}$		A_1	A_2	E							
	Λ, F	1	2	3							
$\boldsymbol{D_{2h}}$		A_g	A_u	B_{1g}	B_{1u}	B_{2g}	B_{2u}	B_{3g}	B_{3u}		
	N	1	2′	2	1′	4	3′	3	4′		
$\boldsymbol{C_{2v}}$		A_1	A_2	B_1	B_2						
	Σ, K	1	2	4	3						
	G, D, S, U, Z	1	2	3	4						

the character tables and find the representations for the translations and for the rotations of the point group $\boldsymbol{C}_{2v}$. In both ways we find

$$\Gamma^{(\mathrm{tr})} = A_1 + B_2 + B_1$$

for the translations and

$$\Gamma^{(\mathrm{rot})} = A_2 + B_2 + B_1$$

for the rotations. Note that the total representations for the translations and for the rotations are always three-dimensional. The irreducible representations for the translations and for the rotations have to be subtracted from the total set given in (8.17). Thus, the pure vibrations are distributed on the three remaining species

$$\Gamma^{(\mathrm{vib})} = 2A_1 + B_2 \ . \tag{8.18}$$

In solid-state physics often Arabic or Greek letters are taken to label the irreducible representations of symmetry points in the Brillouin zone, instead of the Schönflies or international symbols. Table 8.4 compares the different notations of the most important symbols according to [8.2]. Other notations are used as well. The first, second and third column of the table lists the symmetry groups in the Schönflies notation, several selected symmetry points in the Brillouin zone, and the possible irreducible representations in the Schönflies notation. The lower line in the third column contains possible subscripts to the symmetry points to correlate them to the Schönflies symbols. For example, $\Gamma_{25'}$ has F_{2g} symmetry or, Γ_{25} has F_1 symmetry in $\boldsymbol{T}_d$ but F_{2u} symmetry in $\boldsymbol{O}_h$.

8.4 Infinitely Extended Ensembles and Space Groups

Very similar considerations as discussed above apply to infinitely extended ensembles of points in crystals. However, three additional symmetry elements have to be taken into account: translations, screw axes, and glide planes. All SOs are now composed of translations and rotations. Formally they are written as

$$\{R/t_n\} \ ,$$

where R is a symmetry element of the point groups and t_n a translation. t_n is not necessarily a multiple of a primitive translation vector of the lattice. In the case of n-fold screw axes nonprimitive translations $\tau_R = ka/n$ are possible where k can obtain the values 1,2,...n and a is a primitive translation in the direction of the screw axis. As a consequence, translations are always represented by

$$t_n = t_p + \tau_R \ , \tag{8.19}$$

where t_p are the primitive translations of the lattice.

From the symmetry elements for the infinitely extended periodic ensembles again groups can be constructed in a mathematical sense. The total of available elements enables the construction of 230 different *space groups* $\boldsymbol{S}_i$. The order of the space groups is always ∞ since the number of translations is always ∞. The pure translations with the symmetry elements $\{E/t_p\}$ are a subgroup $\boldsymbol{T}$ of the space groups. Space groups which contain only symmetry elements with primitive translations are called *symmorphic.* There are 73 symmorphic and 157 non-symmorphic space groups. Dropping the primitive translations from the space groups yields space-group elements of the form $\{R/\tau_R\}$. They are again elements of a set of groups called the *groups of the unit cell.* These groups are isomorphic to the groups of point symmetry elements which appear in the space-group elements. The latter are assigned as the *crystallographic point groups.* Both types of groups have, of course, finite order. The third type of isomorphic related groups are the *factor groups* defined as $\boldsymbol{S}_i/\boldsymbol{T}$. They are obtained by a group-theoretically defined division of the space groups by the translation groups.

The space groups are again labeled with Schönflies or international symbols. The Schönflies notation counts the space groups belonging to the same point group. $\boldsymbol{D}_{3d}^6$ would be the 6th space group with the point group $\boldsymbol{D}_{3d}$. The international notation uses at first a symbol for the Bravais lattices. (F for fcc, I for bcc, R for rhombohedric, etc.), then a set of the most important symmetry elements follows. The space group $\boldsymbol{D}_{3d}^6$ is labeled as R$\bar{3}$c. All space groups are listed in the International Tables of Crystallography [8.1].

If the analysis of vibrations in crystals is restricted to wave vectors with $q \approx 0$ it is sufficient to study the finite groups related to the space group. In this case one can proceed in a very similar way as demonstrated above for the finite ensembles. The characters for the glide planes and for the screw axes are identical to those for the corresponding rotation axes and mirror planes. For the determination of the stationary atoms N_c space-group elements of the unit cell must be used. This means N_c is always zero for screw axes and glide planes. In addition, an atom is still regarded as stationary if it is moved by a SO from its position in one cell to an identical position in the neighboring cell. As an example in Appendix G.4 a vibrational analysis is performed for the crystal $CaCO_3$.

The irreducible representations of the translations correspond to the acoustic modes since for $q = 0$ the latter are pure displacements. The irreducible representations for the rotations are now real eigenmodes of the crystals since the unit cell is not any more free to move. In molecular crystals the representations for the librational modes (hindered rotations of the molecules) are the irreducible representations of the rotations.

8.5 Quantum-Mechanical Selection Rules

Quantum-mechanical selection rules control the calculation of the transition probability for a system changing from a state i to a state f. The relevant quantity is the matrix element

$$M_{\mathrm{fi}} = \int \psi_{\mathrm{f}}^* \mathbf{P} \psi_{\mathrm{i}} \mathrm{d}x \,, \tag{8.20}$$

where ψ_{i} and ψ_{f} are the wave functions for the initial and the final state, respectively, and $\mathbf{P}$ is the operator driving the transition.

In order to be able to make predictions about the magnitude of M_{fi} it is necessary to generalize the definition for the representations given by (8.9). If we have a set of d linear independent functions $\psi_k(x)$ which are mapped onto themselves by the application of transformations of the form

$$\boxed{\psi_i = \sum_k D_{ik}(R)\psi_k} \tag{8.21}$$

for all SO $R \subset \boldsymbol{G}$ these functions form the basis for a d-dimensional representation $D_{ik}(R)$ of $\boldsymbol{G}$. Since, in particular, the Schrödinger equation

$$\mathbf{H}\psi_k = \epsilon_k \psi_k$$

is invariant versus symmetry operations of the system under consideration the wave functions to a particular eigenvalue are mapped onto themselves and are thus a basis for a (in general reducible) representation. This means the eigenvalues of the system can be classified according to the irreducible representations of the group $\boldsymbol{G}$.

If $\psi_i^{(\varrho)}(x)$ is a basis function for a (irreducible or reducible) representation $\Gamma^{(\varrho)}$ of $\boldsymbol{G}$ the relation

$$\int \psi_i^{(\varrho)}(x)\mathrm{d}x \neq 0 \tag{8.22}$$

is only possible for $\Gamma^{(\varrho)}$ being the trivial representation or containing the trivial representation.

We can always consider $\psi_i^{(\varrho)}$ in (8.22) as the product of basis functions from several representations

$$\psi_i^{(\varrho)}(x) = \prod_{\varrho_k} \psi_i^{(\varrho_k)} . \tag{8.23}$$

Then $\psi_i^{(\varrho)}$ is the basis for the representation obtained from the Kronecker product of the representations $\Gamma^{(\varrho_k)}$. With this assumption we can immediately check under which conditions (8.22) is valid. As discussed in Appendix G.2 the Kronecker product of two different irreducible representations never contains the trivial representation, whereas the Kronecker product of two equal irreducible representations always contains the trivial representation exactly once. Thus, if $\Gamma^{(\varrho)}$ is the Kronecker product of two representations and (8.22) is valid it means two equal representations are contained in this product, and *vice versa.* If two equal representations are contained in the Kronecker product the integral in (8.22) must be finite.

Since all operators transform like particular basis functions we can consider the integral in (8.20) as the triple product of basis functions representing the basis for a representation obtained from the Kronecker product of three representations. Because of the validity of the associative law we can, in this case, at first evaluate the Kronecker product of the first two representations and determine the irreducible contributions. Then the Kronecker product for each of these contributions with the last representation is investigated. If in any of these products the trivial representation is contained the integral in (8.22) is $\neq 0$, and we say the transition from state i to state f is allowed.

The above results can be used immediately to establish selection rules for various processes. If the operator $\mathbf{P}$ is a scalar it transforms according to the trivial representation. In this case only matrix elements for transitions between energy levels with equal symmetry are $\neq 0$.

For IR absorption the operator $\mathbf{P}$ is a dipole moment and transforms therefore according to the coordinates. The corresponding irreducible representations are assigned as x, y, z in the character of Appendix G.1. For the process of IR absorption by phonons or vibrons we can assume the initial state to be the zero-phonon state. This means the vibration is created from the vacuum state and the process of absorption does not depend on the initial electronic states or any other vibronic state. In this case the intial state is characterized by the trivial representation. The final state is a one-phonon state with the symmetry of the phonon characterized by its irreducible representation. Since we need two equal representations to make the integral in (8.22) $\neq 0$ IR absorption is only possible for phonons with an irreducible representation equal to at least one of the representations of the coordinates.

In other words, the irreducible representation assigned in the character tables with x, y, z are exactly the representations for the IR active phonons or vibrations of the corresponding point group. The coordinates given refer to the polarization of the phonon. As an example, in the point group $\boldsymbol{O_h}$ all phonons with irreducible representation F_{1u} but none of the others are IR active.

For the optical absorption $\mathbf{P}$ is the same dipole operator as for IR absorption. However, the initial state is not characterized by the trivial representation. It is rather determined by the symmetry of the electronic ground state represented by $\Gamma^{(\alpha)}$. The integral in (8.20) or in (8.22) is taken over the product of basis functions corresponding to the representation $\Gamma^{(\beta)}\Gamma^{(\text{coord})}\Gamma^{(\alpha)}$. According to the above analysis transitions from state α are possible to all states β for which the irreducible representations $\Gamma^{(\beta)}$ occur in the product $\Gamma^{(\text{coord})}\Gamma^{(\alpha)}$ since then two equal representations are present and the integral in (8.22) will be finite. This statement is expressed by

$$\Gamma^{(\beta)}\left(\Gamma^{(\text{coord})}\Gamma^{(\alpha)}\right) = \Gamma^{(\beta)}\left(\sum n_{\delta_i}\Gamma^{(\delta_i)}\right) ,$$

where $\Gamma^{(\delta_i)}$ are the irreducible representations contained in $\Gamma^{(\text{coord})}\Gamma^{(\alpha)}$. If one of the representations $\Gamma^{(\delta_i)}$ equals $\Gamma^{(\beta)}$ the sum contains $\Gamma^{(\beta)}\Gamma^{(\beta)}$ which contains $\Gamma^{(1)}$.

Let us analyze as an example the possible transitions for the point group $\boldsymbol{O}$ with the irreducible representation $\Gamma^{(\delta_i)} = A_1, A_2, E, F_1, F_2$. The coordinates transform according to F_1. Thus, all products $F_1\Gamma^{(\delta_i)}$ must be analyzed with respect to their irreducible components where $\Gamma^{(\delta_i)}$ is any of the irreducible representations of $\boldsymbol{O}$. A transition from $\Gamma^{(\delta_i)}$ to any of these components will be optically allowed. For the transitions between states with equal symmetry the symmetric product $[\Gamma^{(\delta_i)}]_2$ must be used. The characters for the symmetric product are

$$[\chi^{(\alpha)}(R)]_2 = \frac{1}{2}\left[(\chi^{(\alpha)}(R))^2 + \chi^{(\alpha)}(R^2)\right] . \tag{8.24}$$

If $\Gamma^{(\text{coord})}$ occurs in this product the product $\Gamma^{(\text{coord})}\Gamma^{(\delta_i)}$ contains two equal representations and therefore also the trivial representation. In the case of the group $\boldsymbol{O}$ none of the transitions between states with the same symmetry contains F_1 which means none of them is allowed. The following matrix scheme gives a summary of the allowed and forbidden transitions in $\boldsymbol{O}$. The representation $\Gamma^{(i)}$ in the first column of the scheme is either the representation for the coordinates F_1 (for the off diagonal transitions) or the representation under consideration (for the diagonal transitions). In the latter case the symmetric product must be considered.

	A_1	A_2	E	F_1	F_2
$A_1\Gamma^{(i)}$	0	0	0	M_{14}	0
$A_2\Gamma^{(i)}$	0	0	0	0	M_{25}
$E\Gamma^{(i)}$	0	0	0	M_{34}	M_{35}
$F_1\Gamma^{(i)}$	M_{41}	0	M_{43}	0	M_{45}
$F_2\Gamma^{(i)}$	0	M_{52}	M_{53}	M_{54}	0

For M_{ik} finite the transition is allowed. For example, F_1F_1 contains $A_1 + E + F_2$. Therefore transitions from F_1 to A_1, E, and F_2 are possible but $[F_1]_2$ does not contain F_1. Hence a transition from F_1 to F_1 is forbidden.

Also the splitting of degenerate states by an interaction with an external perturbation can be determined from group theory. This is possible if the symmetry of the perturbation U is lower than the symmetry of the system under consideration, and the symmetry group of U is a subgroup of the symmetry group of the system. The Hamiltonian for the perturbed system has the form

$$\mathbf{H} = \mathbf{H_0} + \mathbf{U} \tag{8.25}$$

with the symmetry of U. The wave functions which represented a basis for an irreducible representation of H_0 are certainly also a basis for a representation of H. However, since the symmetry group of H has less elements than the one of H_0 a smaller set of eigenfunctions may already be mapped onto itself. Thus, a representation of dimension s which is irreducible for the symmetry group of H_0 may be reducible for the group of H. This means the corresponding s-fold degenerated energy levels are split.

Let us consider a threefold degenerated energy level with symmetry F_2 in group $\boldsymbol{T}_d$. The symmetry elements of $\boldsymbol{T}_d$ and the characters for F_2 can be obtained from Appendix G.1 and are listed in Table 8.5 with interchanged order. The system is subjected to a perturbation U with symmetry $\boldsymbol{C}_{3v}$. $\boldsymbol{C}_{3v}$

Table 8.5. Splitting of energy level F_2 from $\boldsymbol{T}_d$ by a perturbation U with symmetry $\boldsymbol{C}_{3v}$.

$\boldsymbol{T}_d$	E	$8C_3$	$6\sigma_d$	$3C_2$	$6S_4$
χ^{F_2}	3	0	1	-1	-1
$\boldsymbol{C}_{3v}$	E	$2C_3$	$3\sigma_v$		
χ^{red}	3	0	1		

has the symmetry elements $E, 2C_3, 3\sigma_v$ also given in Table 8.5. Since only part of the symmetry elements from $\boldsymbol{T}_d$ are available in $\boldsymbol{C}_{3v}$ the truncated representation is reducible in $\boldsymbol{C}_{3v}$. Reduction yields the components $A_1 + E$. Thus the energy level F_2 splits into levels with symmetry $A_1 + E$ in $\boldsymbol{C}_{3v}$.

Finally, we want to study the selection rules if **P** is a tensor. Since a tensor transforms like the product of coordinates we must work out the representation for the product of the coordinates and determine its irreducible components. These irreducible representations determine the allowed transitions.

If we consider as a special case the selection rules for Raman scattering we have to keep the symmetric nature of the Raman tensor in mind. Thus, only the symmetric part of the product representations has to be checked. The characters for this part are evaluated from (8.24). Using as a representation for the transformations of the coordinates the matrices (8.2) the characters for the symmetric product of two (equal) representations is explicitly given by

$$[\chi(R)]_2 = 2\cos\phi(\pm 1 + 2\cos\phi)\,. \tag{8.26}$$

If again like in the case of the IR absorption the representation for the initial state is the trivial representation all irreducible representations occurring in the symmetric product of the coordinates are the representations of the allowed vibrations. So, finally all that needs to be done is to apply the magic counting formula (8.11) to the characters $[\chi(R)]_2$ of (8.26). For the point group $\boldsymbol{C}_{2v}$ these characters have already been evaluated above and are listed in Table 8.3. Reducing with respect to the irreducible representation of $\boldsymbol{C}_{2v}$ yields

$$\Gamma^{[\chi]_2} = 3A_1 + A_2 + B_1 + B_2\,.$$

This means all irreducible representations of $\boldsymbol{C}_{2v}$ are contained in the representation for the symmetric tensor product or all vibrational species are Raman active.

In many character tables the irreducible representations which are contained in the representation for the symmetric tensor products are assigned for all point groups. In the tables of Appendix G.1 this assignment was dropped for simplicity and to avoid duplications. Instead, attention is directed at this point to Appendix H.1 where the various irreducible representations are listed together with the Raman tensors for the various point groups.

Problems

8.1 Prove that the two symmetry elements C_4 and C_4^3 belong to the same class in $\boldsymbol{C}_{4v}$ but not in $\boldsymbol{C}_{4h}$.

(Purpose of exercise: understand the meaning of classes.)

8.2 Show that the trace for the Kronecker product of a matrix A with dimension a and matrix B with dimension b is the product of the traces of matrix A and matrix B.

(Purpose of exercise: understand the Kronecker product.)

8.3 Show explicitly the homomorphic character of the mapping in (8.8) by establishing two multiplication tables.

(Purpose of exercise: understand homomorphism.)

8.4 Consider the coordinates x, y, and z as a basis and find a three-dimensional representation for the point group $\boldsymbol{C_{3v}}$ by using (8.9).

(Purpose of exercise: understand the meaning of a set of coordinates establishing a basis for a representation.)

8.5[a] Derive the magic counting formula from the orthogonality of the characters for different irreducible representations.

(Purpose of exercise: understand orthogonality relations.)

8.6 Find the vibrational modes of a tetrahedron and discuss the degeneracy of the modes.

(Purpose of exercise: perform a simple vibrational analysis and understand degeneracy.)

8.7 Show that for a crystal with crystallographic point group $\boldsymbol{O}$ the transition from a state with symmetry E to a state with symmetry F_2 is allowed but the transition from E to E is forbidden.

(Purpose of exercise: understand selection rules.)

8.8[a] Calculate the Kronecker square for the representations of group $\boldsymbol{D_4}$ and demonstrate its difference to the symmetrized direct products.

(Purpose of exercise: understand the symmetrized Kronecker product.)

8.9[*] Let $\psi_i^{(m)}$ be a basis function of the (reducible) representation $\Gamma^{(m)}$. Show that the relation

$$\int \psi_i^{(m)} \mathrm{d}x \neq 0 \tag{8.27}$$

holds only if $\Gamma^{(m)}$ is the trivial representation or contains the trivial representation. Hint: Show first that the integral disappears if $\psi_i^{(m)}$ is a basis for an irreducible representation which is not the trivial representation. Use: the integral over the full space must remain constant versus any symmetry transformation and $\sum_R D_{ik}^{(\alpha)}(R) = 0$ from the orthogonality of the representation matrices.

(Purpose of exercise: prove a fundamental statement for the application of representation of group theory.)

8.10 Show that for a reduction of the symmetry I_h to T_h a fivefold degenerate vibration decays into a twofold degenerate vibration E_g and a threefold degenerate vibration F_g vibration.

(Purpose of exercise: study lifting of degeneracy by reduction of symmetry.)

9. Light-Scattering Spectroscopy

So far the propagation of light was assumed to be straight or, at most, straight with a discontinuous change in the direction of propagation at a flat boundary between two media with different indices of refraction. Straight propagation of light holds only as long as the medium is homogeneous. If this is not the case and if the inhomogeneities are, in particular, of the size of the light wavelength scattering into arbitrary or well defined directions occurs. For purely geometrical or local inhomogenities with no time dependence the scattering is elastic which means without a change of the light energy. Depending on the size and nature of the optical inhomogeneity the processes are called *Tyndall* scattering, *Mie* scattering, or *Rayleigh* scattering. For time-dependent inhomogeneities the scattering process is inelastic and for inhomogeneities periodic in time sidebands to the excitation line occur. This is the case for the various forms of *Brillouin* scattering and *Raman* scattering. Such scattering experiments give valuable information on the electronic and vibrational states of the material.

9.1 Instrumentation and Setup for Light-Scattering Experiments

In light-scattering experiments the spectral distribution of the scattered light is analyzed relative to the spectrum of the incident light. In the case of Raman or Brillouin spectroscopy the changes in the spectrum are very close in energy to the energy of the incident light but usually many orders of magnitude smaller in intensity. Therefore, a very good suppression is required for the elastically or quasi-elastically scattered light. Lasers are optimum as light sources of excitation, and double monochromators or triple monochromators as they were described in Sect. 4.2 are optimum for analysis. If the expected changes in the spectrum are as close as a few wave numbers to the exciting line single-mode lasers and a Fabry–Perot interferometer are recommended for the excitation and for the analysis, respectively. This is usually the case in Brillouin spectroscopy.

For the purposes of light analysis various optical elements like polarizers, analyzers, $\lambda/4$-platelets, etc., can be inserted into the beam of the incident and of the scattered light. For a quantitative analysis of the scattering intensity the use of a scrambler is recommended in order to account for the very strong difference in spectral sensitivity of the gratings for parallel and perpendicular polarized light, as was described in Sect. 4.2.

For a good resolution in the spectrometer with a simultaneous high light intensity the source of the scattered light must be small. Thus, in general, the laser is concentrated onto the sample by a strongly focusing lens. Figure 9.1 presents several possible geometries. Part (a) is the classical 90°-scattering geometry for transparent crystals. Incident as well as scattered light intensities are increased by a factor of two by the two mirrors M_1 and M_2. Part (b) of the figure represents a 90° scattering geometry for highly absorbing material. Excitation and scattering occurs only close to the crystal surface. A variation to this arrangement is shown in (c) where the scattered light, is observed in 180° backscattering. Because of its well defined geometry with respect to the propagation of the incident and scattered light this arrangement is particularly useful to study selection rules in single crystals. In special cases scattering in the forward direction with scattering angles $\approx 0°$ (Fig. 9.1d) can also be important. The directly transmitted beam must be deflected from the scattered light beam by a small prism or a small mirror, as shown in the figure. Finally, if the crystal under investigation is light-sensitive a cylindrical lense can be used for focusing (Fig. 9.1e). This geometry is appropriate as long as a spectrometer is used with a slit for the light entrance. The local irradiance on the crystal is highly reduced due to the linear spread of the focus but the scattering output remains high if the line focus is properly oriented with respect to the slit.

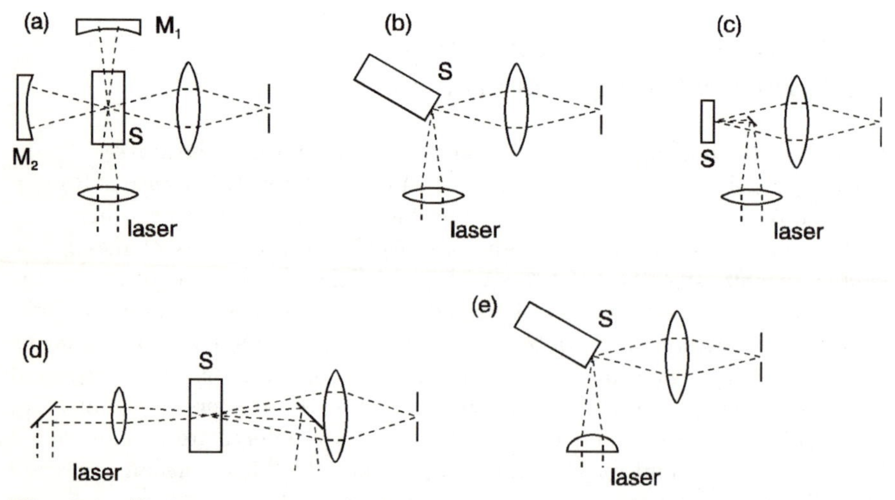

Fig. 9.1. Various geometries for light-scattering experiments: 90° scattering for transparent crystals (**a**), 90° scattering for absorbing crystals (**b**), 180° backscattering (**c**), 0° forward scattering (**d**), and line focus (**e**); (S: sample, $M_{1,2}$: mirrors).

The classical detectors for Brillouin and Raman experiments are the photomultipliers operating as photon counters. Counting rates can be as low as a few counts per seconds. Today, multichannel systems, as they were described in Sect. 5.4, are becoming more and more important.

In the case of light scattering by spatial inhomogenities (Mie scattering) a broad-band spectrum is investigated, in general. This means a broad-band excitation source is appropriate as in the case of optical-absorption spectroscopy, and the spectrometers do not need to have the high resolution required for Raman or Brillouin scattering. Since the light intensity is high, usually no high-sensitivity light detection is needed. Instead, the application of a microscope can be very useful since it may enable the analysis of single scattering centers. If an investigation of the angular distribution of the scattered light is required laser excitation is advantageous even in this type of experiment.

9.2 Raman Spectroscopy

Raman spectroscopy is based on the analysis of inelastically scattered light. Scattering occurs from optical modes of quasi-particles. Classical scatterers are optical phonons but other quasi-particles like optical magnons, plasmons or even electronic excitations provide similar sources for the Raman process.

9.2.1 Fundamentals of Raman Scattering

Raman scattering originates from a change in the polarizability of molecules or the susceptibility of crystals by the excited quasi-particles. The optical phonons are the most often investigated species. In contrast to absorption spectroscopy it is the modulation of the response by the vibrations which is important, rather than the contribution of the vibronic oscillators themselves. The effect is demonstrated in Fig. 9.2 for a two-atom molecule. For an applied field $E(\omega)$ the polarizability α_0 of the orbitals shown leads to a dipole moment $P_\mathrm{D}(\omega) = \alpha_0 E(\omega)$ which acts as a source for the evanescing EM wave. If the molecule is vibrating with frequency Ω the distance between the atoms A and B changes periodically and the polarizability will be modulated. In this case the total dipole moment has the form[1]

$$\boldsymbol{P}_\mathrm{D}(\omega) = (\alpha_0 + \alpha_1 \cos \Omega t)\boldsymbol{E}_\mathbf{0} \cos \omega t \,. \tag{9.1}$$

Application of trigonometric sum rules yields

$$\boldsymbol{P}_\mathrm{D}(\omega) = \alpha_0 \boldsymbol{E}_\mathbf{0} \cos \omega t + (\alpha_1 \boldsymbol{E}_\mathbf{0}/2)[\cos(\omega + \Omega)t + \cos(\omega - \Omega)t] \,. \tag{9.2}$$

[1] Note: The presentation given here is for a single molecule. Thus, α has the units $\mathrm{A\,s\,m^2/V}$.

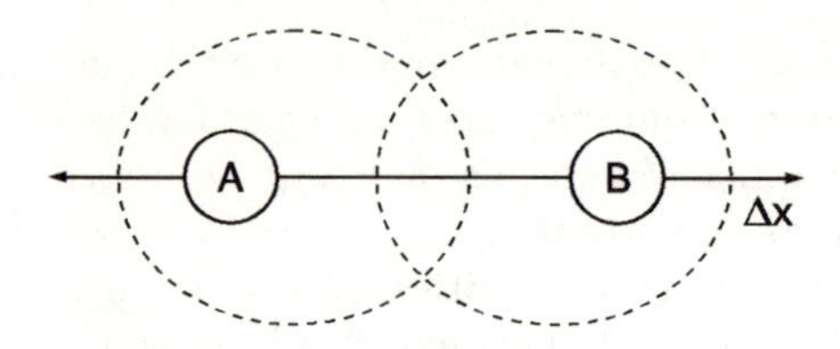

Fig. 9.2. Schematic demonstration of the Raman effect for a two-atomic molecule.

Thus, the evanescing light oscillates not only with the frequency ω but also at sidebands with frequencies $\omega \pm \Omega$. Since ω for visible light is of the order of 20 000 cm^{-1} and the phonon frequencies can be as low as a few cm^{-1} the line shifts can be very small. In addition, α_1 is always many orders of magnitude smaller than α_0, which means that the sidebands are usually very weak. Nevertheless todays spectroscopic techniques enable the measurement of the Raman sidebands for more or less all solid systems under various conditions.

In crystals the situation is more complicated since the phonons have periodic structures and scattering from different parts of the wave will interfere. A constructive interference occurs for the condition

$$2\Lambda \sin(\theta/2) = n\lambda \,. \tag{9.3}$$

Λ and λ are the wavelengths of the phonon and of the light, respectively, n is the index of refraction, and θ the angle between the incident and the scattered beam. Equation (9.3) means that scattering occurs in a well defined direction for a given phonon and for a given wavelength of the incident light.

The classical formulation of the scattering process as given above is easily reinterpreted in a quantum-mechanical picture. In this case the scattering geometry is determined by momentum conservation as shown in Fig. 9.3 and the sidebands are interpreted as emission or absorption of a phonon by the light with conservation of energy. The corresponding mathematical relationships are

$$\begin{aligned} \hbar\omega_{\mathrm{i}} &= \hbar\omega_{\mathrm{s}} \pm \hbar\Omega \,, \\ \hbar\boldsymbol{k}_{\mathrm{i}} &= \hbar\boldsymbol{k}_{\mathrm{s}} \pm \hbar\boldsymbol{q} \,. \end{aligned} \tag{9.4}$$

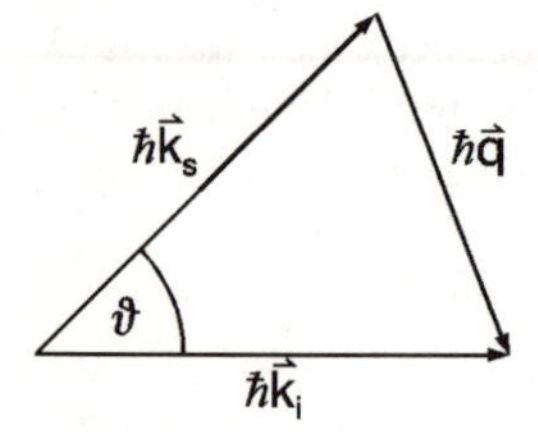

Fig. 9.3. Momentum conservation for a light-scattering process with phonon generation; ($\mathbf{k}_i$, $\mathbf{k}_s$, and $\mathbf{q}$: wave vector for incident and scattered photon and for the phonon, respectively).

The indices s and i refer to the scattered and incident light, respectively. The + sign is for phonon generation and the − sign is for absorption. As can be seen from Fig. 9.3, the direction of the phonon participating in the scattering process depends on the direction of observation. Also, scattering can only occur for quasi-particles with very small values of q as compared to the size of the Brillouin zone. For 180° backscattering the maximum allowed value of q is obtained. It is related to k_i and k_s by

$$q_{max} = k_i + k_s \approx 2k_i \ . \qquad (9.5)$$

Since k_i is of the order of $10^5\,cm^{-1}$ in the visible spectral range only scattering with phonons from the center of the Brillouin zone with $q \approx 0$ is allowed. Depending on whether the quasi-particle is absorbed or emitted, the energy of the scattered light is higher or lower than the energy of the incident light. In the first case we speak about *antiStokes* scattering and in the second case about *Stokes* scattering.

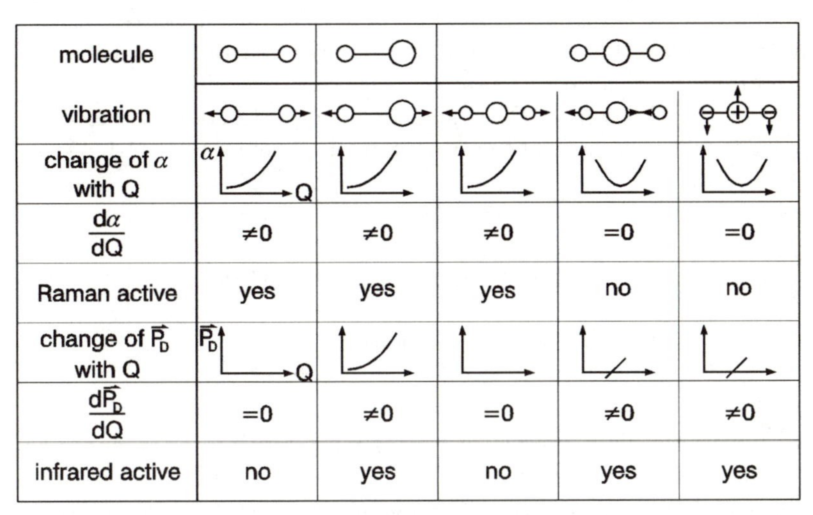

molecule					
vibration					
change of α with Q					
$\frac{d\alpha}{dQ}$	≠0	≠0	≠0	=0	=0
Raman active	yes	yes	yes	no	no
change of $\vec{P}_D$ with Q					
$\frac{d\vec{P}_D}{dQ}$	=0	≠0	=0	≠0	≠0
infrared active	no	yes	no	yes	yes

Fig. 9.4. Selection rules for Raman activity and for infrared activity of vibrations.

A phonon can only contribute to a Raman process if it induces a change in the polarizability. This is not necessarily the case for any vibration but rather depends on the mechanical deformation induced in the molecule or in the unit cell of a crystal. The situation can be demonstrated with the model molecules listed in Fig. 9.4. The figure shows which deformations lead for symmetric diatomic, asymmetric diatomic and symmetric triatomic molecules to a change in the polarizability. In addition, the geometric deformations are indicated which lead to a change of the dipole moment $\boldsymbol{P}_D$ of the molecule. Vibrations of the latter type are infrared active. Column 2 and 3 refer to the change of the polarizability and the dipole moment of diatomic molecules, and column 4 shows the same for three different vibrations of a triatomic molecule. From the symmetry of the vibrations in column 2, 3 and also for the symmetric vibration in column 4 the polarizability is changed by the displacement of the atoms. Therefore these vibrations lead to a Raman effect and are called *Raman active.* This is not so for the asymmetric vibrations of the triatomic molecule. In first order the changes induced by one part of the molecule are compensated by the other part, and the derivative of the polarizability with respect to the normal coordinate Q is zero at $Q = 0$. The oscillations are *Raman inactive.* Similar considerations can be applied with respect to infrared activity. Only the oscillation shown in column 3 and the asymmetric oscillations of column 4 induce a dipole moment and are therefore *infrared active.* Obviously a vibration can be either only Raman active or only infrared active or active to both probes. In fact, it turns out that vibrations can also be inactive or *silent* to both spectroscopic techniques.

The analysis exemplified above relies completely on geometrical considerations. The geometry of the displacement of the atoms must be known. This is in contrast to the analysis of Raman activity and infrared activity discussed in Sect. 8.5. The final results with respect to the activity of the various vibrations is, of course, the same.

The intensity of the light in the sidebands is proportional to the incident intensity. This means the ratio of the two intensities defines a scattering cross section of the form

$$\frac{\mathrm{d}\sigma}{\mathrm{d}\Omega} = \frac{1}{I_\mathrm{i}} \frac{\mathrm{d}\Phi_s}{\mathrm{d}\Omega} \quad \text{(differential scattering cross section), or}$$
$$S = \frac{1}{\mathcal{V}} \frac{\mathrm{d}\sigma}{\mathrm{d}\Omega} \quad \text{(Raman cross section)}\,, \tag{9.6}$$

where $\mathrm{d}\Phi_\mathrm{s}$ means the light power (in watts) scattered into the solid angle $\mathrm{d}\Omega$, I_i the intensity of the incident light (in watts/m^2) and $\mathcal{V}$ the scattering volume. S is a cross section per unit volume and thus a property of the material. Alternatively to (9.6), the derivative of S with respect to the light frequency or spectral energy is often used as a cross section.

The normalization of the Raman cross section to the scattering volume is important for strongly absorbing materials where the scattering volume is determined by the penetration depth of the light. For a comparison of Raman intensities obtained for excitation with different laser lines a correction to the observed results is required. If R and α are the reflectivity and the absorption coefficients of the material, the Raman cross section is given by

$$S = \frac{1}{I_\mathrm{i}} \frac{\mathrm{d}\Phi_\mathrm{s}}{\mathrm{d}\Omega} \frac{(\alpha_\mathrm{i} + \alpha_\mathrm{s})/F}{(1 - R_\mathrm{i})(1 - R_\mathrm{s})}\,, \tag{9.7}$$

where F means the cross section of the laser focus or more generally the cross section of the scattering volume. Equation (9.7) is derived for strongly absorbing materials in 180° backscattering. Modified equations must be used for less strongly absorbing materials [9.1].

9.2.2 Classical Determination of Scattering Intensity and Raman Tensor

For description of the Raman scattering process and for evaluation of the Raman intensity we can proceed analogous to (9.1). Instead of the scalar polarizability α we use the susceptibility tensor χ_{jl} in the case of the crystals. The displacements of the atoms are replaced by the normal coordinates Q_k of the oscillations. The susceptibility can then be expanded with respect to the normal coordinates Q_k and one obtains in analogy to α in (9.1)

$$\chi_{jl} = (\chi_{jl})_0 + \sum_k \left(\frac{\partial \chi_{jl}}{\partial Q_k}\right)_0 Q_k + \sum_{k,m} \left(\frac{\partial^2 \chi_{jl}}{\partial Q_k \partial Q_m}\right)_0 Q_k Q_m + \dots \,, \qquad (9.8)$$

where the sum runs over all normal coordinates. $\partial \chi_{jl}/\partial Q_k$ is a component of the *derived polarizability* tensor. This tensor is also known as the *Raman tensor* and often written as

$$(\chi_{jl})_k \,, \quad \text{or simply as} \quad \chi_{jlk} \,, \quad \text{or} \quad \chi_{jl,k} \,.$$

The intensity of the scattered light is proportional to the square of the Raman tensor. The components of the tensor have three indices. j and l extend over the coordinates 1 to 3 and k runs over the $3N-3$ normal coordinates for the vibrations, where N is the number of atoms in the unit cell. In other words, k runs over all optical modes with wave vector $q = 0$. The Raman tensor which refers to all zone-center vibrations thus has rank three. For an individual mode this tensor is given by a matrix with three rows and three columns determined from the derived susceptibilities. This quantity is called the Raman tensor of a particular mode.

Group theory enables the determination of which vibrational species can have non-vanishing components in their Raman tensor (Sect. 8.5). Since the Raman intensities are described by symmetric tensors exactly those species are Raman active for which the representations are contained in the representation for the symmetric tensors. Group theory can predict even more. It determines which components of the Raman tensors must be zero in the various point groups and which are finite. Thus, the structures of the Raman tensors can be evaluated for all vibrational species and for all point groups. The tensors are tabulated in many books. Appendix H.1 gives a listing following the work of *Poulet and Mathieu* [9.2]. For example, the Raman tensor for a highly symmetric mode (A_g-vibration) in the cubic point group $\boldsymbol{O}_h$ has the form

$$(\chi)_{A_g} = \begin{pmatrix} a & 0 & 0 \\ 0 & a & 0 \\ 0 & 0 & a \end{pmatrix} . \qquad (9.9)$$

As we already saw from Fig. 9.4, vibrations can be Raman active as well as infrared active. For point groups with inversion symmetry Raman activity and infrared activity is mutually excluded.

By expanding the susceptibility into normal coordinates the Raman intensity is obtained quantitatively from the vibration-induced polarization. In the linear approximation the susceptibility is according to (9.8)

$$\chi_{jl} = (\chi_{jl})_0 + (\chi_{jl,k})_0 Q_k \tag{9.10}$$

with

$$Q_k = Q_{k0} \cos \Omega_k t \ .$$

We consider the emission from an elementary oscillator. With a harmonic incident field $E_l^{\mathrm{i}}(t)$ the induced dipole moment which accounts for the emission in the sidebands is

$$P_{\mathrm{D}j}^{\mathrm{s}}(\omega \pm \Omega_k) = \chi_{jl,k} \varepsilon_0 \mathcal{V}_{\mathrm{u}} E_{l0}^{\mathrm{i}} Q_{k0} \cos(\omega \pm \Omega_k)t \qquad (\text{in A s m}) \ , \tag{9.11}$$

where $\mathcal{V}_{\mathrm{u}}$ is the volume of the unit cell. $\chi_{jl,k}$ has the dimension of a reciprocal normal coordinate.

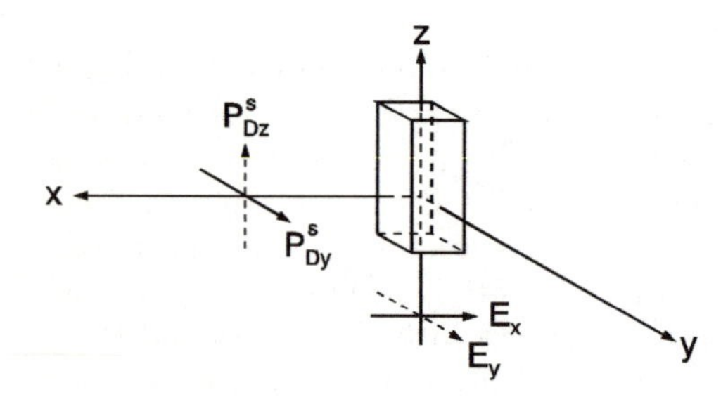

Fig. 9.5. Beam and sample geometry for 90° scattering; (*full drawn arrows*: $(\|, \perp)$-geometry, *dashed arrows*: $(\perp, \|)$-geometry).

For the explicit calculation of the scattering intensity it is convenient to select a special geometry[2]. For the case shown in Fig. 9.5 light is incident in z direction and observed in x direction. Thus, the scattering plane is the xz plane. The polarization for the incident light is in the x direction (full drawn arrow) which means in the scattering plane. The scattered light is y polarized, i.e., perpendicular to the scattering plan. Other allowed orientation for the incident and for the scattered light in this geometry are indicated in the figure as dashed arrows. It is convenient to describe orientations of the incident and scattered light polarization with respect to the scattering plane using the symbols $\|$ and $\perp$. The scattering geometry $E_x P_{\mathrm{D}y}$ given by the full arrows in Fig. 9.5 is then labeled ${}^{\|}\Phi_{\perp}$. Using (2.18) and (2.17) with $\mu = 1$ and $\theta = 90°$ together with (9.11) the time-averaged scattering intensity per elementary unit cell and unit solid angle can be evaluated. For a total scattering volume $\mathcal{V}$, consisting of $N = \mathcal{V}/\mathcal{V}_{\mathrm{u}}$ unit cells, the intensities from elementary oscillators are incoherently superimposed. This yields

[2] Note: Intensities are given as radiance in watts per steradian rather than as watts per area.

$$\frac{\mathrm{d}^{\|}\Phi_{\perp}}{\mathrm{d}\Omega} = \frac{(\omega \pm \Omega_k)^4 \mathcal{V}_\mathrm{u} \chi_{yx,k}^2 \varepsilon_0 E_{x0}^2 Q_{k0}^2 \mathcal{V}}{32\pi^2 c_0^3} . \tag{9.12}$$

It is often more convenient to deal with the intensity I_i of the incident light instead of its field. With (2.8) we obtain

$$\boxed{\frac{\mathrm{d}^{\|}\Phi_{\perp}}{\mathrm{d}\Omega} = \frac{(\omega \pm \Omega_k)^4 \mathcal{V}_\mathrm{u} \chi_{yx,k}^2 Q_{k0}^2 I_\mathrm{i} \mathcal{V}}{16\pi^2 c_0^4} \qquad \text{(in W/ster).}} \tag{9.13}$$

For the Raman scattering cross section (9.6) yields

$$S_{yx} = \frac{(\omega \pm \Omega_k)^4 \mathcal{V}_\mathrm{u} \chi_{yx,k}^2 Q_{k0}^2}{16\pi^2 c_0^4} . \tag{9.14}$$

Similar equations can be derived for other polarizations and for unpolarized light.

Equations (9.12) to (9.14) describe the observed scattering intensity only in a very phenomenological way. For example, the amplitude of the normal coordinates Q_{k0} is not known. It depends strongly on the temperature and is, in fact, the most significant temperature-dependent factor in the equations. This factor can be calculated from a rather simple quantum-mechanical consideration, to be demonstrated in Sect. 9.2.5. The Raman tensor $\chi_{jl,k}$ is also used in a completely phenomenological way. Its calculation needs the evaluation of transition matrix elements similar to those discussed in Chap. 7 but perturbation theory of second order is required.

For the characterization of the scattering geometry the directions of the incident and scattered light as well as the directions of the corresponding polarizations are important. It is convenient to describe these directions with the symbol

$$a(bc)d ,$$

where the letters refer to Cartesian coordinates x, y, and z. a and d give the directions for the incident and scattered light, and b and c the directions for the corresponding polarizations. This assignment is known as *Porto notation.* The Porto notation for the scattering geometry of the full drawn arrows in Fig. 9.5 is $z(xy)x$.

For many experiments it is enough to determine the components of the Raman tensor which contribute to the scattering intensity. For example, to analyze the symmetry species of a vibrational mode it is enough to check which components of the Raman tensor are nonzero. Since from (2.18) the (radiant) scattering intensity is proportional to the square of the dipole moment P_D^s, it is given for the mode k and for a selected direction of polarization $\boldsymbol{e}^\mathrm{s}$ of the scattered light by the absolute square of the projection of the $\boldsymbol{P}_\mathrm{D}^\mathrm{s}$ on $\boldsymbol{e}^\mathrm{s}$.

$$\Phi(k) = C|\,\boldsymbol{e}^\mathrm{s}\,\boldsymbol{P}_\mathrm{D}^\mathrm{s}|^2 = C|\sum_j e_j^\mathrm{s} P_{\mathrm{D}j}^\mathrm{s}(k)|^2 , \tag{9.15}$$

where $P^{\rm s}_{{\rm D}j}$ are the components of the dipole moment induced by the Raman effect. This dipole moment is given according to (9.11) by

$$P^{\rm s}_{{\rm D}j} = \sum_l \chi_{jl,k} e^{\rm i}_l E_l V_{\rm u} \varepsilon_0 Q_k \ .$$

$e^{\rm i}_l$ are the components of the unit vector of the polarization for the incident light. Thus, for a scattering geometry where the polarizations for the incident and scattered light are given by arbitrary vectors $\boldsymbol{e}^{\rm i}$ and $\boldsymbol{e}^{\rm s}$ the scattering intensity is

$$\boxed{\Phi(k) = C'|\boldsymbol{e}^{\rm s}(\boldsymbol{\chi}_k \boldsymbol{e}^{\rm i})|^2 = C'|\sum_{j,l} e^{\rm s}_j \chi_{jl,k} e^{\rm i}_l|^2 E_0^2 \ .} \tag{9.16}$$

If the modes are degenerate, summation of intensities originating from the various Raman tensors corresponding to the same mode is required. Equation (9.16) shows immediately the possibility to select any component of the Raman tensor by properly choosing the polarization of the incident and the scattered light. If the observation for the scattering geometry of Fig. 9.5 is for y polarization but the excitation is for a polarization under 45° to the y direction, the recorded intensity is proportional to $(\chi_{yy,k} + \chi_{yx,k})^2 E_0^2/2$.

Let us consider experimental results for a scattering experiment with calcite ($CaCO_3$). As discussed in Appendix G.4 the crystal has $\boldsymbol{D}_{3d}$ point symmetry with two formula units per unit cell. This yields 27 optical modes distributed over the irreducible representations of $\boldsymbol{D}_{3d}$ as

$$\Gamma^{(3N)} = A_{1g}({\rm R}) + 3A_{2g} + 4E_g({\rm R}) + 2A_{1u} + 4A_{2u}({\rm IR}) + E_u({\rm IR}) \ .$$

The Raman and the IR active modes are assigned with R and IR, respectively. A_{2g} and A_{1u} are silent species. From the table of Raman tensors in Appendix H.1 we find the form of the Raman tensors for the A_{1g} and for the E_g modes

$$A_{1g}: \begin{pmatrix} a & 0 & 0 \\ 0 & a & 0 \\ 0 & 0 & b \end{pmatrix}, \quad E_{g1}: \begin{pmatrix} c & 0 & 0 \\ 0 & -c & d \\ 0 & d & 0 \end{pmatrix},$$

$$E_{g2}: \begin{pmatrix} 0 & -c & -d \\ -c & 0 & 0 \\ -d & 0 & 0 \end{pmatrix}.$$

The twofold degenerate E_g modes have two different Raman tensors. For the scattering geometry $z(xx)y$ the Raman lines of the A_{1g} and of the E_g species can be observed proportional to a^2 and to c^2, respectively. For a scattering geometry $y(zz)x$ only the A_{1g}-modes proportional to b^2 can be observed. In this way the different symmetry species and the individual components of the tensors can be determined experimentally. Figure 9.6 exhibits spectra for different scattering geometries.

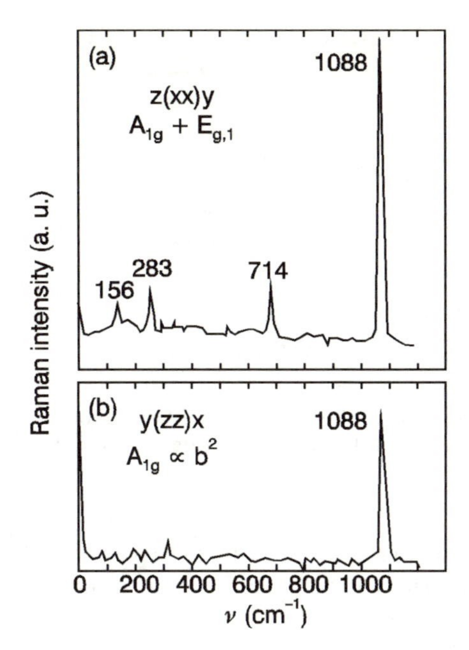

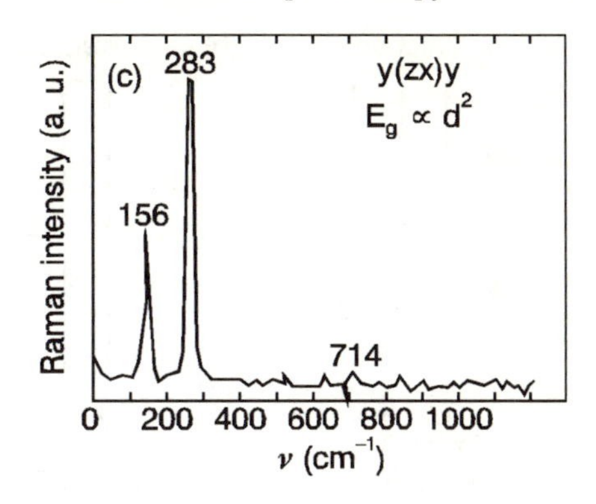

Fig. 9.6a–c. Raman spectra of calcite for different scattering geometries. The four lines in (**a**) can be either A_{1g} or E_g. From (**b**) the mode at 1088 cm^{-1} is A_{1g}, from (**c**) the modes at 156, 283, and 714 cm^{-1} are E_g; after [9.3].

9.2.3 Longitudinal and Transverse Optical Modes

Optical modes with $q \neq 0$ can be longitudinal (LO) or transverse (TO), depending on the direction of the displacement with respect to q. As long as the modes do not carry a dipole moment the LO and the TO components are degenerate. They are electrically inactive and do not contribute to the dielectric function. The situation is different for *polar* modes which carry a dipole moment and are electrically active. The polar modes are the IR-active species. Since the longitudinal electric field contributes much more strongly to the force constants than the transverse field, the LO and TO components are not degenerate any more and the LO frequency is always higher than the TO frequency. This was already evident from the Lyddane–Sachs–Teller relation (6.42) (for $\omega_{\mathrm{L}} = \omega_{\mathrm{LO}}$) in Sect. 6.3. Note that the differentiation between LO and TO modes needs the definition of a direction of propagation. There is no LO-TO splitting for $q = 0$. For the polar modes the direction of polarization is assigned with x, y, and z in the character tables and can be taken from there right away.

Using well defined scattering geometries in a Raman experiment the LO and the TO components of a mode can be studied separately. Figure 9.7 sketches different scattering geometries of the z polarized A_1 species in the point group $\boldsymbol{C}_{2v}$. According to Appendix H.1 the Raman tensor has the form

$$A_1(z) = \begin{pmatrix} a & 0 & 0 \\ 0 & b & 0 \\ 0 & 0 & c \end{pmatrix} .$$

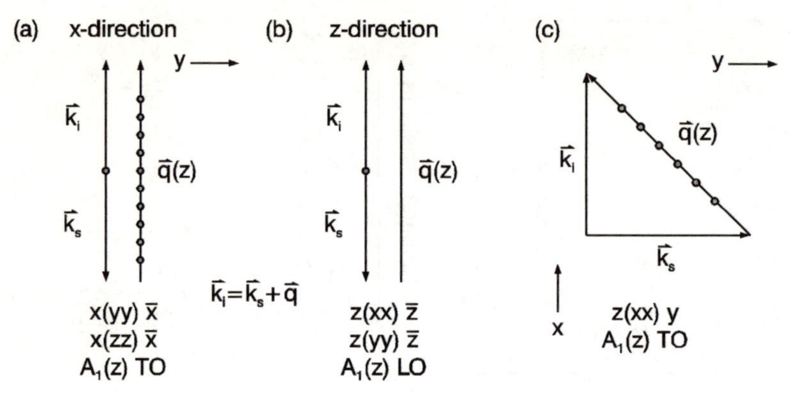

Fig. 9.7. Scattering geometry for the observation of the TO and the LO component for the $A_1(z)$ mode in $\boldsymbol{C}_{2v}$; (**a**) and (**b**) is for 180° backscattering, (**c**) for 90° scattering.

For the geometries in part (a) and (c) of Fig. 9.7 only the TO component is observed whereas for the geometry of part (b) only the LO component is seen.

Difficulties arise if we want to observe the LO component of the $B_1(x)$ species in the same point. It needs a measurement with incident and scattered light propagating in the x direction. The Raman tensor for $B_1(x)$ has the form

$$B_1(x) : \begin{pmatrix} 0 & 0 & e \\ 0 & 0 & 0 \\ e & 0 & 0 \end{pmatrix}$$

which requires a xz polarization. Both conditions cannot be fulfilled simultaneously because of the transverse nature of the light polarization. In transparent crystals the problem can be solved by using a 0° forward scattering geometry. In this case the incident beam is shielded by a beam stop and the

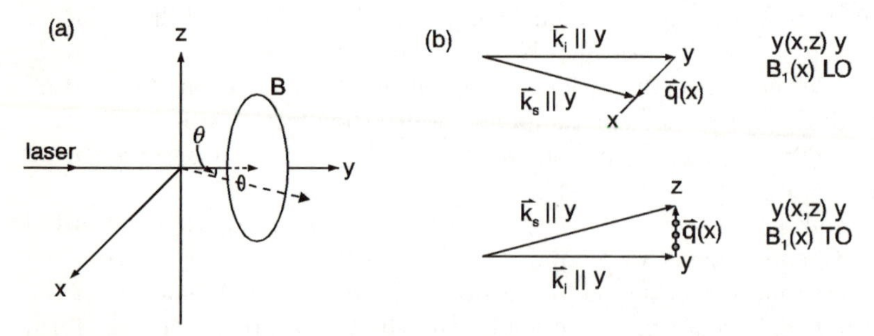

Fig. 9.8. Schematic arrangement for forward scattering to determine the LO and the TO component of a polar mode (**a**). The light is incident in y direction, scattering is for a small angle θ. (**b**) shows the scattering diagrams for LO and TO phonon observation; (B: beamstop).

scattered light is observed through a small pinhole under a very small angle θ in the nearly-forward direction (Fig. 9.8). As the vector diagram in the figure demonstrates, for light incident and observed along the y direction and a xz polarization the LO component of the $B_1(x)$ mode is observed if the xy plane is the scattering plane. Rotating the diaphragm by 90° renders the yz plane as the scattering plane and the TO component of the same mode is observed. In this way all Raman active polar species can be investigated and the LO-TO splitting can be determined. The LO-TO splitting can be between a fraction of a wave number and several tens of wave numbers. It is an important measure for the longitudinal electric field of the mode.

9.2.4 Polaritons

Since the LO-TO splitting does not exist for $q = 0$, the behavior of the modes must change dramatically for approaching the zero wave vector. This is indeed the case. Since the TO component has a transverse polarization which is equivalent to a transverse electric field it has not only mechanical but also electromagnetic character. Thus, the description of the polar modes needs not only lattice dynamics but also Maxwell's equations. With the relationships

$$\boldsymbol{D} = \varepsilon_0 \boldsymbol{E} + \boldsymbol{P}, \qquad \boldsymbol{P} = \boldsymbol{P}_0 \cos(qx - \Omega t), \qquad \boldsymbol{E} = \boldsymbol{E}_0 \cos(qx - \Omega t)$$

the latter yield for the electric field accompanying the wave

$$\boldsymbol{E} = \frac{(\Omega^2/c_0^2)\boldsymbol{P} - \boldsymbol{q}(\boldsymbol{q}\boldsymbol{P})}{\varepsilon_0(q^2 - \Omega^2/c_0^2)} \,. \tag{9.17}$$

From this equation the electric field for the LO mode with $\boldsymbol{P} \| \, \boldsymbol{q}$ and for the TO mode with $\boldsymbol{P} \perp \, \boldsymbol{q}$ is obtained, as expressed by the polarization $\boldsymbol{P}$, from

$$\begin{aligned} \boldsymbol{E}_{\mathrm{LO}} &= \frac{-\boldsymbol{P}}{\varepsilon_0}\,, \\ \boldsymbol{E}_{\mathrm{TO}} &= \frac{\Omega^2 \boldsymbol{P}}{\varepsilon_0(q^2 c_0^2 - \Omega^2)}\,. \end{aligned} \tag{9.18}$$

Since $q^2 c_0^2 \gg \Omega^2$, $E_{\mathrm{TO}} \approx 0$ except for very small values of q. The behavior of the modes for $q \to 0$ can be checked if $\boldsymbol{P}$ is replaced in (9.17) by $\varepsilon_0 \chi \boldsymbol{E}$. This yields a homogeneous equation for the determination of the field $\boldsymbol{E}$. Nontrivial solutions of the equation imply the condition

$$\boxed{\delta_{ij} + \frac{q_i \sum q_k \chi_{kj}(\Omega) - (\Omega^2/c_0^2)\chi_{ij}(\Omega)}{q^2 - \Omega^2/c_0^2} = 0\,.} \tag{9.19}$$

This equation is the dispersion relation for the polar modes. Assuming the phonon parallel to the i direction yields for the ii-component of the equation (LO modes)

$$\boxed{1 + \chi(\Omega) = \varepsilon(\Omega) = 0\,,} \tag{9.20}$$

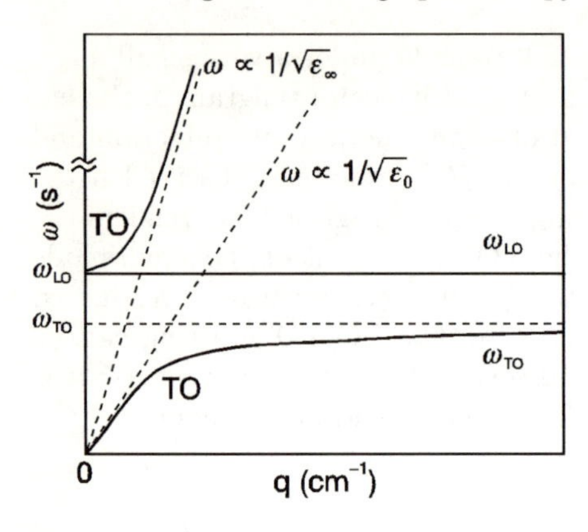

Fig. 9.9. Schematic presentation of LO-TO splitting and polariton dispersion.

or $\Omega = \Omega_{\text{LO}}$. This means, the dispersion for the LO phonons is zero. For the field component perpendicular to the q vector of the phonon (jj-component or TO modes) (9.19) yields

$$1 - \frac{(\Omega^2/c_0^2)\chi(\Omega)}{q^2 - \Omega^2/c_0^2} = 0 \,. \tag{9.21}$$

This equation has two solutions. For small values of q either $\Omega \to \Omega_{\text{LO}}$ (assuming $q = 0$ in (9.21), upper branch of the TO modes) or $q^2 \to \Omega^2\varepsilon(0)/c_0^2$ (for the fraction in (9.21) approaching 1, lower branch for the TO modes). $\varepsilon(0)$ is the static dielectric constant. For large values of q either $\chi(\Omega)$ (lower branch) or Ω (upper branch) must go to ∞ which means Ω becomes either Ω_{TO} or ∞. Between these limits (9.21) yields the proper dispersion. The modes propagating according to this equation are called *polaritons.* Figure 9.9 depicts the dispersion behavior schematically. For small values of q the TO mode exhibits indeed a strong dependence of the frequency on the wave vector and approaches for $q \to 0$ an EM wave with the propagation velocity $c_0/\sqrt{\varepsilon(0)}$. The upper branch of the TO mode approaches the LO frequency for $q \to 0$. For large values of q it corresponds to a true light wave with propagation velocity $c_0/\sqrt{\varepsilon_\infty}$. Polaritons occur in all crystals with polar modes and have partly electromagnetic and partly mechanical character. If the crystal has more than one polar mode the different branches must not intersect. This means only the branch with the highest frequency approaches the light wave. Figure 9.10 shows experimental results for the polariton dispersion in GaP. The values for the frequency and the wave vector were obtained from Raman scattering in forward direction to guarantee small enough wave vectors. For scattering angles smaller than 3° a clear dispersion for the TO mode is observed.

The upper branch of the TO modes can be measured by IR spectroscopy.

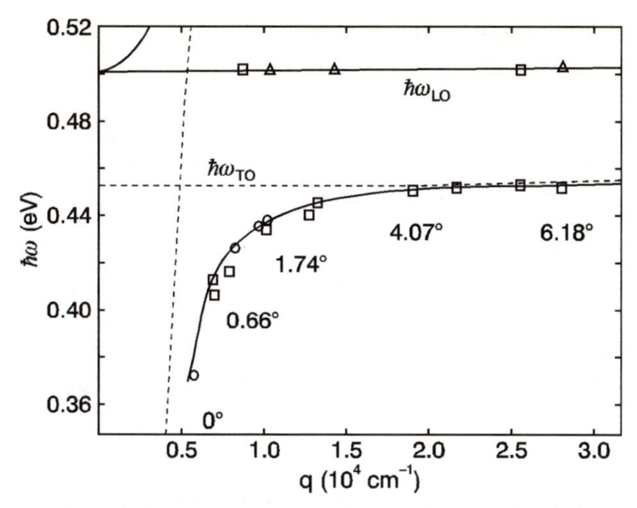

Fig. 9.10. Polariton dispersion of GaP as measured by Raman scattering in the forward direction. The dashed lines give the dispersion for the uncoupled photons and phonons. The angles indicated refer to the scattering geometry; after [9.4].

9.2.5 A Simple Quantum-Mechanical Theory of Raman Scattering

For the quantum-mechanical calculation of the Raman process excitation of an electron by a photon is anticipated. This excitation is followed by a recombination with the simultaneous emission of a photon of different energy. Since energy and momentum must be conserved, the generation or absorption of an additional quasi-particle is required during these two processes. The final state f is reached from the initial state i via an intermediate state z. It differs from the initial state by the generation or absorption of a quasi-particle with energy $\hbar\Omega$ (Fig. 9.11). In the following the quasi-particle is assumed to be a phonon (or vibron) even though it could be any particle for which the selection rules allow a Raman process.

To calculate the probability of the two optical transitions second-order perturbation theory is required, as it is discussed in many special books [9.2, 4.3]. In addition, the generation or absorption of the phonon must be considered. The following description is a simplified version where only the

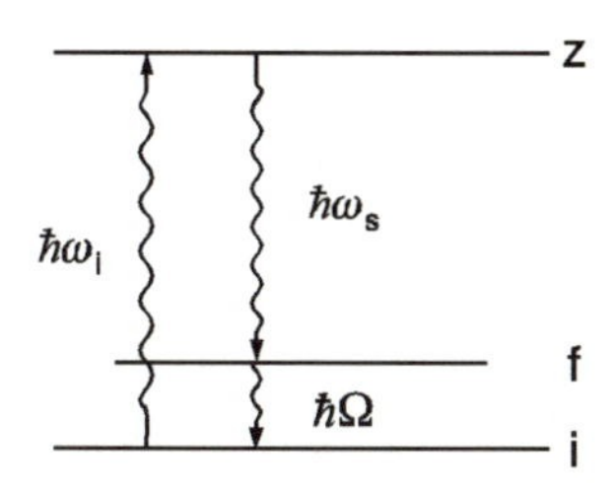

Fig. 9.11. Energy level diagram for a Raman process.

vibrational part is evaluated explicitly. The optical transitions are treated phenomenologically in a way similar to Sect. 7.4 for optical absorption.

To discuss the inelastic scattering processes shown in Fig. 9.11 we must evaluate the matrix element for the transition between state i and f. For the scattering process the transitions are driven by the polarization induced by the light $\boldsymbol{P}= \boldsymbol{\chi}\varepsilon_0\boldsymbol{E}$. The matrix element has therefore the form

$$\boldsymbol{P}_{\mathrm{fi}} = \langle \mathrm{f}|\boldsymbol{P}|\mathrm{i}\rangle = \langle \mathrm{f}|\varepsilon_0\boldsymbol{\chi}\boldsymbol{E}|\mathrm{i}\rangle \ . \tag{9.22}$$

Since $\langle \mathrm{f}|$ and $\langle \mathrm{i}|$ are generalized wave functions the integration runs over all electronic and nuclear coordinates. If the wavelength of the light is much larger than the interatomic distances the electric field can be considered to be constant in (9.22) so that we can extract from the equation a generalized form of the susceptibility known as the *transition susceptibility*

$$[\chi_{mn}]_{\mathrm{fi}} = \langle \mathrm{f}|\chi_{mn}|\mathrm{i}\rangle \ . \tag{9.23}$$

$[\chi_{mn}]_{\mathrm{fi}}$ is a material-specific quantity determined by the electronic orbitals in the crystal. If the final and intial states are both the ground state, it turns into the susceptibility as we have discussed it so far. The situation for Raman scattering is different. We proceed similarly as in Sect. 7.4 by applying the adiabatic approximation (7.23):

$$\boxed{[\chi_{mn}]_{\mathrm{fi}} = \int \varrho_{\mathrm{f}}^*(X)\varphi_{\mathrm{f}}^*(x,X)\chi_{mn}\varphi_{\mathrm{i}}(x,X)\varrho_{\mathrm{i}}(X)\,\mathrm{d}x\,\mathrm{d}X \ .} \tag{9.24}$$

First we consider the integration over the electron coordinates x. Assuming that the final and the intial electronic states are the same renders for this integral the electronic part $\chi_{mn}(X)$ of the transition susceptibility. In contrast to Sect. 7.4 we did not apply the Condon approximation. This means the susceptibility still depends on the nuclear coordinates X. If we introduce normal coordinates Q_k, we can expand the electronic part of the susceptibility with respect to the normal coordinates. Considering only the linear term of the expansion and extracting the expansion coefficient from the integral we obtain

$$\begin{aligned}[\chi_{mn}]_{\mathrm{fi}} =& (\chi_{mn})_0\langle \ldots v_{\mathrm{f}k}\ldots|\ldots v_{\mathrm{i}k}\ldots\rangle \\ &+\sum_k \left(\frac{\partial\chi_{mn}}{\partial Q_k}\right)_0 \langle \ldots v_{\mathrm{f}k}\ldots|Q_k|\ldots v_{\mathrm{i}k}\ldots\rangle \ .\end{aligned} \tag{9.25}$$

The bra and ket symbols represent total vibrational wave functions from the integral in (9.24). They are expressed as the product of harmonic-oscillator wave functions with the occupation numbers $v_{\mathrm{f}k}$ or $v_{\mathrm{i}k}$.

$$\begin{aligned}\langle v_{\mathrm{f}1},\ldots v_{\mathrm{f}k},\ldots v_{\mathrm{f}n}| &= \prod_k \langle v_{\mathrm{f}k}| \\ \langle v_{\mathrm{i}1},\ldots v_{\mathrm{i}k},\ldots v_{\mathrm{i}n}| &= \prod_k \langle v_{\mathrm{i}k}| \ ,\end{aligned} \tag{9.26}$$

where $\langle v_{\mathrm{f}k}|$ and $\langle v_{\mathrm{i}k}|$ are the harmonic-oscillator wave function for occupation numbers $v_{\mathrm{f}k}$ and $v_{\mathrm{i}k}$, respectively, as given in Appendix F.4. Since we do not use the Franck–Condon principle the oscillators are unshifted and the expectation values in (9.24) are

$$\langle v_{\mathrm{f}k}|v_{\mathrm{i}k}\rangle = \begin{cases} 0 & \text{for } v_{\mathrm{f}k} \neq v_{\mathrm{i}k} \\ 1 & \text{for } v_{\mathrm{f}k} = v_{\mathrm{i}k} \end{cases} \tag{9.27}$$

and

$$\langle v_{\mathrm{f}k}|Q_k|v_{\mathrm{i}k}\rangle = \begin{cases} 0 & \text{for } v_{\mathrm{f}k} = v_{\mathrm{i}k} \\ (v_{\mathrm{i}k}+1)^{1/2}\sqrt{\hbar/2\Omega_k} & \text{for } v_{\mathrm{f}k} = v_{\mathrm{i}k}+1 \\ (v_{\mathrm{i}k})^{1/2}\sqrt{\hbar/2\Omega_k} & \text{for } v_{\mathrm{f}k} = v_{\mathrm{i}k}-1\,. \end{cases} \tag{9.28}$$

Because of the orthogonality of the wave functions all expectation values from (9.25) can be factorized into relationships like (9.27) and (9.28). Then, the first term in (9.25) is only different from zero if $v_{\mathrm{f}k} = v_{\mathrm{i}k}$ for all k. This means, the quantum state of the system has not changed. If $(\chi_{mn})_0$ is properly calculated it describes the process of absorption or Rayleigh scattering. The second term is responsible for the Raman process which is evident from the appearance of the derived susceptibility. According to (9.27) and (9.28) it is only nonzero if for all $k' \neq k$ $v_{\mathrm{f}k'} = v_{\mathrm{i}k'}$ and for the mode k $v_{\mathrm{f}k} = v_{\mathrm{i}k} \pm 1$ holds. In this case the transition susceptibility (9.25) has the form

$$[\chi_{mn}]_{v_{\mathrm{i}k}+1,v_{\mathrm{i}k}} = (v_{\mathrm{i}k}+1)^{1/2}\sqrt{\hbar/2\Omega_k}\left(\frac{\partial\chi_{mn}}{\partial Q_k}\right)_0 \tag{9.29}$$

and

$$[\chi_{mn}]_{v_{\mathrm{i}k}-1,v_{\mathrm{i}k}} = (v_{\mathrm{i}k})^{1/2}\sqrt{\hbar/2\Omega_k}\left(\frac{\partial\chi_{mn}}{\partial Q_k}\right)_0 . \tag{9.30}$$

Equation (9.29) and (9.30) obviously describe the Stokes and the antiStokes Raman processes. From a comparison with (9.8) for the classical evaluation of the Raman intensity the equivalence between the tensor for the transition susceptibility and the derived susceptibility multiplied with the amplitude of the normal coordinate is evident. In the quantum-mechanical calculation the amplitude of the latter is replaced by its quantum-mechanical equivalent $\sqrt{\hbar v_k/2\Omega_k}$. We have dropped here the index i, for simplicity, and will do this also in the following equations.

For a comparison with the experimental results attention must be paid to the dependence of the intensities on the vibronic occupation number v_k. Since the latter is determined by a Boltzmann factor

$$W(\epsilon_k) = \frac{\exp(-\epsilon_k/k_{\mathrm{B}}T)}{Z} = \frac{\exp[-\hbar\Omega_k(v_k+1/2)/k_{\mathrm{B}}T]}{\sum_{v_k}\exp[-\hbar\Omega_k(v_k+1/2)/k_{\mathrm{B}}T]} , \tag{9.31}$$

a thermal averaging of the form

$$\sum_{v_k}(v_k+1)W(\epsilon_k)$$

is required to obtain the effective square of the Raman tensor from (9.29) and (9.30). This is similar to the thermal averaging used for the evaluation of the optical absorption from localized wave functions (7.34).

In the case of Stokes scattering the average is $n_k + 1$ where n_k is given by the Bose–Einstein distribution for the mode k

$$n_k = f_E(\Omega_k) = \frac{1}{\exp(\hbar\Omega_k/k_\mathrm{B}T) - 1} \,. \tag{9.32}$$

For antiStokes scattering the average yields n_k. Thus, the scattering intensity per steradian is derived from (9.13) for the incident intensity I_i by replacing Q_{ko}^2 with the square of the quantum-mechenical amplitude $\hbar v_k/2\Omega_k$ and the factor for thermal averiging $(n_k + 1)$.

$$\boxed{\frac{\mathrm{d}^{\|}\Phi_\perp}{\mathrm{d}\Omega} = \frac{\hbar(\omega-\Omega_k)^4\mathcal{V}_\mathrm{u}\chi_{yx,k}^2(n_k+1)I_\mathrm{i}\mathcal{V}}{32\pi^2c_0^4\Omega_k}} \,. \tag{9.33}$$

A corresponding relation is obtained for antiStokes scattering.

9.2.6 Temperature Dependence of Raman Scattering

The temperature dependence of the Raman intensity is immediately obtained from (9.33) by the dependence of n_k on T. A conveniently measured quantity is the ratio between antiStokes and the Stokes intensities of one mode

$$\frac{\Phi'_a}{\Phi'_s} = \left(\frac{\omega+\Omega_k}{\omega-\Omega_k}\right)^4 \exp\left(\frac{-\hbar\Omega_k}{k_\mathrm{B}T}\right) \,. \tag{9.34}$$

The dash at the symbol for intensity stands for the derivation with respect to the solid angle. Equation (9.34) provides a good check on the temperature in the laser focus on the crystal. Figure 9.12a shows experimental results for the Stokes/antiStokes ratio in Si for the optical phonon at 525 cm^{-1} in comparison to the behavior calculated from (9.34).

The widths of the Raman lines also change with temperature. In many cases the line shape is Gaussian or Lorentzian with FWHM Γ. The value of Γ depends on the decay mechanism of the phonons. The most simple but often observed process is a decay of the optical phonon into two longitudinal acoustic modes. In this case the width of the Raman line is

$$\boxed{\Gamma(\Omega_k,T) = \Gamma(\Omega_k,0)\left(1+\frac{2}{\exp(\hbar\Omega_k/2k_\mathrm{B}T)-1}\right)} \,. \tag{9.35}$$

Alternative shapes for the temperature dependence may occur if the phonons couple to rotational modes or to diffuse motions. Since such motions are thermally activated, line shapes have a temperature dependence of the form

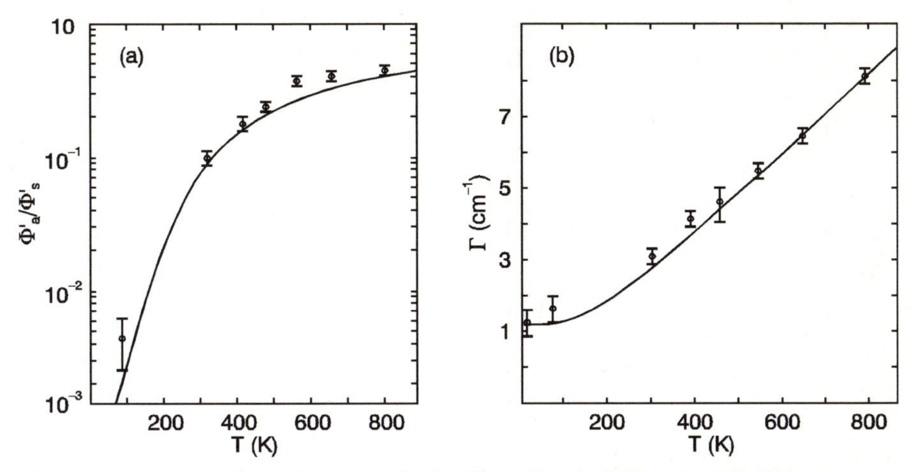

Fig. 9.12. Ratio of antiStokes to Stokes Raman intensities for Si versus temperature; (•): experiment, (—): calculated from (9.34) (**a**), and width of the Raman line in Si versus temperature; (•): experiment, (—): calculated (**b**); after [9.5].

$$\Gamma(\Omega_k, T) = \Gamma(\Omega_k, 0) \exp(-\epsilon_A / k_B T) \; , \tag{9.36}$$

where ϵ_A is the activation energy for the diffuse motion. Figure 9.12b displays the linewidth of the optical mode in Si. The full drawn line has been calculated according to (9.35).

A particularly strong temperature dependence of the Raman lines can be observed if the phonons themselves exhibit a strong temperature dependence. This occurs frequently at structural phase transitions if at least one component of the oscillation coincides with the vector driving the phase transition. For approaching the temperature T_c of the phase transition the energy of the mode vanishes like

$$\hbar\Omega \propto \frac{(T - T_c)^\alpha}{T_c} \tag{9.37}$$

and the mode does not exist above T_c. α is called the *critical exponent* of the *soft mode* Ω. Figure 9.13 exhibits the temperature dependence of two Raman active modes in $SrTiO_3$ for approaching the temperature of the phase transition at 110 K.

9.2.7 Raman Scattering from Disordered Structures

For polycrystalline solids, for disordered polymers, or for molecules in solution a selection of vibrational species is not possible in the sense of Sect. 9.2.2, but even in this case valuable information can be obtained from the observed vibrations. One possibility is to correlate the vibrational features with the

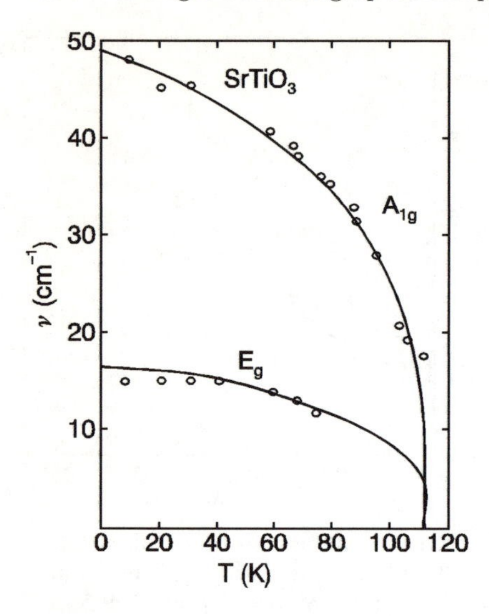

Fig. 9.13. Raman line position for two soft modes in $SrTiO_3$; after [9.6].

structure of the chemical bonds. Characteristic bonds give rise to characteristic frequencies. This is particularly useful for organic compounds where the bonds are well defined. The procedure is similar to the analysis used in IR spectroscopy and will therefore be discussed in more detail in Chap. 10. As compared to IR, classical Raman spectroscopy has the disadvantage of relying on a very delicate instrumentation such as high-power ion lasers and photon-counting systems. With the development of Fourier-transform Raman systems the situation changed and Raman spectroscopy has become as convenient for industrial application as IR spectroscopy. Fourier-transform Raman systems will also be discussed in Chap. 10.

To obtain a relation between the Raman tensors and the scattering intensities even for disordered systems the tensors must be subjected to orthogonal transformations with arbitrary angles, and the results must be averaged over the whole angular space 4π. The transformation of a tensor from a system with the coordinates x', y', z' into a system with the coordinates x, y, z is performed by multiplication of the tensor components with the corresponding product of the direction cosines $\cos(mm')\cos(nn')$

$$\chi_{mn,k} = \sum_{m',n'} \chi_{m'n',k} \cos(mm')\cos(nn') \ . \tag{9.38}$$

To make writing easier the index k will be dropped in the following. It is understood that all equations listed below hold for the derived susceptibility and for each vibration separately.

Since we are dealing with orthogonal transformations the trace $3a$ and the anisotropy τ remain constant with

$$a = (\chi_{11} + \chi_{22} + \chi_{33})/3$$

$$\tau^2 = [(\chi_{11} - \chi_{22})^2 + (\chi_{22} - \chi_{33})^2 + (\chi_{33} - \chi_{11})^2 + 6(\chi_{12}^2 + \chi_{23}^2 + \chi_{13}^2)]/2\,. \tag{9.39}$$

On the other hand, the tensor expressions obtained after transformation and averaging must also be independent of any transformation. This means it must be possible to express the tensors by a and τ^2. This is, indeed, possible for the averaged products of the tensor components (Appendix H.2). The relations can be summarized in the following way:

$$\begin{aligned} \overline{\chi_{11}^2} = \overline{\chi_{22}^2} = \overline{\chi_{33}^2} &= (45a^2 + 4\tau^2)/45\,, \\ \overline{\chi_{12}^2} = \overline{\chi_{23}^2} = \overline{\chi_{31}^2} &= \tau^2/15\,, \\ \overline{\chi_{11}\chi_{22}} = \overline{\chi_{22}\chi_{33}} = \overline{\chi_{33}\chi_{11}} &= (45a^2 - 2\tau^2)/45\,. \end{aligned} \tag{9.40}$$

All other averaged components of the Raman tensor are zero. Using equations of the form (9.40) and (9.13) the Raman intensities can be evaluated for the different scattering geometries from the components of the Raman tensor, even in the case of disordered systems. For the $\|, \perp$ scattering geometry we obtain for example

$$\frac{\mathrm{d}^{\|}\Phi_{\perp}}{\mathrm{d}\overline{\Omega}} = {}^{\|}\Phi'_{\perp} = \frac{\hbar(\omega - \Omega)^4 \mathcal{V}_{\mathrm{u}} \mathcal{V} \tau^2 (n+1) I_i}{240\pi^2 c_0^4 \Omega}\,. \tag{9.41}$$

Looking at Fig. 9.5 we note that for 90° scattering a $\|, \|$ scattering geometry gives the same intensity as $\|, \perp$ or $\perp, \|$.

Since absolute scattering intensities cannot be determined very accurately, often the ratio of the intensities for different scattering geometries is investigated. The corresponding quantities are obtained for each mode from

$$\varrho_{\|} = \frac{{}^{\|}\Phi'_{\perp}}{{}^{\|}\Phi'_{\|}}, \qquad \varrho_{\perp} = \frac{{}^{\perp}\Phi'_{\|}}{{}^{\perp}\Phi'_{\perp}}, \qquad \text{and} \qquad \varrho_n = \frac{{}^{n}\Phi'_{\|}}{{}^{\|}\Phi'_{\perp}}\,. \tag{9.42}$$

The various forms of ϱ are the *depolarization factors* since they determine how much of a polarization is retained for the scattered light after the interaction of the incident light with a phonon. For oriented crystals the depolarization factors are simple ratios of the two components of the Raman tensor and are obtained from equations like (9.13). For the averaged tensors one obtains for 90° scattering

$$\begin{aligned} \varrho_{\|} &= 1 \quad \text{for} \quad \tau \neq 0\,, \\ \varrho_{\perp} &= \frac{3\tau^2}{45a^2 + 4\tau^2}, \quad 0 < \varrho_{\perp} < 3/4\,, \\ \varrho_n &= \frac{6\tau^2}{45a^2 + 7\tau^2}, \quad 0 < \varrho_n < 6/7\,. \end{aligned} \tag{9.43}$$

The limitation of the depolarization factors to the values indicated is a consequence of the possibility for τ and a to become zero. In particular, a line is called *depolarized* if $\varrho_\perp = 3/4$ ($a = 0$) and fully *polarized* if $\varrho_\perp = 0$. From the Raman tensors in Appendix H.1 it is evident that only totally symmetric modes can have a nonzero trace. This means, only for these modes $\varrho_\perp$ can be very small whereas for all others $\varrho_\perp = 3/4$. Thus, at least the totally symmetric modes can be distinguished from the non-symmetric species.

Note: in the case of resonance scattering the depolarization ratios can be dramatically different from the values given above.

9.2.8 Resonance Raman Scattering and Electronic Raman Scattering

So far the Raman process has been described as the excitation of an electron by the incident photon and the subsequent recombination of the electron with the simultaneous emission of another photon. During these processes a phonon is absorbed or emitted. The question whether the intermediate state is an eigenstate of the system has not been considered. For the magnitude of the transition susceptibility this is, however, of crucial relevance. Transitions into eigenstates have a much larger matrix element, and the intermediate states have a much longer lifetime as compared to transitions into virtual states. As a consequence the probability of the generation of a phonon is much larger for the former, and the Raman scattering intensities can be many orders of magnitude larger. A Raman process of this type is called *resonance enhanced.* The lifetime of an excitation into a virtual intermediate state is determined by the Heisenberg uncertainty principle. The closer the intermediate state comes to an eigenstate the less energy conservation is violated and the more stable is the excitation.

For a resonance scattering process discrete excited states or critical points in the density of states are relevant, very similar to optical absorption. Resonances of the first type are frequently observed in molecules but also in molecular crystals and for excitonic excitations in solids. Resonances at critical points in the density of states are frequently found in semiconductors.

Figure 9.14 exhibits Raman spectra for polydiacetylene excited with different lasers. Polydiacetylene is a conjugated polymer which can be prepared in macroscopic single crystals. It has a quasi-one-dimensional band structure with a strong exciton transition close to 2 eV. The more the exciting laser approaches this transition the stronger the light scattering intensity. Raman lines shown in the figure correspond to stretching modes of the C=C bond (1495 cm^{-1}) and of the C≡C bonds (2090 cm^{-1}), and to a deformation vibration of the chain (946 cm^{-1}).

In Fig. 9.15 the scattering cross section for GaP is depicted. In this case the resonance is at a critical point in the density of states. By approaching this point the scattering intensity increases by more than a factor of ten. The maximum of the resonance has two peaks since two critical points exist: One

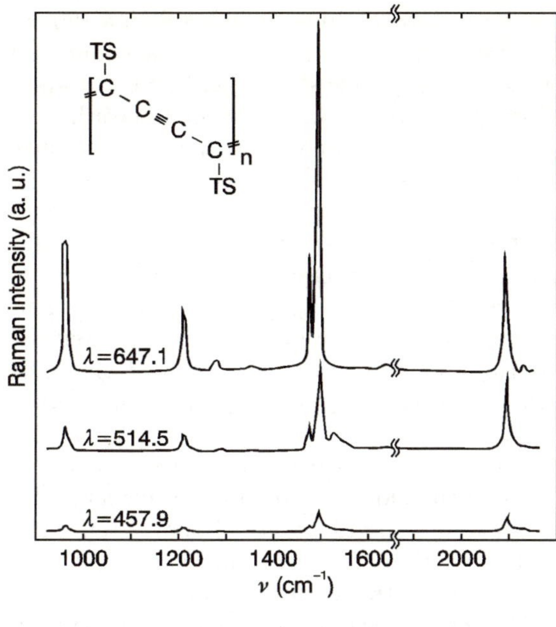

Fig. 9.14. Raman spectra of polydiacetylene-TS as excited with different lasers of equal intensity; after [9.7]. Insert: chemical structure of the polymer.

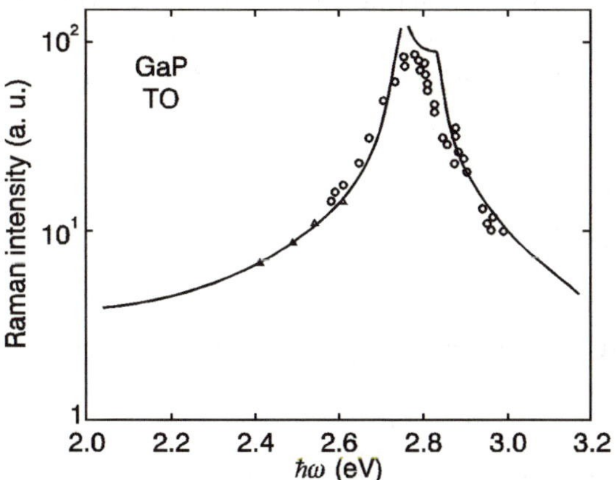

Fig. 9.15. Resonance Raman cross section for the optical mode at 365 cm^{-1} in GaP. The full drawn line is calculated; after [9.8].

for the energy ϵ_0 corresponding to transitions from the valence band to the conduction band, and another for an energy $\epsilon_0 + \Delta_0$ where Δ_0 is the split-off energy from the valence band due to spin-orbit coupling.

The calculation of the resonance Raman intensities is often simpler than the calculation for nonresonant scattering since only one electronic transition has to be considered. The resonance is expressed by a resonance denominator in the transition susceptibility. Resonance Raman scattering allows, in a particularly simple way, to determine the interband electron–phonon coupling constant or the Franck–Condon coupling constant.

A frequently discussed question regards the difference between a Raman process and a luminescence process. In a Raman process the recombination always starts directly from the excited electronic state whereas for the luminescence a strong relaxation from this state always occurs first according to Kasa's rule. In contrast to conventional luminescence, the hot luminescence, where the emission starts from an excited state, is immediately related to the Raman process.

So far only phonons were considered as quasi-particles excited in the Raman process. This limitation is not necessary. Even free electrons or plasmons can interact with the excited state via a Raman process. For a nondegenerate electron gas free carriers obey a Maxwellian velocity distribution. The Doppler effect occurring during the interaction with the light gives rise to a continuous upshift or downshift of the stray light spectrum in the immediate neighborhood of the exciting laser.

If the concentration of the free carriers is large enough longitudinal plasma oscillations can be excited. The frequency for these oscillations coincides for $q = 0$ with the plasma frequency $\omega_{\text{pl}} = \sqrt{ne^2/\varepsilon\varepsilon_0 m^*}$ of the system. The plasmons can be another source for a Raman process and appear as sidebands to the exciting laser line in the spectra. Figure 9.16 displays spectra for GaAs excited with a Nd:YAG laser at 1.06 μm. The carrier concentration of $n = 1.75 \times 10^{17}$ cm^{-3} corresponds to a plasmon mode at $\omega_{\text{pl}} = 8.2 \times 10^{12}$ s^{-1} ($\nu_{\text{pl}} = 130$ cm^{-1}). The spectrum in (a) at the top was taken for $x(yz)y$

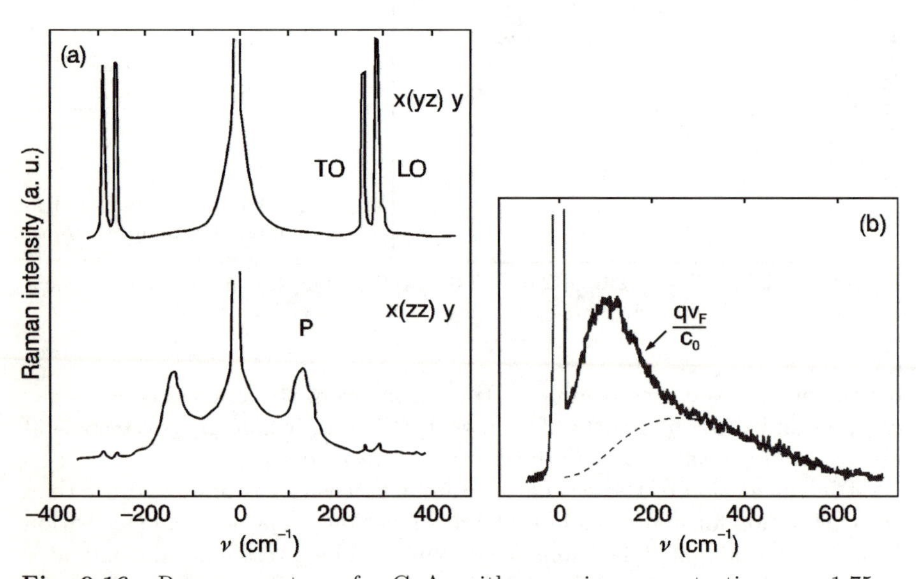

Fig. 9.16. Raman spectrum for GaAs with a carrier concentration $n = 1.75 \times 10^{17}$ cm^{-3} for two different scattering geometries; after [9.9] (**a**), and single-particle scattering continuum for $n = 1.3 \times 10^{18}$ cm^{-3} (**b**). The dashed line in (b) is an estimated contribution from a luminescence emission; after [9.10].

scattering and reveals sharp maxima for the LO and TO components of the threefold degenerate polar mode of the semiconductor as well as the single-particle spectrum close to the exciting laser. For the zz polarization shown at the bottom the excitation of the phonon lines is strongly suppressed and the plasmon at 130 cm^{-1} is clearly observed instead. The plasmon shifts to higher frequencies with $\sqrt{n}$. As soon as it approaches the polar mode it interacts strongly with the LO component and coupled plasmon-phonon modes propagate with the frequencies ω^- and ω^+. Such modes have been observed extensively with Raman scattering [9.9]. An example of this behavior is shown in Fig. 14.4 in Sect. 14.2.3.

In Fig. 9.16b a blown up spectrum for single-particle scattering is shown. The carrier concentration was high enough to shift the Fermi level into the conduction band. Under this condition single-particle scattering extends as far as

$$\boxed{\nu_{\rm el} \approx q v_{\rm F}/c_0 \ ,}$$

where q is the scattering vector and $\nu_{\rm el}$ is given in cm^{-1}. For non-degenerate conduction electrons where the Fermi level is still in the energy gap single-particle scattering extends only as far as

$$\boxed{\nu_{\rm el} \approx q v_{th}/c_0 = (q/c_0)\sqrt{2k_{\rm B}T/m^*} \ .}$$

The scattering vector is in both cases approximately equal to the wave vector of the light.

Interference effects can have a strong influence on the scattering spectra if in a certain range of the spectrum contributions from the single-particle continuum or from any other continuum and the response from discrete phonon lines overlap and if the two systems of quasi-particles interact. The phonon response appears then with a distorted line shape of the Fano type, as was discussed in Sect. 3.2 and demonstrated in Fig. 3.4.

9.3 Brillouin Scattering and Rayleigh Scattering

Brillouin scattering is closely related to Raman scattering. According to a classical definition scattering by optical phonons is a Raman process and scattering by acoustic phonons is a Brillouin process. Definitions from molecular physics according to which Raman scattering originates from molecular vibrations which change the polarizability of the molecules and Brillouin scattering originates from thermodynamic fluctuation of the density of the system is not applicable to solids. A transverse acoustic mode can well contribute to a light scattering process but does not induce density fluctuations.

A basic difference between Raman scattering and Brillouin scattering comes from the dispersion relation of the quasi-particle generated. We talk about a Brillouin process if the frequency is zero for $q = 0$ ($\omega(q = 0) = 0$) and about a Raman process if $\omega(q = 0) \neq 0$.

9.3.1 Fundamentals of Brillouin Scattering

Like in Raman scattering in most experiments visible light is used for the excitation in Brillouin scattering. The frequency range of the probed excitations is 10^{-5}–1 cm^{-1}. This means only long-wavelength modes will be excited where phonons have still the same dispersion as the sound waves. Thus, sound velocities and elastic constants can be determined together with their response to physical processes in the material.

Conservation of energy and momentum must be retained like in Raman scattering (Fig. 9.3) and the same phenomenological description holds for the classical picture. The modulation of the response function creates sidebands. Instead of the susceptibility usually the dielectric function

$$\varepsilon(\omega, q, \Omega) = \varepsilon(\omega, 0) + \Delta\varepsilon(\omega, q, \Omega) \tag{9.44}$$

is considered. The wave vector is now included since we consider "acoustic" modulations which definitely require $q \neq 0$. For classical Brillouin scattering density fluctuations in the material are usually considered as the source of the modulation. The corresponding induced polarization $\boldsymbol{P}^{\mathrm{s}}(\omega)$ which is the source of the scattered light is obtained from

$$\boldsymbol{P}^{\mathrm{s}}(\omega) = \varepsilon_0 \Delta\boldsymbol{\chi}\boldsymbol{E} = \Delta\varepsilon(\omega, q, \Omega)\boldsymbol{E}(\omega)\varepsilon_0 = \frac{\partial\varepsilon(\omega)}{\partial\varrho}\Delta\varrho(q, \Omega)\boldsymbol{E}(\omega)\varepsilon_0 \,. \tag{9.45}$$

For dielectric solids the real displacement field $u_k(r, t)$ or more precisely the Fourier transform of the strain tensor

$$S_{kl}(q, \Omega) = \frac{1}{2}[u_{k,l}(q, \Omega) + u_{l,k}(q, \Omega)] \tag{9.46}$$

has to be considered. In (9.46) $u_{k,l}(q, \Omega)$ is the Fourier transform of the derived displacement $\partial u_k(r, t)/\partial x_l$. The modulation of the DF is then expressed by the *photoelastic* tensor p_{mnkl}. This tensor is defined by the strain-induced change of the index of refraction

$$\frac{1}{n_{mn}^2} - \frac{1}{n_{mn0}^2} = p_{mnkl} S_{kl} \,. \tag{9.47}$$

With this definition we obtain for the modulation of ε

$$\Delta\varepsilon_{ij}(\omega, q, \Omega) = -\varepsilon_{im}(\omega)\varepsilon_{nj}(\omega)(\Delta\varepsilon^{-1}(q, \Omega))_{mn} \,, \tag{9.48}$$

where

$$(\Delta\varepsilon^{-1}(q, \Omega))_{mn} = p_{mnkl} S_{kl}(q, \Omega) \,. \tag{9.49}$$

For a longitudinal acoustic strain in an isotropic medium $S_{kk} = \Delta\varrho/\varrho$. This yields

$$\Delta\varepsilon(\omega, q, \Omega) = \varepsilon^2(\omega) p_{1122} \frac{\Delta\varrho(q, \Omega)}{\varrho} . \tag{9.50}$$

From the induced polarization (9.45) the same relations can be used to calculate the scattered intensity as described in Sect. 2.2 or as given for Raman scattering in (9.12) to (9.14). The Porto notation is applicable as well.

Labeling the scattering geometry by the unit vectors $\boldsymbol{e}^{\mathrm{i}}$ and $\boldsymbol{e}^{\mathrm{s}}$ for the incident and scattered light, respectively, we obtain for the light energy scattered per unit of time and per steradian for a scattering volume $\mathcal{V}$

$$\boxed{\frac{\mathrm{d}^{e^{\mathrm{i}}}\Phi_{e^{\mathrm{s}}}}{\mathrm{d}\Omega} = \frac{(\omega - \Omega)^4 \mathcal{V}_{\mathrm{u}} |\boldsymbol{e}^{\mathrm{s}} \Delta\varepsilon \boldsymbol{e}^{\mathrm{i}}|^2 I_{\mathrm{i}} \mathcal{V}}{16\pi^2 c_0^4}} . \tag{9.51}$$

In a more general treatment of the problem thermodynamic fluctuations must not only be considered for the density ϱ but also for the temperature T. In the case of solids rotational deformations R_{jl} may also add to the modulation of ε.

To proceed further a quantum-mechanical treatment of the electronic and structural transitions is needed. Instead of doing this we will introduce a thermodynamical approach and thus obtain at least the temperature dependence of the scattering process as in the case of the Raman effect.

The energy spectrum of the scattered light is obtained from the space average (ensemble average) of the absolute square of the scattered field $\overline{|\boldsymbol{E}^{\mathrm{s}}(r, \omega \pm \Omega)|^2}$ at the point of observation. This quantity is directly related to the average of the absolute square of the fluctuations. For the case of longitudinal acoustic modes (density fluctuations) in an isotropic material the energy spectrum becomes

$$\frac{\mathrm{d}^2(^{e^{\mathrm{i}}}\Phi_{e^{\mathrm{s}}})}{\mathrm{d}\epsilon \mathrm{d}\Omega} = \frac{(\omega - \Omega)^4 \mathcal{V} |\boldsymbol{e}^{\mathrm{s}} \boldsymbol{e}^{\mathrm{i}}|^2 I_{\mathrm{i}}}{16\pi^2 c_0^4} \left| \frac{\partial\varepsilon(\omega)}{\partial\varrho} \right|^2 S_\varrho(q, \Omega) \tag{9.52}$$

with

$$S_\varrho(q, \Omega) = \frac{1}{2\pi\hbar\tau_{\mathrm{obs}}} \overline{|\Delta\varrho(q, \Omega)|^2} , \tag{9.53}$$

where τ_{obs} is the time of observation. $S_\varrho(q, \Omega)$ is called the spectral density or the *power spectrum* of the fluctuations or alternatively the *dynamic form factor* of the system and relates directly to the power spectrum given by (2.47). It is a very general and very important quantity widely used to study scattering phenomena. It appears likewise in advanced descriptions of Raman scattering, Brillouin scattering, inelastic neutron scattering or electron scattering, and will be used again in a more general context in Chaps. 15 and 17. The Wiener–Khintchin theorem (3.46) relates the power spectrum to the autocorrelation function of the density fluctuations.

To describe the scattering process in crystals the spectral density (9.53) has to be expressed by the dynamical form factor for the displacement gradients

$$S_u(q,\Omega) \equiv \frac{1}{2\pi\tau_{\mathrm{obs}}} \overline{|\sum_{k,l} u_{k,l}(q,\Omega)|^2} \, . \tag{9.54}$$

In this case we have to replace $\Delta\varepsilon$ in (9.51) with (9.48) and (9.49). Note that the dynamical form factors in (9.53) and in (9.54) become independent of the time of observation if τ is long enough since the average of the square of the fluctuations increases linear with the observation time (see also Sect. 5.1).

To proceed further the dynamical form factor must be calculated. This was done for the first time by A. Einstein in 1910 for liquids. The result obtained for the (Ω integrated) dynamical form factor is

$$S_\varrho(q) = \mathcal{V}\varrho^2\beta_{\mathrm{T}}k_{\mathrm{B}}T \, , \tag{9.55}$$

where β_{T} is the isothermal compressibility (in m^2/N). The dynamical form factor for crystals is similarly

$$S_u(q) = \frac{\mathcal{V}k_{\mathrm{B}}T}{2\varrho v_{\mathrm{s}}^2} \, , \tag{9.56}$$

where v_{s} is the appropriate sound velocity. For a particular component of ε and S the expression for $\Delta\varepsilon/S_{ik}$ becomes $\varepsilon^2 p$ which yields finally for the scattering intensity per steradian and a scattering volume $\mathcal{V}$

$$\boxed{\frac{\mathrm{d}\Phi_{\mathrm{s}}}{\mathrm{d}\Omega} = \frac{(\omega-\Omega)^4}{c_0^4 16\pi^2} n^8 p^2 \frac{\mathcal{V}k_{\mathrm{B}}T}{2\varrho v_s^2} I_{\mathrm{i}} \, .} \tag{9.57}$$

Inserting approximate numerical values into (9.57) yields $\Phi'_{\mathrm{s}}/I_{\mathrm{i}} \approx 10^{-8}$–$10^{-10}$, per unit volume.

The Brillouin scattering process is likewise often used to study the propagation of coherent sound waves. If S is the strain amplitude of the sound its intensity is $I_{\mathrm{sound}} = \varrho v_{\mathrm{s}}^3 S^2/2$. The scattering intensity for a sound beam of width B is given in this case by

$$\boxed{\frac{\Phi'_{\mathrm{s}}}{I_{\mathrm{i}}} = \frac{(\omega-\Omega)^2}{4\pi^2 c_0^2 v_{\mathrm{s}}^2} n^6 p^2 B^2 I_{\mathrm{sound}} \, .} \tag{9.58}$$

This is known as *Debye–Sears scattering* or the *Raman–Nath limit.* Since in this case I_{sound} can be very large scattering of light can be very efficient. Such systems are used for light modulation or for sweeping of light beams.

9.3.2 Experimental Results of Brillouin Scattering

Figure 9.17 shows a Brillouin spectrum of SbSI as measured with a multipass Fabry–Perot interferometer. T_1, T_2, and L label the sidebands originating from an inelastic scattering by the two transverse acoustic and the longitudinal acoustic phonon modes. From the scattering geometry the q vector of the phonons can be determined, which enables the calculation of the sound velocities from the shift of the sidebands with respect to the central line. The sidebands which are not assigned in the spectrum are replicas originating from the small value of less than two wave numbers chosen for the free spectral range of the interferometer.

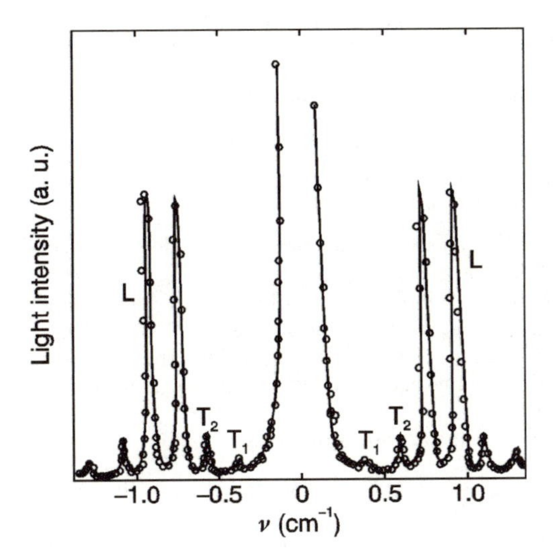

Fig. 9.17. Brillouin spectrum of SbSI as measured with a five-pass Fabry–Perot interferometer. (T_1, T_2: transverse acoustic modes, L: longitudinal acoustic mode); after [9.11].

9.3.3 Rayleigh Scattering

Rayleigh scattering is a diffuse propagation of light with hardly any frequency shift. The temperature-induced fluctuations of the density are responsible for the modulation of the response function. Accordingly Rayleigh scattering is observed very close to the excitation line.

Problems

9.1 Which geometry is required to observe the B_{1g} species in $\boldsymbol{D}_{2h}$?

(Purpose of exercise: learn to use selective geometries for scattering experiments (easy example).)

9.2 The Raman scattering is observed for a calcite crystal (space group $\boldsymbol{D}_{3d}$) in 180° backscattering in the {110} direction and for geometries $(\|, \|)$ and $(\|, \perp)$. Which modes can be seen and which components of the Raman tensor contribute? Hint: For 180° backscattering $\|$ and $\perp$ refer only to the relative orientations of the E vector for the incident and scattered light.

(Purpose of exercise: learn to use selective geometries for the scattering experiment (advanced example).)

9.3* For the same scattering geometry as in Problem 9.2 a polymer with the point group $\boldsymbol{D}_{2h}$ is investigated. The chains are oriented along x by stretching but randomly arranged in the yz plane. Which modes become visible as compared to the fully oriented polymer?

(Purpose of exercise: training with selection rules for analysis of experiments.)

9.4 Using Maxwell's equations show that the electric field of a polar mode is given by (9.17).

(Purpose of exercise: recall the usefulness of Maxwell's equations.)

9.5* Calculate the dispersion relation $\Omega = f(q)$ for polar modes by using the Kramers–Heisenberg dielectric function and evaluate explicitly the behavior of very large and very small q.

(Purpose of exercise: explore the usefulness of a simple DF.)

9.6 Show that the average of the occupation numbers $(v_k + 1)W(\epsilon_k)$ is $n_k + 1$. Hint: Use the relation $a\exp(-ax) = -\mathrm{d}\exp(-ax)/\mathrm{d}x$.

(Purpose of exercise: prove a very important relation.)

9.7 Calculate the transition probability (decays per second) for the decay process of an optical phonon $\Omega(q)$ into two LA modes with frequency $\Omega' = \Omega'' = \Omega(q)/2$. Hint: The probability $\Gamma = \mathrm{d}w/\mathrm{d}t$ is proportional to the difference in the product of the final and initial occupation numbers. These products are $(n_q + N)(n_q' + 1)(n_q'' + 1)$ and $(n_q + N + 1)(n_q')(n_q'')$, respectively. The equilibrium occupation numbers n_q, n_q', n_q'' are given by the Bose–Einstein distribution (9.32). N is a small perturbation to the thermal population of n_q.

(Purpose of exercise: learn to handle occupation numbers.)

9.8 Discuss the depolarization factors $\varrho_\parallel$ and $\varrho_\perp$ for a 180° backscattering geometry.

(Purpose of exercise: note the differences for backscattering and 90° scattering.)

9.9 The carrier concentration in a sample of $YBaCu_3O_{7+\delta}$ is $7\times10^{21}\,\mathrm{cm}^{-3}$. Up to which wave number can scattering from free carriers be expected if the excitation is performed with a green laser?

(Purpose of exercise: estimate experimental conditions for light scattering.)

10. Infrared Spectroscopy

Infrared spectroscopy is one of the most popular spectroscopic techniques in solid-state physics. The simple reason for this is that nearly all materials exhibit a more or less expressed structure of the absorption in the IR spectral range. The origin of these structures has already been discussed to some extent in Chap. 7. Absorption processes due to transitions across the energy gap, from excitons or from impurity states, are found in the visible spectral range as well as in the IR. Important additional sources for absorption and reflection are the IR active phonons or vibrational modes which can give valuable supplementary information to results from Raman scattering. In Chaps. 8 and 9 we have already discussed under which conditions vibrational modes are observable in an IR spectrum and in Sect. 6.3 we have even given a mathematical description of the response function in the form of the Kramers–Heisenberg dielectric function. In the preset chapter particular attention will be paid to special instrumentation not discussed previously, to advanced problems and to several examples from solid-state physics. The first two sections elucidate the characteristic difference between radiation sources, optical components, spectrometers, and detectors for the visible and the IR spectral ranges. To do this it is useful to divide the spectral range into three sections:

- near infrared (NIR) 0.8–10 μm,
- middle infrared (MIR) 10–40 μm,
- far infrared (FIR) 40–1000 μm,

The breakthrough in modern IR spectroscopy was the development of Fourier spectrometers. With this spectroscopic technique it is no longer necessary to disperse the probing light into its spectral components. The whole light energy is always measured simultaneously, and only after the experiment disentangled mathematically into its spectral components. According to its dominating role Fourier spectroscopy will be discussed extensively in a special section. Finally, examples from solid-state physics will be used to demonstrate the power and broadness of applications of this technique.

10.1 Radiation Sources, Optical Components, and Detectors

As in the visible spectral range standard radiation sources are black-body emitters at high temperatures. Even though the maximum of radiation is shifted to shorter wavelengths with increasing temperature (Fig. 3.1), the absolute amount of emission in the IR still increases about linearly with T. This means even here high-temperature sources are an advantage, but for an efficient radiator in the spectral range around 100 μm very good filtering of the emission in the visible is required. A simple source operating on the basis of the black-body radiation is the *glowbar*. As the name indicates, it consists of a SiC rod with the dimensions 2 cm (length) × 0.5 cm (diameter), which is heated by about 5 A to 1450 K. The glowbar is generally used in the spectral range up to 40 μm, in extreme cases even up to 100 μm. Since it is small and extends in one direction, it can easily be imaged onto the entrance slit of a monochromator. The glowbar is operated in vacuum.

For more sophisticated applications or for applications in the FIR gas plasmas must be used as radiation sources. A hot gas plasma emits a long-wavelength continuum, in addition to the characteristic lines of the atoms. This continuum originates from collisions of electrons with ions or neutral particles. The special advantage of plasma radiators is their smaller emissivity in the visible or NIR as compared to a black body at the same temperature. Only in the FIR the emission becomes equivalent so that the ratio between emission in the FIR and the emission in other spectral ranges is much higher than that of a solid black-body radiator.

The plasma emission has a characteristic spectrum. For a fully ionized plasma it is independent of the wavelength but proportional to $n_\mathrm{e}^2 d/\sqrt{T}$, where n_e is the density of the electrons and d the thickness of the plasma. A drop in the emission intensity on the short-wavelength side starts only at $hc/\lambda_\mathrm{T} = \hbar\omega_\mathrm{T} = k_\mathrm{B}T$. On the long-wavelength side the emission starts to drop if either the emission of the black body or the plasma edge at

$$\lambda_\mathrm{p} = 2\pi c_0/\omega_\mathrm{p} = 2\pi\sqrt{c_0^2\varepsilon_0 m_\mathrm{e}/n_\mathrm{e}e^2} \tag{10.1}$$

is reached. The plasma is highly reflecting for wavelengths larger than λ_p and has therefore also a low emissivity. This is an immediate consequence of the Kirchhoff laws. Figure 10.1 shows the calculated emission from two plasmas with different thicknesses as a function of the wavelength. The straight line on the long-wavelength side of the emission represents the black-body radiation. The flat tops of the curves in the central part represent the above-mentioned independence of the emission from the wavelength. The dashed curve is an experimental result.

Even though the plasma emission decreases as $1/\sqrt{T}$ in reality it is advantageous to choose the temperature of the plasma as high as possible. This is due to the fact that the degree of ionization of the plasma for realistic temper-

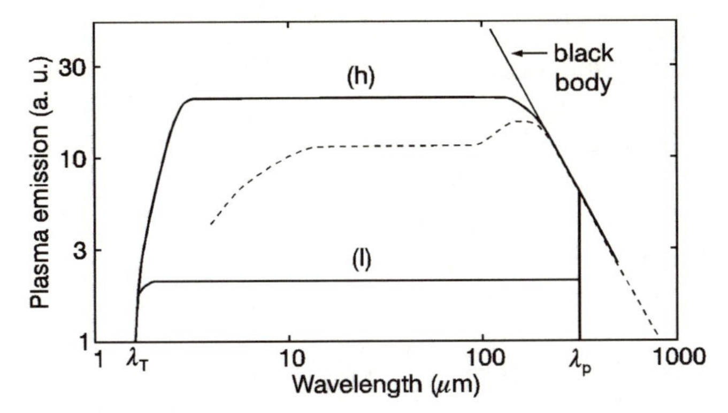

Fig. 10.1. Schematic representation of the emission of a plasma with high thickness (h) and with low thickness (l). The *dashed line* is the observed emission from a mercury arc; after [10.1].

atures is still far from complete. The gain in electron density with increasing temperature more than outweighs the loss due to the factor $1/\sqrt{T}$.

A high-pressure mercury arc lamp in a quartz tube, as discussed in Sect. 3.1, is an appropriate representation of a plasma source. However, even though the temperature reaches 6000 K the degree of ionization is only 1%. Inspite of this, the spectral distribution of the emission follows well the idealized behavior discussed above, as seen from the dashed line in Fig. 10.1.

Besides thermic sources lasers play an important role in the whole IR spectral range. Their dominance is, however, not as clear as was the case for Raman scattering. The lack of the possibility to tune lasers over a sizable spectral range is the big drawback. On the other hand, semiconductor lasers cover a wide spectrum not only in the near but also in the MIR or even in the FIR spectral range if properly constructed. The advantage of such laser systems was already mentioned in Sect. 3.4. Since the absorption lines of several MIR and FIR lasers cover the spectral positions of important gas molecules applications in gas analysis or pollution detection in the atmosphere are widespread. Laser systems on the basis of PbTe or $Cd_xHg_{1-x}Te$ are appropriate. This is demonstrated in Fig. 10.2 where the range of emission for several laser systems is compared with the absorption characteristics of important gases.

For cyclotron-resonance experiments HCN lasers (372 μm) or H_2O lasers (47–220 μm) are important. Finally, the CO_2 laser with an emission around 10 μm is one of the most powerful laser systems altogether, as already discussed in Sect. 3.4. Because of the many other broad-band sources in this spectral range the importance of the CO_2 laser for spectroscopy has decreased.

The possibility to obtain tunable and still highly monochromatic radiation comes from frequency doubling of microwaves. Microwave generators like clystrons or magnetrons can be very powerful which enables the generation of overtones up to the 15th order. This means that from a 58-GHz magnetron

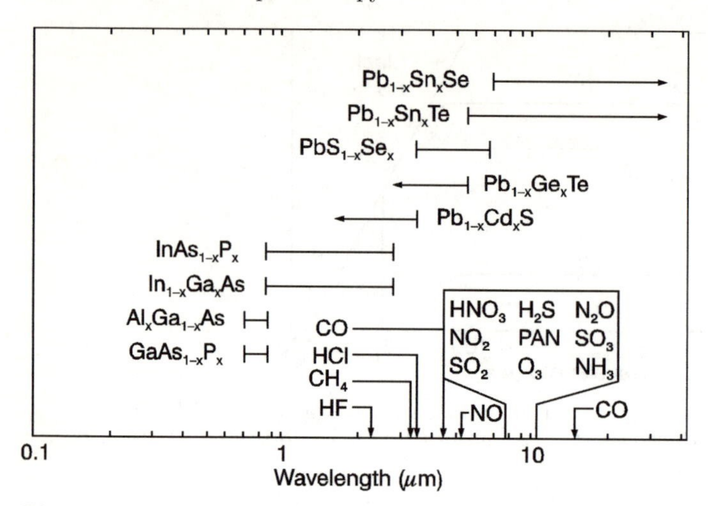

Fig. 10.2. Spectral range of various semiconductor lasers and position of absorption for various gases; after [10.2].

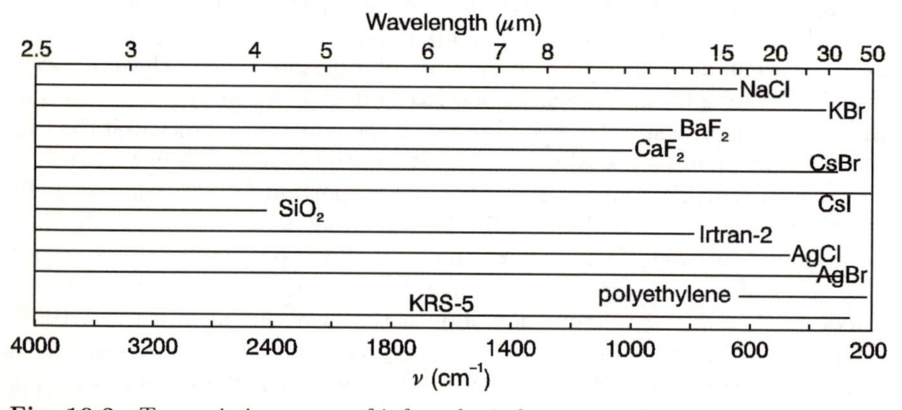

Fig. 10.3. Transmission range of infrared window material.

source radiation with 500-μm wavelength can be obtained. Tuning the magnetron and selection of the appropriate overtone allows a continuous tuning in a wide frequency range.

Using optical components in an IR beam line attention must be paid to the transmission properties of the materials. The longer the wavelength of the radiation the more difficult it becomes to find appropriate materials for windows and lenses. The limitation in the transmission originates from the reststrahlen absorption by lattice vibrations. The heavier the atomic components of the materials the further to the FIR the material can be used. On the other hand, organic materials like polyethylene have no strong absorption lines in the FIR and are therefore well accepted for optical components below

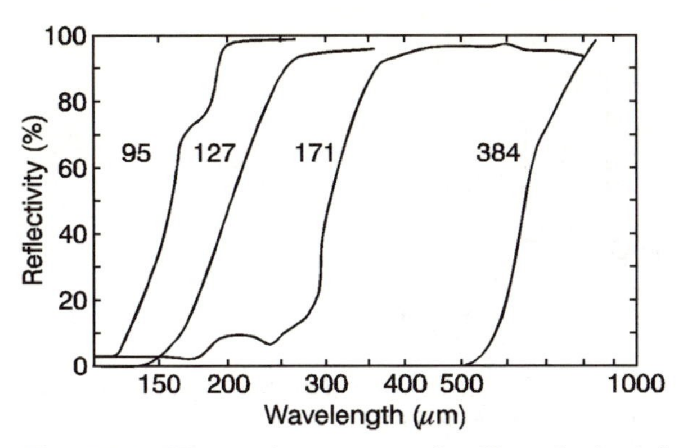

Fig. 10.4. Wire mesh grating as edge filters for far-infrared radiation. The numbers indicate distances between the wires in μm; after [10.3].

about 600 cm^{-1}. Figure 10.3 shows transmission ranges in the IR for various materials.

All materials for which the transmission drops suddenly to zero at a particular wavelength are appropriate for edge filters. In the FIR wire meshes can be used. High reflectivity starts for wire meshes when the wavelength of the light increases beyond the dimension of the meshes. Figure 10.4 is an example. If one component of the mesh is omitted (only linear wires) the system acts as a linear polarizer.

The large value for the wavelength in the FIR allows one to use even simple, highly polished tubes for guiding the light. If the tube is conical the propagating light can be concentrated to a spot.

IR detectors were already discussed in Chap. 5 in connection with photoelectric detectors. Because of the low sensitivity of the latter, particularly in the FIR, and because of the low light powers available in this spectral range other detectors are also frequently used. Examples are various forms of thermal detectors like bolometers or pyroelectric crystals. An often-used system is the *Golay* detector (Fig. 10.5). It operates on a pneumatic principle. The incident IR light is absorbed by a thin film. The generated heat increases the pressure in the gas chamber, which drives a mirror. The mirror is part of an optical system which images a grating onto itself. Any small motion of the mirror leads to a change in the overlap between grating and image and thus gives a signal to the detector. As compared to low-temperature bolometers the Golay detector has a low detectivity of only $5\times 10^9\,W^{-1}$ and is rather slow. On other hand, it is simple to operate and can be used up to 1000 μm. The detector can only be employed for alternating signals, at best between 3 and 10 Hz.

Another simple but often used detector works on the basis of the pyroelectric effect. Crystals with a permanent electric dipole moment respond to a sudden change in the dipolar order with the generation of compensating

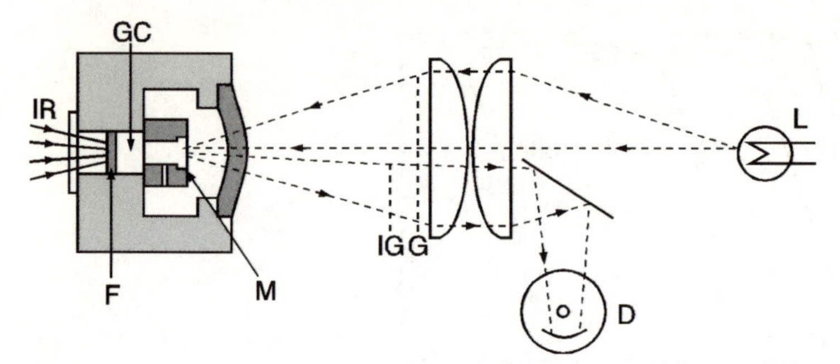

Fig. 10.5. Schematic arrangement of a Golay detector. (IR: far-infrared beam, F: film, GC: gas chamber, M: mirror, G: grating, IG: image of grating, L: light source (visible), D: detector); after [10.4].

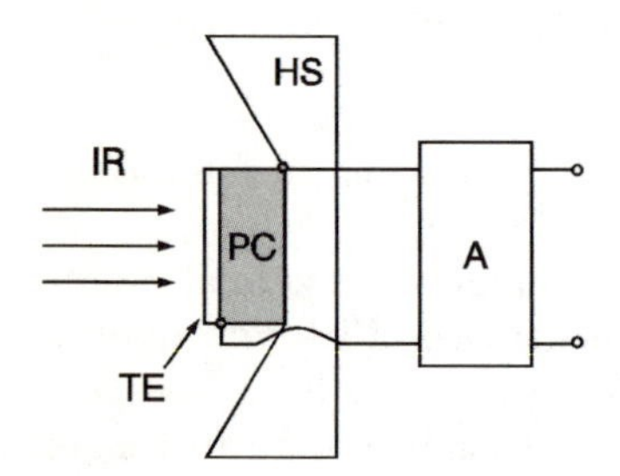

Fig. 10.6. Basic components of a pyroelectric detector. (IR: IR beam, PC: pyroelectric crystal, TE: transparent electrode, HS: heat sink, A: amplifier).

surface charges. IR or heat pulses can be the origin of such induced disorder. The voltage accompanying the compensation charges can be used to detect the IR or heat pulse. A well known crystal for such detectors is triglycine sulphate (TGS). Figure 10.6 shows the basic components. A good heat sink on the backside electrode is important. Even though the detector relies on a thermic effect it can be rather fast. The time constant is given by the ratio between heat capacity H and heat conductance G to the heat sink. Time constants can be as short as 10^{-5} s with a sensitivity of 100 V/Watt and a detectivity of 10^8–10^9 W^{-1}.

More sophisticated but also more elaborate detectors are the low temperature Ge bolometers. Their operation is based on the temperature-induced change in the conductivity of a Ge crystal cooled to 4.2 K. Figure 10.7 illustrates the basic construction. The IR light hits the Ge crystal through a cold filter which is supposed to reduce the background radiation from the environment. The change of the current through the crystal is measured by the voltage drop across a cooled resistor. A widely used dopand for the Ge crystal is Ga. In a broad spectral band around 100 μm BLIP condition is reached, which means a detectivity of $10^{12}\,\mathrm{W}^{-1}$. For a spectral range of 1000 μm even $10^{14}\,\mathrm{W}^{-1}$ can be obtained. This value already approaches the detectivity of photomultipliers in the visible. However, the signal level is very low which means low noise amplifiers are required to reach high detectivity. This limits the bandwidth for the measurement to 10 Hz. By counterdoping

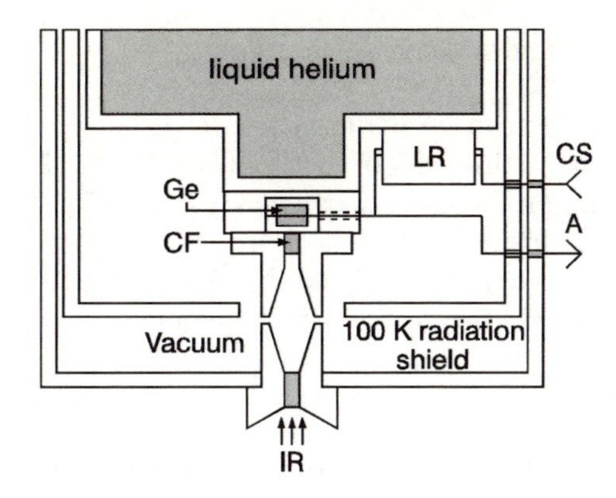

Fig. 10.7. Construction of a Ge bolometer. (Ge: Ge crystal, CF: cold filter, LR: load resistor, CS: current supply, A: amplifier).

with Sb the detectivity is slightly reduced but the bandwidth increases to 500 Hz.

Really fast experiments with time constants of the order of 10^{-8} s can only be performed with detectors based on photoconduction. Ge doped with Cu, Zn, or Ga leads to flat impurity levels with ionization energies between 10 and 40 meV. Such detectors can be operated at 4.2 K up to 150 μm (60 cm^{-1}). In InSb detectors the strong temperature dependence of the carrier mobility rather than the change of the carrier concentration is used to probe the heating of the sample. Such systems are known as InSb-transformer detectors. Figure 10.8 compares various detectors which are particularly useful in the FIR.

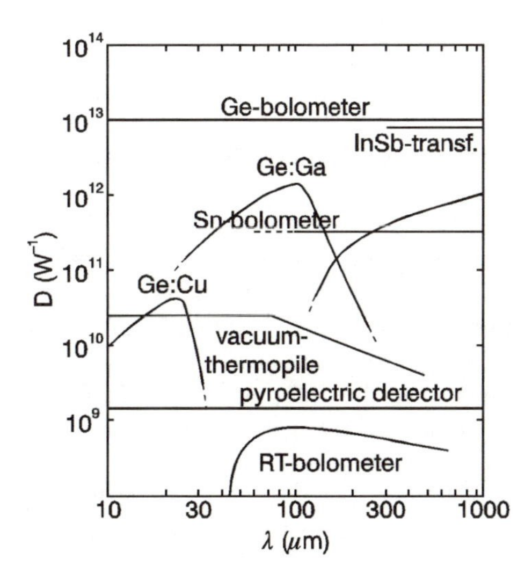

Fig. 10.8. Detectivity for far infrared detectors; after [10.5].

10.2 Dispersive Infrared Spectroscopy

Like optical spectroscopy IR spectroscopy can be performed with dispersive instruments. Prism spectrometers and grating spectrometers from Chap. 4 are appropriate. Both spectrometers are problematic, particularly in the FIR. Away from the absorption range the prisms have a very low dispersion which leads to very low light intensities for a required bandpass $\delta\lambda$. According to (4.6) the power in the spectral bandpass $\delta\lambda$ incident on the detector is

$$P_\lambda \delta\lambda = T I_0 E \delta\lambda = T I_0 \frac{A^2}{F^2} W H \delta\lambda \, , \tag{10.2}$$

where according to (4.19) $\delta\lambda = (\mathrm{d}\theta/\mathrm{d}\lambda)^{-1}(W/F)$. Using W from this relation yields

$$P_\lambda \delta\lambda = T I_0 \frac{A^2}{F} H (\delta\lambda)^2 \frac{\mathrm{d}\theta}{\mathrm{d}\lambda} \, . \tag{10.3}$$

This means for small values of $\mathrm{d}\theta/\mathrm{d}\lambda$ the energy available for a predefined bandpass becomes very small. A prism which is very good at $180\,\mathrm{cm}^{-1}$ because it has a high dispersion can be very bad at $1000\,\mathrm{cm}^{-1}$. Another disadvantage of the prisms is the limited applicability in the FIR because of *reststrahlen* absorption as already discussed in Sect. 4.2.2. Appropriate crystalline materials with their frequency limit in parentheses are NaCl ($650\,\mathrm{cm}^{-1}$), KBr ($400\,\mathrm{cm}^{-1}$), CsBr ($280\,\mathrm{cm}^{-1}$), and CsI ($180\,\mathrm{cm}^{-1}$).

Using (10.3) and the definition for the etendue in (4.5) together with (4.6), (4.10), and (4.13) we can find this value for the prism spectrometers from

$$\begin{aligned} E^{\mathrm{Pr}} &= \frac{A^2}{F} H \delta\lambda \frac{\mathrm{d}\theta}{\mathrm{d}\lambda} = \frac{A R_0 H}{F} \delta\lambda \\ &= \frac{A R_0 H}{F \nu^2} \delta\nu \, . \end{aligned} \tag{10.4}$$

Apparently, this is formally the same result as was derived for the grating spectrometers in (4.21).

Due to the larger wavelength of the IR light for grating spectrometers the requirements on the gratings are not as severe as they are in the visible spectral range. On the other hand the gratings must be much larger to accommodate the required number of lines for a good interference. Dimensions of $30 \times 30\,\mathrm{cm}^2$ and 100 lines per grating are quite common.

To avoid the distribution of light intensity on the different orders of diffraction, gratings which are *blased* for a particular wavelength (*echellet* gratings) are in use. As shown in Fig. 10.9, such gratings have a stepped surface. For these gratings the intensities are no longer distributed to all orders of diffraction, rather they are concentrated to the order which comes closest to the specular reflection of the long part of the step. For a constructive interference the grating equation (4.14) and simultaneously also the relationship $\theta_{\mathrm{i}} + \alpha = \theta_{\mathrm{d}} - \alpha$ must be satisfied as can be inferred from Fig. 10.9. Thus the equation for the echellet grating is

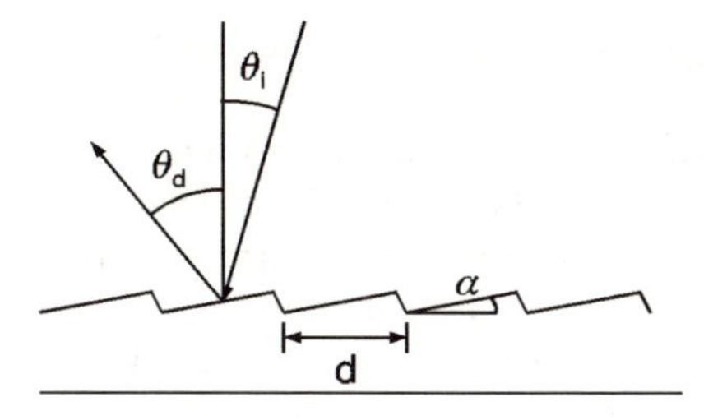

Fig. 10.9. Diffraction of an echellet grating.

$$m\lambda = 2d\sin\alpha\cos(\theta_\mathrm{i} - \alpha)\;. \tag{10.5}$$

To record spectra fixed slits and a rotating grating are used as in optical spectroscopy. The light path is limited to an angular spread of $\theta_\mathrm{i} + \theta_\mathrm{d} = 2(\theta_\mathrm{i} + \alpha) \approx 20°$. Since measurements are usually performed in first order, (10.5) defines an optimum wavelength λ_b termed the *blase* wavelength. Radiation with wavelength λ_b can be analyzed in an optimum way in first order. If λ becomes much smaller, e.g., equal to $\lambda/2$ also light from second-order diffraction will hit the detector, or more generally, radiation with wavelength λ/m will be observed in the m-th order. This means using echellet gratings only small ranges of the spectrum of the order of

$$\frac{2}{3}\lambda_\mathrm{b} \leq \lambda \leq \frac{3}{2}\lambda_\mathrm{b}$$

can be analyzed. The rest must be cut off with filters. This is not a serious problem in the visible since with $\lambda_\mathrm{b} \approx 0.6\,\mu\mathrm{m}$ the whole visible spectral range is covered. To cover, on the other hand, the spectral rang from 5–240 cm^{-1} five gratings blased for 1300, 650, 325, and 80 μm are required for given values of α and $\theta_\mathrm{i}+\theta_\mathrm{d}$. In addition, for each grating the other spectral ranges must be filtered. This means that spectroscopy becomes very laborious and difficult. Accordingly spectrometers in this spectral range are only operated down to 30 cm^{-1}.

10.3 Fourier Spectroscopy

It is obvious from the above discussion that in the MIR and FIR spectral range non-dispersive spectrometers have a great advantage. As a consequence of the high capacity and speed of today's computers this non-dispersive or interferometric spectroscopy is already entering the visible spectral range as well and is known as *Fourier* spectroscopy. In the following the basic principles and the operation conditions for this very important spectroscopic technique will be discussed.

10.3.1 Basic Principles of Fourier Spectroscopy

Fourier spectroscopy is based on the Michelson interferometer shown in Fig. 10.10. The white light from the source located at the focus of lens L_1 is separated into two parts by the beam splitter. The reflected part is focused onto the detector D after reflection from the stationary mirror M_1 and after a second split by the beam splitter. The transmitted part of the light is also focused onto the detector after it was reflected from the mirror M_2 and split again by the beam splitter. The mirror M_2 is mobile and can glide a distance Δx. In this way interference fringes develop at the detector. Their intensity $I(x)$ depends on the position x of the mirror M_2. $I(x)$ is termed *interferogram function.* In contrast to the multiple beam interference occurring with diffraction from the gratings or from Fabry–Perot plates, here the interference is only between two beams.

If the incident wave is monochromatic of the form $E(x,t) = E_0 \cos(kx - \omega t)$ the field E_D at the detector is

$$E_\mathrm{D} = \frac{1}{2}\{E_0 \cos(\omega_0 t - k_0 x) + E_0 \cos[k_0(x + 2\Delta x) - \omega_0 t]\} , \tag{10.6}$$

where $2\Delta x$ is the optical path difference between the two beams (twice the shift of the gliding mirror). Choosing $x = 0$ and replacing $2\Delta x$ by $2x$ yields from (2.45) for the intensity at the detector

$$I(x) = c_0 \varepsilon_0 \langle E^2 \rangle = \frac{c_0 \varepsilon_0}{4} E_0^2 [1 + \cos(4\pi\nu_0 x)] , \tag{10.7}$$

where we have replaced k_0 by $2\pi\nu_0$. Rewriting this equation by using a spectral intensity $I(\nu) = \varepsilon_0 c_0 E_0^2 \delta(\nu - \nu_0)/2$ yields

$$I(x) = \frac{1}{2}\int_0^\infty I(\nu)[1 + \cos(4\pi\nu x)]\mathrm{d}\nu . \tag{10.8}$$

Generalizing this equation to an arbitrary intensity spectrum $I(\nu)$ yields the basic relationship for Fourier spectroscopy

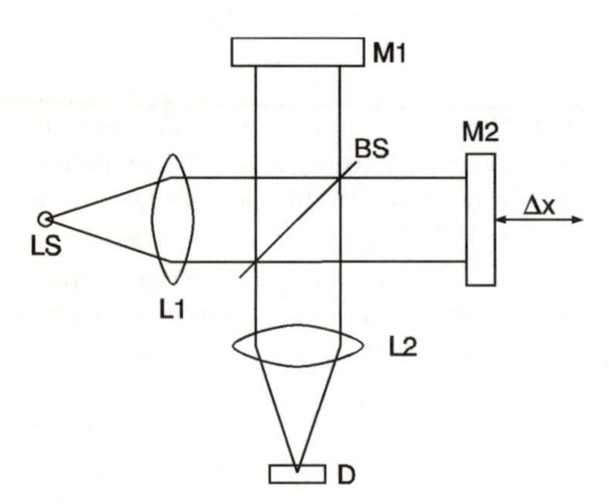

Fig. 10.10. Optical path in a Michelson interferometer; (LS: light source, L: lenses, M: mirrors; BS: beam splitter).

$$I'(x) = I(x) - \frac{1}{2}\int_0^\infty I(\nu)\mathrm{d}\nu = \frac{1}{2}\int_0^\infty I(\nu)\cos(4\pi\nu x)\mathrm{d}\nu\,. \tag{10.9}$$

The interferogram function $I(x)$ or $I'(x)$ contains the whole information about the spectrum $I(\nu)$. In fact, $I'(x)$ is the Fourier transform of $I(\nu)$ performed with a cosine function. The observed intensity $I(x)$ oscillates around an average intensity $\int I(\nu)\mathrm{d}\nu/2 = I_0/2$ which is exactly half of the original total intensity of the beam. For $x = 0$ it reaches its maximum value of I_0 as immediately seen from (10.8). This position corresponds to zero optical path difference. It is called the *white lightposition.* For $x \to \infty$ the coherence of the radiation is lost. According to Fig. 10.10 the intensity at the detector becomes then $I_0/2$. A Fourier transformation of $I'(x)$ yields for $x = y/2$

$$\begin{aligned}&\int I'(y/2)\cos(2\pi\nu' y)\mathrm{d}y\\ &\quad = \frac{1}{2}\int_0^\infty I(\nu)\mathrm{d}\nu \int \cos(2\pi\nu y)\cos(2\pi\nu' y)\mathrm{d}y = \frac{I(\nu')}{2}\,,\end{aligned} \tag{10.10}$$

since the integration over y gives $\delta(\nu - \nu')$.

This equation means we obtain the spectral components of the light directly from the interferogram by Fourier transformation, without any spectral dispersion. As compared to dispersive spectroscopy the procedure has two basic advantages, known as *energy advantage* and *multiplex advantage.* The energy advantage originates from the fact that during the whole period of measurement nearly always the total beam intensity hits the detector. This means the detection operates on a high signal level which improves the signal-to-noise ratio, particularly for weak radiation sources. The multiplex advantage originates from the simultaneous measurement of the full spectrum during the whole period T of detection. In contrast in dispersive spectroscopy N parts of widths $\Delta\nu$ of the spectrum will be measured successively so that for each part only the time $T' = T/N$ is available. The signal-to-noise ratio would be smaller by $1/\sqrt{N}$. The need for a mathematical process in the form of a Fourier transformation after registration of the interferogram is not really a drawback. The dramatic developments in computer capacity enables the transformations to be carried out in a very short time even for very large data sets. Another great advantage of the Fourier spectrometers is their higher brightness as derived from (10.15).

Figure 10.11 displays the interferogram for a transmission spectrum of a polyethylene film. Of the 4096 points measured for every 0.632 μm only 950 points are plotted. As expected the signal has a maximum at the white-light position in the center of the interferogram and reaches $I_0/2$ for large distances x. The corresponding distribution of the IR light is shown in Fig. 10.12. The upper curve is a single-beam spectrum obtained immediately after the

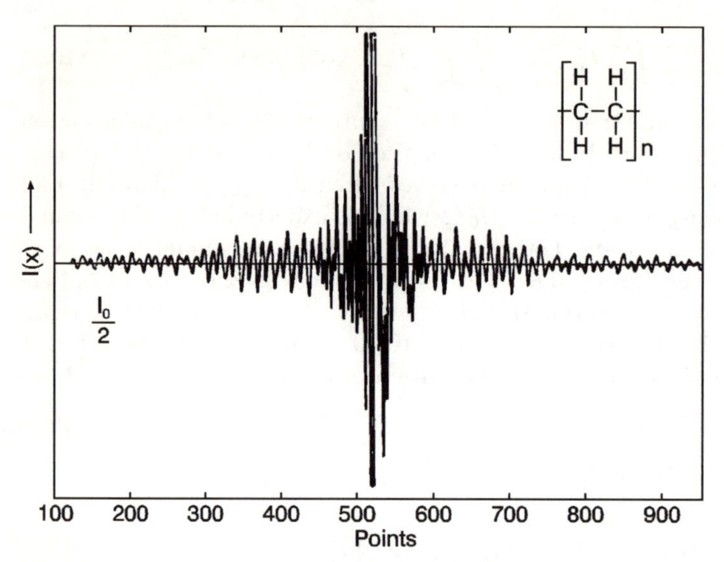

Fig. 10.11. Interferogram of transmission through a polyethylene film. The insert shows the structure of the polymer.

Fourier transformation of the interferogram. The strong absorption lines at 2900, 1450, and 730 cm^{-1} are characteristic vibrational frequencies of the polymer. They correspond to a CH-stretching, a CC-deformation, and a CH-deformation oscillation in the plane of the polymer backbone. The last two lines are split by interchain interaction. This is not seen in the figure since the spectral resolution is too low.

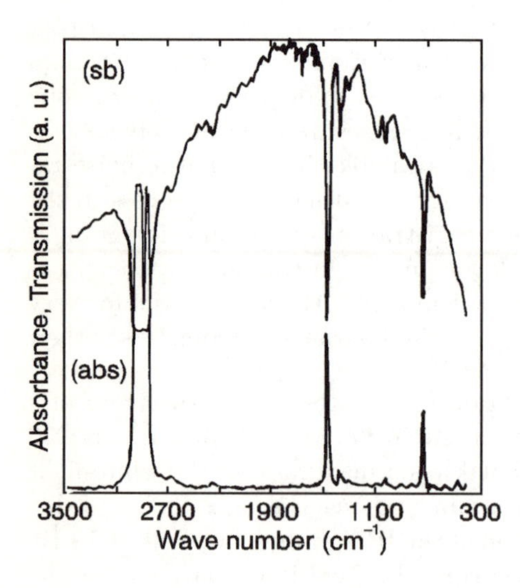

Fig. 10.12. Single-beam spectrum (sb) and absorbance (abs) for the interferogram of Fig. 10.11.

The overall shape of the curve (sb) in Fig. 10.12 is mainly determined by the characteristics of the light source and by the characteristics of the optical elements in the light path, including the detector. If the spectrometer is operated under ambient conditions additional absorption lines appear from CO_2 and H_2O in the atmosphere. Since these lines have nothing to do with the properties of the solid the single-beam result is usually divided by the background spectrum. The logarithm of this ratio is plotted versus the light energy (in wave numbers or meV). The quantity obtained in this way is the *absorbance*. It gives a much better characterization of the absorption lines as compared to the single-beam spectrum. The absorbance is plotted in Fig. 10.12 as the lower spectrum. The procedure anticipates that the observed transmission is directly related to the absorption according to the Lambert law (6.8), and reflections have only a minor influence on the spectrum.

10.3.2 Operating Conditions for Fourier Spectrometers

As for the dispersive spectrometers the resolution of the Fourier spectrometers can be expected to depend on deviations from ideal conditions of operation. Two facts are crucial: the finite value for the shift of the mirror and the limited number of registrations during the scan.

Since the mirror cannot be shifted from $-\infty$ to ∞ the Fourier transformation is truncated. For a shift from the positions $-x_{\max}$ to $x_{\max}$ and a monochromatic light with $I(\nu) = I_0\delta(\nu - \nu_0)$ the calculated spectral intensity $I'(\nu')$ yields from (10.10)

$$\begin{aligned} I'(\nu') &= \frac{I_0}{2}\int_0^{\infty}\int_{-x_{\max}}^{x_{\max}} \delta(\nu-\nu_0)\cos(4\pi\nu x)\cos(4\pi\nu' x)\,\mathrm{d}x\,\mathrm{d}\nu \\ &= \frac{I_0 x_{\max}}{2}\frac{\sin[4\pi x_{\max}(\nu_0-\nu')]}{4\pi x_{\max}(\nu_0-\nu')}\,, \end{aligned} \tag{10.11}$$

instead of $I_0\delta(\nu' - \nu_0)/2$. Thus, for a monochromatic light wave the resulting intensity spectrum is not a δ function but a function of the form $\sin x/x$ as depicted in Fig. 10.13. Note that the evaluated intensity can have negative values. This is, of course, a consequence of the approximation. The spectral resolution is given by the distance $\delta\nu$ between the central maximum at $\nu' = \nu_0$ and the first minimum at $\nu'_{\min}$. From (10.11) the minimum is found at

$$\delta\nu = \nu_0 - \nu'_{\min} \approx \frac{0.7}{2x_{\max}}\,. \tag{10.12}$$

At this value of $\nu - \nu'$ the calculated intensity $I'(\nu')$ is negative. A good estimate and an easy-to-remember number for the resolution is the inverse of the total scan width $2x_{\max} = \Delta x$ or

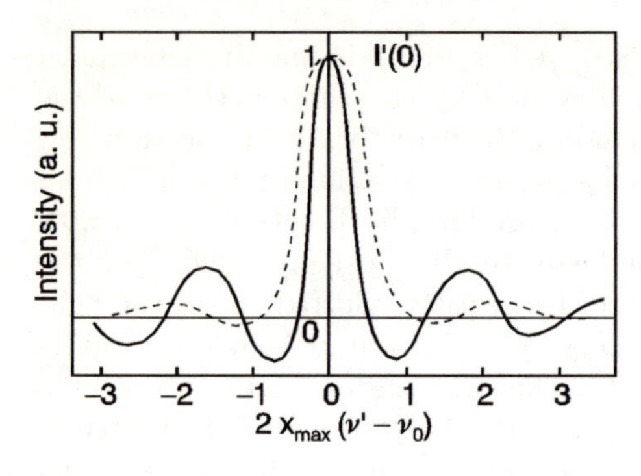

Fig. 10.13. Spectral intensity evaluated for monochromatic light with a Fourier transformation over a limited range in space; (—) without apodization, (− − −) with apodization.

$$\boxed{R_0 = \frac{\nu}{\delta\nu} = \Delta x \nu \ .} \tag{10.13}$$

As expected the resolution increases directly with the scanning distance during registration.

Optimum resolution is only obtained if the size of the entrance pinhole to the interferometer and the focal length F of lens L_1 in Fig. 10.10 are properly tuned. From a theoretical analysis the required relationship is $r = F\sqrt{2/R_0}$ where r is the radius of the entrance pinhole and R_0 is the resolution. This yields for the brightness (etendue) of the interferometer

$$E^{\mathrm{Fu}} = \frac{r^2 \pi A^2}{F^2} = \frac{2\pi A^2}{R_0} = \frac{2\pi A^2}{\nu} \delta\nu \ , \tag{10.14}$$

where A is the beam diameter in the spectrometer. Comparing this value with the brightness of the grating spectrometer from (4.21) gives convincing evidence of the advantage of Fourier spectrometers

$$\boxed{\frac{E^{\mathrm{Fu}}}{E^{\mathrm{Gr}}} = \frac{2\pi F}{H} \ ,} \tag{10.15}$$

where we have used $R_0 \approx A/d \approx A/\lambda$ like in (4.21). Since H and F were the height of the slit and the focal length of the grating spectrometer, respectively, the ratio in (10.15) is of the order of 500. This enhancement of the brightness of Fourier spectrometers over the grating and prism spectrometers is known as the *Jacquinot* advantage.

Besides the finite width of the line at the center position also the side maxima of the $\sin x/x$ curve are disturbing. They originate from the abrupt cutoff of the interferogram at $-x_{\max}$ and $x_{\max}$. To reduce this problem the jump can be smoothed by multiplication of the interferogram with $1 - |x/x_{\max}|$ or any other appropriate function before subjecting it to the Fourier transformation. Even though some of the resolution is lost by this process the side maxima of the $\sin x/x$ curve are truncated, as shown by the dashed line in Fig. 10.13. Because of the cutting of the sidebands the process is called *apodization*.

Another problem of the analysis is the lack of an analytical function for $I(x)$. Since the interferogram is only known pointwise the Fourier transformation can only be performed with finite Fourier sums. For each frequency ν' the intensity is obtained from

$$I(\nu') = x_0 \sum_{m=-n}^{n} I'(mx_0)\cos(4\pi\nu' mx_0) \ . \tag{10.16}$$

The continuous variable x is replaced by the discrete values mx_0 where x_0 is the sampling distance. Equation (10.16) can yield a wrong result if the sampling distance and the band pass of the incident light are not properly correlated. An intensity which is evaluated for a frequency ν' holds also for

$$\nu'' = \nu' \pm \frac{l}{2x_0} \, , \qquad l = 1, 2, \ldots \ . \tag{10.17}$$

since

$$\cos[4\pi(\nu' \pm l/2x_0)mx_0] = \cos(4\pi\nu' mx_0 \pm 2\pi lm) = \cos(4\pi\nu' mx_0) \ .$$

Thus, the solution is only single valued for the frequency interval $0 \leq \nu' < 1/2x_0$. If the bandwidth of the incident light is characteristically larger than $1/2x_0$, the contribution from low wave numbers can lead to a wrong value for $I(\nu')$ for high values of ν' and the other way round.

Looking into the details the situation is even worse. The results from (10.16) are the same for $\nu' = l/4x_0 \pm \nu$. Thus, the function $I(\nu')$ reflects about $\nu = l/4x_0$ and a particular spectrum appears repeatedly in the analysis of increasing wave numbers with a folding frequency

$$\nu_{\mathrm{F}} = \frac{1}{4x_0} \ .$$

This means the free spectral range is not $0 \leq \nu' \leq 1/2x_0$ but only $0 \leq \nu' \leq 1/4x_0$. If x_0 is not chosen small enough the part of the spectrum for low wave numbers will overlap the spectrum for high wave numbers and thus cause an irritation. The described behavior is called *aliasing* derived from the Latin word *alius* which means *another*. It is the reason why Fourier spectroscopy becomes more and more difficult, the more the visible or even the ultraviolet spectral range is approached. If ν_{F} is the highest frequency in the incident light a probe interval $x_0 \leq 1/4\nu_{\mathrm{F}}$ must be selected in order to avoid aliasing. This means the number of measurements N to reach a spectral resolution $\delta\nu$ for this spectrum is

$$N = \frac{\Delta x}{x_0} = \frac{4\nu_{\mathrm{F}}}{\delta\nu} \ . \tag{10.18}$$

In the FIR with $\nu_{\mathrm{F}} = 500$ this number is 2000 for a resolution of $\delta\nu = 1$ cm^{-1}. For the same resolution in the visible with $\nu_{\mathrm{F}} = 20\,000$ the number is already 80 000. Such large sets of data make the Fourier transformation more elaborate. Since, however, today's computer capacity is nearly unlimited

Fourier spectrometers are already commercialized up to the frequency range of the near ultraviolet. On the other hand, aliasing can be used to extend a narrow frequency range at high frequencies by folding down into a lower spectral range. To avoid aliasing a spectral range appropriate for the chosen value of x_0 must be selected by filtering.

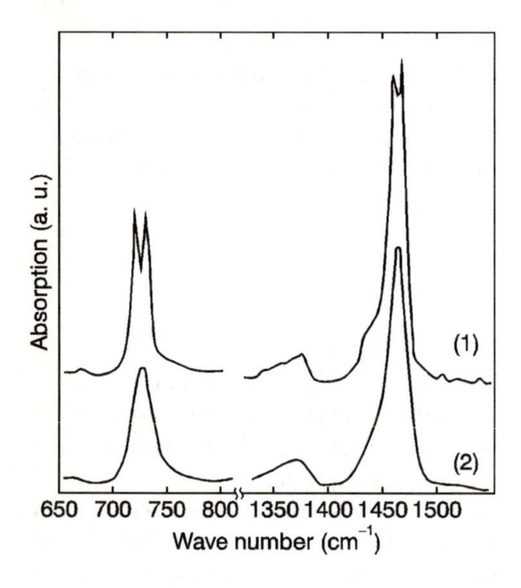

Fig. 10.14. Resolution of a Fourier spectrometer demonstrated for polyethylene. Spectra (1) were recorded for a distance Δx four times larger than for spectra (2).

Figure 10.14 demonstrates the influence of the maximum shift Δx on the resolution of the Fourier spectrometer. The spectra (1) are the response of the absorption of the CH deformation and of the CC deformation for polyethylene, as already shown in Fig. 10.12, but with an extended scale for the frequency. The splitting of the lines is now well observed. If instead of the 4096 probes only 1024 probes are recorded for the same distance x_0 the spectra (2) are obtained where the splitting is obviously not resolved any more.

Particular attention must be paid to the beam splitter. Since it usually consists of a thin and transparent polymer film like mylar multiple-beam interferences occur as discussed in Sect. 4.3.1. The multiple-beam interference contributes crucially to the transmission properties of the beam splitter. Using the results of the plane parallel plate the intensity at the detector is obtained from

$$I(d) = \frac{8R(1-R)^2(1-\cos\phi)}{(1+R^2-2R\cos\phi)^2} I_0 \qquad \text{with} \qquad \phi = 4\pi n d\nu \cos\theta' \, . \quad (10.19)$$

R is the reflectivity of the interface, d the thickness of the film, n its index of refraction, and θ' the angle of incidence (inside the beam splitter). For $\cos\phi = 1$ all light is reflected and cannot reach the detector. The power observed is thus a critical function of ϕ and consequently also of the thickness d. The latter must be selected in a way to transfer optimum power to the detector for the spectral range under investigation. The transmission is plotted in Fig. 10.15 for three different beam splitters. The broader the spectral distribution under investigation the lower the values of d required for the beam splitter.

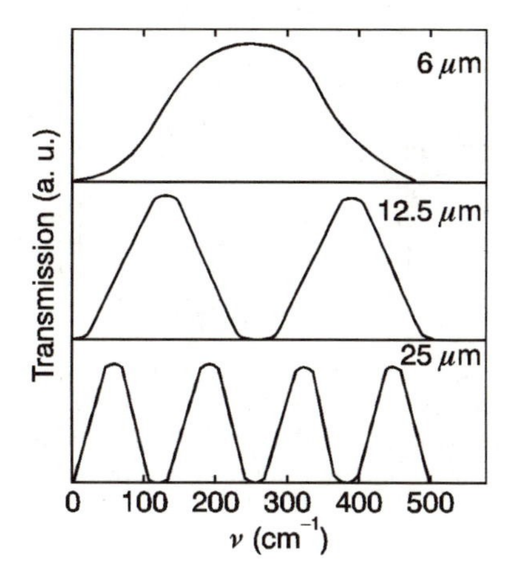

Fig. 10.15. Transmission through beam splitters with three different thicknesses for the spectral range from 0–500 cm^{-1} as calculated from (10.19).

Figure 10.16 sketches the layout of a commercial Fourier spectrometer. It consists essentially of four chambers: the light-source chamber, the beam-splitter chamber, the sample chamber, and the detector chamber. Advanced spectrometers operate with two or even three different detectors which are computer-controlled and available for the different spectral ranges. The same holds for different beam splitters and light sources. The latter can even be external. For measurements below 1000 cm^{-1} it is necessary to evacuate all chambers or to flush them at least with dry nitrogen to get rid of any traces of moisture. The vibrational and rotational lines of water will otherwise fully cover up the investigated spectra.

10.3.3 Fourier-Transform Raman Spectroscopy

Since Fourier spectroscopy is an excellent tool to analyze weak spectra it was certainly tempting to apply it also to Raman spectroscopy. As we saw, the brightness of the Michelson interferometer can be up to a factor 500 higher than for good grating spectrometers with the same cross section of the beam. This has basically to do with the much smaller focal length of the former with respect to the latter. Unfortunately the multiplex advantage of conventional Fourier spectroscopy turns into a multiplex disadvantage in the Raman experiments due to stray light. Since in Fourier spectroscopy the whole spectrum is always measured simultaneously the large amount

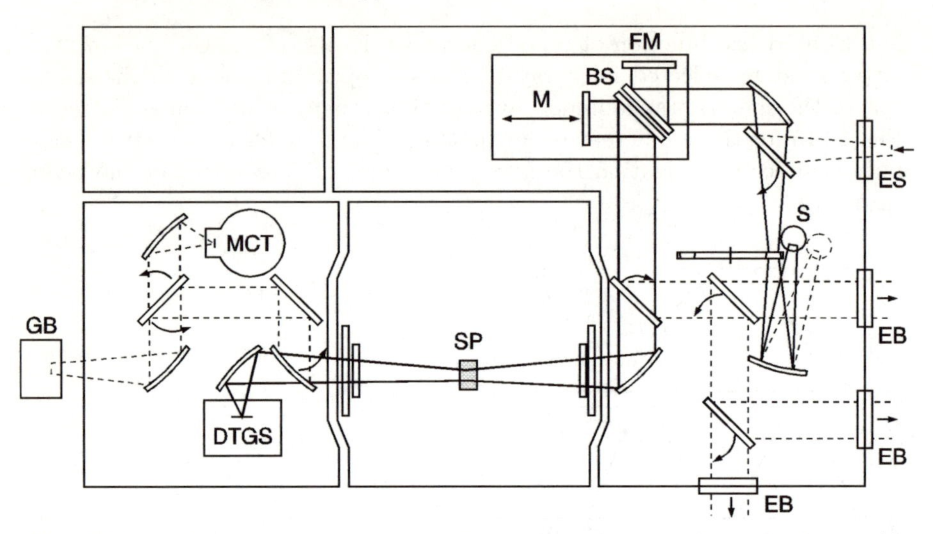

Fig. 10.16. Construction of a vacuum Fourier spectrometer for multiple use. (S: light source, M: gliding mirror, FM: fixed mirror, BS: beam splitter, SP: sample, DTGS, MCT, GB: detectors, EB: external beam, ES: external source). The layout corresponds to the spectrometer IFS66v from Bruker. The arrows indicate possible computer-controlled flipping of the mirrors to select different light paths.

of quasi-elastic scattered light induces a strong noise proportional to $\sqrt{I(t)}$ even on top of the weak part of the spectrum. With the development of very efficient and very sharp cutting filters it was possible to stop the largest part of the elastically scattered light and to use the Fourier technique even in the case of Raman scattering. Today such experiments can be performed down to $80\,\mathrm{cm}^{-1}$ in the spectrum for excitation with a Nd:YAG laser at 1064 nm. Fourier-transform Raman spectroscopy has several advantages like a less critical sample alignment, excitation at a much lower light energy and a large area sample illumination. Several companies offer Raman equipment as an additional supply to conventional Fourier spectrometers.

10.4 Intensities for Infrared Absorption

As the response of matter to IR radiation can have a purely electronic or a purely vibronic character the approximations for a quantitative treatment of the interaction processes are different for the two cases and will be discussed separately in the following.

10.4.1 Absorption for Electronic Transitions

The electronic transitions in the IR are fully analogous to the transitions in the UV-vis spectral range. Therefore the relationships derived in Chaps. 6

and 7 hold for the IR spectral range likewise. The response to the radiation is particularly well described by a dielectric function of the form (6.34). Introducing the oscillator strength of (6.38) we obtain for ε

$$\varepsilon = \varepsilon_\infty + \frac{S\omega_{\mathrm{T}}^2}{\omega_{\mathrm{T}}^2 - \omega^2 + \mathrm{i}\omega\gamma}\,, \tag{10.20}$$

where $\hbar\omega_{\mathrm{T}}$ is the transition energy. The oscillator strength is calculated from the transition matrix element, as discussed in Chaps. 7 and 8. Absorption and reflectivity are obtained as in Chap. 6.

10.4.2 Absorption for Vibronic Transitions

The response of the IR radiation to polar modes is described by the Kramers–Heisenberg dielectric function as well. For an evaluation of the oscillator strength a treatment very similar to the case of Raman scattering is possible. This was already indicated in the discussion of Fig. 9.4. Whereas the Raman intensity was given by the dependence of the field-induced dipole $\chi\varepsilon_0 E\mathcal{V}_{\mathrm{u}}$ on the normal coordinate the IR intensity is given by the dipole moment induced by the normal coordinate. Following the traditional assignments we will use the symbol μ instead of P_{D} for the molecular or microscopic dipole moment. To proceed like in the case of Raman scattering we expand $\mu(Q)$ in a Taylor series

$$\mu = \mu_0 + \sum \frac{\partial\mu}{\partial Q_k} Q_k + \dots\,. \tag{10.21}$$

The square of the second term in the expansion is proportional to the IR absorption intensity.

For the simple quantum-mechanical calculation the perturbation Hamiltonian is $-\mathbf{pA}$. Using harmonic-oscillator wave functions we obtain the transition dipole moment in analogy to the case of Raman scattering

$$[\boldsymbol{\mu}]_{\mathrm{fi}} = \langle \dots v_{k\mathrm{f}} \dots |\boldsymbol{\mu}| \dots v_{k\mathrm{i}} \dots \rangle\,. \tag{10.22}$$

With the Taylor expansion from (10.21) the transition dipole moment for the absorption of one phonon becomes

$$[\boldsymbol{\mu}]_{\mathrm{fi}} = \langle v_k + 1| \left(\frac{\partial\boldsymbol{\mu}}{\partial Q_k}\right) Q_k |v_k\rangle = \left(\frac{\partial\boldsymbol{\mu}}{Q_k}\right)\left(\frac{\hbar}{2\Omega_k}\right)^{1/2} (v_k + 1)^{1/2}\,. \tag{10.23}$$

Averaging over the occupation numbers with the weight $W(\epsilon_k)$ from (9.31) yields

$$\boxed{\alpha_k \propto \left|\frac{\partial\boldsymbol{\mu}}{\partial Q_k}\right|^2 \frac{\hbar}{2\Omega_k}(n_k + 1) \approx \left|\frac{\partial\boldsymbol{\mu}}{\partial Q_k}\right|^2 \frac{\hbar}{2\Omega_k}\,.} \tag{10.24}$$

The approximation adopted in the equation is good for $\hbar\Omega_k > k_{\mathrm{B}}T$ where Ω_k is the frequency for the transverse optical phonon. The expression on the right-hand side of (10.24) is a good value for the oscillator strength in a Kramers–Heisenberg DF, such as (10.20), for the polar mode.

The use of the square of the transition matrix element in (10.20) is not fully self-consistent. The denominator in the equation was obtained from a very simple model whereas the numerator is now calculated quantum-mechanically. In a self-consistent description the damping of the electric field $E(t)$ is considered. The absorption is then given from the Fourier transform of the field. The explicit result for α is obtained from [10.6]

$$\alpha_k = \frac{4\pi}{3c_0\hbar}|\langle \dots v_{kf} \dots |\boldsymbol{\mu}| \dots v_{ki} \dots \rangle|^2 \frac{\gamma_k}{(\omega - \Omega_k)^2 + \gamma_k^2} \, . \tag{10.25}$$

This compares to the imaginary part of (10.20) which is proportional to the absorption. From the evaluation the denominator becomes $(\Omega_k^2 - \omega^2)^2 + \omega^2\gamma_k^2$ in slight contrast to the denominator in (10.25). This difference has no physical meaning. It is a simple consequence of the different approaches.

In a more general and very fundamental treatment the IR absorption can be evaluated from the dipole autocorrelation function $\langle \boldsymbol{\mu}(0)\boldsymbol{\mu}(t) \rangle$. As demonstrated in Appendix I.1 this yields a very important and very useful relationship between the absorption coefficient α and the derived dipole moment in the form

$$\boxed{\left|\frac{\partial \boldsymbol{\mu}}{\partial Q_k}\right|^2 = \frac{12\Omega_k c_0 \varepsilon_0}{n_\mathrm{d}} \int \frac{\alpha_k(\omega)}{\omega} \mathrm{d}\omega \, .} \tag{10.26}$$

n_d is the number of dipole moments per unit volume. Equation (10.26) is often utilized to determine the derivative of the dipole moment with respect to the normal coordinate from a measurement of the absorption coefficient.

10.5 Examples from Solid-State Spectroscopy

Solid-state physics has a large number of problems where IR spectroscopy is very informative. They cover simple vibrational spectroscopy in polycrystalline material or the determination of bonding in organic solids as well as electronic properties like bandgaps, impurity levels, or cyclotron resonance in semiconductors. Even for the analysis of metallic systems and superconductors IR spectroscopy is used. Some examples will be described in the following sections.

10.5.1 Investigations on Molecules and Polycrystalline Material

As discussed extensively in Sect. 9.2, lattice modes or molecular vibrations can be classified as Raman-active, IR-active and polar, or silent. The IR-active species and the direction of their polarization can be looked up in character tables. Since characteristic bonds have characteristic frequencies

Table 10.1. Frequency range of important modes of organic material in the IR and Raman spectrum; (vs: very strong, s: strong, m: medium, w: weak, vw: very weak).

Stretching vibrations				Finger print range			
Vibration	Range $[\mathrm{cm}^{-1}]$	Intensity IR	Intensity Raman	Vibration	Range $[\mathrm{cm}^{-1}]$	Intensity IR	Intensity Raman
νCH	3100–2800	s	w	δCH_3	1370–1470	m	m
νNH	3500–3300	m	w	νC-C arom.	1600	m	s
νOH	3650–3000	s	w		1500	ms	m
νSiH	2250–2100	m	s		1000	w	vw
νPH	2440–2275	m	s	νC-C	1100–650	m	s
νSH	2600–2550	w	s	νN-N	865–800	w	s
νC≡C	2250–2100	w	s	ν_{as}COC	1150–1060	s	w
νC≡N	2255–2220	w	m	ν_sCOC	970–800	w	s
νC=C	1900–1500	w	vs	νS-S	550–430	w	s
νC=O	1820–1680	vs	w	νC-F	1400–1000	s	w
νC=N	1680–1610	m	s	νC-S	800–600	m	s
νN=O	1590–1530	s	m	νC-Cl	800–550	s	s
				νC-Br	700–500	s	s
				νC-I	660–480	s	s
				$\delta_s CF_3$	740	m	s
				$\delta_{as} CF_3$	540	m	w

IR spectroscopy is widely used to analyze chemical bonding and thus to analyze structures or identify materials and components in mixed systems. For inorganic materials the frequencies are concentrated in a rather narrow range between 50 cm^{-1} and 800 cm^{-1}, which makes analysis difficult. In organic material the frequencies extend up to 3500 cm^{-1} and the vibrational character of many special atomic groups is very well known. Since IR spectroscopy is easily performed such problems are now investigated in a special research field of computer-assisted IR material analysis. For identification stretching modes between the atoms are particularly useful since their frequencies are characteristically different for the various bonds. This analytical method will be increasingly supported in the future by Fourier-transform Raman spectroscopy. In Table 10.1 the most important and characteristic frequencies are listed together with their visibility in IR and Raman spectra. It is common to classify the vibrations into two groups, one above and one below about 1500 cm^{-1}. The high-frequency group contains only stretching modes, the lower-frequency group is called the *finger print range* since it covers the modes of the molecular stage. For the characterization of normal coordinates the following symbols are used:

- ν for stretching modes,
- δ for in-plane bending modes,
- γ for out-of-plane bending modes, and
- τ for rocking or torsion modes.

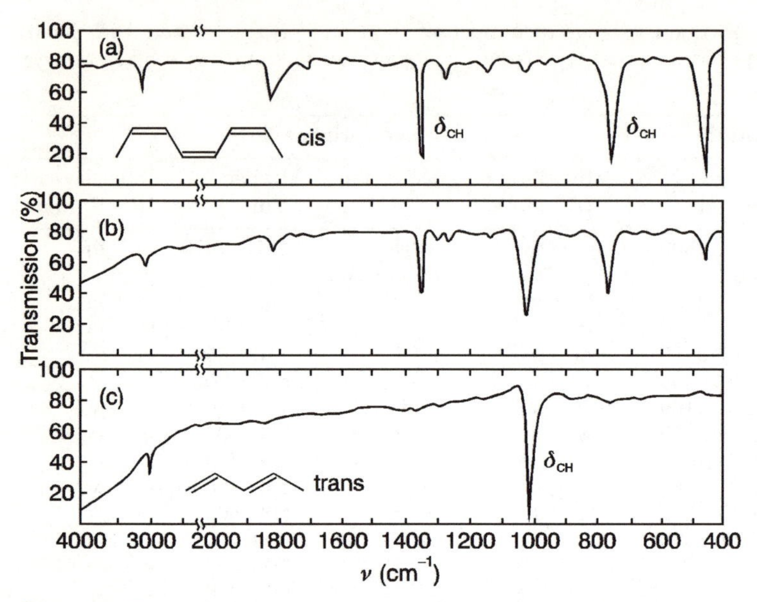

Fig. 10.17. Infrared transmission of cis-polyacetylene during the phase transition into the trans-form. cis-polyacetylene (**a**), intermediate phase (**b**), trans-polyacetylene (**c**); after [10.7].

For the experiments the material to be analyzed is usually mixed and ground together with a carrier which is IR-inactive in the frequency range to be investigated. The mixture is then compressed into pellets and analyzed in the spectrometer relative to a pellet of the pure carrier. For the NIR and MIR range KBr is a standard carrier material. Problems arise for the FIR since nearly all materials absorb strongly in this spectral range. One possibility is powdered polyethylene since this compound has only one rather weak line at 72 cm^{-1} in the FIR.

Besides the analysis of materials the documentation of thermodynamic or chemical processes like phase transitions, phase separations, or oxidation and reduction processes is a wide field in IR spectroscopy. Since the recording of spectra with the Fourier technique may require only several seconds, time-resolved spectroscopy is possible and often applied. Figure 10.17 shows the modulation of the IR spectra during a cis-trans phase transition of polyacetylene. By watching the two bending modes at 740 and 1329 cm^{-1} in the cis-form and the bending mode at 1050 cm^{-1} in the trans-form the dynamics of the isomerization process can be studied very well.

10.5.2 Infrared Absorption and Reflection in Crystals

Infrared absorption and reflection are basically determined by the complex dielectric function $\varepsilon(\omega)$ from Sect. 6.3. To investigate the contributions of the lattice modes the Kramers–Heisenberg model function of the form (6.34) or (6.39) is appropriate. The reflection or the absorption is evaluated from equations like (6.12), (6.19), (6.21), or (6.20).

Considering the Kramers–Heisenberg dielectric function in more detail interesting relations can be stated between the frequencies measured in the Raman spectra and the IR intensities. For a real crystal with N atoms per primitive unit cell the dielectric function must be derived in tensorial form. Then, x and E are vectors, and M, ε and the effective charge e^* are tensors. It is convenient to include the mass M in an *effective charge parameter* Z of the form

$$Z = \frac{e^*}{(\varepsilon_0 M)^{1/2}} \qquad \text{(in 1/s)} .$$

Z is now a $3N\times$ three-dimensional tensor with the components $Z_{k\alpha}$ where k and α run over 1 to $3N$ and 3, respectively. Since e^* and M are given per unit volume Z is given in s^{-1}. Equation (6.34) has then the form

$$\varepsilon_{\alpha\beta} = \varepsilon_{\alpha\beta}(\infty) + \sum_{k=1}^{3N} \frac{Z_{k\alpha} Z_{k\beta}}{\omega_{k\mathrm{TO}}^2 - \omega^2 + \mathrm{i}\omega\gamma_k} . \tag{10.27}$$

The sum in (10.27) runs over all phonon branches including the acoustic modes. For crystals with orthorhombic or higher symmetry only diagonal elements of ε are non zero. This yields

$$\varepsilon_{\alpha\alpha} = \varepsilon_{\alpha\alpha}(\infty) + \sum_{k=1}^{3N} \frac{Z_{k\alpha}^2}{\omega_{k\mathrm{TO}}^2 + \mathrm{i}\omega\gamma_k - \omega^2} . \tag{10.28}$$

If the sum is taken only over the $3N-3$ optical modes the dielectric function for constant strain $\varepsilon_{\alpha\beta}^{\mathrm{s}}$ is obtained. Introducing the oscillator strength $S_{k\alpha}$ of the form

$$S_{k\alpha} = \frac{Z_{k\alpha}^2}{\omega_{k\mathrm{TO}}^2} \tag{10.29}$$

(10.28) becomes (for $\gamma = 0$)

$$\varepsilon_{\alpha\alpha}^{\mathrm{s}} = \varepsilon_{\alpha\alpha}(\infty) + \sum_{k=1}^{3N-3} \frac{\omega_{k\mathrm{TO}}^2 S_{k\alpha}}{\omega_{k\mathrm{TO}}^2 - \omega^2} . \tag{10.30}$$

Introducing into the sum a common denominator and expressing the resulting polynomial in the numerator by its zero positions $\omega_{k\mathrm{LO}}$ yields $\varepsilon_{\alpha\alpha}^{\mathrm{s}}$ in its factorized form

$$\varepsilon_{\alpha\alpha}^{\mathrm{s}} = \varepsilon_{\alpha\alpha}(\infty) \prod_{k=1}^{3N-3} \frac{\omega_{k\mathrm{LO}}^2 - \omega^2}{\omega_{k\mathrm{TO}}^2 - \omega^2} \tag{10.31}$$

which obviously represents a generalized form of the Lyddane–Sachs–Teller relation (6.42). A similar relationship is obtained if the damping is included which means (10.31) enables the calculation of the dielectric function and thus the reflectivity spectrum if the all LO and TO frequencies are known. If the polar modes are Raman-active the latter can be determined from Raman spectroscopy. This means, the IR reflectivity can be fully evaluated from a simple measurement of Raman frequencies. Figure 10.18 displays results for the reflectivity of the $A_1(z)$ species for the orthorhombic crystal $LiGaO_2$ with space group $Pna2_1$. The dashed line was measured with the IR spectrometer. The solid line was calculated using the LO and TO mode frequencies determined by Raman scattering. No fitting parameter is used for the comparison. The very good agreement between the measured and calculated result is good evidence for the relationships derived above.

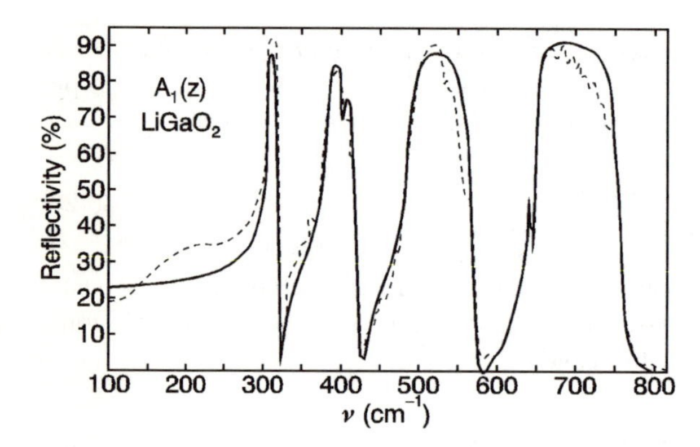

Fig. 10.18. Infrared reflectivity for the $A_1(z)$ species in $LiGaO_2$; experimental result (– –) and calculated from the LO and TO frequencies of a Raman spectrum (—); after [10.8].

A comparison between (10.30) and (10.31) enables, in addition, the oscillator strengths $S_{k\alpha}$ and with (10.29) also the effective charge parameters to be explicitly determined. For the former the equation

$$S_{k\alpha} = \varepsilon_{\alpha\alpha}(\infty)\frac{\omega_{k\mathrm{LO}}^2 - \omega_{k\mathrm{TO}}^2}{\omega_{k\mathrm{TO}}^2}\prod_{l=1,k\neq l}^{3N-3}\frac{\omega_{k\mathrm{TO}}^2 - \omega_{k\mathrm{LO}}^2}{\omega_{k\mathrm{TO}}^2 - \omega_{l\mathrm{TO}}^2} \tag{10.32}$$

is obtained. With this all parameters for the determination of the dielectric function are known. A detailed analysis reveals that the obtained parameters are even sufficient to determinate the nonlinear optical constants of the crystal [10.8].

10.5.3 Attenuated Total Reflection

From the many possibilities to use IR spectroscopy for the analysis of molecular and lattice vibrations one special technique should be mentioned where the total reflection at a crystal boundary plays the fundamental role. The method is known as *attenuated total reflection* since it makes use of special properties of the dielectric function for the situation of total reflection. The geometry of the experiment is sketched in Fig. 10.19. Light from the spectrometer hits the surface of the crystal with refraction index $n_2 + \mathrm{i}\kappa_2$ after passing the medium on top of it with the refractive index $n_1 + \mathrm{i}\kappa_1$. Let the angle of incidence be α. Material 1 can be either air or vacuum or another transparent crystal with a hemispheric form. The reflection depends critically on the angle α and on the related component k_x of the light wave vector. Two alternative experiments are possible. Either the electric field or the magnetic field of the light can be perpendicular to the plane of the beams. In the first case the geometry is called *transverse electric* (TE) and in the second case it is called *transverse magnetic* (TM). In both cases essentially the same results are obtained. The reflectivity R is determined from the Maxwell equations and by considering the steady transition of the tangential component of the electric field through the interface as was done for the derivation of the Fresnel formulae in Appendix E.1. For the TE geometry one finds for non-dissipative media

$$R = \frac{\boldsymbol{E}^r \times \boldsymbol{H}^r}{\boldsymbol{E}^i \times \boldsymbol{H}^i} = \left| \frac{(\omega/c_0) n_1 \cos\alpha - \sqrt{(\omega/c_0)^2 \varepsilon_{2y}(\omega) - k_x^2}}{(\omega/c_0) n_1 \cos\alpha + \sqrt{(\omega/c_0)^2 \varepsilon_{2y}(\omega) - k_x^2}} \right|^2 . \tag{10.33}$$

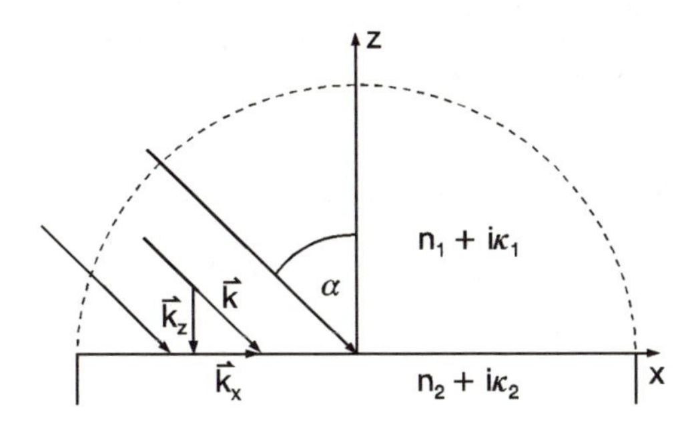

Fig. 10.19. Geometry for attenuated total reflection experiments.

ε_{2y} and k_x are the y component of the dielectric tensor of the crystal and the x component of the wave vector of the incident light, respectively. For very small values of α (perpendicular incidence) this relation yields (6.19) for $\kappa = 0$. For α large conventional total reflection as it is known for a smooth dielectric function is possible if $\sqrt{\varepsilon_{2y}} < n_1$. Then the expression under the square root can become zero or even negative and the reflectivity will be 1.

If the dielectric function diverges for certain values of ω the reflectivity (10.33) becomes a complicated function of α and ω. It depends again critically on whether the square root is real, imaginary, or dominates the numerator and the denominator altogether. In the last two cases the numerator and the denominator become equal, and $R \approx 1$ is expected. Starting the reflectivity measurement for small enough angles α and ω the reflectivity will be conventional and low. As soon as the frequency of light has reached the TO phonon ε_{2y} becomes very large. In this case the square root dominates the fraction in (10.33), and R becomes 1. This holds even for negative values for ε_{2y} at least up to the LO frequency where the dielectric function becomes zero. Only when ε_{2y} becomes positive again the square root can turn to real and the reflectivity will be reduced. When this will happen depends critically on k_x and thus on the angle of incidence α. Consequently, the drop in reflectivity must be shifted to higher frequencies with increasing α. As shown in Fig. 10.20, this is indeed the case.

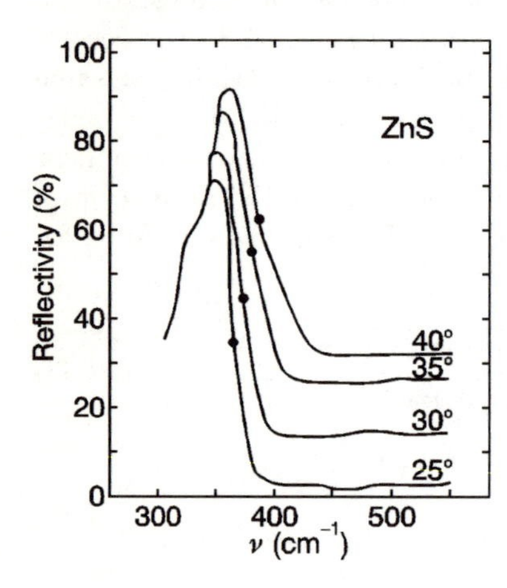

Fig. 10.20. Reflectivity of ZnS as measured for a geometry of attenuated total reflection for various angles of incidence; after [10.9].

The physical interpretation of these results reveals that one can use such experiments to observe the upper branch of the polariton dispersion discussed in Sect. 9.2.4. The first expression under the square root in (10.33) is a wave vector of magnitude

$$q^2 = \left(\frac{\omega}{c_0}\right)^2 \varepsilon_{2y} \ .$$

According to (9.21) this corresponds exactly to the dispersion for the upper branch of the polariton. The drop in reflectivity occurs exactly when the expression under the square root becomes zero. This means that the drop occurs when the x component of the incident light equals the wave vector of the polariton.

$$q_{\text{pol}} = 2n_1 \sin\alpha \ . \tag{10.34}$$

Since the corresponding frequency is obtained from the experiment according to Fig. 10.20, the experimental relation between q and ω is available.

10.5.4 Applications in Semiconductor Physics

Except for general vibrational analysis application to semiconductors is the most important field of IR spectroscopy. To keep the volume of this textbook limited only a few characteristic examples will be described, and reference will be given to special summarizing reports.

Because of the low value of the energy gap fundamental absorption appears in semiconductors often in the MIR or even in the FIR. In general, this absorption is an intrinsic behavior of the crystals but for solid solutions of two systems it can also depend on the concentration of the components. In special cases ϵ_g can even drop to zero and finally re-increase for a continuous variation of the concentration in a mixed-crystal system. This happens when the valence band and the conduction band are mutually interchanged at the two ends of the mixed system. Figure 10.21 gives an example. $Pb_{1-x}Sn_xTe$ can be grown as mixed crystals for a wide range of x. The energy gap for PbTe is 0.22 eV at 77 K. The gap for SnTe is 0.26 eV at the same temperature but the valence band and the conduction band have interchanged. Accordingly, with increasing x the gap decreases until it becomes nearly zero for $x = 0.28$. For x as large as 1 the gap has considerably increased again.

In degenerate semiconductors where the Fermi energy has shifted into the conduction band (or into the valence band) a transition into the lowest (highest) band states is not possible since they are occupied. This leads to an apparent upshift of the absorption band with increasing carrier concentration n of

$$\delta\epsilon_g = \epsilon_F(n) - 4k_BT \, , \tag{10.35}$$

where $\epsilon_F(n)$ is the distance of the Fermi energy from the band edge. This phenomenon is known as the *Burstein shift* and has been observed frequently with IR spectroscopy.

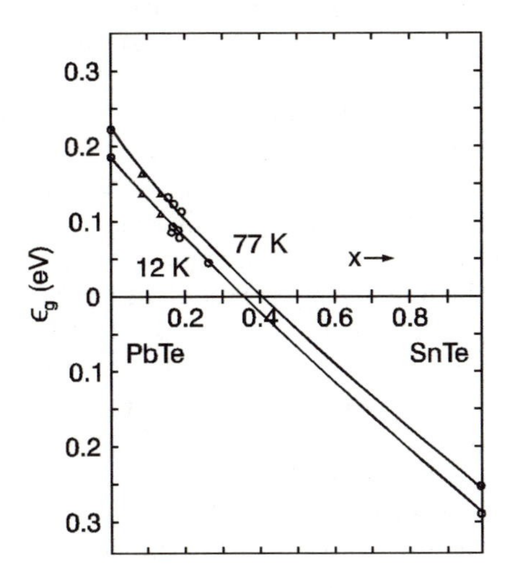

Fig. 10.21. Energy gap at 12 K and 77 K for the mixed crystals $Pb_{1-x}Sn_xTe$ versus concentration x; after [10.10].

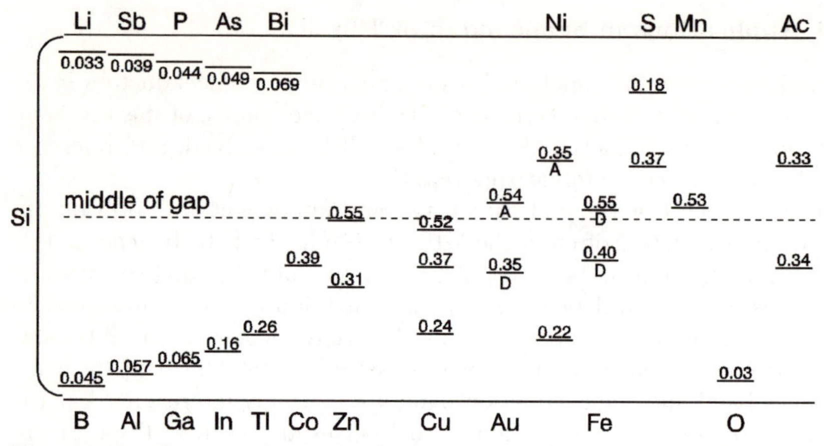

Fig. 10.22. Impurity levels for Si. The numbers are energetic distances to the corresponding band edges in meV; after [10.11].

Impurity levels in the gap are another very large field in semiconductor IR spectroscopy. Figure 10.22 lists impurity levels for Si. Levels close to the band edges are called *shallow*, those appearing more towards the center of the gap are called *deep*. Shallow impurity levels exhibit a hydrogen-like bonding for the electrons. The energy levels ϵ_β are independent of the impurity under consideration and given by

$$\epsilon_\beta = \frac{m^* e^4}{32\pi^2(\varepsilon\varepsilon_0)^2\hbar^2} \,. \tag{10.36}$$

In addition, series of excited states exist and can be populated like in hydrogen. The chemical nature of the impurity atom plays a minor role. It can lead to a small deviation in energy from (10.36) known as *chemical shift*.

To investigate the impurity levels the induced photoconductivity can be used rather than the absorption process itself. This is a particularly sensitive probe since it does not need an extra IR detector. The sample itself is the detector. The incident radiation must not necessarily fully ionize the impurity level. It is enough if this occurs with thermal assistance. Figure 10.23 shows how an electron can be excited from an impurity level and how it may reach the conduction band. In part (a) of the figure the quantum energy of the light is large enough to excite the electron directly into the conduction band. In part (b) a thermal assistance of the amount $\hbar\Omega_q$ is required for full ionization, and in part (c) only an excitonic state is reached. In contrast to the first two examples the electron recombines before it can contribute to the photoconductivity. Processes according to the mechanism (a) and (b) give rise to characteristic maxima in the spectrum of photoconductivity versus IR energy. Since photothermic ionization does not work for very low temperatures an unusual result is obtained. The photoconductivity spectra can show a detailed fine structure for elevated temperatures which disappears on cooling. This is demonstrated in Fig. 10.24 for Sb-doped Ge. The spectrum observed for 4.2 K has no low-energy structures and photoconduction starts only at around 10 meV. By raising the temperature to 10 K several well defined sharp structures appear below the ionization energy of the impurity state. They correspond to the excited levels of the state induced by Sb.

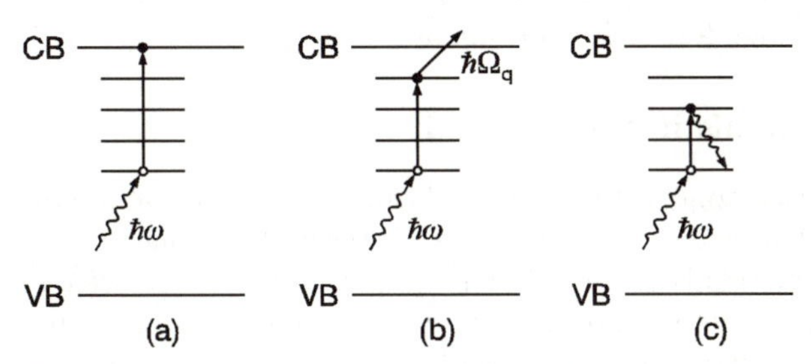

Fig. 10.23. Excitation processes for electrons during photoabsorption from impurities.

Plasma reflection was discussed in Sect. 6.3. Since plasma oscillations occur in semiconductors usually in the IR spectral range investigations for the position and width of the plasma edge are an important issue in IR spectroscopy. The determination of ω_p and τ are alternative methods to obtain the carrier concentration and the carrier mobility. An example of a plasma reflection has already been given in Fig. 6.6.

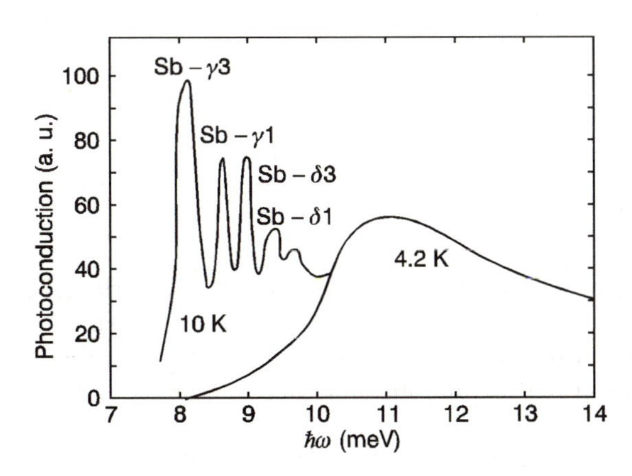

Fig. 10.24. Low temperature extrinsic photoconductivity for Sb-doped Ge at 10 K and at 4.2 K. The symbols assign the excited levels of the Sb-induced impurity state; after [10.12].

Finally, IR investigations proved very important for magneto-transport experiments. If a semiconductor with free carriers is accommodated in a magnetic field the carriers are forced on circular orbits determined by their thermal velocity and the Lorentz force. The frequency of circulation is

$$\omega_c = \frac{eB}{m^*} . \qquad (10.37)$$

known as the *cyclotron frequency.* If light with the same frequency is irradiated onto the semiconductor it is resonantly absorbed provided the cyclotron frequency is larger then the carrier collision frequency $1/\tau$. This phenomenon is called *cyclotron resonance* and very useful to determine the effective masses of carriers and thus the band structure in semiconductors. Since the condition $\omega_c \gg 1/\tau$ requires high enough frequencies for resonance absorption FIR is an appropriate spectral range for such experiments.

10.5.5 Properties of Metals in the Infrared

Because of the high carrier concentration the plasma frequency for metals is above the visible spectral range. Only in the UV, beyond the plasma frequency, transparency of metals is observed. The quality of the reflectivity in the visible and in the IR is determined by the carrier concentration n and by the collision time τ. It is very well described by the dielectric function for free carriers given by (6.47). For frequencies much smaller than the collision frequency ($\omega\tau \ll 1$) this relationship can be approximated as

$$\begin{aligned} \varepsilon_r &= \varepsilon_\infty - \omega_p^2\tau^2 = \text{constant} , \\ \varepsilon_i &= \frac{\omega_p^2\tau}{\omega} = \frac{\sigma_0}{\varepsilon_0\omega} . \end{aligned} \tag{10.38}$$

σ_0 is the dc conductivity. Since ε_i increases with decreasing frequency finally $\varepsilon_i \gg \varepsilon_r$ is reached. Optical constants and reflectivity are then only determined by the conductivity σ_0 as

$$\boxed{\begin{aligned} n \approx \kappa &\approx \sqrt{\frac{\sigma_0}{2\varepsilon_0\omega}} , \\ R &= 1 - 2\sqrt{\frac{2\varepsilon_0\omega}{\sigma_0}} . \end{aligned}} \tag{10.39}$$

The corresponding spectral range is called *Hagen–Rubens* range. The Hagen–Rubens relationship is often used to extrapolate reflectivity measurements of metallic systems to zero frequency if the spectra are to be used for a Kramers–Kronig transformation and results are not available down to low-enough frequencies.

The reflectivity of good metals is of the order of 99.9%. This high value can make it difficult to determine the deviation from 100%. A possibility to avoid this problem is to measure the emissivity of the metal in comparison to a black body. According to the Kirchhoff law the emissivity is identical to the absorptivity and for vanishing transmission we have $A = 1 - R$.

On cooling to low temperatures many metals exhibit a 2nd-order phase transition to a superconducting state. This state is characterized by a complete loss of resistivity and by a diamagnetic behavior in a magnetic field.

The generally acknowledged theory for a microscopic description of this phenomenon was developed by L. Bardeen, L.N. Cooper, and J.R. Schrieffer and is known as BCS theory. The theory claims a pairing of two electrons on the Fermi surface in the strain field of a phonon. Simultaneously with the pairing process a reordering of the electronic states occurs in a way that for the excitation spectrum a gap ϵ_g opens up at the Fermi surface. According to the BCS theory the gap energy and the transition temperature T_c are connected by the famous relationship

$$\epsilon_g = 3.5 k_B T_c \ . \tag{10.40}$$

Since for the conventional superconductors T_c is in the range of 0–20 K, ϵ_g is expected in the meV region. FIR is therefore an appropriate technique for the determination of this energy gap. In contrast to the picture which we have developed for semiconductors the superconducting metals are not transparent even if the quantum energy of the light is lower than the gap energy! Due to the loss of resistivity a high current is excited which is 90° phase shifted to the field and thus does not dissipate energy. The radiation is reflected to 100%. Only for quantum energies of light higher than ϵ_g breaking of the electron pairs is possible and radiation can be absorbed.

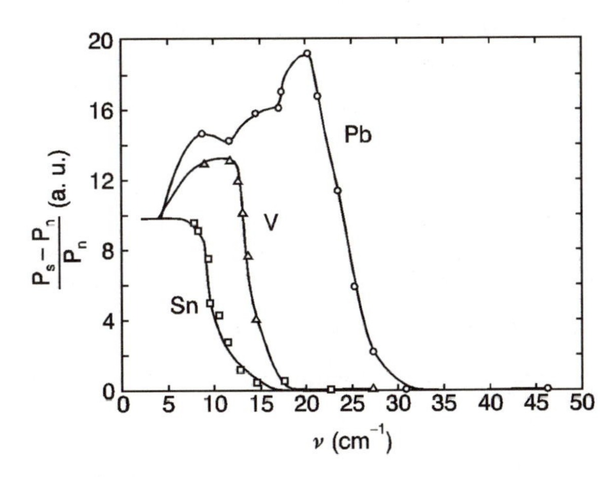

Fig. 10.25. Change of infrared absorption of metals in the superconducting state as a function of the quantum energy of the radiation after [10.13].

For the experimental observation of the gap in superconductors IR radiation is guided to a cavity made of the material under investigation. A very small but very sensitive bolometer is fixed on the wall in order to register the radiation in the cavity. If the cavity is in the superconducting state the power incident on the bolometer is high as long as the energy of the radiation is too small to break the pairs. In the normal state or if the quantum energy of the light reaches the gap energy the radiation is absorbed and less power reaches the bolometer. Figure 10.25 exhibits experimental results for the three metals Pb, V, and Sn. Plotted is the relative difference of the bolometer signal in the superconducting and in the normal state. Switching between superconducting and the normal conducting state was performed by a magnetic field. The difference remains large for low light energies but goes to zero as soon as the gap energy is reached. From the position at which the signal begins to drop the values 20 cm^{-1}, 13 cm^{-1}, and 9 cm^{-1} are obtained for Pb, V, and Sn, respectively. This corresponds to energies of 2.5, 1.62, and 1.13 meV. Using (10.40) the corresponding transition temperatures are 8.3 K, 5.37 K, and 3.75 K. Values for T_c from conductivity measurements are 7.19 K, 5.38 K, and 3.72 K. The agreement is particularly good for V and Sn which are thus considered to be typical BCS superconductors.

In cuprate superconductors like $YBa_2Cu_3O_7$ or BiSrCaCuO the energy gap is at least a factor ten higher which allows conventional reflectivity measurements for the determination of the energy gap. From such experiments the relation (10.40) was found to have a different scaling factor. Instead of 3.5 a factor of 5–6 is frequently observed. This suggests that the high-temperature oxidic superconductors are not BCS-like.

Another special application of IR radiation in the field of superconductivity is found for Josephson junctions. These junctions consist of a tunneling transition between two superconducting metals which are separated by a thin insulator (see also Sect. 15.2). The current-voltage characteristic of the junctions is changed by irradiating the contact with IR light. This can be used for very sensitive detection of IR radiation.

Problems

10.1 Discuss the emission of a black-body radiation source at 50, 100, 1000, and 6000 centigrade. How much light is emitted in the MIR (10–40 μm) spectral range? What is the ratio of the intensities for this emission to the emission in the visible spectral range?

(Purpose of exercise: study the efficiency of IR sources.)

10.2 Show explicitly that the folding frequency for a discrete Fourier transformation is $1/4x_0$ where x_0 is the sampling distance.

(Purpose of exercise: understand the phenomenon of aliasing.)

10.3 Using the results of Sect. 4.3.1 show that the intensity at the detector of a Fourier spectrometer is given by

$$I/I_0 = \frac{8R(1-R)^2(1-\cos\phi)}{[1+R^2-2R\cos\phi]^2} .$$

Hint: Use the Airy formulae and consider that the two beams arriving at the detector are coherent. Thus, the fields must be superimposed, not just the intensities.

(Purpose of exercise: understand the mechanism of the beam splitter.)

10.4 Show that the dielectric function for several LO-TO split phonon branches can be written in the factorized form of (10.31).

(Purpose of exercise: work with LO-TO splitting.)

10.5 Evaluate the Hagen–Rubens approach from the dielectric function for free carriers in the limit $\omega\tau \ll 1$.

(Purpose of exercise: use the free carrier dielectric function.)

10.6 Copper has a conductivity of $6\times10^4\,\Omega^{-1}\,\mathrm{cm}^{-1}$, an electron mobility of $1\,\mathrm{cm}^2/\mathrm{V\,s}$ and $\varepsilon_\infty = 1$. For which wavelengths is the Hagen–Rubens approach valid?

(Purpose of exercise: understand the Hagen–Rubens relation.)

11. Magnetic Resonance Spectroscopy

The electronic and vibronic states investigated so far were obtained from the Schrödinger equation where the Hamiltonian included only electronic and lattice contributions. Whether the atoms on the lattice sites have a magnetic moment or not was irrelevant since the latter does not contribute to the total energy of the system. The situation is different if a magnetic field is applied or if the interaction of the magnetic moments with the internal magnetic field is considered. In this case magnetically degenerate electronic states will split and new transitions become possible. As a consequence the response function will change. For magnetic fields in the conventional range the new transition energies are below the energy of FIR light, which means microwaves or even high-frequency fields are appropriate for the excitation. Since the magnetic moments of the atoms or molecules are immediately related to their spin or to the paramagnetism the investigation of these transitions is called *magnetic resonance* or *spin resonance* spectroscopy. The magnetic moments of interest can originate either from the electrons or from the nuclei.

Like many spectroscopic techniques, the magnetic resonance method has developed recently into one of the most important techniques for the investigation of new materials. This holds, in particular, for nuclear magnetic resonance which has become an excellent tool for chemical analysis.

After a short introduction to the physical properties of magnetic moments of atoms and nuclei this chapter will discuss fundamental procedures and applications of magnetic resonance spectroscopy. For a good part of this description a differentiation between *electron spin resonance* (ESR) and *nuclear magnetic resonance* (NMR) is not required. Only in the last three sections specific spin resonance techniques will be discussed explicitly. For details of more sophisticated applications special references must be considered [11.1, 11.2, 11.3].

11.1 Magnetic Moments of Atoms and Nuclei

All atoms have more or less strong magnetic moments. They originate either from the angular momentum of the electrons (electron spin, orbital angular momentum, or induced angular momentum if a magnetic field is applied) or from the nuclear spin. The magnetic moments of the electrons determine the magnetic properties of the crystals. The nuclear spin or the magnetic moment of the nuclei are the sum of the contributions from the neutrons and protons. They are much smaller than the magnetic moments of the electrons. Magnetic moments and angular moments are connected for the electrons as well as for the nuclei by the magnetogyric ratios γ defined as

$$\boldsymbol{\mu}_\mathrm{e} = -\gamma_\mathrm{e}\hbar\boldsymbol{J} \qquad \text{and} \qquad \boldsymbol{\mu}_\mathrm{N} = \gamma_\mathrm{N}\hbar\boldsymbol{I} \,, \tag{11.1}$$

respectively. $\boldsymbol{\mu}_\mathrm{e}$ and $\boldsymbol{\mu}_\mathrm{N}$ are the magnetic moments of electrons and nuclei. $\hbar\boldsymbol{J}$ and $\hbar\boldsymbol{I}$ are the angular moments. The magnetic moment of the electrons is conveniently measured in units of the Bohr magneton μ_B

$$\begin{aligned} \mu_\mathrm{B} = \frac{e\hbar}{2m_0} &= 9.27\times10^{-24}\,[\,\mathrm{A\,m^2}] \\ &= 5.79\times10^{-5}\,[\,\mathrm{eV/T}]\,. \end{aligned} \tag{11.2}$$

In this unit it can be calculated in a simple way from the quantum number J and the dimensionless g-factor

$$\frac{\mu_\mathrm{e}}{\mu_\mathrm{B}} = g_\mathrm{e}J\,. \tag{11.3}$$

The magnitude of the magnetogyric ratio γ_e as expressed by g_e becomes

$$\gamma_\mathrm{e} = \frac{g_\mathrm{e}\mu_\mathrm{B}}{\hbar}\,[\mathrm{s^{-1}/T}]\,. \tag{11.4}$$

For electrons in a (free) atom the g-factor is obtained from the quantum numbers J, S, and L by the well known Lande formula

$$g_\mathrm{e} = g_\mathrm{LSJ} = 1 + \frac{J(J+1)+S(S+1)-L(L+1)}{2J(J+1)}\,. \tag{11.5}$$

For the free electron $J = S = 1/2$, so that $g_\mathrm{el} = 2$ and the gyromagnetic ratio for the free electron becomes from (11.4)

$$\gamma_\mathrm{el} = \frac{2\mu_\mathrm{B}}{\hbar} = 1.76\times10^{11}\,[\mathrm{s^{-1}/T}]\,.$$

For nuclear angular moments the magnetogyric ratio is much smaller because of the larger mass of the nucleons. The magnetic moments are counted in units of the *nuclear* magneton μ_n which is defined by

$$\mu_\mathrm{n} = \frac{e\hbar}{2M_\mathrm{p}} = 5.05\times10^{-27}[\mathrm{A\,m^2}]\,, \tag{11.6}$$

where M_p is the mass of the proton. Hence for an arbitrary nucleous N we define a g-factor and a magnetogyric ratio by

$$\frac{\mu_N}{\mu_n} = g_N I \qquad \text{and} \qquad \gamma_N = \frac{g_N \mu_n}{\hbar} .$$

The definition of the nuclear mangneton is arbitrary in a sense that it is neither equal to the magnetic moment of the proton nor the magnetic moment of the neutron. For the nuclear magneton and $I = 1/2$ we obtain $g_n = 2$ and

$$\gamma_n = \frac{g_n \mu_n}{\hbar} = 0.956 \times 10^8 \, [\mathrm{s}^{-1}/\mathrm{T}] . \tag{11.7}$$

For the proton the magnetic moment μ_p is

$$\mu_p = 2.79 \mu_n = 1.41 \times 10^{-26} \, [\mathrm{A\,m}^2] \tag{11.8}$$

and its gyromagnetic ratio is obtained from (11.1)

$$\gamma_p = \frac{2\mu_p}{\hbar} = 2.675 \times 10^8 \, [\mathrm{s}^{-1}/\mathrm{T}] . \tag{11.9}$$

The magnetic moment for the neutron is

$$\mu_{ne} = -1.913 \mu_n = -9.663 \, [\mathrm{A\,m}^2] . \tag{11.10}$$

11.1.1 Orientation of Magnetic Moments in a Field and Zeeman Splitting

Applying a magnetic field $\boldsymbol{B}$ to the dipole $\boldsymbol{\mu}$ yields the potential energy U from

$$U = -\boldsymbol{\mu}\boldsymbol{B} . \tag{11.11}$$

Maximum and minimum values for this energy are

$$U_{\max} = \mu B, \qquad \text{and} \qquad U_{\min} = -\mu B$$

for antiparallel and parallel orientation of the dipoles to the field. According to quantum mechanics the magnetic dipoles can be aligned only in $2J + 1$ or $2I + 1$ well defined directions between the two limiting orientations of the field. From this the possible energetic states are

$$\boxed{U_{J_z} = J_z g_e \mu_B B, \qquad \text{and} \qquad U_{I_z} = I_z g_N \mu_n B} \tag{11.12}$$

for the electronic and nuclear dipoles, respectively. J_z and I_z can take the values $J, J-1, \ldots 0, \ldots -J$ and $I, I-1, \ldots 0, \ldots -I$. This behavior of the energy levels in a magnetic field is known as *Zeeman* splitting. Spin transitions are allowed between neighboring values of U. This means $\Delta J_z = 1$ or $\Delta I_z = 1$. From (11.12) the transition energies are

$$\Delta U_{J_z} = g_e \mu_B B = \hbar \gamma_e B, \quad \text{and} \quad \Delta U_{I_z} = g_N \mu_n B = \hbar \gamma_N B \,. \tag{11.13}$$

As a consequence of the splitting electromagnetic radiation will be resonantly absorbed if its energy matches the energy difference of the Zeeman levels given in (11.13). Thus, the frequency for resonance absorption is

$$\omega_0 = 2\pi f_0 = \gamma B \,. \tag{11.14}$$

For a field of 1 T, and the values of γ_e and γ_p as given above, resonance frequencies of 28 GHz and 42.6 MHz follow. Thus, microwaves or radio-frequency waves are the appropriate energy range in the electromagnetic spectrum.

In general in solids the quantum-mechanical expectation value for the spin quantum number must be used for the description of the magnetization, rather than simply the quantum number. This means the magnetic moments of an atom or nucleus with magnetic quantum number J, or I, respectively, are

$$\mu_e = \gamma_e \hbar \sqrt{J(J+1)} \qquad \text{or} \qquad \mu_N = \gamma_N \hbar \sqrt{I(I+1)} \,. \tag{11.15}$$

In contrast to this energy levels and transitions in magnetic resonance spectroscopy are assigned directly by the quantum numbers.

For finite temperatures the magnetic moments do not line up completely in a magnetic field. For spin 1/2 systems the average magnetization M along the field direction is given by $M_0 = n\langle \mu_e \rangle$ where n is the number of magnetic moments per unit volume. As in the theory of magnetism this quantity is obtained from

$$M_0 = n\mu_e L(x) = n\mu_e(\coth x - 1/x) \tag{11.16}$$

with $x = \mu_e B / k_B T$. For small x $L(x)$ can be approximated by $x/3$ which yields for M_0

$$M_0 = \frac{n\mu_e^2 B}{3k_B T} = nJ(J+1)\frac{g_e^2 \mu_B^2 B}{3k_B T} \,. \tag{11.17}$$

For large fields and low temperatures $L(x) \approx 1$, and the magnetization saturates at $M_0 = n g_e \mu_B J$.

In practice, the g-factors and magnetogyric ratios are more frequently used to describe the magnetic behavior of material as compared to the magnetic moments μ_e or μ_N themselves.

11.1.2 Magnetic Moments in Solids

In solids atoms are not free and the orbital quantum number L is not a well defined quantity. The magnetic moments are mainly determined by the spin quantum number S. This is a consequence of the anisotropy of the electrostatic crystal field. However, the quenching of the orbital moments is not complete due to spin orbit coupling. A small correction remains for the determination of the magnetic moments from the g-factor and from the spin quantum number. Thus the g-factor has an anisotropic spin–orbit correction Δ_{ik}. This means it must be written as a tensor of the form

$$g_{ik} = g_e(\delta_{ik} + \Delta_{ik}) , \tag{11.18}$$

where δ_{ik} is the Kronecker symbol.

Table 11.1. Magnetic moments for various ions in units of μ_B.

Ion	Configu-ration	Basic level	$\mu_e = g_{LSJ}\sqrt{J(J+1)}$	$\mu_e = g_{el}\sqrt{S(S+1)}$	μ from experiment
Ce^{3+}	$4f^1 5s^2 p^6$	$^2F_{5/2}$	2.54	1.73	2.4
Pr^{3+}	$4f^2 5s^2 p^6$	$^3H_4$	3.58	2.8	3.5
Sm^{3+}	$4f^5 5s^2 p^6$	$^6H_{5/2}$	0.84	5.9	1.5
Eu^{3+}	$4f^6 5s^2 p^6$	$^7F_0$	0	2.0	3.4
Gd^{3+}	$4f^7 5s^2 p^6$	$^8S_{7/2}$	7.94	7.94	8.0
Dy^{3+}	$4f^9 5s^2 p^6$	$^6H_{15/2}$	10.63	5.9	10.6
Ti^{3+}	$3d^1$	$^2D_{3/2}$	1.55	1.73	1.8
Co^2+	$3d^7$	$^4F_{9/2}$	6.63	3.87	4.8
Ni^{2+}	$3d^8$	$^3F_4$	5.59	2.83	3.2
Cu^{2+}	$3d^9$	$^2D_{5/2}$	3.55	1.73	1.9

With respect to the electronic g-factor an exception of the above rule holds for the ions of the three-valent rare-earth atoms which are well known for their particular magnetic properties. Table 11.1 lists effective magnetic moments in units of μ_B for various rear-earth ions in comparison to ions of transition metals. The columns "Configuration" and "Basic level" show the electron configuration and the ground state of the ions. In the symbol ^iA_k A means the orbital quantum number ($S, P, D, F, \ldots$ for $L = 0, 1, 2, 3, \ldots$), k the total quantum number J, and i the spin multiplicity. This yields, e.g., for the Sm^{3+} ion $L = 5, J = 5/2$, and $S = 5/2$. Columns 4 and 5 show the magnetic moments considering the total quantum number J and only the spin quantum number S, respectively. The last column is the experimental result. Obviously, for transition metal ions the orbital quantum number is quenched and the magnetic moments are indeed only determined by the spin of the atoms.

Important other magnetic moments in solids originate from nearly all defects like color centers or isolated metal atoms in a matrix.

The magnetic moments of the nuclei are composed from the magnetic moments of the protons and the neutrons. The resulting quantum numbers are, in general, between 0 and 5. For nuclei with an even number of protons and neutrons (g,g-nuclei) they are always zero. For g,u-nuclei they are always an uneven multiple of 1/2. Table 11.2 lists important data for some nuclei used in NMR experiments. Other nuclei data can be looked up in, e.g., [11.1]. The use of the nuclear quadrupole moment for resonance experiments will be discussed in Sect. 11.9.

Table 11.2. Properties of nuclei of various isotopes used in NMR and in nuclear quadrupole resonance experiments

Nucleus	Abundance [%]	n_P	n_N	Spin	γ_N [γ_P]	NMR intensity [relat. to ^1H]	Quadrupole moment [10^{-24} cm^2]
^1H	99.98	1	0	1/2	1	1	–
^2H	0.01	1	1	1	0.1535	9.65E–3	2.8E–3
^{10}B	18.83	5	5	3	0.1075	1.99E–2	≈9.4E–2
^{11}B	81.17	5	6	3/2	0.3208	0.165	3.6E–2
^{13}C	1.10	6	7	1/2	0.2514	1.59E–2	–
^{14}N	99.63	7	7	1	0.0722	1.01E–3	≈4.5E–2
^{15}N	0.36	7	8	1/2	0.1013	1.04E–3	–
^{19}F	100.00	9	10	1/2	0.9408	0.833	–
^{31}P	100.00	15	16	1/2	0.4048	6.63E–2	–
^{63}Cu	69.17	29	34	3/2	0.2651	9.31E–2	−0.15
^{111}Cd	12.90	48	63	3/2	0.2120	0.12	0.83

In addition to electron spin resonance or electron paramagnetic resonance (EPR) and nuclear magnetic resonance, various other magnetic resonance techniques are used. Such are, e.g., *spin wave resonance* (SWR), *nuclear quadrupole resonance* (NQR), *optically detected magnetic resonance* (ODMR). Even double-resonance techniques are common like *dynamic nuclear polarization* (DNP) or *electron nuclear double resonance* (ENDOR) where nuclear and electronic spins are excited simultaneously.

The following sections will discuss only the basic magnetic resonance processes in some detail. NQR and double-resonance experiments will be touched briefly.

11.2 Kinematic Description of Spin Resonance

As in Sect. 11.1 the following description is valid for NMR as well as for ESR. For simplicity, the indices to γ, μ, etc., have been dropped in this and in the following two sections.

11.2.1 Motion of Magnetic Moments and Bloch Equations

We start by considering the equation of motion for a magnetic moment μ in a magnetic field B parallel to z. Let the magnetic moment be accompanied by a mechanical angular moment $\hbar J$. Since the time derivative of this quantity equals the acting torque $\boldsymbol{\mu}\times\boldsymbol{B}$ we have

$$\hbar\frac{\mathrm{d}\boldsymbol{J}}{\mathrm{d}t} = \boldsymbol{\mu}\times\boldsymbol{B}\,. \tag{11.19}$$

Introducing the magnetization as magnetic moment per unit volume $\boldsymbol{M} = n\langle\boldsymbol{\mu}\rangle$ and (11.1) yields (for electrons)

$$\boxed{\frac{\mathrm{d}\boldsymbol{M}}{\mathrm{d}t} = \gamma(\boldsymbol{B}\times\boldsymbol{M})\,.} \tag{11.20}$$

Equation (11.20) describes an undamped precession of $\boldsymbol{M}$ around z. It is called the *free Bloch equation.* A relaxation of $\boldsymbol{M}$ to its equilibrium position $\boldsymbol{M} = (0, 0, M_0)$ with $M_0 \propto B/T$ from (11.17) is not possible. This will only happen if relaxation processes are included. If this is the case, the individual magnetic moments precessing around the z axis will gradually relax to the equilibrium position. Assuming an exponential behavior for the relaxation process it can be described by

$$\frac{\mathrm{d}M_z}{\mathrm{d}t} = \frac{M_0 - M_z}{T_1}, \qquad \frac{\mathrm{d}M_x}{\mathrm{d}t} = -\frac{M_x}{T_2} \qquad \frac{\mathrm{d}M_y}{\mathrm{d}t} = -\frac{M_y}{T_2}\,. \tag{11.21}$$

Since the relaxation of the z component changes the energy of the system it must proceed by an interaction with other quasi-particles like phonons or electrons. The time constant T_1 is therefore called the *longitudinal* or *spin–lattice* relaxation time. The x and y components of the magnetic moment $\boldsymbol{M}$ relax to zero without a change in energy. They already become zero when the individual magnetic moments loose their phase coherence, e.g., by spin–spin interaction or by inhomogeneities in the field. Thus, T_2 is different from T_1 and called the *transverse* or *spin–spin* relaxation time. In general $T_2 < T_1$ holds.

The combination of (11.20) and (11.21) yields the Bloch equations for the motion of the magnetization including relaxation

$$\boxed{\begin{aligned}\frac{\mathrm{d}M_z}{\mathrm{d}t} &= \gamma(\boldsymbol{B}\times\boldsymbol{M})_z + \frac{M_0 - M_z}{T_1}\,,\\ \frac{\mathrm{d}M_{x,y}}{\mathrm{d}t} &= \gamma(\boldsymbol{B}\times\boldsymbol{M})_{x,y} - \frac{M_{x,y}}{T_2}\,.\end{aligned}} \tag{11.22}$$

They describe the full time-dependence of the magnetic moments and thus also the time dependence of the spins in the magnetic field.

11.2.2 The Larmor Frequency

From the equation of motion without relaxation [$T_1 = T_2 = \infty$ in (11.22)] the frequency of precession can be obtained directly. It is, however, here and for many other problems of spin precession very useful to study the kinematic of the magnetic moments not only in the laboratory system (x, y, z) but also in a system (x', y', z') rotating with a frequency ω around the z axis. The geometries and the precession of the magnetization $\boldsymbol{M}$ are demonstrated in Fig. 11.1. The connection between the change of the magnetization in the laboratory system $\mathrm{d}\boldsymbol{M}$ and in the rotating system $\mathrm{d}'\boldsymbol{M}$ is given by

$$\mathrm{d}'\boldsymbol{M} = \mathrm{d}\boldsymbol{M} - \boldsymbol{\omega}\mathrm{d}t \times \boldsymbol{M} . \tag{11.23}$$

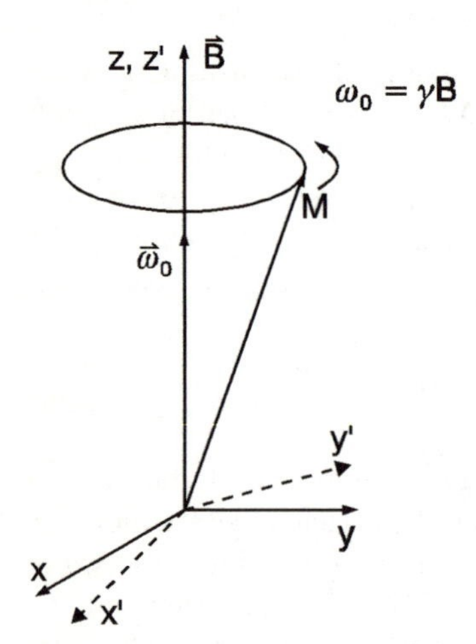

Fig. 11.1. Rotating frame (x', y', z') and rotation of a magnetic moment $\boldsymbol{M}$ in the laboratory frame (x, y, z).

This follows from the general relation between vectors in a laboratory system and in a rotating system (Appendix J.1). $\boldsymbol{\omega}$ in (11.23) is a vector $(0,0,\omega)$ pointing in the z direction. Thus, in the rotating system the equations of motion without relaxation have the form

$$\boxed{\frac{\mathrm{d}'\boldsymbol{M}}{\mathrm{d}t} = (\gamma\boldsymbol{B} - \boldsymbol{\omega}) \times \boldsymbol{M} .} \tag{11.24}$$

The equation shows that the magnetic field is effectively reduced to the value

$$B_{\mathrm{eff}} = B - \frac{\omega}{\gamma} . \tag{11.25}$$

If the dashed coordinate system rotates with the precession frequency $\boldsymbol{M}'$ will be constant and thus $\mathrm{d}'\boldsymbol{M}/\mathrm{d}t = 0$. This means from (11.24) that the precession frequency or *Larmor* frequency of the magnetic moments is related to $\boldsymbol{B}$ by

$$\boxed{\boldsymbol{\omega} = \boldsymbol{\omega}_0 = \gamma\boldsymbol{B}} \tag{11.26}$$

and the effective field becomes zero.

A comparison with the transition energy for resonance absorption from (11.13) shows that the magnetic system absorbs just at the Larmor frequency. This is the reason why the search for the resonance positions can either be performed by checking the induction of the rotating magnetic moments into a sensor coil or by measuring the energy absorbed by the system.

11.3 Induction into a Sensor Coil

The detection of the excited magnetic system with a sensor coil is the classical technique of NMR since in this case the frequency range is covered by RF equipment. The excitation occurs via an ac magnetic field in the x direction and the bias magnetic field in the z direction. The signal is detected with a pickup coil in the y direction[1]. A schematic setup is depicted in Fig. 11.2. x is pointing into the plane of the paper. The RF source can be tuned through the resonance frequency.

Since any linearly polarized field can be separated into two circularly polarized fields with opposite directions of circulation, the following description is given for circularly polarized fields. In fact, for such fields the motion of the magnetic moments is much more transparent as compared to linear excitation.

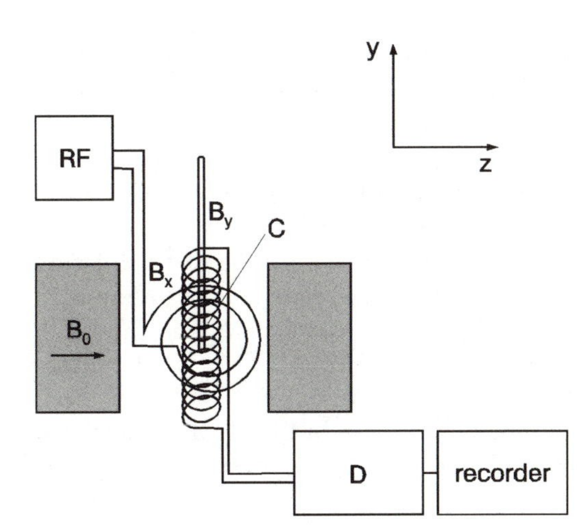

Fig. 11.2. Schematic representation of a setup for the detection of magnetic resonance by pickup coils. (B_0: static field, B_x: coil for excitation, B_y: coil for detection, RF: RF source, D: detector). The inset shows the orientation.

The excitation field B_1 is assumed circularly polarized in the xy plane. Then, the magnetization $\boldsymbol{M}$ precesses not only around z with the Larmor frequency $\omega_0 = \gamma B_0$ but also around the momentary direction of the field B_1 with frequency $\omega_1 = \gamma B_1$. This is a rather complicated motion. Its analysis

[1] In modern spectrometers excitation and detection are performed with a single coil in the x or y direction.

turns out to be much more simple if performed first in a coordinate system rotating with the excitation field B_1 as shown in Fig. 11.3. In this system B_1 is constant and oriented, e.g., in the x' direction. Since the static field along z is reduced to the effective field of (11.25) the resulting reduced but static field in the rotating system is

$$\boldsymbol{B}_{\text{eff}}^* = B_1\boldsymbol{e}_x' + B_{\text{eff}}\boldsymbol{e}_z' = B_1\boldsymbol{e}_x' + (B_0 - \omega/\gamma)\boldsymbol{e}_z' \tag{11.27}$$

with the magnitude

$$\boxed{B_{\text{eff}}^* = \sqrt{B_1^2 + (B_0 - \omega/\gamma)^2}\ .}$$

$\boldsymbol{e}_x'$ and $\boldsymbol{e}_z'$ are unit vectors in the rotating system. The magnetization rotates in the dashed system about B_{eff}^* as shown in Fig. 11.3a. Its components oscillate there with the frequency γB_{eff}^*. The oscillation amplitudes are determined by B_1/B_{eff}. The amplitudes of the M_y and M_z components increase with increasing ratio $B_1/B_{\text{eff}} = \sin\theta$ until they reach a maximum value M_0 for $B_{\text{eff}}^* = B_1$. Since they determine the signal in the pickup coil the latter depends critically on $B_{\text{eff}} = B_0 - \omega/\gamma$. For returning to the laboratory system we just have to rotate the x' axis in Fig. 11.3a with the frequency ω about $z' = z$.

The situation becomes particularly simple if the frequency ω is tuned to the Larmor frequency γB_0. In this case B_{eff} becomes zero and $B_{\text{eff}}^* = B_1$. For $B_{\text{eff}} = 0$ the angle θ will be 90° so that the magnetization rotates on a circle in the $y'z'$ plane with the component $M_x' = 0$ and the rotation frequency γB_1. This is sketched in Fig. 11.3b. In the laboratory system the motion of $\boldsymbol{M}$ is twofold: it rotates rapidly about z with frequency $\omega_0 = \gamma B_0$ and slowly with frequency γB_1 about the momentary direction of B_1. This yields the spiral motion on the surface of a sphere shown in Fig. 11.3c. The components M_x and M_y are double-modulated. They are subjected to a rapid modulation from the rotation about z and to a slower "amplitude" modulation with frequency γB_1. For the fully tuned case the amplitudes for M_x and M_y reach $|\boldsymbol{M}|$. Figure 11.3d displays the time dependence for M_x. M_y is 90° phase-shifted.

The mathematical form for the time dependence in Fig. 11.3d is obtained by solving the unrelaxed Bloch equations for the rotating system. This yields for the resonance conditions

$$M_x' = 0, \qquad M_y' = M_0\cos(\omega_1 t), \qquad M_z' = M_0\sin(\omega_1 t)\ ,$$

and after back transformation to the laboratory system

$$\begin{aligned} M_x &= -M_0\cos(\omega_1 t)\sin(\omega_0 t)\ , \\ M_y &= M_0\cos(\omega_1 t)\cos(\omega_0 t)\ , \\ M_z &= M_0\sin(\omega_1 t)\ . \end{aligned} \tag{11.28}$$

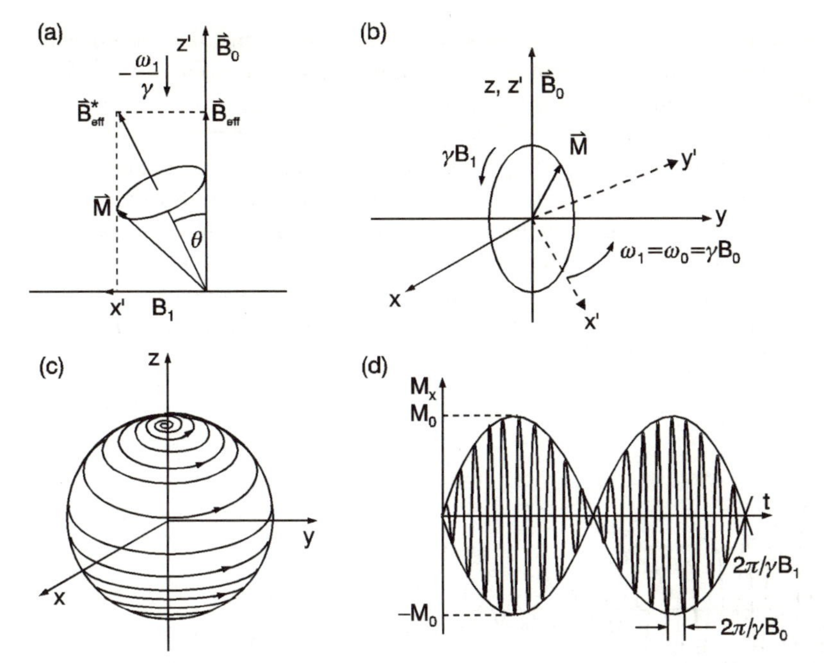

Fig. 11.3. Motion of the magnetization $\boldsymbol{M}$ in a static magnetic field $B_0||z$ and a field B_1 circularly polarized in the x, y plane, as observed in the rotating system (**a**); same motion in the laboratory system for resonance condition $\omega_0 = \gamma B_0$ (**b**); topography of the endpoint of $\boldsymbol{M}$ in real space (**c**); and the time dependence of component M_x at resonance condition (**d**).

Obviously M_x and M_y are phase-shifted by 90°. M_y or M_x is the signal which is detected by the pickup coil. The central frequency for the observation is $\omega_{\text{obs}} = \omega_0 = \gamma B_0$.

The oscillations of (11.28) as they are depicted in Fig. 11.3 are only good for a relaxation free motion. If T_1 and T_2 are finite a steady-state result is obtained in the rotating frame as will be discussed in the next section. Before we do this we will demonstrate the usefulness of the above ideas for a very important class of resonance experiment in which relaxation is already included.

The detection of relaxation is possible in a straightforward way by applying a pulsed field which pulls the magnetization away from its original z orientation and observing the relaxation back to the intial state after termination of the pulse. If, for example, a field B_1 is applied at resonance for the time period $t = \pi/2\gamma B_1$ the magnetization has just arrived at the xy plane when the field is turned off. After this 90° or $\pi/2$ *pulse* the magnetization relaxes back to the z direction. For the z component this happens with the longitudinal relaxation time T_1 and for the x and y components with the transverse relaxation time T_2. The transverse components already disappear when the individual magnets or spins have lost their mutual coherence by inhomogeneities in the sample or by spin–spin interaction. With the pickup coil so-called *free induction decay* (FID) is observed. It appears as an exponentially damped sine wave of frequency γB_0 with a decay constant $1/T_2$. The time dependence of the pickup signal is shown in Fig. 11.4. The mathematical expression for free induction decay is again readily obtained from solving the Bloch equations first for the rotating system with $B_1 = B_{\text{eff}} = 0$ and then transforming the result back to the laboratory system.

90°-Pulses and several other more sophisticated pulse sequences are often employed in magnetic resonance spectroscopy to determine T_1 and T_2. This holds, in particular, for NMR.

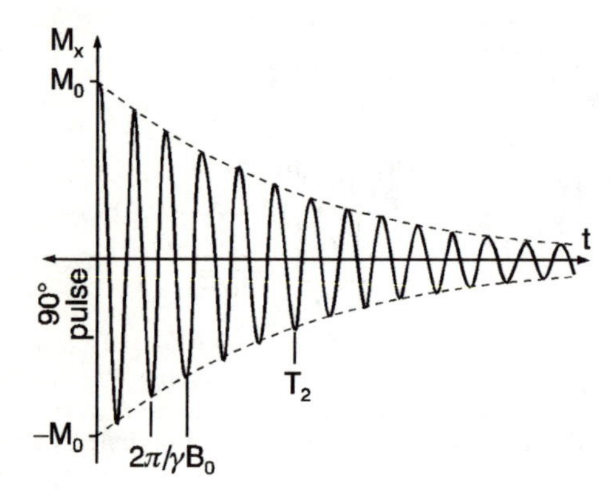

Fig. 11.4. Free induction decay of the signal in the pickup coil after a 90° pulse.

11.4 Continuous-Wave Experiments with Relaxation

ESR as well as NMR are often performed in CW configuration. This holds for experiments with induction in a pickup coil as well as for the absorption measurements in a microwave cavity. Thus, we have to work out the kinetics for this case, including relaxation.

11.4.1 Tuning the Resonance

To find the absorption either the field B_0 or the frequency ω has to be tuned. The speed of this tuning has an important consequence on the absorption behavior. It is measured with respect to the characteristic time constants of the experiment such as $1/\omega_0, T_2$, and T_1.

The change in the tuning parameter is called suddenly if it occurs faster than the inverse Larmor frequency. This means no precessional motion occurs during the change. An example would be a rapid switch of direction of the magnetic field B_0 by 90°. If the switch is faster than ω_0 no energy will be dissipated. Energy dissipation occurs only afterwards during the free induction decay.

The contrary are very slow or *adiabatic* changes. In this case the system is at any time in equilibrium. To reach this equilibrium energy must be dissipated. The adiabatic processes can either be fast or slow, depending on whether they are faster or slower than the time constants for transverse or longitudinal relaxation. The $\pi/2$-pulse described in Sect. 11.3 is an adiabatic fast process provided $\gamma B_1 \gg 1/T_2$. The process of the free induction decay or any CW resonance absorption experiments are adiabatic slow processes.

11.4.2 Susceptibility and Absorption of Power

To describe absorption it is convenient to introduce a magnetic susceptibility. We consider again the situation of a strong static (but tunable) field B_0 in z direction and a small rotating field B_1 perpendicular to it. The magnetization rotates around the effective field B^*_{eff} of Fig. 11.3. In the frame rotating with B_1 it has a component M'_x and M'_y, both proportional to B_1 but not necessarily equal in magnitude. We define two ratios χ' and χ''

$$\boxed{\chi' = \frac{M'_x \mu_0}{B_1} \qquad \text{and} \qquad \chi'' = \frac{M'_y \mu_0}{B_1}\ .} \tag{11.29}$$

These ratios are not very useful for an unrelaxed motion of the magnetization, since both, χ and M'_x, M'_y are time-dependent. This is different for the case where a relaxation of the magnetization is included. Then, the values for M'_x and M'_y become stationary in the rotating frame and (11.29) defines a susceptibility. Evidence of this change in kinetic behavior will be given below.

Since $\boldsymbol{M}$ and $\boldsymbol{B}_1$ rotate with the same frequency in the laboratory system in this case, the two relations from (11.29) can be expressed in the latter by

$$\mu_0 M = \chi B_1\ ,$$

where a complex notation for M, B_1 and $\chi = \chi' + \mathrm{i}\chi''$ was introduced. Note that $\boldsymbol{M}$ and $\boldsymbol{B}_1$ are not necessarily in phase, so that M may be complex even for real values of B_1.

The orientation of $\boldsymbol{M}$ in the rotating frame depends on the type of process considered and must be evaluated in each case. We will restrict ourselves in the following to the most widely used case of adiabatic slow processes which includes continuous-wave experiments.

The quantitative description of the kinetics of the magnetization is deduced from the Bloch equations. Represented in the rotating frame they read in components and with the use of (11.22) and (11.27) for the effective field B^*_{eff}

$$\begin{aligned}
\frac{\mathrm{d}M'_x}{\mathrm{d}t} &= -(\gamma B_0 - \omega)M'_y - \frac{M'_x}{T_2}\,,\\
\frac{\mathrm{d}M'_y}{\mathrm{d}t} &= (\gamma B_0 - \omega)M'_x - B_1 M'_z - \frac{M'_y}{T_2}\,,\\
\frac{\mathrm{d}M'_z}{\mathrm{d}t} &= \gamma B_1 M'_y + \frac{M_0 - M'_z}{T_1}\,.
\end{aligned} \tag{11.30}$$

ω is assumed independent of t. This means that ω is either constant or at most tuned adiabatically slow. Under these conditions (11.30) is a set of inhomogeneous linear differential equations of first order for $\boldsymbol{M'}$. The general solution for the homogeneous equation is an exponentially decaying function which dies out for $t \to \infty$. Thus, the particular solution for the inhomogeneous equation is the only steady-state solution. A particular solution is possible if all derivatives with respect to time are zero on the left side of (11.30). This means that $\boldsymbol{M'}$ is stationary (time-independent) in the rotating frame. With this we obtain from (11.30) a set of algebraic equations for M'_x, M'_y, and M'_z. The straightforward solution for $\boldsymbol{M}$ gives the complex susceptibilities

$$\begin{aligned}
\chi' &= \frac{M'_x \mu_0}{B_1} = \frac{\gamma(\gamma B_0 - \omega)T_2^2 M_0 \mu_0}{1 + (\gamma B_0 - \omega)^2 T_2^2 + \gamma^2 B_1^2 T_1 T_2}\,,\\
\chi'' &= \frac{M'_y \mu_0}{B_1} = \frac{\gamma T_2 M_0 \mu_0}{1 + (\gamma B_0 - \omega)^2 T_2^2 + \gamma^2 B_1^2 T_1 T_2}\,.
\end{aligned} \tag{11.31}$$

For small enough values of B_1 which means $\gamma^2 B_1^2 T_1 T_2 \ll 1$ this approaches

$$\begin{aligned}
\chi' &= \frac{\gamma(\gamma B_0 - \omega)T_2^2 M_0 \mu_0}{1 + (\gamma B_0 - \omega)^2 T_2^2}\,,\\
\chi'' &= \frac{\gamma T_2 M_0 \mu_0}{1 + (\gamma B_0 - \omega)^2 T_2^2}\,,
\end{aligned} \tag{11.32}$$

where χ' and χ'' describe the linear response M_x and M_y to B_1. Note that susceptibilities derived for a linearly polarized field B_1 are a factor of two smaller in comparison to the results from (11.32).

In the approach of (11.32) the susceptibility has the typical shape of a response function of an oscillator. The complex part is Lorentzian with a FWHM of $2/T_2$ and the real part has a zero crossing at resonance.

The change of energy density in the spin system per unit volume and per unit time is given by

$$\frac{\mathrm{d}W}{\mathrm{d}t} = \frac{\mathrm{d}}{\mathrm{d}t}(-\boldsymbol{M}\boldsymbol{B}) = -\boldsymbol{M}\frac{\mathrm{d}\boldsymbol{B}}{\mathrm{d}t} - \boldsymbol{B}\frac{\mathrm{d}\boldsymbol{M}}{\mathrm{d}t}\,. \tag{11.33}$$

The first part on the right side of (11.33) originates from the time-varying field and the second part describes the energy delivered to the lattice. Since for an adiabatic slow process no energy can be stored in the spin system the balance of (11.33) must be zero. In other words, the second part and the first part are of equal magnitude but opposite sign. Then the first part can be used to calculate the power dissipated to the lattice. From (11.23) at resonance $\mathrm{d}\boldsymbol{B}/\mathrm{d}t = \boldsymbol{\omega_0} \times \boldsymbol{B_1} = \omega B_1 \boldsymbol{e_y}$ has only a y component with magnitude ωB_1. With $M_y = \chi'' B_1/\mu_0$ the dissipated power is

$$W' = \left(\frac{\mathrm{d}W}{\mathrm{d}t}\right)_{\mathrm{diss}} = \chi'' \omega B_1^2/\mu_0 \,. \tag{11.34}$$

In combination with (11.31) this relation reveals that the resonance absorption decreases like $1/T$ as a consequence of the temperature dependence of M_0. Also, W' saturates for large values of B_1. The reason for this saturation is immediately evident from a statistical concept of the resonance absorption. Through the absorption process more and more spins arrive in the upper state until equal occupation is obtained. Under these conditions absorption is saturated since an equal number of spins is excited by the radiation from the lower level to the upper level and from the upper level to the lower level (see also Sect. 11.4.4).

For practical reasons in experiments often $\mathrm{d}W'/\mathrm{d}B$ is determined instead of W'. Alternatively the complex impedance can be measured which enables χ' and χ'' to be determined separately. In analogy to optics χ' and χ'' are called the *dispersive* and the *absorptive* part of the susceptibility.

11.4.3 Resonance Absorption

To measure the absorbed power of the radiation directly a resonance cavity or the immersion of the sample into a waveguide is appropriate. This technique corresponds more closely to the picture of transitions between the Zeeman levels. It is used, in particular, for ESR experiments as they are conducted in the spectral range of microwaves and the explicit handling of the electromagnetic field is much more elaborate in this case. Measurements of incoming and outgoing power is easier but still requires waveguides and resonator cavities. Reflex klystrons and bolometers or diode crystals are the standard radiation sources and detectors, respectively. Figure 11.5 shows a schematic experimental setup. The samples are inserted into a microwave cavity after they have been sealed in a very thin quartz or pyrex tube. The tube material must be completely free from paramagnetic impurities. To find the resonance positions the magnetic field is usually tuned by some extra windings on the static magnet used for the Zeeman splitting.

11.4.4 The Resonance Excitation as an Absorption Process

From the above discussion and from the description of the spin resonance as the transition between two Zeeman levels one would expect that the quantum-mechanical model, as developed in Sect. 7.1 for the optical absorption, applies

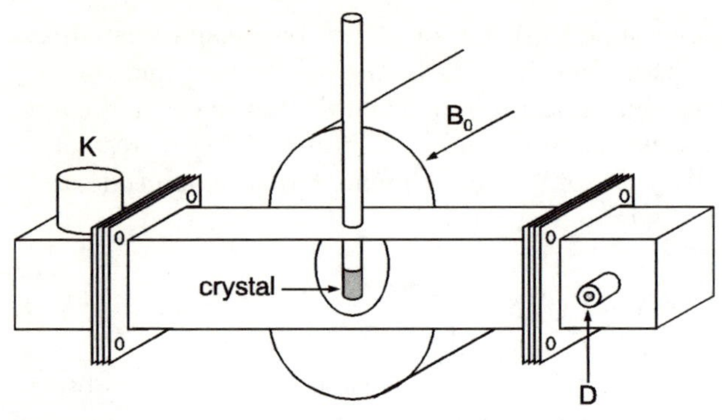

Fig. 11.5. Schematic setup for microwave absorption in magnetic resonance; (K: klystron, B_0: magnet, D: detector).

here as well. This is indeed true. The golden rule for the transition probability per unit time between two Zeeman levels becomes

$$P_{\Delta U_z} = \frac{2\pi}{\hbar} |H'_{\Delta U_z}|^2 \delta(\Delta U_z - \hbar\omega) \; . \tag{11.35}$$

The matrix element must be evaluated for transitions between the lower and upper Zeeman states which we labeled l and u in the following. There is, however, one important difference to the case of optical absorption. Since the difference in energy between the two states is only a few meV in ESR and only a few μeV in NMR the thermal population of the states is nearly equal. This means for the evaluation of the overall transition rate that the probability for the induced transitions from l to u as well as for the transitions from u to l must be considered. The matrix elements for both transitions are equal. [The square of the transition matrix elements corresponds to the Einstein coefficients B_{12} and B_{21} from Appendix C.4, equation (C.12)].

More details about the quantum-mechanical description of resonance absorption must be left to special textbooks [11.2]. In the following we will rather discuss the consequences of the nearly equal population of the upper and lower Zeeman levels on the rate equations.

The ratio between the population of energy levels l and u is given by the Boltzmann factor

$$\frac{n_\mathrm{l}}{n_\mathrm{u}} = \mathrm{e}^{g\mu B/k_\mathrm{B}T} \; . \tag{11.36}$$

Since for not too low temperatures and for not too high fields the exponent in (11.36) is $\ll 1$ the ratio of the two populations is well approximated by $(1 + g\mu B/k_\mathrm{B}T)$. Introducing the total number of spins per unit volume as $n = n_\mathrm{l} + n_\mathrm{u}$ the concentrations in the upper and lower levels are straightforwardly evaluated as

$$n_\mathrm{u} \; = \; (n/2)(1 - g\mu B/2k_\mathrm{B}T)$$

$$n_\mathrm{l} = (n/2)(1 + g\mu B/2k_\mathrm{B}T) . \tag{11.37}$$

This yields for the relative difference in the population

$$\frac{\Delta n}{n} = \frac{n_\mathrm{l} - n_\mathrm{u}}{n} = \frac{g\mu B}{2k_\mathrm{B}T} . \tag{11.38}$$

Since for protons $g_\mathrm{p}\mu_\mathrm{p}$ is only $4 \times 10^{-26}\,\mathrm{A\,m^2}$ the difference in magnetic energy $g_\mathrm{p}\mu_\mathrm{p}B$ is only $2.4\times10^{-4}\,\mathrm{meV}$ for a field of 1 T. At room temperature the thermal energy is, on the other hand, 27 meV so that the two populations differ only by a fraction of 10^{-5}.

We are interested in the energy absorbed by the spin system per unit time and per unit volume. This rate is given by

$$\frac{\mathrm{d}W}{\mathrm{d}t} = n_\mathrm{l}P_\mathrm{ul}(\epsilon_\mathrm{u} - \epsilon_\mathrm{l}) + n_\mathrm{u}P_\mathrm{lu}(\epsilon_\mathrm{l} - \epsilon_\mathrm{u}) = \Delta n(t)P\Delta\epsilon , \tag{11.39}$$

where $P = P_\mathrm{ul} = P_\mathrm{lu}$ and $\Delta n(t)$ are the transition probability from (11.35) and the difference $(n_\mathrm{l} - n_\mathrm{u})$ in the population of the levels l and u, respectively. The time dependence for $\Delta n(t)$ is readily obtained from the rate equation for the change of the population in the levels l or u. For level l we have

$$\frac{\mathrm{d}n_\mathrm{l}}{\mathrm{d}t} = n_\mathrm{u}P_\mathrm{lu} - n_\mathrm{l}P_\mathrm{ul} = P(n_\mathrm{u} - n_\mathrm{l}) \tag{11.40}$$

which yields by combining it with a similar equation for n_u

$$\frac{\mathrm{d}\Delta n}{\mathrm{d}t} = -2P\Delta n(t) . \tag{11.41}$$

The solution of (11.41) is an exponentially decaying function which implies that absorption goes to zero for increasing time. This is, of course, due to the equal rate of up and down transitions as soon as the population of the upper level and the lower level become the same. As in the case of the precessing magnetization we obtain a different result if energy relaxation from the spin system to the lattice is allowed. In this case the upper energy level is not only depopulated by the induced transitions but also by the conversion of magnetic energy to lattice energy. Using an exponential term for the relaxation as in (11.22) or (11.30) we can rewrite (11.41) as

$$\frac{\mathrm{d}\Delta n}{\mathrm{d}t} = -2P\Delta n(t) - \frac{\Delta n(t) - \Delta n(0)}{T_1} , \tag{11.42}$$

where $\Delta n(0)$ is the population difference at thermal equilibrium. For the steady-state solution we immediately obtain

$$\Delta n(t) = \frac{\Delta n(0)}{1 + 2PT_1} \tag{11.43}$$

and thus from a combination with (11.39) the absorbed power

$$\boxed{\left(\frac{\mathrm{d}W}{\mathrm{d}t}\right)_\mathrm{diss} = \Delta n(0)\Delta\epsilon\frac{P}{1 + 2PT_1} .} \tag{11.44}$$

Since P is proportional to the incident power of the electromagnetic field, a power for which $2PT_1 \gg 1$ leads to a saturation of the absorption. This is completely equivalent to the result of (11.34) with (11.31) for χ'' and large values of B_1. Note the difference for small values of B_1 where T_2 limits the absorption.

11.5 Relaxation Times and Line Widths for Magnetic Resonance

As seen from (11.34) in combination with (11.32) the linewidth of the resonance absorption is determined by T_2 even though the final relaxation of the absorbed energy to the lattice is given by T_1. It is the dephasing of the individual magnetic moments after the time T_2 which limits further absorption. What are the physical processes to determine the relaxation behavior?

The width of χ'' in a solid is to a first approximation given by the dipole–dipole interaction of the individual spins. The magnetic field of a dipole is well known from magnetostatics.

$$\boldsymbol{B}_\mu = \frac{3\boldsymbol{r}(\boldsymbol{\mu}\boldsymbol{r}) - \boldsymbol{\mu}r^2}{4\pi r^5}\mu_0 , \tag{11.45}$$

where $\boldsymbol{r}$ is the vector pointing from dipole μ to the fieldpoint. The interaction energy between two magnetic moments μ_i and μ_k with the distance r_{ik} is therefore

$$\epsilon_{\mathrm{d}}(ik) = -\boldsymbol{\mu}_i\boldsymbol{B}_k = \frac{\boldsymbol{\mu}_i\boldsymbol{\mu}_k}{4\pi r_{ik}^3}\mu_0 - \frac{3(\boldsymbol{\mu}_i\boldsymbol{r}_{ik})(\boldsymbol{\mu}_k\boldsymbol{r}_{ik})\mu_0}{4\pi r_{ik}^5} , \tag{11.46}$$

where $\boldsymbol{B}_k$ is the field of the dipol $\boldsymbol{\mu}_k$ at the position of the dipole $\boldsymbol{\mu}_i$. $\boldsymbol{r}_{ik}$ is the vector pointing from dipole $\boldsymbol{\mu}_i$ to dipole $\boldsymbol{\mu}_k$.

In general, the field at dipole $\boldsymbol{\mu}_i$ is the sum of all other randomly distributed dipoles in the system. Thus, the local field is subjected to random fluctuations. The width of the resonance lines is determined by these fluctuations. Its evaluation needs an averaging over the statistically distributed moments. A simple but physically instructive approximation is obtained by evaluating the maximum energy of one dipole in the field of an equal dipole at distance a_0. With $\mu = g\mu_{\mathrm{B}}J$ this yields from (11.46)

$$\epsilon_{\mathrm{d}} = \mu B_\mu \approx \frac{\mu_0 g^2 \mu_{\mathrm{B}}^2 J^2}{2\pi a_0^3} , \tag{11.47}$$

and a related equation for the nuclear spins. This quantity divided by $\hbar$ is a good measure for the linewidth (in ω) from the dipole–dipole interaction. For J or $I = 1/2$ and n spins per unit volume it yields, except for a constant factor of the order of 1 resulting from averaging procedures,

$$\boxed{\delta\omega = \delta(\gamma B) = \frac{\mu_0\gamma^2\hbar n}{8\pi}} . \tag{11.48}$$

As an example we consider the nuclear spin of ^{19}F in Ca F_2. ^{19}F has $I = 1/2$ and $\gamma_N = 2.5 \times 10^8$ s^{-1} T^{-1}. With $n = 5 \times 10^{28}$ F^- ions/m^3 we obtain a linewidth of $\delta\omega = 1.5 \times 10^4\,\mathrm{s}^{-1}$ or 0.06 m T. Line widths for electron spins are usually much larger because of the larger magnetogyric ratio γ_e.

Even though the linewidth from dipole–dipole interaction is a good basis linewidths are frequently observed to be much smaller than evaluated from (11.48). Obviously there are mechanisms which lead to a narrowing of the resonance lines. Two of them are well known: motional narrowing and exchange narrowing.

Motional narrowing of lines broadened by inhomogeneities results from a cancellation of broadening interactions by spin diffusion or molecular rotation. Since T_2 for inhomogeneous broadening is determined by a local variation in the precession frequency of the magnetic moments in space or time, the phase coherence is pertained for a longer time if the spins diffuse around. In this case spins which are advanced in their phase may arrive at a position with lower local field and will thus be retarded. Other spins which are behind may be accelerated in their precession if they arrive at a spot with higher field. Thus, on average the magnetic moments stay longer in phase and T_2 will increase. In such a case the dephasing will not proceed linear in time but diffusive with the square root of the time. For a field with a fluctuation amplitude $\pm\delta B$ and a characteristic time constant τ for one step in the spin diffusion the phase difference after n steps is

$$\Delta = \sqrt{n\tau}\,\gamma\delta B\,. \tag{11.49}$$

Assuming T_2 is obtained for a dephasing of 1 rad the number of steps needed is $n = 1/(\delta B)^2\gamma^2\tau^2$ steps. Hence, T_2 is given from

$$\boxed{1/T_2 = 1/n\tau = (\delta B)^2\gamma^2\tau\,.} \tag{11.50}$$

Without diffusion the phase difference would increase linearly in time. This means $T_2(0) = 1/\gamma\delta B$. Thus, for $\tau^{-1} \gg \gamma\delta B$ (11.50) describes a remarkable increase of T_2 with a correlated narrowing of the resonance line.

Exchange interaction is a quantum-mechanical phenomenon which causes an additional interaction energy with no counterpart in classical physics (see also Appendix J.2). Phenomenologically it is described by an exchange constant K characterizing the probability of the exchange of two electrons from two atoms or the exchange of two spins from two lattice sites. If the width of resonance absorption is determined by a dipole–dipole interaction and by exchange interaction T_2 is obtained from

$$\frac{1}{T_2} = \frac{\pi}{2\hbar}\frac{\epsilon_d^2}{K}\,. \tag{11.51}$$

For $K \gg \epsilon_d$ the half width $1/T_2$ can become very small. The physical background for the exchange narrowing is a diffusion of the spins between states split by exchange interaction. If N spins interact the resonance splits into N states between which the spins can diffuse. In this sense exchange narrowing is also based on motional narrowing. For spin concentrations higher than 100 ppm considerable exchange narrowing can be expected.

To relax the z component of the exited magnetization energy must be transferred to the lattice. This energy can either be dissipated to the phonons or to the electrons. For dissipation to phonons a direct process is possible. The energy of the excited spin system is directly converted into a phonon. Since the spin energy is rather low acoustic phonons are generated. The situation is very similar to non-radiative recombination in semiconductors. The relaxation rates are proportional to the temperature T like

$$\frac{1}{T_1} = K_{\mathrm{ph}}\omega_0^2 T \,, \tag{11.52}$$

where ω_0 is the transition frequency and K_{ph} is the spin–lattice interaction constant.

Other decay channels are possible and known as Raman or Orbach processes. Figure 11.6a shows the two processes together with the direct relaxation in an energy scheme. According to the Raman mechanism an incident phonon is scattered and gains energy like in an antiStokes Raman process. Since it is a scattering process all phonons independent of their energy can participate. As the phonon density of states increases with Ω a Raman process can be as likely as a direct process even though it is a four-particle event.

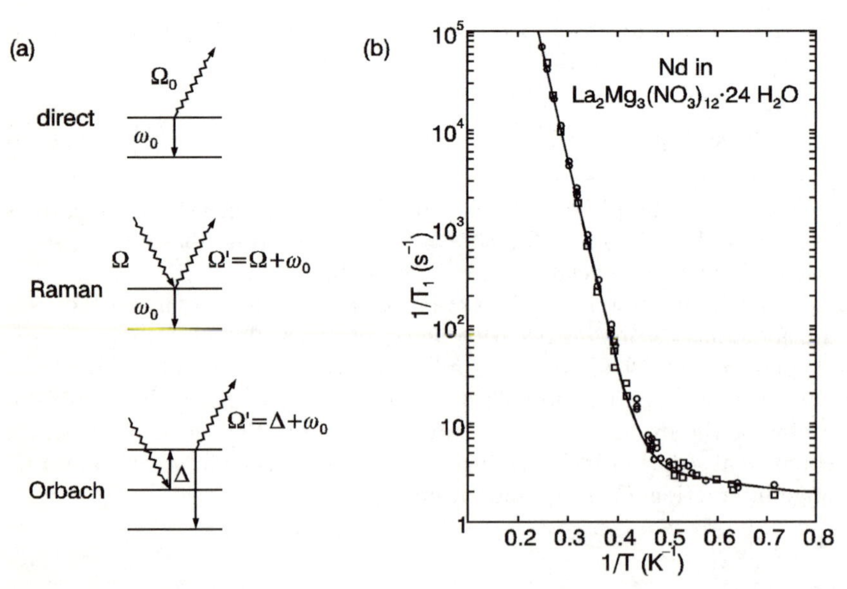

Fig. 11.6. Spin–lattice relaxation via phonons. Energy diagram for a direct, a Raman, and an Orbach process (**a**), and inverse of longitudinal relaxation time T_1 on a logarithmic scale versus inverse temperature for the electron spin of Nd in $La_2Mg_3(NO_3)_{12}{\cdot}24\ H_2O$ at 9.37 GHz (**b**); after [11.4].

The relaxation rate from a Raman process is proportional to T^7 or T^9. In the Orbach process the spin from the upper level is fist excited thermally to a third level and relaxes only from there to the ground state by the emission of a phonon. According to the thermal activation involved relaxation by the Orbach process is proportional to $\exp(-\Delta/k_\mathrm{B}T)$. Figure 11.6b shows the relaxation rate for the spins of Nd impurities in a crystalline matrix. The high temperature part of the curve shown is clearly exponential and thus indicates the dominance of an Orbach process. Relaxation rates via lattice deformations can be particularly strong close to structural phase transitions.

In systems with free carriers like metals longitudinal relaxation can also occur by dissipation of energy to the electrons (or holes). This is known as *Korringa relaxation*. Korringa relaxation originates from the coupling of spins to the fluctuating field the electrons generate at their position. In the case of nuclear spins a Korringa relaxation is thus immediately related to the hyperfine interaction and to the Knight shift $K = (\omega_\mathrm{K} - \omega_0)/\omega_0$ of the resonance line to be discussed in Sect. 11.8. The relaxation energy is transferred from the spin system of the nuclei to the spin system of the free electrons. As a consequence of the Pauli principle, only the fraction $k_\mathrm{B}T/\epsilon_\mathrm{F}$ of electrons around the Fermi energy can participate. This leads to the famous Korringa relation

$$TT_1K^2 = \frac{4\chi_\mathrm{P}^2}{\pi k_\mathrm{B}\gamma_\mathrm{e}^2\gamma_\mathrm{N}^2\hbar^3 g_\mathrm{v}^2(\epsilon_\mathrm{F})\mu_0^2}\,, \tag{11.53}$$

where $g_\mathrm{v}(\epsilon_\mathrm{F})$ is the density of states per unit volume at the Fermi level, χ_P is the Pauli susceptibility, and K is the Knight shift. For a Sommerfeld electron gas (11.53) can be simplified using

$$g_\mathrm{v}(\epsilon_\mathrm{F}) = \frac{3n}{2\epsilon_\mathrm{F}} \tag{11.54}$$

and

$$\chi_\mathrm{P} = \mu_\mathrm{B}^2 g_\mathrm{v}(\epsilon_\mathrm{F})\mu_0 = \frac{(\gamma_\mathrm{e}\hbar)^2}{4} g_\mathrm{v}(\epsilon_\mathrm{F})\mu_0\,, \tag{11.55}$$

where n is the electron density. This yields the standard formulation of the Korringa relaxation

$$\boxed{\frac{1}{T_1T} = K^2S_0 = K^2\frac{4\pi k_\mathrm{B}}{\hbar}\left(\frac{\gamma_\mathrm{N}}{\gamma_\mathrm{e}}\right)^2 .} \tag{11.56}$$

In metallic systems the Korringa relaxation dominates over the phonon relaxation. The lack of any temperature dependence for the product T_1T is therefore often used to check the metallicity of a spin system for low temperatures. This is possible even though a direct spin–lattice relaxation yields also TT_1 independent of temperature. However, in metals the direct spin–lattice relaxation as well as the other spin-phonon relaxation mechanisms are always much weaker than the Korringa relaxation.

The Korringa relation also holds for the interaction of localized electron spins with free carriers. In this case the coupling is by exchange interaction.

11.6 The Effective Spin Hamiltonian

Since for magnetic resonance the origin of the electronic states of the crystals is not the central problem the relevant Hamiltonian is not the one for the electrons and nuclei in the crystal as we used it in the previous chapters. What is essential are the interactions of the magnetic moments with an external field or with the internal field generated by themselves. These interactions are the origin of the different types of splitting observed. The Hamiltonian which describes the most important interaction processes is known as the *effective spin Hamiltonian* and has the form

$$\boxed{\mathbf{H} = \mu_{\mathrm{B}}\mathbf{B}\,\mathbf{g}_{\mathrm{e}}\,\mathbf{S} + \mathbf{S}\,\mathbf{D}\,\mathbf{S} + \mathbf{S}\,\mathbf{A}\,\mathbf{I} - \mu_{\mathrm{n}}\mathbf{B}\,\mathbf{g}_{\mathrm{N}}\,\mathbf{I} + \mathbf{I}\,\mathbf{Q}\,\mathbf{I} + \dots\,.} \tag{11.57}$$

$\mathbf{S}$ and $\mathbf{I}$ are the operators for the electronic and nuclear spin. According to the comment given above about the quenching of the orbital quantum number J was assumed equal to S. $\mathbf{D}$, $\mathbf{g}_{\mathrm{e}}$, $\mathbf{A}$, $\mathbf{g}_{\mathrm{N}}$ and $\mathbf{Q}$ are operators describing the influence of the crystal field (fine structure, $\mathbf{D}$), the Zeeman splitting of electronic states ($\mathbf{g}_{\mathrm{e}}$), the electron-nuclear spin interaction (hyperfine structure $\mathbf{A}$), the Zeeman splitting for the nuclear spins ($\mathbf{g}_{\mathrm{N}}$) and the nuclear quadrupole interaction ($\mathbf{Q}$). With the various resonance techniques the different interactions are probed. In principle, many other interaction Hamiltonians like the one for spin–lattice interaction, etc., can be added. In the following section we will discuss the interactions without considering the operator character of the coupling terms. All operators will be replaced by their eigenvalues and treated as coupling constants. However, the tensor character of these constants has to be retained for crystalline material as well as for molecules.

11.7 Electron Spin Resonance

The simplest case of a magnetic resonance experiment is a study of the resonance absorption of electron spins at the Zeeman energy. Only the first contribution to the spin Hamiltonian from (11.57) is used. From (11.13) the field for resonance absorption is

$$B_{\mathrm{R}} = \frac{\hbar\omega}{g_{\mathrm{e}}\mu_{\mathrm{B}}}\,, \tag{11.58}$$

where the selection rule $\Delta S = 1$ was applied. For a field (given in tesla) and for $g_{\mathrm{e}} = 2$ this yields

$$f_R = g_e \mu_B B / 2\pi\hbar = 28 \times 10^9 B \, [\mathrm{Hz}] \, . \tag{11.59}$$

Except for cubic crystals the tensor character of g_e must be considered. The effective g-value is then given by the components of the g-tensor and the orientation of the field with respect to the crystal axes. This orientation is conveniently described by the direction cosines l, m, n of the angles the field makes with the crystal axes. For Bravais lattices with trigonal or higher symmetry the g-tensor is diagonal and the effective g-factor is obtained from

$$g = \sqrt{l^2 g_x^2 + m^2 g_y^2 + n^2 g_z^2} \, . \tag{11.60}$$

Figure 11.7a shows the position of resonance absorption for the organic crystal $Qn(TCNQ)_2$ (quinolinium-tetracyanoquinodimethane) as a function of the orientation of the magnetic field relative to the crystal axes. Qn and TCNQ are two large organic molecules accommodated in a trigonal bravais lattice. Thus, the tensorial character of g is retained for any electron spins and the Zeeman splitting depends on the orientation of the field. The spin in the system originates from a charge transfer of one electron between the two molecules which renders TCNQ as a positive charged ion with one unpaired electron. This electron gives rise to the ESR signal with $S = 1/2$. The diagonal components of the g-tensor were found to be 2.00236, 2.00279, 2.00356.

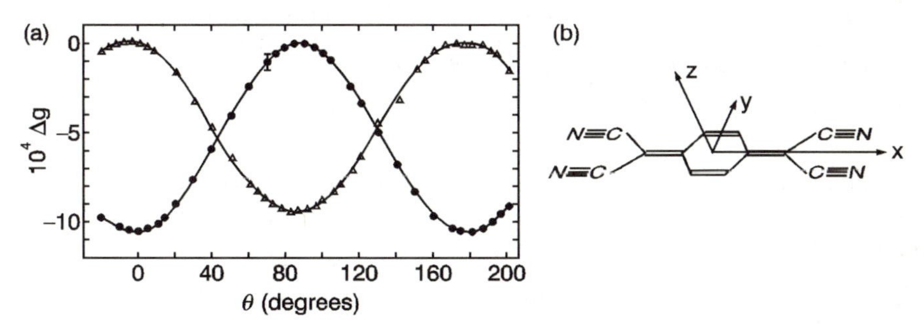

Fig. 11.7. Dependence of electron spin resonance position expressed as Δg of the organic crystal $Qn(TCNQ)_2$ on the orientation of the field in the crystal. The resonance was measured versus a reference to increased accuracy; (•) field in yz plane, (△) field in xy plane (**a**); (**b**) shows the TCNQ molecule with the coordinates for g_{ik}; after [11.7].

Even in cubic systems the position of the energy levels for the spin states depend very often on the orientation of the magnetic field with respect to the crystallographic axes. This is, in particular, so if the spin is larger than 1/2. In this case higher-order terms change the diagonal elements and introduce nondiagonal contributions to the g-tensor.

Besides the pure Zeeman splitting the system described by

$$H = SDS + \mu_B g_e SB \tag{11.61}$$

is the simplest case, in particular if the term SDS is small. The crystal field modifies the Zeeman levels. This does not only lead to a shift of the resonance absorption but also to an additional splitting if $S > 1/2$. The splitting is a consequence of the dependence of the shift on the individual Zeeman states. Figure 11.8 shows the splitting of the ESR line of Cr^{3+} in $AlCl_3$ as a function of the orientation. θ is the angle between [100] direction and a magnetic field in a (011) plane. The threefold degeneracy of the transition between the four Zeeman levels of the $S = 3/2$ spin of Cr^{3-} is lifted.

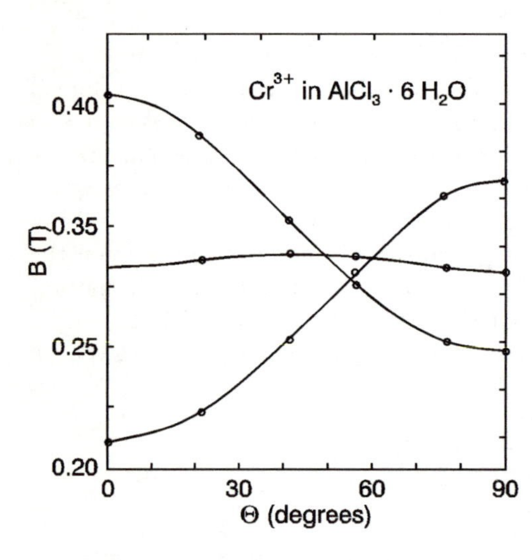

Fig. 11.8. Splitting and anisotropy of electron spin resonance of Cr^{3+} with S=3/2 in $AlCl_3$; after [11.6].

The most important and thus also the most often investigated splitting of the resonance lines originates from hyperfine interaction. The corresponding term in (11.57) is

$$H = g_e \mu_B SB + ASI . \tag{11.62}$$

Because of the tensorial character of g and A the hyperfine interaction cannot be read directly from (11.62). It must be calculated again for the different orientations of the field from the direction cosines l, m, n. As in (11.13) the energy levels are given by the possible z components of the spins. For diagonal A and g tensors the result is

$$U_{S_z,I_z} = \epsilon = g_e \mu_B B S_z + a S_z I_z \tag{11.63}$$

with

$$a = \frac{1}{g}\sqrt{A_x^2 g_x^2 l^2 + A_y^2 g_y^2 m^2 + A_z^2 g_z^2 n^2}$$

and

$$g = \sqrt{g_x^2 + g_y^2 + g_z^2} \, .$$

The splitting of the levels for a system with $S = 1/2$ and $I = 1$ is displayed in Fig. 11.9. The electronic levels split into $2S + 1 = 2$ Zeeman components and each of them splits into $2I + 1 = 3$ nuclear Zeeman levels with the distance $|(aS_z/2) \times 2|$ between two levels. Since the selection rules for transitions are again $\Delta S_z = 1$ the resonances occur for

$$\hbar\omega = g_e \mu_B B + a I_z \tag{11.64}$$

and the resonance field is at

$$\boxed{B_R = \frac{\hbar\omega}{g_e \mu_B} - \frac{a I_z}{g_e \mu_B}} \, . \tag{11.65}$$

A strong signal is only obtained for $\Delta I_z = 0$. Thus the spin resonance line splits into $2I + 1$ components separated by a constant field $a/g_e\mu_B$. Equation (11.64) is easily verified from Fig. 11.9.

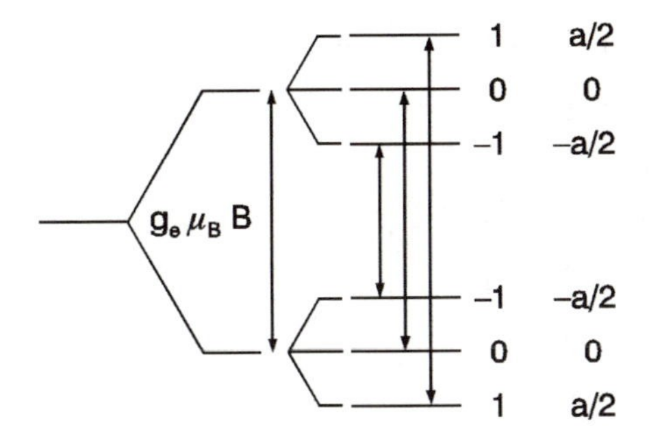

Fig. 11.9. Splitting of energy levels by hyperfine interaction in a system with $S = 1/2$ and $I = 1$.

The splitting obtained for $\Delta S_z = 0, \Delta I_z = 1$ is the nuclear Zeeman effect.

The hyperfine splitting is particularly simple for the electron in hydrogen. From $I = 1/2$ for the proton a splitting into two lines results in a difference in resonance frequency of $a/2\pi\hbar = 1420\,\text{MHz}$ or $50.7\,\text{mT}$. This corresponds to an energy difference of $6\times10^{-6}\,\text{eV} \approx 6\,\mu\text{eV}$.

Figure 11.10 exhibits the hyperfine splitting of Mn^{2+} in ZnS. The spectrum is represented by the derivative of the absorped power $\mathrm{d}W'/\mathrm{d}B$. Mn^{2+} has an electron spin and a nuclear spin of 5/2 each. Sweeping the field B_R we expect six resonance lines from (11.65) as shown in the figure. The Zeeman splitting for the electronic spins gives six levels with equal distance. Thus, all electronic Zeeman transitions give the same hyperfine pattern.

The arrangement of the energy levels in Fig. 11.9 enables interesting double-resonance experiments. While exciting the spins with microwaves between the electronic Zeeman levels an RF frequency can be applied to induce

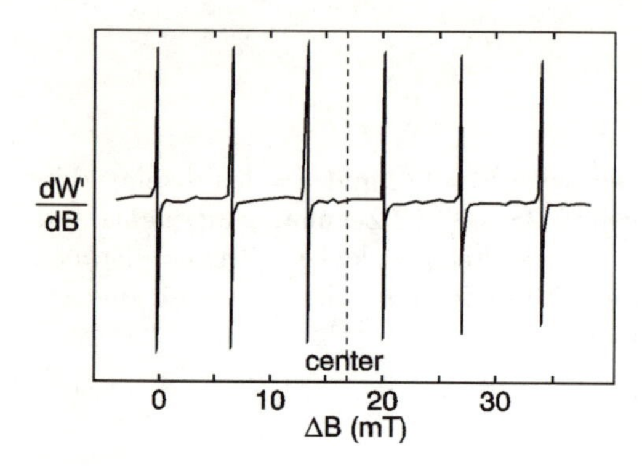

Fig. 11.10. Hyperfine splitting of electron spin resonance for Mn^{2+} in ZnS; after [11.7].

transitions between the levels of the nuclear spin. Since both excitations start or end at the same levels there is a strong interaction between the two resonance transitions. Such experiments are known as electron-nuclear double resonance (ENDOR) or dynamic nuclear polarization (DNP).

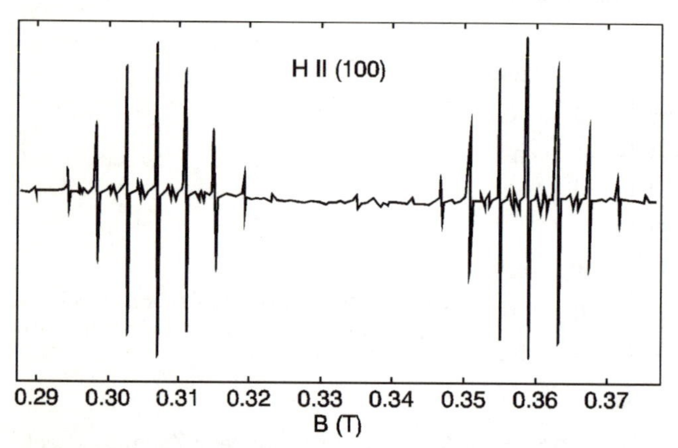

Fig. 11.11. Electron spin resonance of hydrogen on a cubic interstitial site in CaF_2; after [11.8].

A multiple splitting of the ESR signal occurs also if the electron spin interacts with several *equivalent* nuclear spins I. Since the nuclear spins can be added to yield an effective nuclear spin with maximum value NI, $2NI + 1$ possible lines follow where N is the number of equivalent nuclear spins. Figure 11.11 shows the ESR spectrum of hydrogen in CaF_2. In this crystal the F^- ions are accommodated on the edges of a cube. In every other center of the cubes are the Ca^{++} ions. The free cubes can accommodate the hydrogen atoms which has thus 8 F^- neighbors. Since ^{19}F has spin 1/2 the resonance splits into nine lines. However, the absorption of the lines do not have the same intensity since the probability for generating a particular effective spin is not the same for the different configurations. The maximum effective total spin is obtained by only one configuration where all equivalent nuclear spins are parallel. Smaller effective total spins are obtained by several or even many spin configurations. Therefore, the central lines of the splitting are the strongest. Away from the center the lines become weaker and weaker.

The hyperfine interaction between electron and nuclear spin of the hydrogen leads in CaF^2, in addition to the ninefold splitting to a doubling of the absorption lines where both components have the same intensity. The two components interact independently with the eight nearest neighbor nuclear spins of ^{19}F.

If the wave function of the spin is extended over several nuclei the discrete lines of the hyperfine splitting are washed out and a broad line centered around $I = 0$ appears. Motional narrowing or exchange interaction can again reduce the width of this line but the splitting is lost. Figure 11.12 depicts the behavior for electron–proton interaction. Spectrum (1) is the response for a localized electron spin with a splitting of $a/2\mu_B$ and spectrum (2) is for a delocalized spin. In spectrum (3) the resonance line appears narrowed from spin diffusion. The hyperfine interaction is fully averaged out.

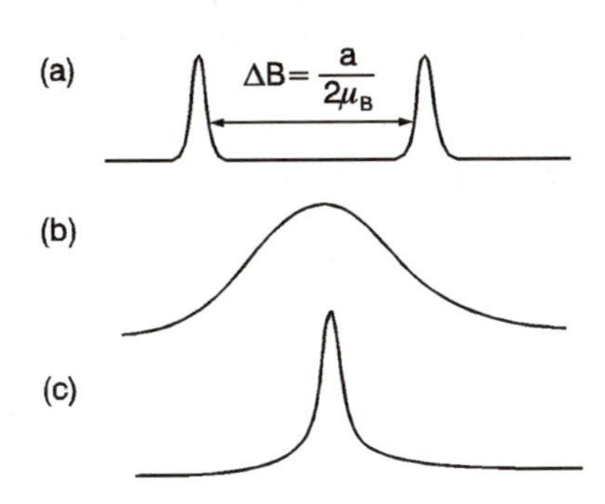

Fig. 11.12. Lineshape of the electron spin resonance with hyperfine interaction; localized spins (1), delocalized spins (2), delocalized spins with motional narrowing (3).

11.8 Nuclear Magnetic Resonance

From the value for μ_n a resonance frequency for NMR is in the range 10–100 MHz for a static field of the order 1 T. More precisely (11.58) yields (for B in tesla) and for protons

$$f_R = g_p \mu_p B / 2\pi\hbar = 42.58 \times 10^6 B \, [\text{Hz}] \,. \tag{11.66}$$

Nuclear magnetic resonance experiments are usually carried out with a setup shown in Fig. 11.2. The frequency is tuned instead of the field until all positions of resonance absorption are detected.

11.8.1 The Chemical Shift

Considering that the magnetic moments are only given by the spin of the protons and the neutrons in the nucleus one would expect the same NMR signal for all atoms or at least definitely the same signal for a particular atom in various materials. This is not the case. Looking at the NMR signal of diethylamine (Fig. 11.13) we find three resonances with a relative intensity (area of lines) of 6:4:1, even though only the hydrogens contribute to the resonance (proton NMR). The reason for this unexpected result originates from the accommodation of the protons in three different chemical environments. The field in (11.66) is not the externally applied field but the effective local field B_{loc} which is given by the applied field and by the contributions from the chemical environment. Since for each type of bonding a different chemical environment is expected several resonance frequencies will be observed for one nucleus. This effect is known as *chemical shift*. In diethylamine the hydrogens are located in three different chemical environments which explains the three different lines. Moreover, the ratio of the numbers of hydrogens in the different chemical bonds is 6:4:1, exactly like the intensities of the resonance absorption. Each nucleus contributes the same to the total oscillator strength. This means that the NMR spectrum of a compound is a fingerprint of the distribution of the hydrogen atoms or other NMR active nuclei in the material. NMR spectroscopy is therefore a very important tool for the analysis or characterization of new compounds, particularly in organic chemistry.

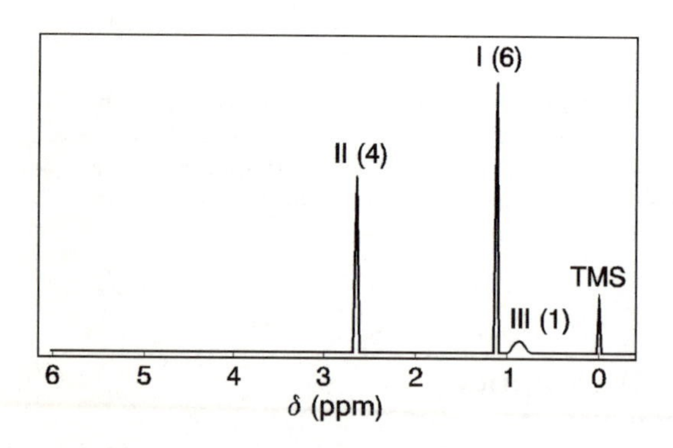

Fig. 11.13. NMR signal for diethylamine $(CH_3CH_2)_2NH$. TMS is the calibration signal from tetramethyl silane.

The line assigned as TMS in Fig. 11.13 originates from protons in tetramethylsilane [TMS, $(CH_3)_4Si$]. It serves as calibration. The chemical shift is often given in relative units as

$$\delta = \frac{\Delta f}{f_0} \times 10^6 \qquad \text{(in ppm)} . \tag{11.67}$$

Δf is the distance of the resonance from the TMS signal and f_0 the excitation frequency.

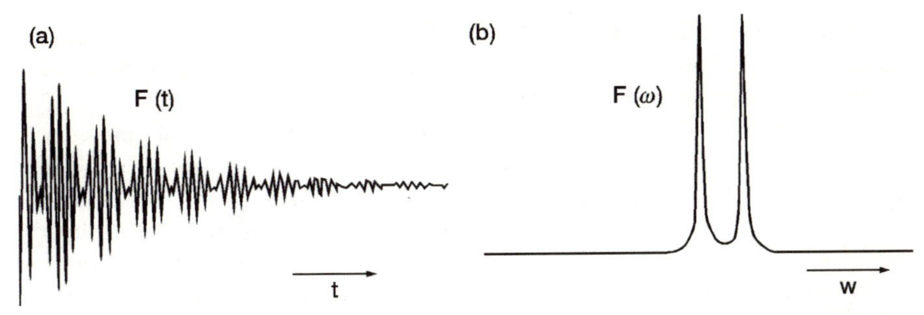

Fig. 11.14. NMR signal of a spin system with two resonance positions after a $\pi/2$-pulse (**a**), and absorption spectrum evaluated by Fourier transformation (**b**); after [11.9].

In general, the chemical shift at a particular position in a molecule or in a crystal is also anisotropic. This means the shift depends on the orientation of the field with respect to the molecular axis. Thus, in polycrystalline samples the chemical shift anisotropy leads to a line broadening.

NMR has been used very successfully in chemical analysis of compounds for more than 40 years. Although new fields of application are ever appearing, fundamental applications in solid-state physics and in medicine have been developed. Examples are the different forms of pulsed NMR, artificial narrowing of resonance lines, two-dimensional NMR, and NMR tomography.

11.8.2 Pulsed Nuclear Magnetic Resonance

In the case of pulsed NMR the free induction decay is observed after the spin system has been excited by an RF pulse as described in Sect. 11.2. Detection is with a sensor coil. From the drop in the signal the transverse relaxation time is immediately obtained. Since the time dependence of the damped oscillation consists of a superposition of eigenfrequencies of the precessing moments it contains all information about the resonance positions of the system. Thus, from a Fourier transformation of $F(t)$ of the pickup signal the full NMR spectrum can be determined. This is known as Fourier transform NMR spectroscopy. Figure 11.14a shows the measured FID of a spin system with two resonances. In part (b) of the figure the NMR spectrum is displayed as obtained from a Fourier transformation.

Besides the simple Fourier analysis of the FID, several procedures exist in which spins are excited with a well defined sequence of RF pulses of varying duration. Figure 11.15 shows two possibilities. In case (a) the spin system is first excited by a $\pi/2$-pulse. Within or immediately after the transverse relaxation time but definitely well within T_1 a second pulse is applied which rotates the magnetization by 180° (π-pulse). After the first pulse and the subsequent decay of the phase coherence of the individual spins they are still well inclined with respect to the z axis. The second pulse inverts the direction of precession of the spins so that their phase coherence increases again. If the π pulse is applied after a time τ the original signal is regained as an echo after the time 2τ, though with somewhat lower intensity. The $\pi/2 - \pi$ sequence is known as a *Hahn* sequence. The signal of the echo is decreased by $\exp(-2\tau/T_2)$ from the original pulse. For the other case shown in Fig. 11.15b the second pulse is again $\pi/2$ and applied after a time interval τ with $T_2 < \tau < T_1$. Since the individual magnetic moments are already dephased but not yet returned to the z axis the observed free induction decay after the second pulse is smaller. From a measurement of the signal height just after the second pulse for different values of τ the longitudinal relaxation time can be determined.

A large number of other pulse sequences have been studied and are presently used in NMR. Details can be looked up in special references [11.10].

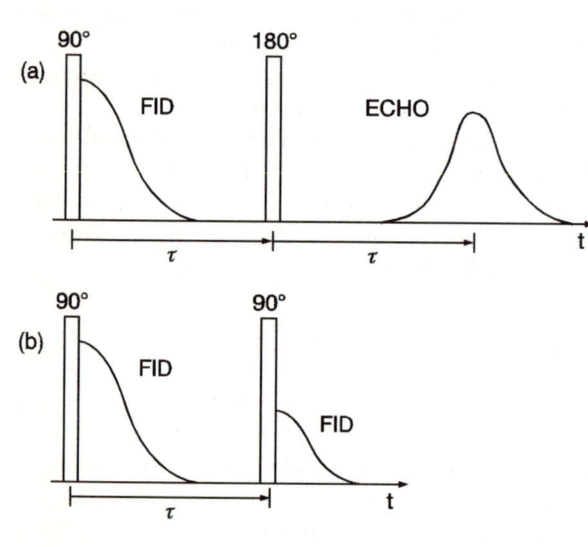

Fig. 11.15. Pulse sequence for a time resolved NMR spectroscopy; spin echo sequence (**a**), and pulse sequence for the determination of T_1 (**b**).

11.8.3 Magic-Angle Spinning NMR

The use of high resolution NMR in solid-state physics was delayed initially since the linewidths were very large as a consequence of the anisotropy of the spin–spin interaction and the chemical shift. Motional narrowing as in liquids is observed in solids only at high temperatures. Since the spin–spin interaction is a dipole–dipole coupling and thus well defined in its geometry a trick can be used to bring the effective spin–spin interaction to zero.

If all spins are oriented in one direction by a magnetic field the spin–spin interaction becomes zero if the vector $\boldsymbol{r}_{ik}$ pointing from one spin to the other makes a *magic angle* of $54° \, 44'$ to the orienting field. For this orientation the

z component of the field of magnetic moment μ_i at the position of moment μ_k

$$B_{ik,z} = \frac{\mu_i}{4\pi r_{ik}^3}(3\cos^2\Theta_{ik} - 1)\mu_0 \tag{11.68}$$

becomes zero. The geometrical situation for the quenching of the spin–spin coupling is demonstrated in Fig. 11.16.

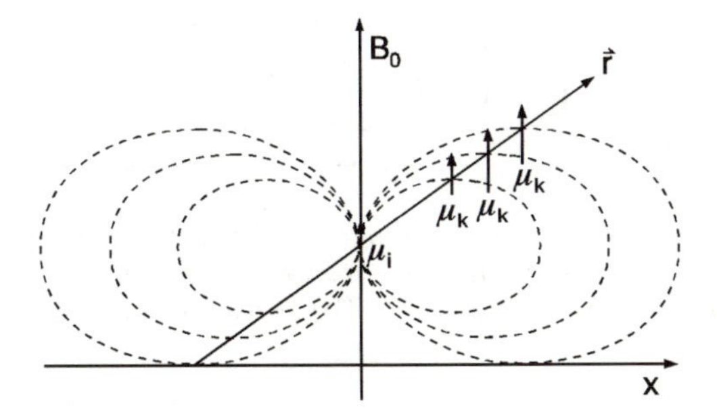

Fig. 11.16. Geometric arrangement of magnetic moments μ_i and μ_k for zero interaction.

Even though in reality the spins are distributed on the lattice sites a very fast rotation of the crystal around the magic angle of 54° 44′ with respect to the field B_0 generates a geometry where the connection lines between the spins are on average oriented under 54° 44′ to B_0. The rotation must be faster than the inverse of the spectral widths to be investigated. Then the interactions from the different positions relative to B_0 obtained successively during the rotation cancel on average and the spins appear as non-interacting. The situation is shown in Fig. 11.17a. The cancellation holds also for broadening from the chemical shift anisotropy as a consequence of a motional narrowing. Thus, the resonance linewidths become very narrow even for solids. An example is depicted in Fig. 11.17b. Spectrum (1) is a conventional NMR response. The lines are very broad and not very useful for analysis. In contrast spectrum (2) is obtained if the sample is rapidly rotated around the magic angle. The gain in resolution is obvious from the figure. In praxis spinning speeds of several KHz are used.

11.8.4 Knight Shift

The hyperfine interaction between electron spin and nuclear spin can also be observed in NMR. It appears as the influence of the field of the magnetic moments of the electrons at the position of the nuclei. A shift in resonance position and a splitting of lines can occur. The former is very well investigated in metals and known as the *Knight shift.*

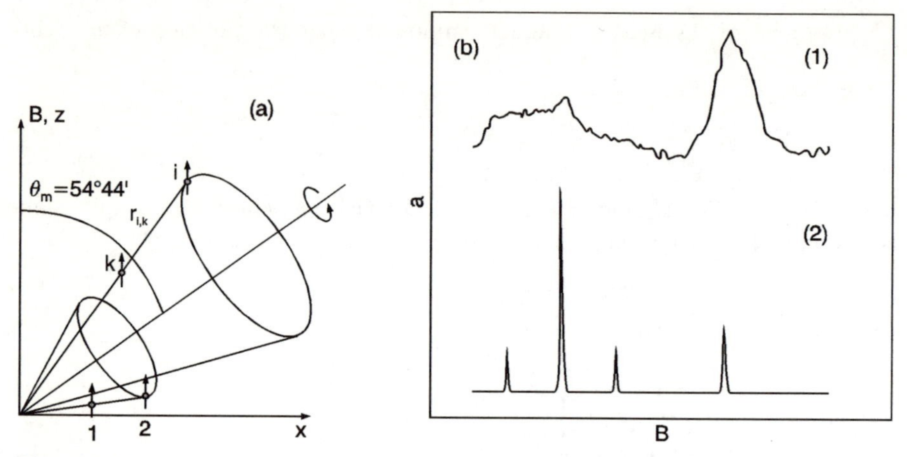

Fig. 11.17. Spinning of nuclear moments about the magic angle. 1,2 and i, k are arbitrarily selected nuclear magnetic moments (**a**). Line narrowing by spinning about the magic angle 54° 44′ for a 200-MHz proton NMR (**b**); without spinning (1) and with spinning (2); after [11.9].

The appropriate Hamiltonian is again the one for hyperfine interaction with the electron Zeeman interaction replaced by the nuclear Zeeman interaction, i.e.,

$$H = -g_N\mu_n B I_z + a I_z S_z = -g_N\mu_n I_z \left(B - \frac{aS_z}{g_N\mu_n}\right) . \tag{11.69}$$

The second contribution in parentheses on the right hand side is the effective local field produced by the electrons at the position of the nucleus. In reality, only the mean value of S_z is effective. Thus, the shift in the field of resonance is given by

$$\Delta B = \frac{a\langle S_z\rangle}{g_N\mu_n} . \tag{11.70}$$

The quantity $\langle S_z\rangle$ is a thermal average and can be calculated from the bulk magnetic susceptibility χ' of the electrons since

$$M_0 = \chi'_{el} B/\mu_0 = n g_e \mu_B \langle S_z\rangle .$$

Thus, the relative shift in the resonance field (or frequency) is

$$\boxed{\frac{\Delta B}{B_{res}} = \frac{\chi_{el} a}{n\mu_0 g_e\mu_B g_N\mu_n} .} \tag{11.71}$$

The hyperfine coupling constant a can be related to the local density ϱ_N of the electron spin at the position of the nucleus by

$$a = \frac{2}{3} g_e\mu_B g_N\mu_n\mu_0\varrho_N . \tag{11.72}$$

This finally yields the Knight shift

$$\boxed{K = \frac{\Delta B}{B_{\text{res}}} = \frac{2}{3}\chi_{\text{el}}\frac{\varrho_{\text{N}}}{n}\,.} \tag{11.73}$$

The classical case of a Knight shift in metals is obtained if the Pauli susceptibility χ_{P} is used for χ_{el}. It enables the determination of the electron concentration ϱ_{N} at the position of the nucleus.

The Knight shift can be observed explicitly by studying the NMR for a metal atom in a metallic crystal and in a nonmetallic compound. For ^{63}Cu the Knight shift is 0.237%. Since $\gamma_{\text{N}}/2\pi$ for ^{63}Cu is $11.28\times10^6\,\text{Hz}\,\text{T}^{-1}$ the line shifts by 26.73 KHz for a field of 1 T.

11.8.5 Two-Dimensional NMR and NMR Tomography

For more complex pulse schemes in NMR experiments the detected magnetization can be a function of several independently varying time scales. For example, in a two-dimensional NMR experiment the magnetization would be a function of the form $M(t_1, t_2)$. Fourier transformation with respect to t_1 and t_2 yields a two-dimensional spectrum $S(\omega_1, \omega_2)$.

A typical example of a two-dimensional NMR spectrum is the detailed analysis of a pulsed NMR experiment. The procedure can be divided into three parts. The first is the preparation period, the second the development period, and the third the detection period. In the first period the spins are prepared, e.g., by applying a $\pi/2$-pulse. In the second period the spin system develops for various times given by t_1. In the third period the magnetization is detected as a function of time $t_2 = t - t_1$. t_1 acts as a parameter. If the number of experiments with different values for t_1 is big enough the magnetization can be considered as a two-dimensional function $M(t_1, t_2)$.

Another multi-dimensional NMR spectroscopy refers to the spatial resolution of the distribution of NMR-active nuclei. In this case a magnetic field with a constant gradient is applied to the sample. This means a 1:1 relation between resonance frequency and position in space is established. Then, from the observed intensity for each frequency the distribution of the active nuclei can be determined. If this is done for several nuclei and for several directions a computer can calculate the distribution and plot it as a three-dimensional image. This procedure is known as *NMR tomography* and widely used in medical applications. It enables pictures to be made of the interior of biological systems in vivo without the use of the dangerous x rays. Figure 11.18 exhibits the NMR tomogram of a human head.

11.9 Nuclear Quadrupole Resonance

The last contribution in the spin Hamiltonian (11.57) describes the interaction of an electric nuclear quadrupole moment with an electric field gradient.

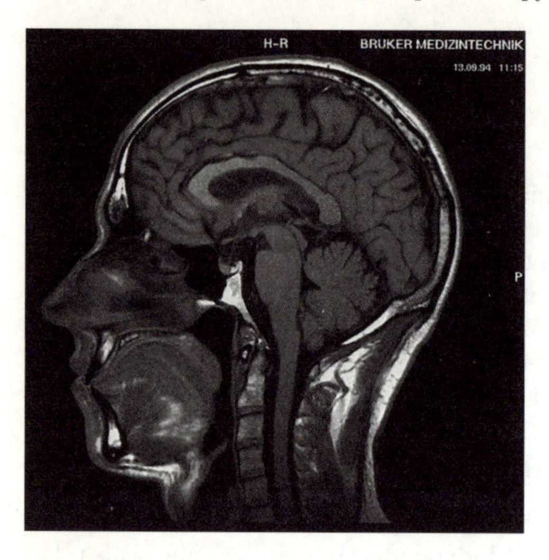

Fig. 11.18. NMR tomogram of a human head; after [11.11].

The field gradient is generated at a particular nucleus from the surrounding valence electrons and from the other nuclei. This means nuclear quadrupole resonance (NQR) spectroscopy does not need an external field. The nuclear quadrupole moment P_{Q} is given by the distribution of the charges in the nucleus. According to (B.12) it has the form

$$P_{\mathrm{Q}} = \int (3z^2 - r^2)\varrho(\boldsymbol{r})\mathrm{d}^3x \qquad (\text{in A s m}^2)\,. \tag{11.74}$$

Only nuclei with $I > 1/2$ can have a quadrupole moment. If U is the local potential energy the field gradient is a second-rank tensor with the components $\partial^2 U/\partial x_i x_k$. If the tensor has axial symmetry only the component U_{zz} contributes to the quadrupole splitting. From a first order perturbation calculation the resonance transition occurs for

$$\boxed{\hbar\omega_{\mathrm{Q}} = \frac{3P_{\mathrm{Q}}U_{zz}}{2I(2I-1)}}\,. \tag{11.75}$$

Table 11.2 lists quadrupole moments for various nuclei. Since nuclear quadrupole moments are known for most nuclei U_{zz} at the position of the nucleus can be determined. Its value depends strongly on the shielding effect of the valence electrons. The quadrupole splitting of the nuclear magnetic levels can be larger or smaller then the Zeeman splitting for conventional fields. For iodine in ICN the splitting due to the field gradient is 2.42 GHz.

Problems

11.1 Show that the magnetic moment for an electron in the first Bohr orbit of hydrogen is $-e\hbar/2m_0$.

(Purpose of exercise: recall connection between circular motion of charge and magnetic moments.)

11.2 Evaluate the g-factor and the magnetogyric ratio γ_{ne} for the neutron.

(Purpose of exercise: remember important definitions in NMR.)

11.3 Solve the relaxation-free Bloch equations and evaluate from this the Larmor frequency.

(Purpose of exercise: recall the description of circular motion.)

11.4 Show that the z component returns to its original value M_0 after a 90°-pulse with an exponential function and a time constant T_1. Similarly, find the expression for the x and y components.

(Purpose of exercise: practice transformation between rotating and laboratory coordinate system.)

11.5 Use a PC to calculate the anisotropy of the Zeeman splitting for TCNQ with the values given for the g-tensor in Sect. 11.7.

(Purpose of exercise: get familiar with anisotropic systems.)

11.6 Calculate the hyperfine coupling constant a for hydrogen from the Bohr orbital of the electron. Why is the result not very good?

(Purpose of exercise: remember the shortcommings of the Bohr model.)

11.7 Find the energy levels and the line splitting due to hyperfine interaction for a system with $S = 3/2, J = 5/2$.

(Purpose of exercise: understand the multiplicity of the transitions.)

11.8 Evaluate the population for the Zeeman levels for $\mu\gamma B \ll k_{\mathrm{B}}T$. At which temperature is this approximation no longer appropriate for electrons in a 1-T field?

(Purpose of exercise: get a feeling for the difference in population of Zeeman levels in ESR.)

11.9 Show explicitly that the dipole–dipole interaction is zero if the line connecting the two dipoles makes an angle of $54°\,44'$ with the magnetic field.

(Purpose of exercise: understand the magic angle condition.)

11.10* An electronic spin interacts with eight equivalent nearest-neighbor nuclear spins of 1/2. Calculate the relative intensities for the resulting resonance lines. Hint: The intensities are proportional to the number of representations for each of the $2NI + 1$ allowed orientations of the total spin.

(Purpose of exercise: become familiar with composition of spins.)

12. Ultraviolet and X-Ray Spectroscopy

The higher the energy of the electromagnetic radiation the deeper we can look into the atoms and into the band structure of solids. Thus, for ultraviolet radiation and x rays a new area opens up for spectroscopic investigations. It becomes possible to analyze the structure of the valence band and of lower bands down to the core levels.

The classical application of x rays is elastic diffraction used for the analysis of crystal structures and x-ray fluorescence for chemical analysis. Since elastic diffraction is not a spectroscopy in the sense of an energy analysis, it is not included into this textbook. X-ray fluorescence is a standard technique for the elemental analysis of compounds. It is also known as *electron dispersive x-ray analysis* (EDX). Also, *photoelectron spectroscopy* (PES) in its various forms has gained fundamental importance since the technical development of good electron spectrometers.

Other spectroscopic techniques to be discussed here are *inverse photoemission* (IPES), *extended x-ray absorption fine structure* (EXAFS), and *near-edge x-ray absorption fine structure* (NEXAFS).

Since PES of the core levels is very sensitive to the chemical environment of the atoms it is an appropriate technique for the chemical analysis of material. Hence it is often discussed under the acronym ESCA which means *electron spectroscopy for chemical analysis.*

At present strong efforts are being made to apply synchrotron radiation to inelastic scattering experiments for the detection of phonons or other quasi-particles. The technique is similar to the Raman or Brillouin experiments discussed in Chap. 9 but does not suffer from the limiting conditions $q \approx 0$. The problem with these experiments is the energy resolution. Since the excitation energy for quasi-particles in the solid is several meV and the x-ray energy is several KeV, a resolution of the order of 10^{-6} is required while the primary light is still ten orders of magnitude higher than the scattered light. Nevertheless, experiments are already successful on an academic level.

12.1 Instrumentation for Ultraviolet and X-Ray Spectroscopy

Ultraviolet and x-ray spectroscopy needs special instrumentation which is available today with a very high degree of perfection. This refers to the light sources as well as to spectrometers and detectors. For the light x-ray tubes are the classical sources. However, synchrotron radiation as it was described in detail in Chap. 3 is gaining more and more importance. Spectrometers and detectors must be designed for the analysis of x rays as well as for the analysis of electrons.

12.1.1 X-Ray Sources and X-Ray Optics

The conventional x-ray sources are x-ray tubes where high energy electrons are decelerated when they hit a metal anode. Two types of radiation are emitted as a consequence. From the deceleration of the electrons a continuous spectrum is generated. If the energy of the incident electrons was several KeV the emitted light will be in the x-ray range. Alternatively the incoming electrons can eject electrons from core orbitals of the target material by an ionization process. As a consequence characteristic x-ray lines are emitted from the recombination between the generated holes and electrons from higher orbitals.

From the discussion in Sect. 2.2 we know that the emitted power of a decelerated charge is proportional to the square of the absolute value of the deceleration $\dot{v}$ and given by (2.23). Since the deceleration occurs along a very short distance of the order of 1 μm after the electron has entered the anode, $\dot{v}$ is very large and high-energy light can be emitted. The high-energy limit for the emitted light is given by

$$\hbar\omega_{\text{max}} = \frac{m_0 v_0^2}{2} , \tag{12.1}$$

where v_0 is the velocity the electrons have gained from the acceleration in the tube. The emission of the x-ray light should be highly oriented in a well defined direction and polarized as we know from the emission characteristics of dipoles (Fig. 2.1). This is usually not the case since the deceleration proceeds in several steps while the electron undergoes a diffuse motion. If the acceleration voltage is higher the 30 KV relativistic velocities are obtained and the emission characteristic turns to the forward direction as for the particles in synchrotron radiation. For a simplified model where the deceleration occurs continuously with constant $\dot{v}$ antiparallel to v Sommerfeld derived a relation for the total intensity observed at a distance r under an angle θ from the direction of the velocity. For relativistic particles this intensity is

$$I_{\text{tot}}(\theta) = \frac{e^2 \dot{v} \sin^2\theta}{16\pi^2 \varepsilon_0 c_0^3 r^2} \int_0^{\beta_0} \frac{c_0 \mathrm{d}\beta}{(1-\beta\cos\theta)^5}$$

$$= \frac{e^2 \dot{v} \sin^2 \theta}{64\pi^2 \varepsilon_0 c_0^3 r^2 \cos\theta} \left(\frac{1}{(1-\beta_0 \cos\theta)^4} - 1 \right) , \quad (12.2)$$

where $\beta_0 = v_0/c_0$ is the normalized velocity of the electrons at the beginning of the deceleration. For $\beta_0 \ll 1$ $I(\theta)$ is proportional to $\sin^2\theta$. With increasing energy of the electrons the emission turns more and more to the forward direction and finally reaches the relativistic limit of Fig. 3.6.

The frequency spectrum of the emitted light depends strongly on the energy of the decelerated electrons. Figure 12.1 plots calculated total intensities versus wavelength for different acceleration voltages applied to the x-ray tube. The similarity of the curves to those of Figs. 3.1 and 3.8 is noteworthy but not really surprising. It is in all cases the deceleration (or the acceleration) of electrons which is the origin of the emission. A semi-empirical description of the frequency spectrum expressed in wavelengths of the emitted radiation is

$$\boxed{I_{\text{tot}}(\lambda) = CZ \frac{(\lambda - \lambda_{\min})}{\lambda_{\min}\lambda^3} .} \quad (12.3)$$

Z is the nuclear charge of the anode material, C a constant and $\lambda_{\min}$ is obtained from (12.1).

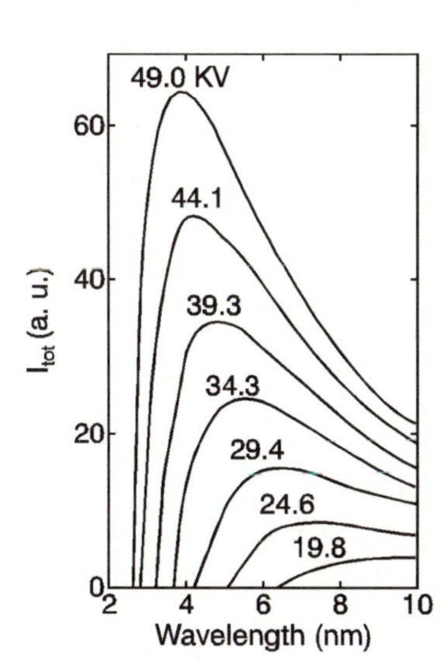

Fig. 12.1. Spectral intensity I_{tot} for the emission of the x-ray continuum from a tungsten anode. The parameter for the curves is the acceleration voltage in KV; after [12.1].

The other type of radiation emitted from the anode is a narrow line characteristic for the material of the cathode. The cascades of recombination with one particular hole are known as the x-ray emission series and assigned with the Greek letters α, β, γ, etc. Depending on the position of the hole the emission is called $K-, L-, M-, N-, \ldots$ series. The transitions are schematically shown in Fig. 12.2. Since for shells with main quantum number $n > 1$ the

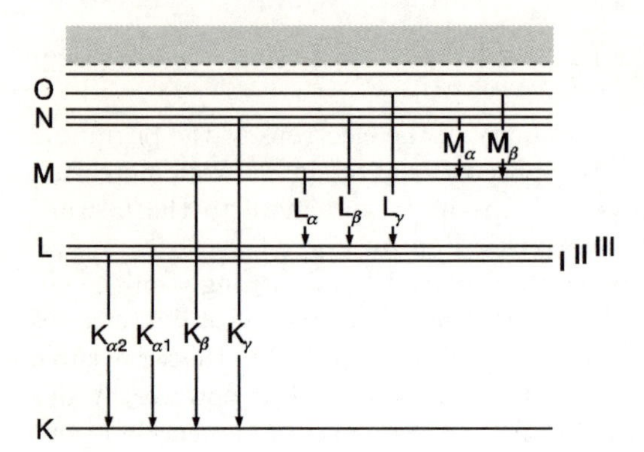

Fig. 12.2. Characteristic x-ray emission series of atoms.

orbitals are split according to different quantum numbers l the characteristic emission lines show a fine structure. For example, the L-shell orbitals are split into three sublevels L(I), L(II), and L(III) where the orbital quantum number is 0 for L(I) and 1 for the other two. Since the selection rules for the transitions are

$$\Delta n \neq 0, \qquad \Delta l = \pm 1, \qquad \Delta j = 0, \pm 1 ,$$

this leads to the well known splitting of the lowest line in the K-series into two components $K_{\alpha 1}$ and $K_{\alpha 2}$.

The energy of the emission depends on the nuclear charge Z of the material selected for the anode and on the particular transition

$$\boxed{\epsilon_{\mathrm{fi}} = \hbar\omega = \mathrm{Ry}(Z - \sigma_{\mathrm{f}})^2 \left(\frac{1}{n_{\mathrm{f}}^2} - \frac{1}{n_{\mathrm{i}}^2} \right) .} \tag{12.4}$$

n_{f} and n_{i} assign the initial and the final shell for recombination, σ_{f} is a screening parameter, and Ry is the Rydberg constant

$$\mathrm{Ry} = \frac{m_0 e^4}{64\varepsilon_0^2 \pi^3 \hbar^3} = 13.607\,[\mathrm{eV}] .$$

For $n_{\mathrm{f}} = 1$ the K-series is obtained. The linear scaling of the square root of the emission frequency with the nuclear charge Z is known as *Moseley's* law. The energy in (12.4) extends from the hard UV to the hard x-ray spectral range. A list of representative lines is given in Table 12.1. The linewidths are typically 0.5–0.8 eV for the x-ray range and one to two orders of magnitude smaller in the UV range.

With the wide range of energies for the emission lines spectroscopy can be carried out from the outer valence band range down to the lowest core levels, as shown schematically in Fig. 12.3.

Todays x-ray tubes operate in high vacuum with a tungsten filament for the emission of the electrons. Tube currents are in the range of several 100 μA to several 10 mA with acceleration voltages of several 10 KV. The photon

Table 12.1. Characteristic ultraviolet and x-ray lines used in PES. (λ [nm]$\times\epsilon$ [KeV] = 1.23985; for a complete listing of lines see [12.2])

Atom	ϵ [KeV]	λ [nm]	Atom	ϵ [KeV]	λ [nm]	Atom	ϵ [KeV]	λ [nm]
He I	0.021	58.428	Cr $K\alpha_1$	5.4055	0.2294	Mo $K\alpha_1$	17.4795	0.0709
He II	0.041	30.373	Cr $K\alpha_2$	5.4148	0.2290	W $K\alpha_1$	8.4330	0.1470
Y $M\zeta$	0.132	9.3715	Fe $K\alpha_1$	6.4039	0.1936	W $L\beta_3$	9.8189	0.1263
Al $K\alpha$	1.487	0.8340	Cu $K\alpha_1$	8.0478	0.1541	W $L\gamma_2$	11.6081	0.1068
			Cu $K\alpha_2$	8.0279	0.1544			

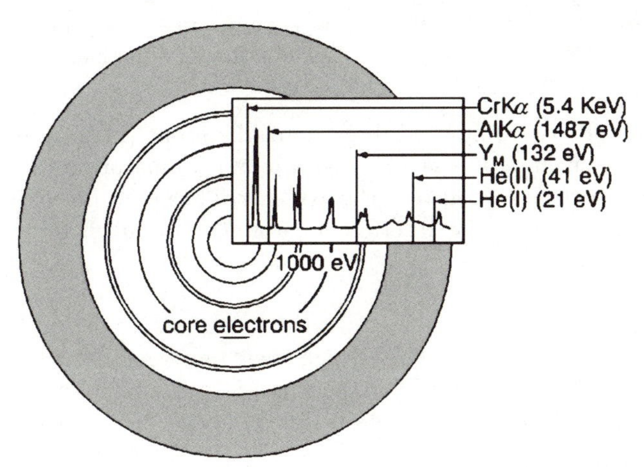

Fig. 12.3. Schematic representation of the excitation of atoms with hard electromagnetic radiation.

efficiency is very low. Usually less than 1% of the electrical power used can be transformed into light. A 100-μA current gives about 10^{12} photons/s. The power consumption is high. 50 KV acceleration at 20 mA current yields 1000 W input power. Thus, efficient water cooling or rotating anodes are required.

An appropriate window material for the x-ray range is beryllium because of its mechanical stability and low atomic mass. The thickness of the windows can be as low as 30 μm to allow for optimum transmission. Filters can be made from various metals which usually have very sharp and well expressed edge structures in the absorption. Al has proven to be an excellent filter and window material for the VUV spectral range. The transmission for a thin Al film is displayed in Fig. 12.4. The low-energy cutoff is given by the plasma reflection from the free carriers. On the high-energy side the first core-level absorption (Al $2p$ level) limits further transmission.

12.1.2 X-Ray and Electron Spectrometers

X-ray spectroscopy needs spectrometers for the high-energy photons and for the generated photoelectrons. The photons from the x-ray tubes are, in gen-

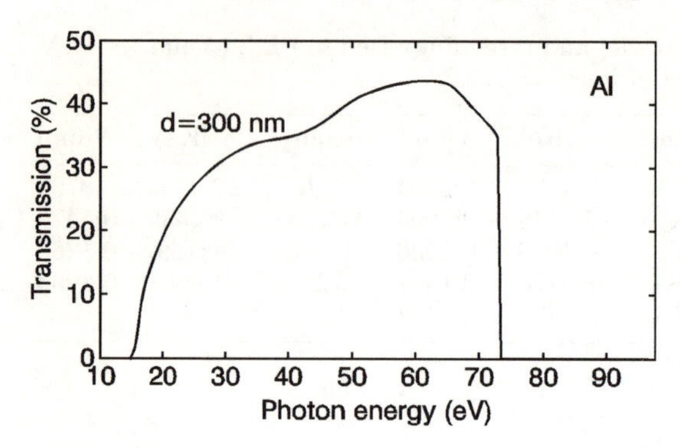

Fig. 12.4. Transmission of a 300-nm film of aluminum.

eral, sufficiently monochromatic to be used directly on the input side of the experiment. X rays from synchrotron sources need a monochromator on the input side and, for x-ray spectroscopy, also on the exit side. The monochromators work on the basis of a constructive interference from crystal diffraction, as given by the Bragg equation

$$n\lambda = 2d \sin\theta \ . \tag{12.5}$$

d is the spacing of the lattice planes, and θ the Bragg angle measured between the incident ray and the deflecting planes. For the simplest case of one plane analyzing crystal the angular dispersion is obtained from (12.5) for $n = 1$ by

$$\frac{\mathrm{d}\theta}{\mathrm{d}\lambda} = \frac{1}{2d\cos\theta} = \frac{\tan\theta}{\lambda} \ . \tag{12.6}$$

As in the case of optical spectroscopy, two lines are considered as resolved if their diffracted beams at the distance L from the crystal surface are separated by more than the width W of the exit slit. If the analyzing crystal must be rotated by $\delta\theta = W/2L$ to obtain glancing-angle diffraction for the two beams the resolution of the spectrometer is

$$\boxed{\frac{\lambda}{\delta\lambda} = \frac{2L\tan\theta}{W}} \ . \tag{12.7}$$

For a slit width of 0.01 cm, $\theta = 20°$, and a spectrometer length of $L = 100$ cm a resolution of 7×10^3 is obtained. With this the α_1 and α_2 components of Cu $K\alpha$ are easily resolved.

For advanced x-ray spectrometers two or even four crystals are used or the crystals are bent to obtain a focusing effect (Fig. 12.5). The additive mode of Fig. 12.5a gives a higher dispersion and thus a higher resolution but the direction of the beam changes during scanning. For the subtractive modes the beam direction is constant and the brightness is higher as a consequence of the smaller dispersion. The spectrometer presented in Fig. 12.5a at the bottom has the advantages of (1) and (2) but still low brightness. The spectrometer in Fig. 12.5b uses a thin bent crystal in transmission. The diffracted light of

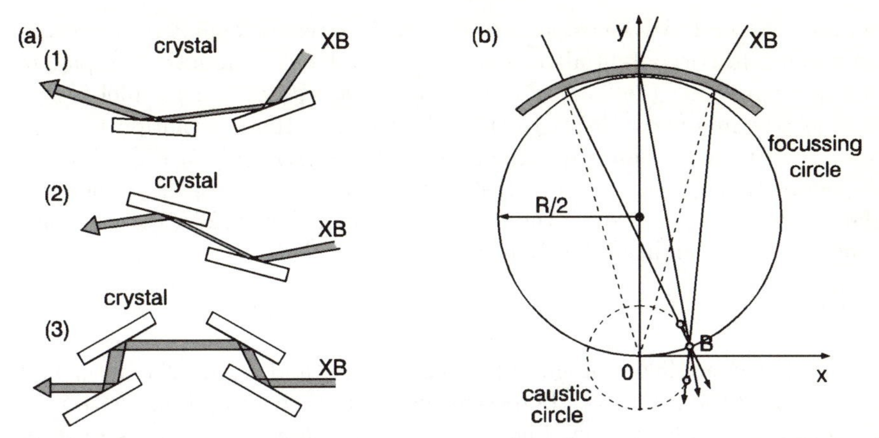

Fig. 12.5. Various crystal monochromators: double monochromator (**a**) in the additive mode (1), in the subtractive mode (2), and in two subtractive but pairwise additive arrangements; (**b**) exhibits a bent crystal spectrometer. (XB: x-ray beam, O: central focus, B: focus for the off-axis beam).

different wavelengths can be shown mathematically to be more or less focused on a circle with radius $R/2$ where R is the bending radius of the crystal. This more or less means the diffracted light arrives at the focus within a small *caustic* circle.

For an energy analysis of the emitted electrons dispersive or non-dispersive spectroscopy is possible. Non-dispersive spectrometers use a vacuum tube in which the electrons are slowed down by an electric potential barrier between two grids. The potential is tuned until no more electrons can pass. The corresponding voltage directly gives the kinetic energy of the electrons. The spectral resolution of these systems is only of the order of 0.2 to 0.3 eV.

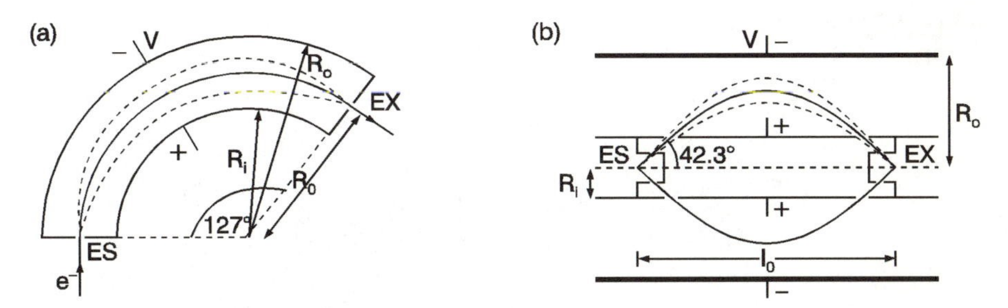

Fig. 12.6a,b. Energy analyzers for electron spectroscopy. Cylinder deflection analyzer (**a**) and cylinder mirror analyzer (**b**); (ES: entrance slit, EX: exit slit, R_i, R_o: radii of curvature, V: applied voltage, l_0: length of spectrometer).

Dispersive spectrometers use either an electric field or a magnetic field oriented perpendicular to the motion of the electrons. Electric spectrometers are most common. Advanced systems employ concentric cylinders connected to different electric potentials to deflect the electrons. The latter propagate either perpendicular (*cylindrical deflection analyzer, CDA*) or parallel (*cylindrical mirror analyzer, CMA*) to the cylinder axis. High resolution but not a very good transmission is obtained for the CDA geometry of Fig. 12.6a. In this geometry the incoming electrons can be focused (with respect to their angular spread) to an exit slit if the sector of the cylinder is 127°. For electrons of energy ϵ_0 the condition for focusing is

$$2eV = \epsilon_0 \frac{R_\mathrm{o}}{R_\mathrm{i}} ,$$

where V is the applied voltage difference, and R_o and R_i are the outer and inner radius of the cylinders. High transmission but low resolution is obtained for the CMA geometry from Fig. 12.6b. In this case the electrons have to enter the space between the cylinders from the inner cylinder under an angle of 42.3° in a 2π geometry and propagate in the direction of the cylinder axis. For

$$eV = 1.3\epsilon_0 \ln\left(\frac{R_\mathrm{o}}{R_\mathrm{i}}\right) \qquad \text{and} \qquad l_0 = 6.1 R_\mathrm{i}$$

they are refocused at a distance l_0 from their entrance. The required angle of incidence is established by electric lenses.

For experiments of photoemission with synchrotron radiation combined spectrometers must be used. Figure 12.7 illustrates a setup. The spectrometer utilizes a bent Ge crystal Ge1 in reflection to focus the beam of a synchrotron

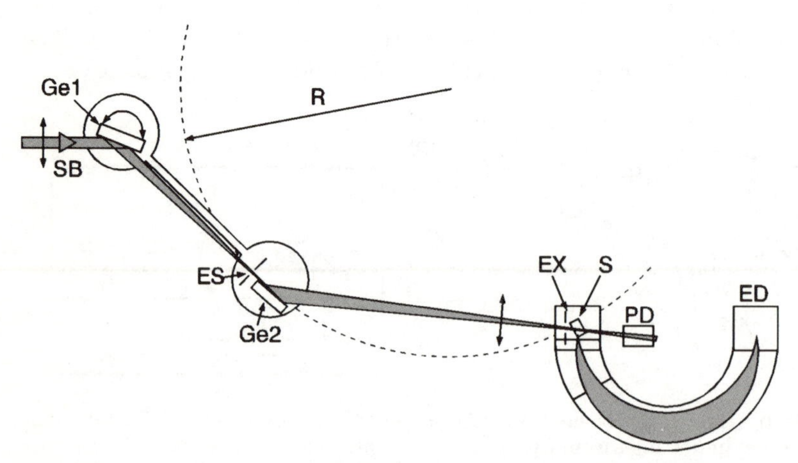

Fig. 12.7. Crystal monochromator for synchrotron radiation and electron spectrometer for the analysis of photoelectrons; (SB: synchrotron beam, Ge1, Ge2: bent germanium crystals, R: radius of focusing circle, ES: entrance slit, EX: exit slit, S: sample, PD: photon detector, ED: electron detector).

onto the entrance slit (ES) of the x-ray spectrometer. The spectrometer operates with another bent crystal Ge2. Light of a given wavelength is now focused to the position EX on the focusing circle of the crystal Ge2. This is the position of the exit slit of the spectrometer and the entrance slit of the electron spectrometer. The sample to be investigated is mounted exactly there. The two Ge crystals enable the selection of x rays with varying wavelengths from the synchrotron beam. The photoelectrons are subsequently analyzed with respect to their energy in an electron spectrometer.

12.1.3 X-Ray and Electron Detectors

The emitted electrons as well as the x rays must be detected. For electrons the simplest way is to use an electron collector and observe the current generated by the external photoelectrons.

If the emission rates are very small the use of an electron amplifier is appropriate. The electron amplifier operates like a photomultiplier but without a photocathode. The electrons hit the dynodes directly and induce secondary electron emission. The dynodes are continuous and consist of curved channels with glassy side walls from where the secondary electrons are released. Such amplifier tubes are called *channeltrons*. The detection speed for channeltrons is about an order of magnitude higher than for photomultipliers. The channeltron operates similarly to the channelplates of the image amplifier in Fig. 5.9.

For the detection of x rays several systems are possible. Classical techniques are photographic films, scintillation counters or proportional detectors.

Scintillation counters are solid-state detectors where the high-energy quantum excites a luminescence process in a scintillation crystal. In this way the energy of the quantum is transformed to a lower level in the near-UV or blue spectral range. This light can be detected subsequently and with high efficiency by a photomultiplier. Common scintillator crystals are NaI activated with Tl. The heavy I^- ion is good for x-ray (and γ-ray) absorption. Up to 0.02 nm wavelength nearly 100 % efficiency can be obtained in the absorption process. The excitation energy is transferred from the I^- ion to the Tl atom from where 410 nm radiation is emitted. This means for one quantum of $\mathrm{Cu}K\alpha_1$ radiation at 0.1541 nm $410/0.1541 = 2670$ blue light quanta are generated. NaI(Tl) scintillation counters are only good for light energies higher than 6 KeV.

In addition to the scintillation counters on the basis of NaI(Tl) plastic scintillators are common.

Proportional counters employ discharge chambers filled with gas such as $Ar\text{-}CH_4$. The applied voltage is high enough to allow for impact ionization. Each electron generated by ionization from the absorption of a light quantum can generate other electrons. The number of secondary electrons generated is proportional to the voltage applied. Thus, if the quantum energy of the light and the ionization energy of the gas are $\hbar\omega$ and U_i, respectively, the charge arriving at the electrode wire per incident light quantum is

$$N = A\frac{\hbar\omega}{U_\mathrm{i}} = A'V\frac{\hbar\omega}{U_\mathrm{i}} \,. \tag{12.8}$$

A is the gain factor of the gas which can be tuned with the applied voltage V over several orders of magnitude. Even though the ionization energy of the gases in use is rather high (several 10 eV) and therefore the number of primary charges generated is not very large, finally a strong signal can be detected at the electrode. Proportional counters can be utilized for energies as low as 100 eV.

Both the scintillation counter and the proportional counter do not only allow to determine the number of incoming high-energy light quanta but also, though in general with low resolution, the energy of the quanta. In both cases the signal height is proportional to the quantum energy of the light. The counters are therefore called *dispersive.* Proportional counters have an energy resolution of about 10% for a 6-KeV radiation.

Very good x-ray and γ detectors are available on the basis of semiconductor p-n junctions. p-i-n diodes from Si and Ge are in use where the intrinsic part is obtained by compensation. The diodes are biased in reverse direction. The light quanta absorbed in the intrinsic region give rise to a photocurrent. In a Si(Li) detector about 3.8 eV are needed to generate one electron–hole pair. This means from one quantum of Cu $K\alpha$ radiation about 3250 electron–hole pairs can be created. The Si(Li) detectors must be kept at liquid nitrogen temperature throughout to prevent the Li from diffusing out of the intrinsic zone in the diode.

The sensitivity of the various detectors depends strongly on the spectral range. This is demonstrated in Fig. 12.8. The quantum efficiency used in Fig. 12.8 is defined as the ratio between the number of primary excitations generated per unit of time and the number of incident light quanta per unit of time. ($\eta = 1$ if each light quantum makes a contribution to the excitation).

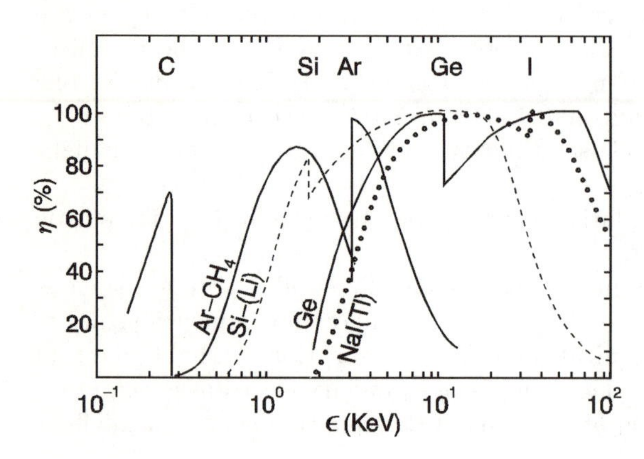

Fig. 12.8. Quantum efficiencies for various detectors. The K-absorption edges are marked with symbols for the elements. The response for the Si(Li) detector is shown for a 220-Å gold layer and a 7.5-μm Be window; after [12.3].

Table 12.2. Detectors for x and γ radiation.

Detector	Resolution Energy $[\Delta\epsilon/\epsilon]$	Space [µm]	Time [ns]	Amplification	Remark
Photographic films		0.3			Sensitivity decreases with ϵ
Ionization chamber				1	for high intensities
Proportional counter	0.1		20	E5	for small ϵ
Plastic scintillator	0.6		1	E8	High sensitivity
Crystal scintillator		20	20	E8	High sensitivity
p^+-nSi-, Ge-diodes	0.05		10	1	Cooling required
p-i-n Si(Li) diodes	0.02		10	1	Cooling required
p^+-nSi array Detectors	0.1	5	10	1	High electronic effort
CCD from MOS diodes	0.1	5	10	1	$\epsilon <$ several KeV required

The number of incident light quanta is the incident power divided by the quantum energy of the light.

Finally, Table 12.2 compiles some details and technical data about x-ray and γ-ray detectors. Since detectors are often utilized for dispersionless spectral analysis the resolution in energy, space, and time is included in the table. CCD detectors and array detectors operate in the same way as described in Sect. 5.4.4.

12.2 X-Ray Fluorescence

X-ray fluorescence is the most traditional spectroscopic technique in the x-ray regime. The fluorescence spectra are excited by photoionization of an electron from an inner shell. The hole created is refilled by a recombination process with electrons from higher levels. The recombination occurs either radiative with the emission of a light quantum with an energy equal to the recombination energy, or by an Auger process. The energy-level diagram for the fluorescence process is illustrated in Fig. 12.9a. The radiative recombination of outer-shell or valence electrons with an inner-shell hole is dominated by a dipole transition. The same relationships apply as they were derived in Sect. 7.1. The transition probability between shell n_i and shell n_f per unit time is

$$\boxed{P_\mathrm{fi} = \frac{\omega_\mathrm{fi}^3}{3\pi\varepsilon_0\hbar c_0^3}|\boldsymbol{M}|^2} \tag{12.9}$$

with the dipole matrix element $(M_j)_\mathrm{fi}$ from (7.6). For an element with nuclear charge Z the magnitude of the matrix element is proportional Z^2. This quadratic dependence on the nuclear charge is the reason for the rather low efficiency of x-ray luminescence in low atomic weight material.

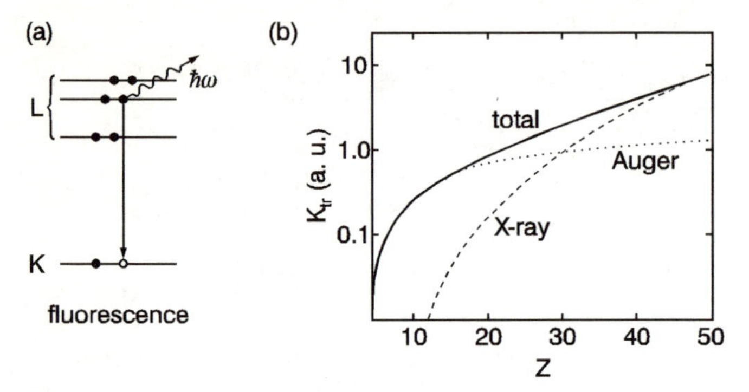

Fig. 12.9. Light emission in x-ray fluorescence (**a**), and calculated transition rates K_{tr} to the K-shell for the two competing effects of Auger emission and fluorescence, versus nuclear charge Z (**b**); [(....) Auger transitions, (––) fluorescence, and (—) sum of both transitions].

The shape for the core-emission lines is Lorentzian or Gaussian with a half width determined by the lifetime of the excited state. From the proportionality of the intensity to Z^4 one can expect the lifetime of the hole state to be proportional to Z^{-4}. This means for a lifetime-determined linewidth a scaling of the latter with Z^4 is expected. This is indeed in good agreement with experimental results for $Z > 30$.

For the light elements the process of luminescence has to compete with the Auger process according to which recombination is nonradiative. The lifetime is not only determined by the matrix elements for the radiative transitions but also by the transition due to the Auger process. This holds likewise for the intensity of the emission. Figure 12.9b compares the dependence of the transition rates on the nuclear charge Z of the anode material. Obviously the Auger process is important for material with nuclear charges smaller than 30. The efficiency of a radiative recombination was given in (7.35). Applied to the competition between luminescence and the Auger process this means

$$\eta_x = \frac{P_{fi}}{P_{fi} + P_{Aug}} = \frac{bZ^4}{bZ^4 + a}, \tag{12.10}$$

where P_{Aug} is the transition probability for the Auger process. For $Z > 30$ this probability becomes nearly independent from Z and was approximated by the constant a in (12.10), in agreement with the dotted line in Fig. 12.9b. Thus, for large Z the quantum efficiency for radiative emission becomes 1. This result is also well confirmed experimentally.

Since the fluorescence emission is characteristic for each element of the periodic table, x-ray fluorescence is widely applied in chemical analysis. Sophisticated techniques have reduced the detection limit to several 10^{-15} g of material. The hole in the core shell can be generated either by irradiation with x rays or with electrons. Excitation with x rays has the advantage of a higher penetration of the light and allows irradiation at ambient conditions. With electron excitation a very high spatial resolution of less than 1 μm can be obtained for the analysis but the irradiation needs ultrahigh vacuum. Figure 12.10 shows a core-level fluorescence of several impurities on a Si waver after excitation with 10-KeV radiation. From the position and intensity of the peaks the type and concentration of the radiating atoms can be determined. The mass for the Ni metal observed in Fig. 12.10 is 10 pg. The detection limit for Ni on Si was 13 fg. Because of the factor Z^4 in the emission intensity this technique becomes less appropriate for elements lighter than oxygen.

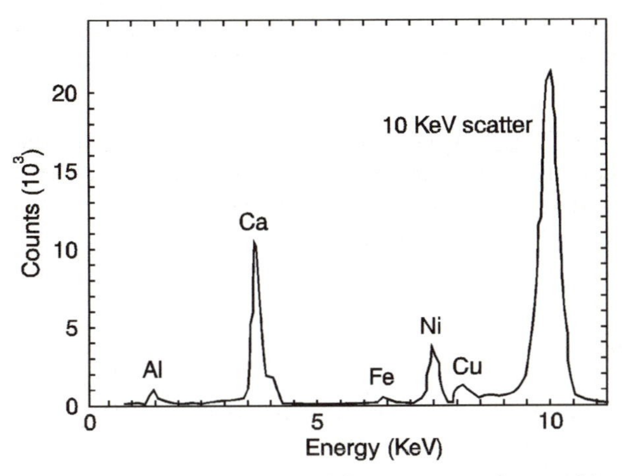

Fig. 12.10. X-ray core level fluorescence of several impurities on Si for excitation with 10-KeV synchrotron radiation and a 10-μm aluminum film as a detector filter; after [12.4].

For not too deep core levels the exact energy of the core hole depends on the chemical environment. Thus, in this case even more details about the chemical structure can be obtained if the energy resolution of the spectrometer is high enough. This phenomenon holds likewise for photoemission and will be discussed in detail in the next section.

12.3 X-Ray Electron Spectroscopy

In x-ray electron spectroscopy electrons in the solid are excited by x rays or by other electrons from an external source. The excitation energy must be high enough to allow for the electrons to leave the solid either by an Auger process or by external photoemission. The kinetic energy of the electrons is

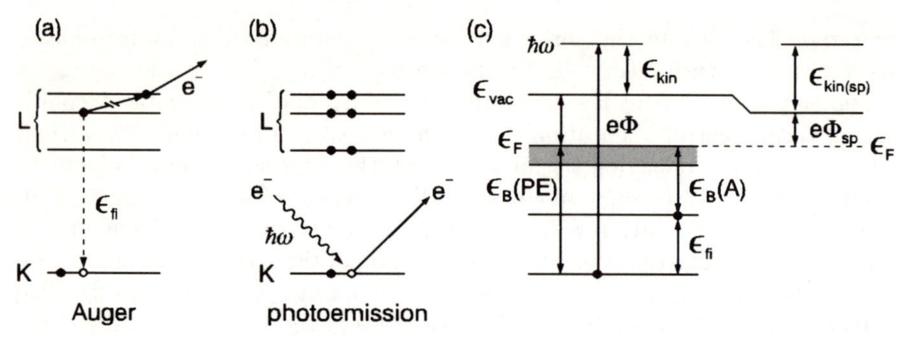

Fig. 12.11. Process of Auger electron emission (**a**), photoelectron emission (**b**) and energy level diagram for the emission processes into vacuum or into the spectrometer (**c**). ($e\Phi$, $e\Phi_{sp}$: work functions with respect to the vacuum level and with respect to the spectrometer, $\hbar\omega$: total energy available for emitted photoelectron, ϵ_{fi}: total energy available for Auger electron, ϵ_F: Fermi level, $\epsilon_B(PE)$: binding energy of photoelectron, $\epsilon_B(A)$: binding energy of Auger electron, ϵ_{vac}: vacuum level).

subsequently measured in an electron spectrometer. Figure 12.11 shows the two emission processes schematically together with an energy level diagram for the energy balance. In order to get the electron out of the solid its energy has to be raised only to the spectrometer level in both cases. This is in general less effort than raising it to the vacuum level.

12.3.1 Auger Spectroscopy

In the Auger process the energy of the electron which is subjected to the recombination process is transferred to another electron which is emitted. The energy of the latter is equal to the recombination energy but reduced by the sum of its binding energy $\epsilon_B(A)$ and the work function of the solid with respect to the spectrometer level. It is thus obtained from

$$\epsilon_{Aug} = \epsilon_{fi} - [\epsilon_B(A) + e\Phi_{sp}] . \tag{12.11}$$

As demonstrated in Fig. 12.11c $e\Phi_{sp}$ is measured from the Fermi level to the spectrometer level and $\epsilon_B(A)$ is measured from the Fermi level to the level from where the recombination started. The process is only possible if the resulting kinetic energy ϵ_{Aug} is finite. The electron which is emitted can come from any of the shells equal or higher than the shell of the hole. Thus a signal can be observed for all binding energies. An often used nomenclature to describe the Auger process is FIY. F labels the shell of the hole, I the shell from where the recombination starts, and Y the shell from where the Auger electron is emitted. Thus, processes in which electrons from the L-shell recombine with a hole in the K- shell are called KLY. The special process depicted in Fig. 12.11a is assigned KLL. If the initial and the final hole state are on the same shell (but in different orbitals), like $L_{\mathrm{I}}L_{\mathrm{III}}M_{\mathrm{I}}$, the process is said to be of the *Coster–Kronig* type. For $L_{\mathrm{I}}L_{\mathrm{III}}M_{\mathrm{I}}$ an electron recombines from L_{III} to L_{I} and the energy is transferred to an electron in M_{I} from where it is emitted. Note that the atom is left in a state with two holes after the Auger process.

For Auger emission satellite lines are often observed in addition and close to the main lines. Such lines originate from a general disturbance of the valence electrons as a consequence of the dramatic change in electronic structure from the emission of one core electron. Such lines are known as *shake-up satellites.*

12.3.2 Photoelectron Spectroscopy

Like in x-ray fluorescence and Auger spectroscopy in PES a hole is created at a certain level of the core or valence orbitals. Instead of detecting the radiation emitted from a recombination process or the Auger electron the kinetic energy of the emitted electrons is directly analyzed. The basic process is shown in Fig. 12.11b.

Photoelectron spectroscopy was originally used mainly for the determination of the bonding in molecules. Today it is the dominating technique to analyze the electronic structure in solids. The technique benefitted strongly from the development of synchrotron-radiation sources.

The kinetic energy of the electron has to be measured again with respect to the spectrometer level $e\Phi_{\mathrm{sp}}$. Its value is therefore

$$\epsilon_{\mathrm{pe}} = \hbar\omega - [\epsilon_{\mathrm{B}}(\mathrm{PE}) + e\Phi_{\mathrm{sp}}] \; . \tag{12.12}$$

Like in Auger spectroscopy $e\Phi_{\mathrm{sp}}$ is a spectrometer constant which has to be calibrated with a well known sample such as Au or Cu. The similarity between equation (12.12) and (12.11) for the Auger electrons is evident.

The experimental requirements for PES are those discussed in Sects. 12.2 and 12.3.1. In particular, the primary electron can be generated either by light quanta or by electrons but in any case UHV conditions are required.

Since the number of photoelectrons emitted at a particular energy is given by the density of states of the electrons in the crystal, core level photoemission and valence band photoemission give different types of spectra. For core level emission high energy x rays are required and the spectra consist of sharp lines. This is known as *x-ray photoelectron spectroscopy* (XPS). In contrast, for the emission of valence electrons only soft x rays or hard UV radiation in an approximate energy range 3 eV $\leq \hbar\omega \leq$ 100 eV is required and the spectra are broad with characteristic structures. In this case we talk about *ultraviolet photoelectron spectroscopy* (UPS). XPS can be used to study core levels as well as valence band structures.

The basic property to be determined in the XPS and UPS experiments are the binding energies of the electrons. These energies are well known for the free atoms. Table 12.3 gives selected examples for frequently used atoms. In molecular or solid compounds the energies are subjected to a chemical shift according to the various chemical environments. Similar to NMR this allows not only the concentration of atomic species but also their chemical bonding and many other details about a material to be determined.

Table 12.3. Binding energies in eV for core levels of selected free atoms. VE means valence orbital. (A complete set of binding energies can be obtained from [12.5].)

Atom	$1s$	$2s$	$2p_{1/2}$				
^{1}H	16.0						
^{2}He	24.59						
^{3}Li	54.7	VE					
^{6}C	284.5	VE	VE				
^{7}N	409.9	37.3	VE				
^{8}O	543.1	41.6	VE				
^{9}F	696.7	VE	VE				
Atom	$1s$	$2s$	$2p_{1/2}$	$2p_{3/2}$	$3s$	$3p_{1/2}$	$3p_{3/2}$
^{11}Na	1071.4	63.6	30.6	30.4	VE		
^{13}Al	1562.3	118.5	72.7	72.3	VE	VE	VE
^{14}Si	149.6	99.2	99.8	VE	VE	VE	
Atom	$2s$	$2p_{1/2}$	$2p_{3/2}$	$3s$	$3p_{1/2}$	$3p_{3/2}$	$3d_{3/2}3d_{5/2}4s$
^{23}V	627.2	521.1	513.4	66.4	37.2		VE
^{26}V	848.7	720.4	707.2	91.6	53.0		VE
^{29}Cu	1097.5	952.3	932.7	122.4	77.2	14.9	VE

The advantage of PES over x-ray fluorescence for the analysis of electronic structures is due to the easier handling of the photoelectrons and the better resolution in energy.

12.3.3 X-Ray Photoemission

In XPS experiments core levels as well as valence states are analyzed. The width of the lines observed for the core levels is given by the lifetime of the intermediate state during the emission process or by broadening due to phonons and not by the density of states itself. The lines often have a slightly asymmetric form. This asymmetry is mainly observed for metals and results from a screening effect by the conduction electrons. If ϵ_0 is the center of the energy distribution curve for a core line it can often be described by a *Doniach–Sunjic* lineshape

$$I(\epsilon - \epsilon_0) = \frac{\Gamma(1-\alpha)\cos[\pi\alpha/2 + (1-\alpha)\arctan(\epsilon-\epsilon_0)/\gamma]}{[(\epsilon-\epsilon_0)^2+\gamma^2]^{(1-\alpha)/2}} , \tag{12.13}$$

where α and γ are two parameters and Γ is the Γ function. The electrons escaping from the core levels will interact in addition with the valence electrons and excite plasma oscillations. Accordingly, in solids core lines are often accompanied by satellites which represent the energy loss by excitation of plasmons. Figure 12.12 shows a core line for the Na $2s$ level with plasmon satellites. The plasmon energy is 6 eV. The small shoulders next to the plasmon lines originate from surface plasmons.

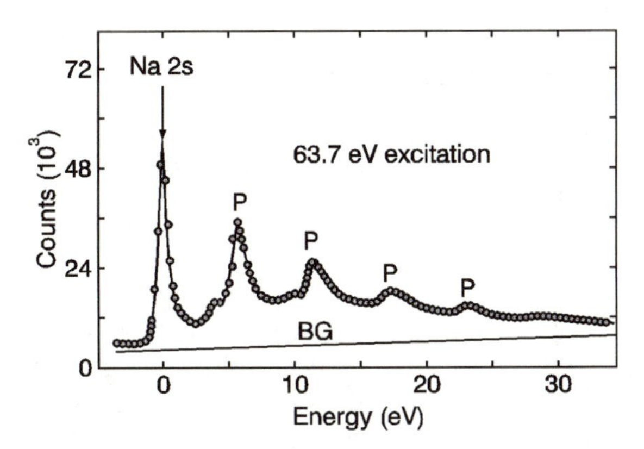

Fig. 12.12. Core line for the Na $2s$ level and plasmon satellites. The energy axis is scaled to the core line. The full drawn line is as calculated with the Doniach–Sunjic formula. BG is the background used for the fit; after [12.6].

A very instructive XPS spectrum is presented in Fig. 12.13. The response of core levels, valence-band levels and Auger electrons are presented for Mg_2Sn after excitation with Al $K\alpha$ at 1.486 KeV. To the left, which means at very high kinetic energies of the electrons, the Fermi edge and the continuing valence band structure is seen. The sharp lines originate from core levels according to the given assignment. Lines assigned with P are plasmon satellites to the preceeding core line. The stepwise increase in the background with decreasing electron energy originates from higher order scattering processes and shake-ups. The steps always start at the position of the core lines. In the Auger part of the spectrum we see two valence band replica, for the recombination of the L_{II} and L_{I} electrons, respectively.

Emission intensities from the top of the valence band are often rather weak as is evident from Fig. 12.13. Blown up and expanded representations of such spectra are very informative about the nature of the valence band. In Fig. 12.14 spectra for several simple metals are shown. The top two presentations are for Ag and Ni. The full drawn line represents a calculated density of states multiplied with the Fermi function. The valence band for Ag consists of ten $4d$ and one $5s$ derived orbital. The $5s$ electrons give rise to the shoulder just below the Fermi edge. For Ni the top orbitals are occupied with two s electrons so the shoulder is pulled up. Figure 12.14c shows explicitly that

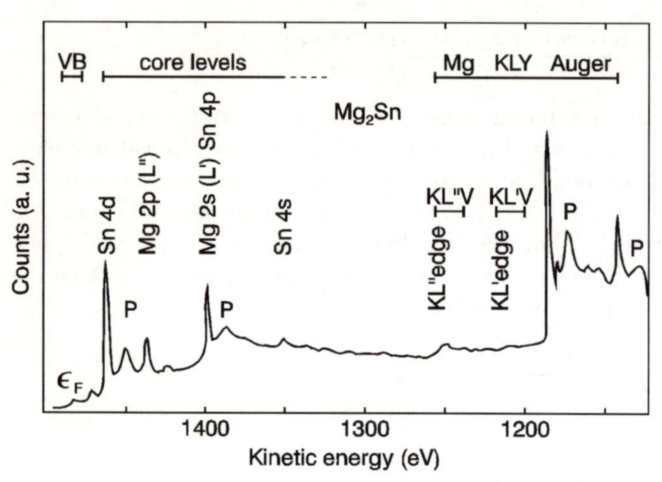

Fig. 12.13. XPS and Auger spectrum for Mg_2Sn after excitation with 1.486-KeV x rays. Left part: XPS spectrum, right part Auger spectrum. The symbols in the Auger spectrum correspond to the definition given in Sect. 12.3.1; after [12.7].

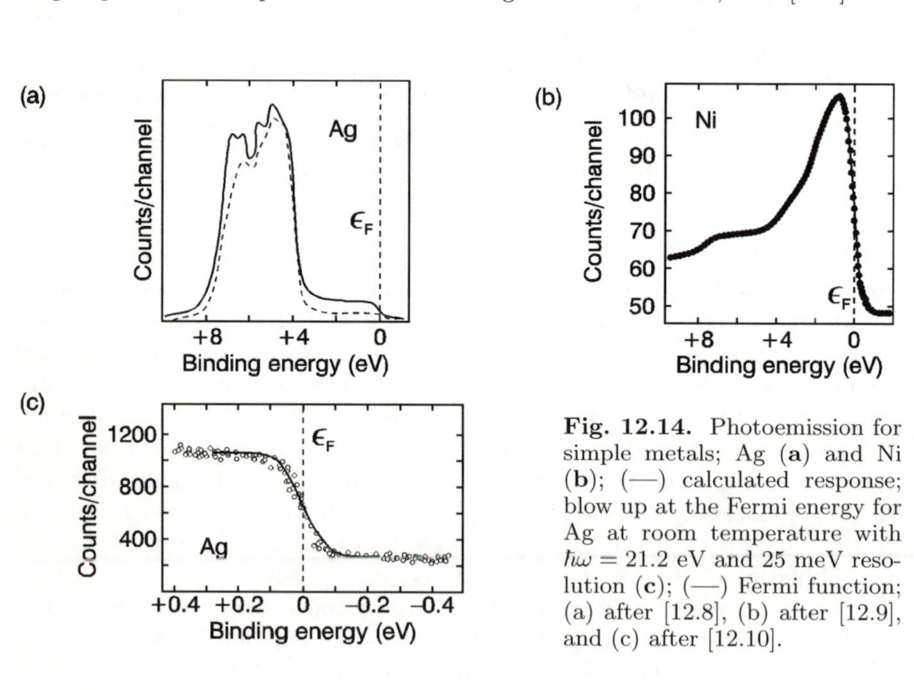

Fig. 12.14. Photoemission for simple metals; Ag (**a**) and Ni (**b**); (—) calculated response; blow up at the Fermi energy for Ag at room temperature with $\hbar\omega = 21.2$ eV and 25 meV resolution (**c**); (—) Fermi function; (a) after [12.8], (b) after [12.9], and (c) after [12.10].

emission is only possible from below the Fermi energy, except for a small tail from thermal excitation given by the Fermi function.

An important application of core-level spectroscopy is the analysis of structural properties from the chemical shift. This shift is well expressed for not too deep core levels. Figure 12.15 exhibits the chemical shift for the $2p$ and $3d$ core levels due to the generation of GeO. Since the spectra come mainly from the surface, the core lines from pure Ge as well as those from Ge in GeO are observed. Due to the reaction with oxygen the binding energy for the $3d$ level is shifted from 28.7 eV to 31.1 eV. For the $2p$ electrons the shift is from 1217.5 eV to 1219.8 eV, respectively. Interestingly, the relative intensity between the line from Ge and GeO is different for the two levels. For the $2p$ electrons the GeO dominates the spectrum whereas the Ge is dominating for the $3d$ electrons. The reason is the different energy of the electrons and the corresponding difference in the escape depth. The $2p$ electrons have a higher binding energy and therefore a lower energy when they escape from the surface. Thus, they are more strongly absorbed and originate on average from a lower depth with a higher concentration of GeO.

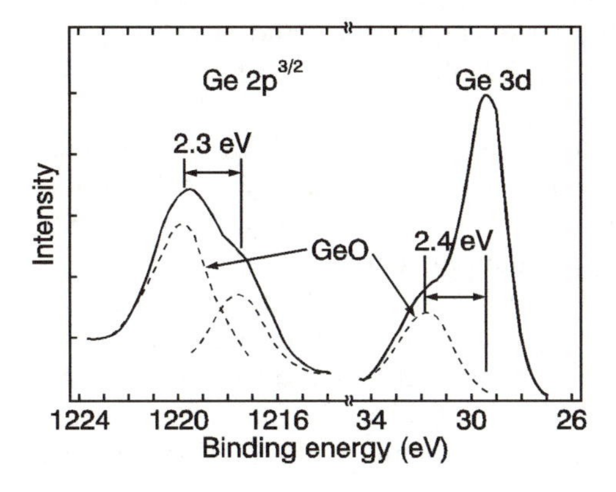

Fig. 12.15. Binding energy for $2p$ and $3d$ electrons in pure Ge and in GeO; after [12.7]

Another useful example for the application of core level spectroscopy is the observation of phase transitions. The crystal KC_{60} has a rock salt structure at temperatures higher than 400 K, where C_{60} is a large highly symmetric molecule consisting of 60 carbon atoms which are arranged in a football-like shape. The K atoms are accommodated at the octahedral interstitial lattice sites. The K $2p$ core-level lines are observed for binding energies of 297.3 and 294.6 eV as demonstrated in Fig. 12.16 curve (1). The splitting originates from a spin–orbit interaction for the $2p^{1/2}$ and $2p^{3/2}$ electrons. For temperatures below 400 K a pair of split lines is observed as indicated by the arrows in the figure. This means the potassium is now accommodated in two different chemical environments. These environments are provided by the octahedral and the two tetrahedral interstitial sites of the fcc lattice. One can conclude the system KC_{60} undergoes a phase transition into a new structure

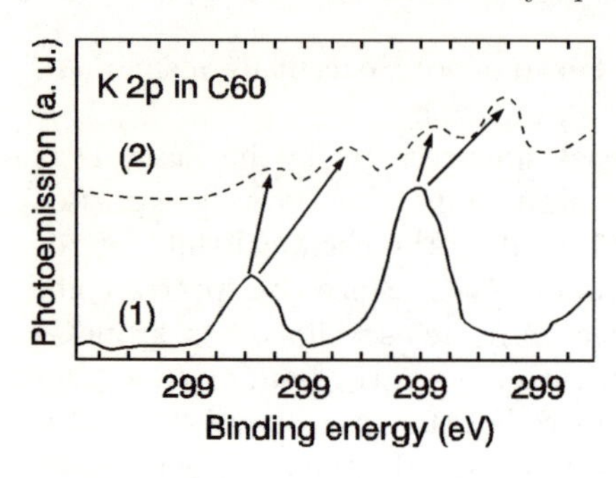

Fig. 12.16. XPS spectra for the $2p$ electrons of K in $K_xC_{60}, x \approx 1$, for two different temperatures; above 400 K (1), below 400 K (2); after [12.11].

where some of the K atoms have moved from octahedral to tetrahedral sites on cooling to below 400 K. The process is fully reversible. For heating back up to 400 K the spectrum (1) reappears.

12.3.4 Ultraviolet Photoelectron Spectroscopy

Ultraviolet photoelectron spectroscopy has been employed very successfully to study the valence band structure or the conduction band of crystals. An example has already been given in Fig. 12.14. In general only the density of states as a function of the energy can be measured. A full analysis of the electronic states in k space is not possible except for low dimensional systems. Since UPS is very sensitive to surface states, very clean surfaces have to be used. Also the experimental result should always be compared with calculations to confirm clean experimental conditions.

An example of the analysis of band structures is presented in Fig. 12.17. The calculated band structure, the density of states derived from it, a density of states folded with the resolution of the spectrometer, and the experimental result are shown for three semiconductors Ge, GaAs, and ZnSe, respectively. The good agreement between experiment and calculation for all three crystals confirms the clean condition for the experiments. In the case of ZnSe the core level of the Zn $3d$ electrons appears in the experiment. It is missing in the calculated spectra, since the calculation was performed without contributions from core levels.

To obtain explicit information on the $\epsilon(k)$ relation angle-resolved recording is required. With this technique at least the component of the k vector parallel to the surface can be analyzed. If the k vector is confined to a crystal plane as is the case in layered structures, like MoS_2 or $TaSe_2$, etc., $\epsilon(k)$ can be measured directly. The situation is even more simple in quasi-one-dimensional structures where the k vector is confined to a certain direction. From a determination of the energy dependence of the photoelectrons on the energy of the incident photons the band structure $\epsilon(k)$ is directly obtained.

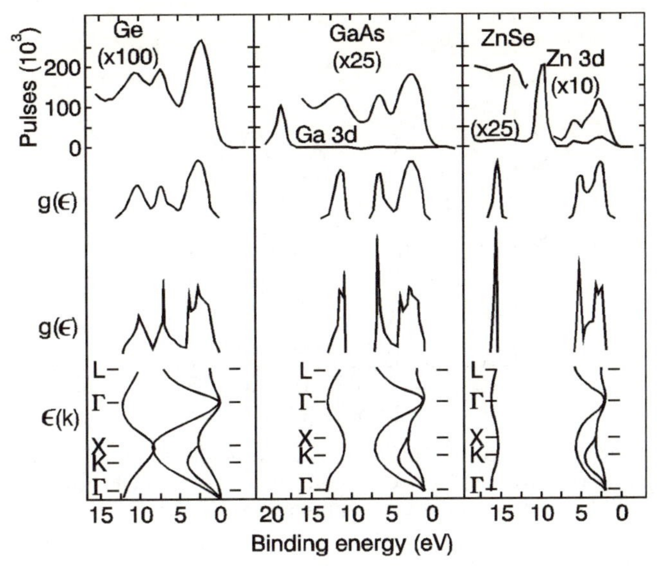

Fig. 12.17. Calculated band structure $\epsilon(k)$, density of states $g(\epsilon)$, $g'(\epsilon)$ as derived from it and after folding it with the instrumental resolution, and observed UPS spectrum for three semiconductors; after [12.12].

Such experiments can be performed with synchrotron radiation where a continuous tuning of the energy is possible for the incident light. Figure 12.18 shows the band structure of crystalline polyethylene. The full drawn lines are from a calculation [12.13].

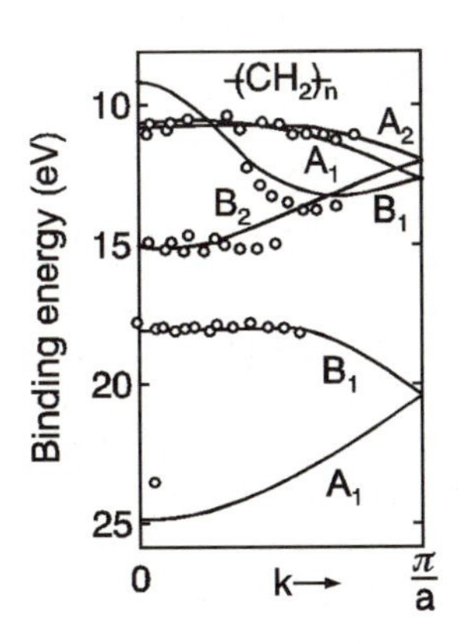

Fig. 12.18. Band structure of crystalline polyethylene as measured (o) and as calculated (—). The symbols A and B assign the irreducible representations of the different bands for $k = 0$; after [12.14] and [12.13].

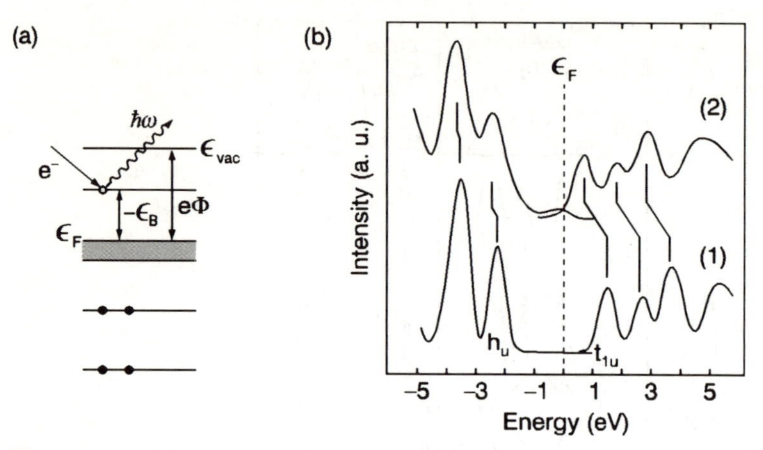

Fig. 12.19. Electronic levels and recombination processes for inverse photoemission (**a**) and photoemission and inverse photoemission for C_{60} (1) and K_3C_{60} (2) (**b**); The Mullikan symbols h_u and t_{1u} label the symmetry of the bands; (b) after [12.15].

12.4 Inverse Photoemission

Photoemission only has access to states below the Fermi energy, since occupation of the states is required. The empty states in the conduction band or in even higher bands cannot be studied. The investigation of such states is possible with inverse photoemission. As demonstrated in Fig. 12.19a in IPES the crystal is irradiated with electrons which can settle at the empty orbitals of the bands above the Fermi level. If this happens the total energy of the incoming electron is increased by the work function Φ of the crystal plus the (negative) binding energy at the particular level. This total energy is emitted as a light quantum $\hbar\omega$.

$$\hbar\omega = \epsilon_{\text{kin}} + (e\Phi + \epsilon_{\text{B}}) \,. \tag{12.14}$$

In early work electrons of varying energy were used for the excitation and only one photon energy was detected. This technique is known as *bremsstrahlung-isochromate spectroscopy or BIS*. In contrast in real IPES work excitation is with monochromatic electrons and the spectrum of the emitted photons is detected. This spectrum is again a replica of the density of states but now for the unoccupied higher bands. Figure 12.19b depicts the photoemission on the left side of the Fermi energy and the inverse photoemission on the right side of the Fermi energy for the two materials C_{60} (1) and K_3C_{60} (2). All peaks on the left side of ϵ_F originate from the occupied bands, all peaks on the right side of ϵ_F originate from the empty bands. For C_{60} the density of states at the Fermi level is zero since the system is an insulator. The energy difference between the peaks labeled h_u and t_{1u} is the gap energy. K_3C_{60} is a metal. Thus PES and IPES overlap and the t_{1u} band appears in both PES and IPES.

12.5 X-Ray Absorption Fine Structure

The development of powerful x-ray sources with a broad-band spectrum triggered the development of additional spectroscopic techniques which are based on x-ray absorption. X-ray absorption decreases with increasing energy of the radiation. A phenomenological description of this decrease is given by the *Victoreen formula*

$$\alpha(\epsilon) = \frac{a}{\epsilon^3} + \frac{b}{\epsilon^4} , \tag{12.15}$$

where a and b are parameters. The absorption exhibits characteristic upward-directed edges whenever the energy becomes large enough to ionize a lower laying shell. A detailed analysis of the upper part of the edge reveals a characteristic fine structure. Figure 12.20a shows the x-ray absorption for metallic copper close to the K-edge. The investigation of this fine structure gives information on the interaction between next neighbors and short range order in the lattice. This is extended x-ray absorption fine structure (EXAFS) or near edge x-ray absorption fine structure (NEXAFS) spectroscopy. NEXAFS is restricted to structures observed within the first 30 to 50 eV to the edge [12.17] with a 50-meV resolution, whereas EXAFS extends far beyond this range.

The fine structure originates from details of the absorption process which we have not considered so far. To emit the photoelectron it must not only be released from its orbital but it must also travel through the crystal until it finally reaches the vacuum (or spectrometer) level, or ends up in a higher

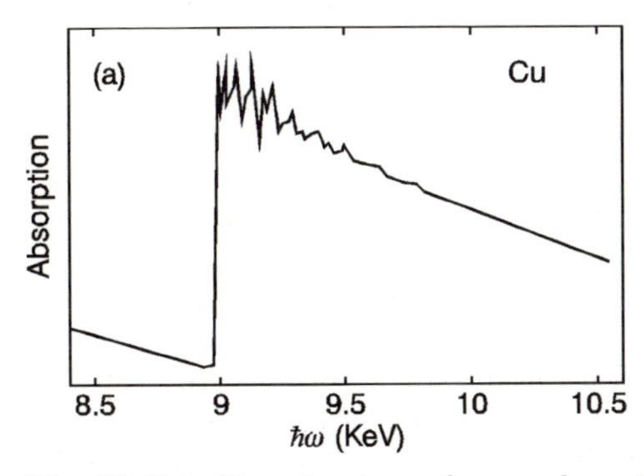

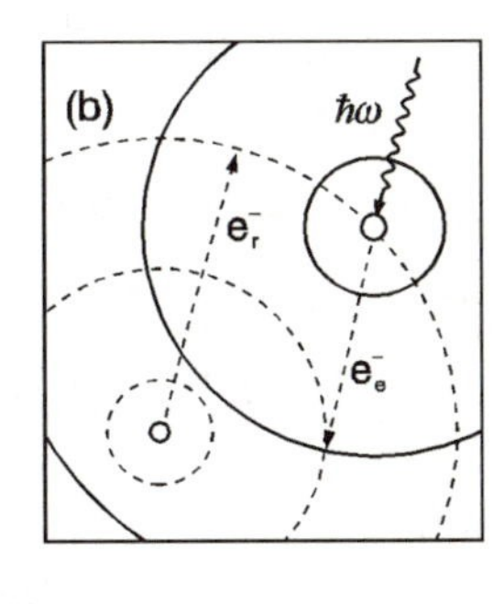

Fig. 12.20. Fine structure of x-ray absorption at 77 K for metallic copper at the K-edge, after [12.16] (**a**) and demonstration of the generation of the fine structure (**b**). (—) evanescing and (– – –) reflected electron wave.

unoccupied state. Even without inelastic scattering the wave function pattern of the electron will be influenced by elastic scattering from neighboring atoms and resulting interference phenomena. To calculate the matrix element for the absorption process the total wave function must be used. Since both the wave function of the released electron and the partial wave functions from its reflected components depend on the energy of the incident photon the matrix element for the absorption varies with the photon energy, particularly in the interference region close to the edge. This behavior is illustrated in a simplified form in Fig. 12.20b. The full drawn circles represent the zeros for the evanescent wave of the released electron. The dashed circles are the zeros for the wave reflected from the neighboring atoms. The interference between evanescent and reflected waves leads to a modulation of the absorption edge.

The experimental results are usually described by the dimensionless relative modulation of the absorption after subtraction of the main background in the form

$$\chi(\epsilon) = \frac{\alpha(\epsilon) - \alpha_0}{\Delta\alpha_{\mathrm{edge}}} \,. \tag{12.16}$$

α_0 is the smooth absorption without interference and $\Delta\alpha_{\mathrm{edge}}$ is the jump in absorption at the edge, i.e., the contribution from the new absorption channel opened up at the edge. Instead of representing χ as a function of the energy in general the wave vector k_{e} of the evanescing electron is used as an argument. The two quantities are related by

$$k_{\mathrm{e}}^2 = \frac{2m_{\mathrm{e}}}{\hbar^2}(\epsilon - \epsilon_0) \,, \tag{12.17}$$

where ϵ_0 is the photon energy for which the kinetic energy of the electron is zero (the vacuum level or the edge energy). The modulation can be expressed by the influence of the next nearest neighbors j on the atom i. As a simplifying assumption only one type of atom may surround the activated atom i. If the distribution of the neighboring atoms is described by a radial function $p_{ij}(r)$ with one peak at an average nearest neighbor distance r_i the modulation $\chi(k)$ can be expressed by

$$k_{\mathrm{e}}\chi_{ij}(k_{\mathrm{e}}) = \frac{2\sqrt{2\pi}}{r_i^2}\mathrm{Re}\{P_{ij}(k_{\mathrm{e}})\Lambda_{ij}(k_{\mathrm{e}}, r_i)\} \,. \tag{12.18}$$

$P_{ij}(k_{\mathrm{e}})$ and $\Lambda_{ij}(k, r)$ are the Fourier transform of the radial distribution

$$P_{ij}(k) = \frac{1}{\sqrt{2\pi}} \int_0^\infty \mathrm{d}r \mathrm{e}^{\mathrm{i}2kr} p_{ij}(r)$$

and the influence of the interacting electrons on the absorption, respectively. It is important to note that the Fourier pairs in this case are k_{e} and $2r$ in contrast to conventional elastic scattering where the pairs are k and r. For a Gaussian distribution of N_i atoms with a width σ_i around a mean distance r_i the simplified form for $k_{\mathrm{e}}\chi(k_{\mathrm{e}})$ is obtained in the form

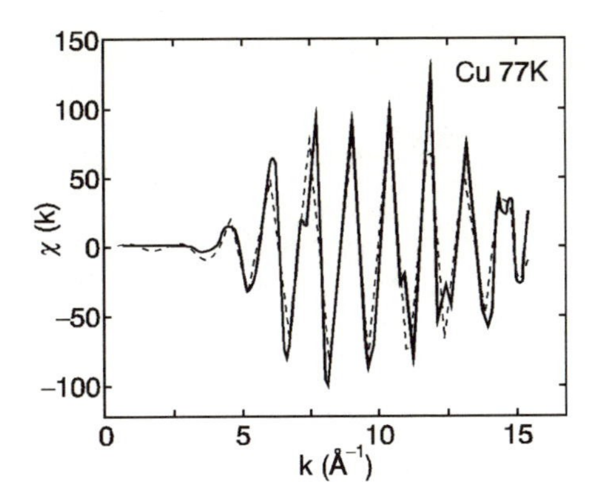

Fig. 12.21. Fine structure of the K-edge absorption of Cu. (– – –) as measured, (—) as calculated; after [12.18].

$$k_{\mathrm{e}}\chi(k_{\mathrm{e}}) = \frac{N_i}{r_i^2}|\Lambda(k_{\mathrm{e}}, r_i)|\mathrm{e}^{-2(\sigma_i k_{\mathrm{e}})^2} 2\cos(2k_{\mathrm{e}} r_i + \Phi_\Lambda) , \tag{12.19}$$

where Φ_Λ is the phase of $\Lambda(k_{\mathrm{e}}, r_i)$. For $|\Lambda| \approx$ constant this is a function oscillating in k_{e} with the period $2r_i$ and decaying with a Gaussian profile for $k_{\mathrm{e}}^2 \propto \epsilon - \epsilon_0 > 0$.

Figure 12.21 shows the experimentally determined spectrum $\chi(k_{\mathrm{e}})$ for the K-edge absorption Cu. The full drawn line is as calculated for a simple nearest neighbor response.

Problems

12.1 Discuss the emission characteristic of an electron which is decelerated with a constant value for $\dot{v}$ as a function of the electron energy.

(Purpose of exercise: study the relativistic effect on emission.)

12.2 Show that for the wave functions of a hydrogen model and the $L_{\mathrm{III}} \rightarrow K$ transition that $P_{\mathrm{fi}} \propto Z^4$.
Hint: Use standard wave functions for the hydrogen atom and remember that ω_{fi} enters to the 3rd power into the transition probability.

(Purpose of exercise: practice evaluation of matrix elements.)

12.3 Show that the Doniach–Sunjic lineshape converts to a Lorentzian line for $\alpha \rightarrow 0$.

(Purpose of exercise: study various lineshape functions.)

12.4 The Fermi energy of Ag is 5.5 eV. Using a PC compare the PE spectrum from the Fermi edge at 400 K with the emission from a system with a conduction band width of 0.5 eV, filled to 1/3 with electrons at the same temperature.
Hint: Use a free electron model in both cases.

(Purpose of exercise: recognize strong temperature effects on narrow metallic bands.)

13. Spectroscopy with γ Rays

Increasing the quantum energy beyond the x rays leads to the spectral range of γ radiation. This radiation originates from a relaxation of excited and quantized states of the nuclei. Even with these very hard light quanta spectroscopy in solids can be very useful. The subject was triggered in 1958 when *R. Mößbauer* reported for the first time recoil free resonance absorption for γ rays. This technique is now known as *Mößbauer spectroscopy*. Recently additional spectroscopic methods were developed on the basis of an angular correlation of γ rays which enable information about the chemical environment of special nuclei inserted into the lattice to be obtained.

13.1 Mößbauer Spectroscopy

The importance of Mößbauer spectroscopy extends over wide areas of physics. For example, it was used to determine for the first time the mass of light quanta. For the discovery of this spectroscopic method Mößbauer was awarded the Nobel prize in 1961.

13.1.1 Fundamentals of Mößbauer Spectroscopy

The principle of Mößbauer spectroscopy can be explained from Fig. 13.1. The left part of the figure refers to a macroscopic picture, the right part to the atomistic model.

In the macroscopic model the test person cannot succeed to jump from one boat to the other since he looses recoil energy when he jumps off. If there are strong waves he will be successful from time to time since the boat can support him with an additional momentum by chance. If the tow boats are frozen in ice the test person will always be successful (unless the ice is too thin and the recoiled boat breaks the ice).

The atomistic model is indeed similar. If the γ quantum is emitted from a nucleus in an excited state with energy ϵ_0 above the ground state part of the energy available from returning to the ground state must be transferred to the nucleus as a recoil energy ϵ_{R}. This means the emission energy is smaller than ϵ_0. From conservation of momentum it follows

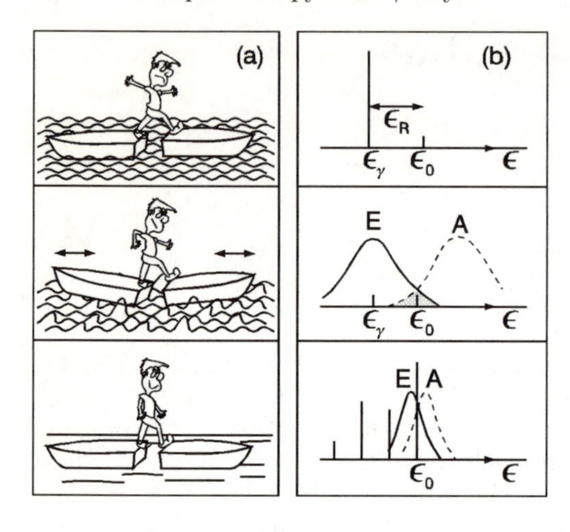

Fig. 13.1. Phenomenological description of recoil-free nuclear resonance absorption; macroscopic model (**a**), atomistic model with energy relations (top), emission and absorption for $T > 0$ (center) and quantized vibronic states (bottom), (**b**); (E: emission, A: absorption)

$$\boxed{\epsilon_{\rm R} = \frac{p_{\rm M}^2}{2M} = \frac{p_\gamma^2}{2M} = \frac{\epsilon_\gamma^2}{2Mc_0^2} \approx \frac{\epsilon_0^2}{2Mc_0^2} .} \tag{13.1}$$

M, $p_{\rm M}$, p_γ, and ϵ_γ are the mass of the nucleus, the momentum of the nucleus, and the momentum and the energy of the γ quantum, respectively. Since

$$\epsilon_\gamma = \epsilon_0 - \epsilon_{\rm R} \approx \epsilon_0 \left(1 - \frac{\epsilon_0}{2Mc_0^2}\right) < \epsilon_0 \tag{13.2}$$

the emitted quantum cannot be reabsorbed by another nucleus at $T = 0$ (Fig. 13.1b, top). If the atom is under thermal motion the emission lines and the absorption lines are subjected to a Doppler broadening. From the overlap between the emission and the absorption a certain probability arises for the reabsorption of the γ quantum (Fig. 13.1b, center). According to this model the effect of resonance absorption or resonance fluorescence should increase with increasing temperature. The opposite is observed in the experiment, at least for solids. Mößbauer explained the unexpected result by a partial recoil-free and thus particularly efficient resonance absorption at low temperatures. The atoms are frozen due to the quantization of the lattice energy and the γ quantum emitted from one nucleus can be absorbed by an equivalent nucleus. It can jump from one nucleus to the other if the recoil energy is smaller than the vibronic energy (Fig. 13.1b, bottom).

To describe the freezing of the atoms the quantum nature of the lattice motions must be considered in detail. For a classical oscillator of frequency $\Omega_{\rm p}$ the total energy ϵ is given by

$$\epsilon = k_{\rm B}T = M\langle x^2\rangle \Omega_{\rm p}^2 , \tag{13.3}$$

where $\langle x^2\rangle$ is the average of the squared oscillation amplitude which can accept any value. According to quantum mechanics this is not so. The energies can only have the discrete values from (7.28)

$$\epsilon_\alpha = \hbar\Omega_p(\alpha + 1/2) ,$$

where α is the quantum number of the oscillation.

The discussion of the Mößbauer effect is simplest for the Einstein model of lattice oscillators. This means there is only one frequency Ω_E and the energy scale has only discrete values with distances $\hbar\Omega_E$. The lowest energy level is $\hbar\Omega_E/2$ obtained for $\alpha = 0$. If $\epsilon_R \gg \hbar\Omega_E$ the emission of the γ quantum is accompanied by the excitation of a large number of oscillations and the loss by recoil energy will be non negligible. For $\epsilon_R < \hbar\Omega_E$ recoil-free emission occurs. In general for a finite temperature only a certain fraction f of the quanta will be emitted without recoil. This fraction can be determined for $\epsilon_R \ll \hbar\Omega_E$ in a first approximation from the average energy of the oscillator given by its averaged square amplitude in the lowest quantum-mechanically allowed state of energy $\hbar\Omega_E/2$. The average transferred recoil energy is under these conditions

$$\langle\epsilon_R\rangle = (1 - f)\hbar\Omega_E .$$

Using (13.1) for ϵ_R and $\langle x^2\rangle = \hbar/2M\Omega_E$ for the classical amplitude of the lowest oscillator Einstein oscillator we obtain for f

$$f = 1 - \frac{\epsilon_R}{\hbar\Omega_E} = 1 - \frac{\hbar^2k^2}{2M\hbar\Omega_E} = 1 - k^2\langle x^2\rangle , \tag{13.4}$$

where k is the wave vector of the γ quantum. A more exact value of f derived for an Einstein oscillator and temperatures larger than the Debye temperature is

$$\boxed{f(T) = \exp(-k^2\langle x^2\rangle) = \exp\left(-\frac{k^2k_BT}{M\Omega_E^2}\right) .} \tag{13.5}$$

$f(T)$ corresponds physically and also formally exactly to the Debye–Waller factor used for the description of diffraction intensities of x ray patterns. Only those quanta contribute to the interference pattern for which no phonon excitation was involved in the scattering process. Equation (13.5) shows that the fraction of recoil-free emitted quanta increases with increasing mass and frequency of the oscillator and with decreasing temperature and wave vector of the γ quantum. If a Debye model is used to describe the lattice modes (see Appendix K.1) more complicated formulas are obtained but the physical meaning and the dependence of f on ϵ_R, Ω and T remain the same.

The natural linewidth Γ for the absorption is determined from the lifetime τ for the excited nuclear state

$$\Gamma = \frac{2\pi\hbar}{\tau} = 2\pi\hbar\frac{\ln 2}{t_{1/2}} , \tag{13.6}$$

where $t_{1/2} = \tau \ln 2$ is the half time of the excitation. If the Mößbauer atoms are in thermal motion the linewidth is still intrinsic but broadened from the Doppler effect. If the thermal energy is of the same order or larger than the recoil energy a good value for it is

$$\Gamma \approx 2\sqrt{\epsilon_{\text{th}}\epsilon_{\text{R}}}, \qquad \epsilon_{\text{th}} = \frac{3k_{\text{B}}T}{2} . \tag{13.7}$$

Since ϵ_{R} is 1.9 meV for ^{57}Fe the linewidth for this source is intrinsic.

13.1.2 Radiation Sources and Detectors

For experiments with resonance absorption appropriate radiation sources, absorber, and detectors are required. To tune the frequency of the radiation to the resonance transition the Doppler effect is used. Either the source or the absorber is shifted with a constant velocity until resonance absorption occurs. Since the shift in energy from the Doppler effect is $\Delta\epsilon = \epsilon v/c_0$, the energy balance for the radiation is

$$\boxed{\epsilon_\gamma = \epsilon_0 - \epsilon_{\text{R}} + \epsilon_{\text{D}} \qquad \text{with} \qquad \epsilon_{\text{D}} = \epsilon_0 \frac{v}{c_0} .} \tag{13.8}$$

The spectral distributions for the emission of a γ quantum and for the absorption of the quantum are well described by Lorentzian lines of width Γ. Hence, in a resonance experiment where a Lorentzian emission line is shifted across the position of absorption with equal width and shape the response has the form

$$\sigma = \sigma_0 \frac{(\Gamma/2\pi)^2}{(\Gamma/2)^2 + (\epsilon - \epsilon_0)^2} . \tag{13.9}$$

σ_0 is evaluated to

$$\sigma_0 = \frac{2\pi c_0^2 \hbar^2 (2I_{\text{e}} + 1)}{\varepsilon_0 (I_{\text{g}} + 1)(\alpha + 1)} .$$

Equation (13.9) is the result of the folding of two Lorentzian lines. I_{e} and I_{g} are nuclear spins for the excited state and for the ground state and α is the total conversion coefficient for the nuclear transition. The finally observed absorption for a particular nucleus K is

$$\alpha_{\text{K}}(T) = \sigma_{\text{K}} f(T) n_{\text{K}} a_{\text{K}} d , \tag{13.10}$$

where n_{K}, a_{K}, and d are the number of nuclei per unit volume, the isotope concentration, and the thickness of the sample, respectively. $f(T)$ is the fraction of recoil-free emitting nuclei from (13.5) or the corresponding value from a Debye model.

The Doppler energy for tuning can be superimposed on the resonance process either by a mechanical or by an electromechanical or piezoelectric shift. The source or the absorber may be moved. Depending on the Mößbauer source shifts are required from several µm/second to several m/second.

The detector is a γ counter which must be adapted to the energy range of the radiation used. Proportional counters, scintillation counters or p-i-n semiconductor detectors are appropriate as discussed in Sect. 12.1 and as

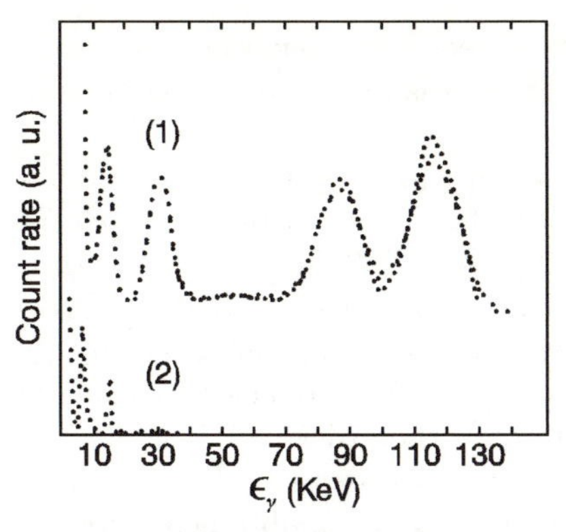

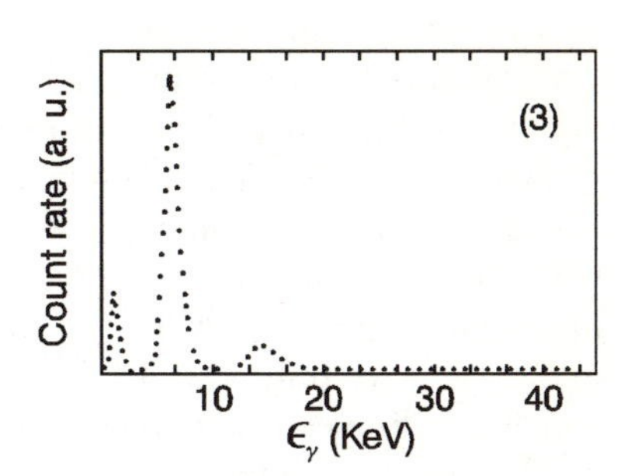

Fig. 13.2. Sensitivity of various γ detectors between 0 and 130 KeV. The emission is for ^{57}Co. The registration is for a scintillation counter (1), a proportional counter (2), and for a Si(Li) detector (3); after [13.1].

they are compiled in Table 12.2. Figure 13.2 shows the response of the various detectors to ^{57}Co radiation. The spectral distribution is obtained directly from the detector by using a pulse height analyzer. Obviously for very high quantum energies only the scintillation counter is useful. For low energies the proportional counter and the Si(Li) detector are more appropriate because of their better efficiency.

For a Mößbauer experiment three components are needed: a Mößbauer source, an appropriate absorber, and a detector. Possible arrangements and the reaction steps during the resonance absorption experiment are shown in Fig. 13.3. As a starting material for the source a nucleus is appropriate, which becomes only Mößbauer active after a nuclear reaction from a mother nucleus. The resonance transition in the absorber can either be studied by a change in the transmission (resonance absorption) or by observation of a resonance scattering (resonance fluorescence). In the latter case the incident γ quantum excites the absorber but is afterwards reemitted in an arbitrary

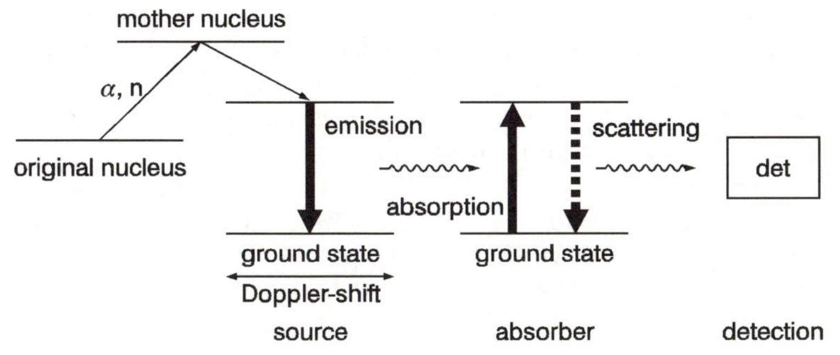

Fig. 13.3. Schematic representation of nuclear transitions in a Mößbauer experiment.

direction. Mößbauer experiments in scattering geometry do not need a sample preparation in the form of very thin films which is an advantage compared to absorption experiments.

The best known example of a Mößbauer source is ^{57}Fe with a γ-ray component at 14.4 KeV. The component is part of a decay pattern for ^{57}Co as shown in Fig. 13.4. The Co nucleus has a lifetime of 270 days versus a transition to ^{57}Fe by electron capture. The resulting Fe nucleus has three radiative transitions in the γ range. The one appropriate for the Mößbauer experiments has a half time $t_{1/2}$ of 97.7×10^{-9} seconds. This corresponds to a natural linewidth Γ of only 4.6×10^{-9} eV. Since shifts of the order of 1% of the half width of a line can be detected relative shifts of the order $10^{-2}\Gamma/\epsilon_\gamma \approx 10^{-13}$ can be observed. This is a resolution which has never been obtained before in spectroscopy. For many nuclei this resolution is several orders of magnitude higher than their hyperfine interaction energy or their quadrupole interaction energy. Thus, these phenomena which are introduced by the crystal field (or more generally by the electrons in their bonding orbitals) can be investigated. This makes Mößbauer spectroscopy another member of experiments where the analysis of the chemical shift yields microscopic information on the materials.

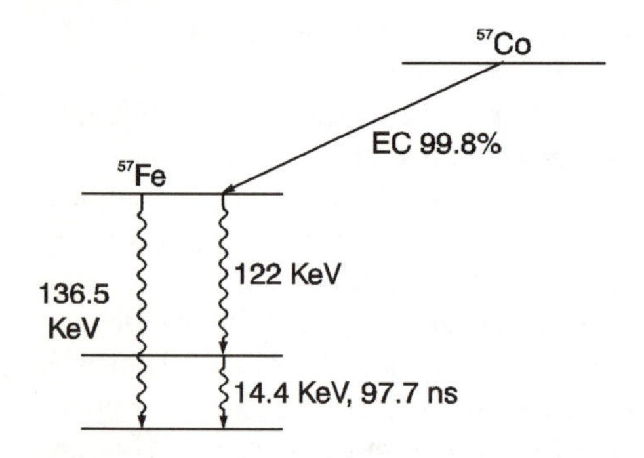

Fig. 13.4. Decay scheme of ^{57}Co into ^{57}Fe by electron capture (EC).

A review on parameters essential for Mößbauer spectroscopy as well as their energy range is shown in Fig. 13.5. If the Debye energy is larger than the recoil energy the system is appropriate for Mößbauer experiments. This condition is well satisfied for ^{57}Fe. Also the hyperfine splitting is two orders of magnitude larger than the natural linewidth.

Table 13.1 compiles several important Mößbauer sources and their characteristic parameters. The linewidth is given in units of mm/s as it refers to experiments where the energy is tuned by a Doppler shift. The linewidth in eV is obtained from (13.8) by $2\Gamma\,[\text{eV}] = \epsilon_0\,[\text{eV}] 2\Gamma\,[\text{mm/s}]/c_0$.

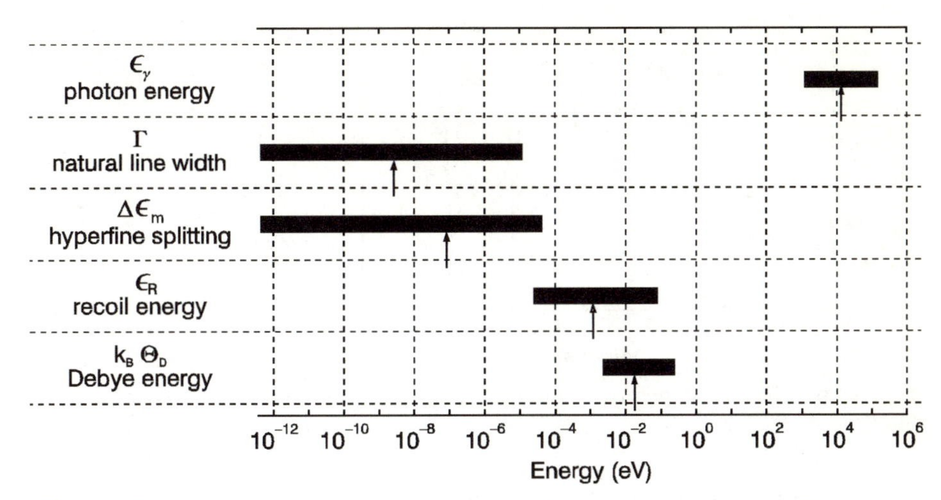

Fig. 13.5. Characteristic energy ranges for Mößbauer spectroscopy. The arrows assign the values for ^{57}Fe.

13.1.3 Results of Mößbauer Spectroscopy

Because of its very high resolution in energy and its local character Mößbauer spectroscopy is particularly useful for probing the local environment of the active nucleus as a function of temperature, pressure, defects, phase transitions, etc. As for IR spectroscopy from defects, for magnetic resonance, or for photoemission the chemical shift of the Mößbauer absorption lines gives information on the local structure. The electron density at the position of the nucleus is important. This is in particular so for the often investigated hyperfine splitting of the energy levels in the nuclei. Figure 13.6 shows Mößbauer

Table 13.1. Mößbauer sources and their characteristic data; adapted from [13.1]. (a: isotope abundance, ϵ_0: transition energy, $t_{1/2}$: half time, $I_{e,g}$: nuclear spins for the excited and ground state, σ_0: cross section, 2Γ: linewidth, ϵ_R: recoil energy, Ω_D: Debye frequency).

Isotope	a [%]	ϵ_0 [KeV]	$t_{1/2}$ [ns]	I_e	I_g	σ_0 [E–20 cm^2]	2Γ [mm/s]	ϵ_R [meV]	Ω_D [cm^{-1}]
^{40}K	0.011	29.4	4.26	3	4	28.97	2.18	11.6	62
^{57}Fe	2.19	14.41	97.81	3/2	1/2	256.6	0.19	1.95	312
^{61}Ni	1.25	67.40	5.06	5/2	3/2	72.12	0.80	39.99	295
^{67}Zn	4.11	93.31	9150	3/2	5/2	10.12	3E–5	69.78	213
^{73}Ge	7.76	13.26	4000	5/2	9/2	361.2	5E–4	1.29	295
^{73}Ge	7.76	68.752	1.86	7/2	9/2	22.88	2.13	34.77	295
^{119}Sn	8.58	23.87	17.75	3/2	1/2	140.3	0.64	2.57	136
^{121}Sb	57.25	37.15	3.5	7/2	5/2	19.70	2.10	6.12	123
^{184}W	30.64	111.19	1.26	2	0	26.04	1.95	36.08	238

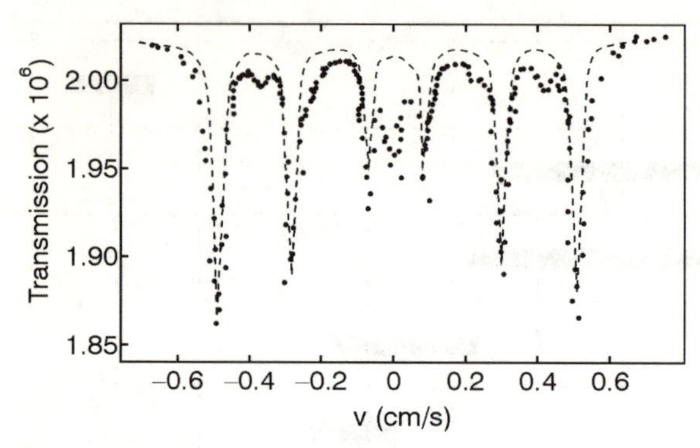

Fig. 13.6. Mößbauer spectrum of pure Fe (−−) and of carbon rich Fe after quenching from 830° (...); after [13.2].

spectra for pure Fe and for Fe with 4.2% carbon after quenching from 850°. The six equidistant lines originate from the hyperfine splitting of the Fe nuclei. The central line for the quenched sample comes from the austenitic phase which is retained during the quenching process. (The austenitic phase is fcc and only stable at high temperatures.) The concentration of this phase retained after cooling is an important control parameter for steel production.

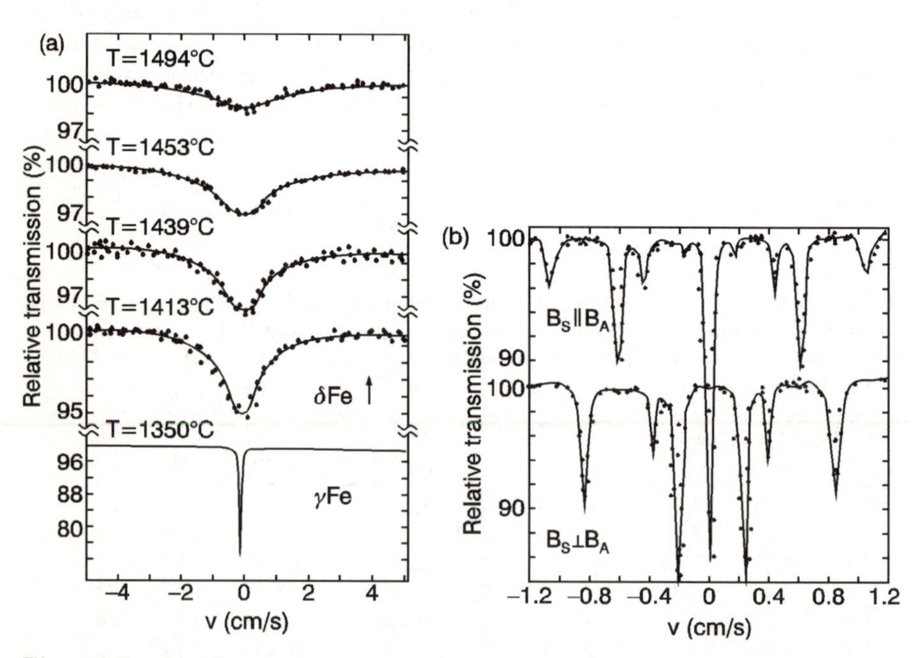

Fig. 13.7. Lineshape of resonance absorption in δ-Fe at various temperatures as indicated and after cooling to the austenitic phase (γ-Fe) (**a**); after [13.3], and absorption for ^{57}Fe with polarized γ quanta (**b**). The magnetic fields at source and absorber are parallel and perpendicular to each other for the two displayed spectra; after [13.1].

The strong temperature dependence of the resonance absorption as discussed above is demonstrated in Fig. 13.7a. It refers to the intensity via the Debye–Waller factor as well as to the linewidth Γ. For the high temperatures indicated the latter is intrinsic and caused by a Doppler shift as a consequence of the diffuse motion of the atoms. It is therefore related to the diffusion coefficient D as

$$\Gamma \propto \frac{\hbar}{\tau_{\mathrm{D}}} = \frac{6\hbar D}{a^2} ,$$

where $1/\tau_{\mathrm{D}}$ and a are the jump rate and jump width for the diffusion process, respectively. Accordingly, such experiments are appropriate to determine the diffusion coefficient on a microscopic level [13.4].

Mößbauer spectroscopy can also be performed with polarized γ radiation. The possibility to obtain polarized γ quanta is based on the anisotropy of the emission process. The emission depends on the orientation of the nuclear spin. It is maximum perpendicular to the spin direction. If the spins are oriented by a magnetic field B the emission is perpendicular to B. The polarization is either perpendicular to B (for transitions with $\Delta I_z = 0$) or parallel to B (for transitions with $\Delta I_z = \pm 1$). Depending on the orientation of the field in the absorber relative to the field in the emitter either transitions with $\Delta I_z = 0$ or transitions with $\Delta I_z = \pm 1$ are observed. Figure 13.7b shows experiments with polarized γ radiation for which the different behavior of the two geometries is clearly seen.

13.2 Perturbed Angular Correlation

Another microscopic method for the probing of the neighborhood of a particular atom in a lattice uses the subsequent emission of two γ quanta in a cascade. The direction of emission is perturbed by local electric and magnetic fields. Thus, the technique is called the *perturbed angular correlation* (PAC) of γ radiation. The analysis of the emission characteristic yields the information on the local fields.

13.2.1 Basic Description of the Perturbed Angular Correlation

The basic idea of experiments using perturbed angular correlation originates from the anisotropy of the emission of γ quanta with respect to the orientation of the nuclear spins. This anisotropy is not observed right away since the nuclei are not oriented and any anisotropy is averaged out. If the nuclei are oriented by the application of a magnetic field B observation of the anisotropy for the γ emission is possible. Such experiments are well known in nuclear physics. The emission of polarized γ radiation as it was discussed in the last section for Mößbauer quanta is a good example. The procedure in the case of PAC is different. According to Fig. 13.8 nuclei are used which emit

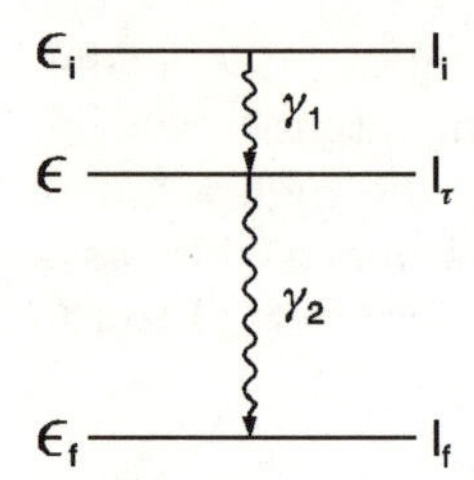

Fig. 13.8. Energy levels and transitions for perturbed angular correlation; I_{i}, I_{τ} and I_{f}: nuclear spins in the initial, intermediate and final state, respectively.

a cascade of at least two quanta in a series. From the direction of detection for the first quantum a certain manifold of spins with a particular direction is selected. The anisotropy of the emission for the second quantum is now investigated by an angular correlated coincidence experiment. If the nuclear spins interact with local magnetic fields they precess around the local field and the anisotropy of the emission for the second γ quantum precesses likewise. Even though the probability of emission of the second quantum decays with the lifetime of the intermediate state it can still be significant for many revolutions. The precession is sustained by thermal activation of the spins from the ground state. In the latter all spins are aligned along the local field. The activation is still relevant to below 1 K.

Figure 13.9 demonstrates the basic features of the emission processes. The dashed line represents the anisotropy $W(\theta)$ for the emission of γ quanta. The emission is strongly reduced along the direction of the spin but independent of the azimuth angle ϕ. By the detection of γ_1 with detector D_1 (Fig. 13.9a) the initial position of a particular nucleus is defined. With time the spin moves from this position by precession around the local field. For simplicity the latter is assumed in the figure as perpendicular to the spin direction. After a time t when the second quantum is observed with detector D_2 the spin has rotated by an angle $\theta = \omega t$ and the orientation for observation has become less favorable (Fig. 13.9b). Finally, after a time period $t = \pi/\omega$ the spin has returned to an orientation for optimum detection with respect to the counter

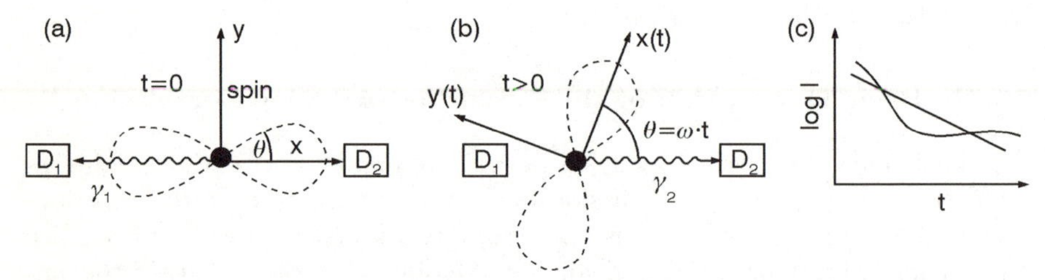

Fig. 13.9. Time delayed coincidence measurement for the investigation of the perturbed angular correlation; starting geometry (**a**), geometry at the time when the second quantum is emitted (**b**) and time dependence of the intensity for the detection of the second quantum (**c**). The field in (a) and (b) is assumed perpendicular to the plane of the paper.

D_2. This means a signal oscillating with the delay time of the coincidence will be detected at the counter D_2. The oscillation frequency is the precession frequency of the spins in the local field. Thus, the anisotropy $W(\theta)$ of the γ emission is studied for a constant geometry in space but varying angle θ in time. This is in contrast to the classical experiments of unperturbed angular correlation in nuclear physics where $W(\theta)$ is studied explicitly for $t = 0$ with the angle θ varying in space and no precessing spins. Because of the finite lifetime τ of the intermediate state the overall signal decays exponentially with time. This yields for the signal observed at D_2 a time dependence of the form

$$I(\theta, t) = I_0 \mathrm{e}^{-t/\tau} W(\theta(t)) \tag{13.11}$$

with

$$W(\theta(t)) \approx W(\theta_0 + \omega t) \, .$$

Thus, on a logarithmic scale the overall intensity decreases linearly with t with a superimposed modulation as shown in Fig. 13.9c. $W(\theta)$ is given by the Legendre polynomials in $\cos\theta$ and depends on the spins in the initial, intermediate, and final state. For I_i and $I_\mathrm{f} = 0$ and $I = 1$ one obtains

$$W(\theta) = 1 + \frac{1}{2} P_2(\cos\theta) = 1 + \frac{1}{4}(3\cos^2\theta - 1) \, . \tag{13.12}$$

For the time-dependent experiments θ has to be replaced by $\theta_0 + \omega t$ where θ_0 is the angle between the two detectors. This means the oscillation in (13.11) follows a variation in time as $\cos^2(\omega t)$.

According to the severe constraints on the character of the emission not so many nuclei are available for the experiments as, for example, in the case of Mößbauer spectroscopy. The most important mother/daughter pairs are Pd/Rh, In/Cd, and Hf/Ta. The decay diagram and the total spectrum of γ radiation emitted are depicted for the first two pairs in Fig. 13.10. In both cases the nucleus for the PAC experiments is obtained from the mother nucleus by electron capture. The actual nuclear spins are indicated by the numbers on the left side of the decay diagram. The energies for the emitted γ cascade and the lifetime of the intermediate state is shown on the right side of the diagram. From the many emitted lines in Fig. 13.10b only two are acceptable for the PAC experiments.

The mother nucleus must be accommodated in the crystal on a lattice site. It acts there as a microscopic probe. 10^{11}–10^{12} nuclei are enough to obtain an activity of the order of 10 μCi as is required for the experiment.

For the detection of the γ quanta at least two, but better four, detectors are required. The quanta must be resolved according to their energy (for identification) as well as according to their time of arrival. Appropriate detectors are again scintillation counters on the basis of NaI(Tl). For a properly designed electronic each detector can be used for the detection of the quanta

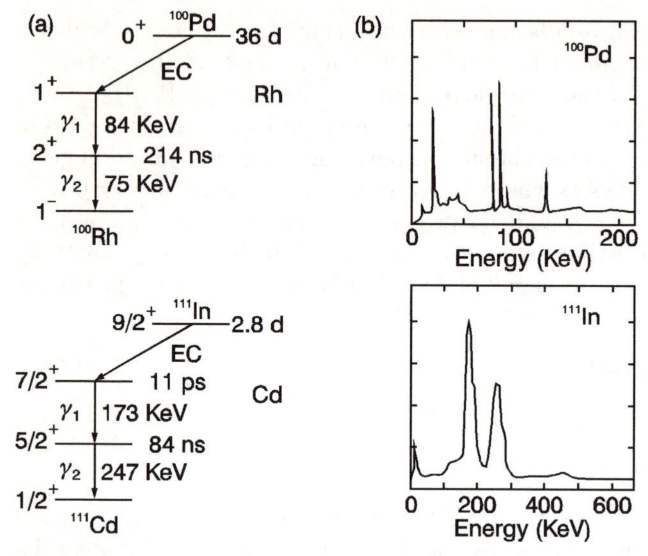

Fig. 13.10. Mother/daughter pairs for use in PAC experiments (**a**) and observed γ spectra (**b**); after [13.5].

γ_1 and γ_2. If $I_{ij}(\theta_0, t)$ is the coincidence spectrum between two detectors i and j out of four detectors on a circle, the ratio

$$\frac{I_{13}(180,t)I_{24}(180,t)}{I_{14}(90,t)I_{23}(90,t)} = \left(\frac{W(180,t)}{W(90,t)}\right)^2$$

is a good measure for the change of the orientation of the nuclei. In the expression the initial intensity I_0 as well as the exponential decay $\exp(-t/\tau)$ have cancelled. Another commonly used expression for the oscillating part of the signal is

$$R(t) = \frac{2}{3}\left(\frac{W(180,t)}{W(90,t)-1}\right) . \tag{13.13}$$

$R(t)$ is known as the *time spectrum* of the PAC experiment.

13.2.2 Experimental Results from Perturbed Angular Correlation

From the time spectrum and a subsequent Fourier transformation the local magnetic field or the local electric field gradient at the position of the nucleus is obtained from the precession frequencies. Experiments can be performed with an applied magnetic field or without field. In the first case the change in time spectra with changing local field conditions is studied. In the second case the local field is monitored directly. Figure 13.11 shows results for a rhodium nucleus in Cu. Part (a) represents the directly measured coincidence rate for two directions + and − of an applied magnetic field.

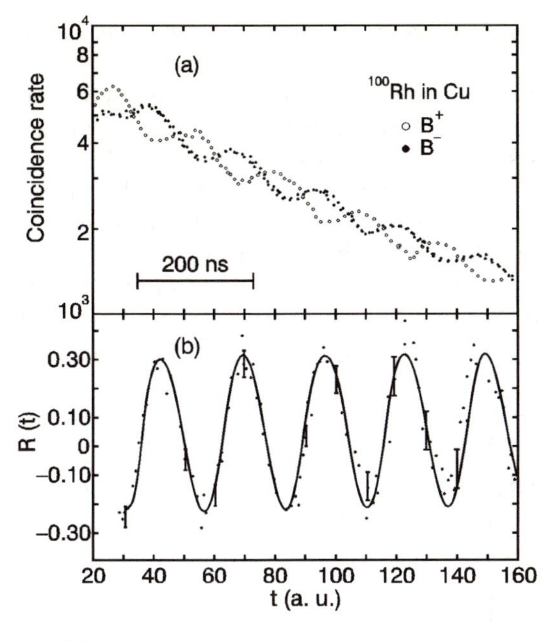

Fig. 13.11. Coincidence spectra of ^{100}Rh in Cu for two magnetic fields in opposite directions (**a**) and ac part of the spectrum according to (13.14) (**b**); after [13.6].

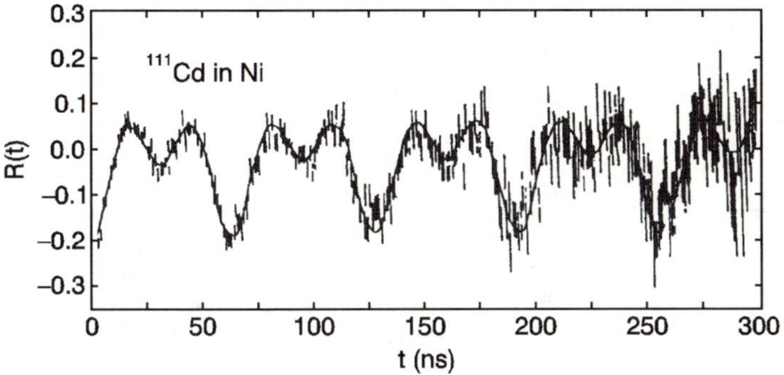

Fig. 13.12. Time spectrum for ^{111}Cd in Ni; after [13.7].

The result corresponds exactly to the schematic presentation in Fig. 13.9c. From the two coincidence measurements the ratio

$$R(t) = \frac{I_+ - I_-}{I_+ + I_-} \tag{13.14}$$

can be constructed in which the initial intensity I_0 as well as the exponential time dependence has cancelled as in (13.13). Accordingly an oscillation in time such as $\cos^2 \omega t$ is obtained as shown in Fig. 13.11b.

Without external field the hyperfine interaction of the electrons at the position of the nucleus is sufficient to excite the nuclear spins to a precession. Figure 13.12 shows a time spectrum for ^{111}Cd in polycrystalline ferromagnetic Ni. The Fourier analysis yields two Larmor frequencies ω_0 and $2\omega_0$. The observation of two frequencies is a consequence of the statistic disorder of the magnetic fields in the sample.

The precession of the nuclear spins can also originate from an interaction of the nuclear quadrupole moment with the electric field gradient at the position of the nucleus. This is shown in Fig. 13.13 from time spectra for ^{111}Cd in different noncubic metals of the 5th period. It is immediately evident from the figure that the field gradients are very different for the three metals. Since the quadrupole moment for ^{111}Cd is known to be 0.83×10^{-24} cm^2 one can even determine their value quantitatively using (11.75) from Chap. 11. From the spectra shown for Cd, Sn, and In the values 5.45, 1.5, and 0.74 V/cm^2 are obtained.

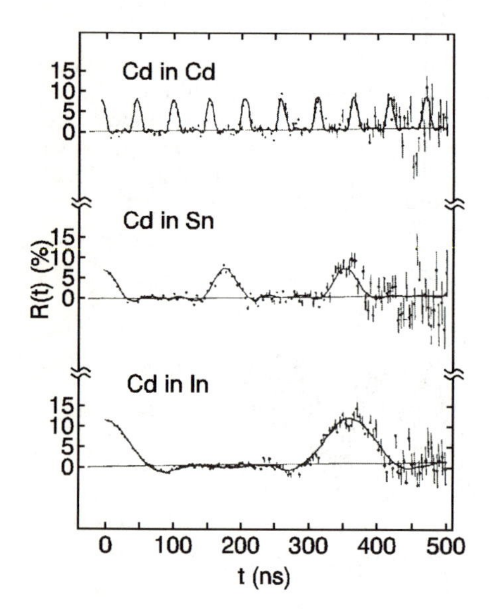

Fig. 13.13. Time spectra for ^{111}Cd in various metals. The spin precession originates from an interaction of the quadrupole moment of ^{111}Cd with an electric field gradient; after [13.8].

Problems

13.1[a] Compare the recoil-free emission of γ quanta for the Debye model and the Einstein model using a personal computer.

(Purpose of exercise: observe the difference between Debye model and Einstein model.)

13.2 Find an approximation for the fraction of recoil-free emission of an Einstein system at low temperatures and discuss the result for the case of ^{40}K. What is the result for f from the Debye system under the same conditions?

(Purpose of exercise: study the possibilities to observe a Mößbauer effect at very low temperatures.)

13.3 Show that the correlation of two Lorentzian lines is different to the convolution of two Lorentzian lines.

(Purpose of exercise: remember difference between correlation and convolution.)

13.4 For which temperature would the thermal absorption of γ quanta equal the recoil-free absorption?
Hint: use numerical data for Ge from Table 13.1.

(Purpose of exercise: note the two different types of absorption; one of them is dominating.)

13.5 Calculate the field gradients in Cd, Sn, In by using the results of Fig. 13.13.
(Purpose of exercise: evaluate a fundamental parameter of the electronic structure from an experiment.)

14. Generalized Dielectric Function

A generalized dielectric function must consider all types of excitations which can occur in a solid. Since we will discuss in the following chapters not only excitations by electromagnetic radiation, a description is needed which goes beyond the electromagnetic response. The description will still be phenomenological as a general formulation of the linear response theory is beyond the frame of this textbook. However, some fundamental concepts of the latter are summarized in Appendices L.1 to L.5. More details can be found in special textbooks like [14.1].

The model dielectric functions discussed in Sect. 6.3 are only useful for simple systems. They do not, for example, consider finite values of the q vector of the excitations or interactions between various quasi-particles. The following description takes care of some of these shortcomings. It concentrates on systems with free carriers.

14.1 The Momentum Dependence of the Dielectric Function

Since quasi-particles in a solid have well defined energy and momentum, the generalized response functions such as $\chi(q, \omega)$ or $\varepsilon(q, \omega)$ must depend on both q and ω. As a consequence of the particle dispersion ω itself is a function of q. The type of quasi-particles excited depends on the nature of the exciting perturbation. For example, the transverse nature of electromagnetic radiation does not allow the excitation of longitudinal polarized plasmons. In contrast, plasmons are excited with a high probability by an electron beam but both types of probes can interact with phonons of well defined symmetries as discussed in Sect. 10.4 and Sect. 15.1.

In Chap. 6 the dielectric response was investigated for $q = 0$ and $\omega \neq 0$. Here we start the discussion with a particular case which is alternative to these conditions. We choose $\omega = 0$ and $q \neq 0$. These conditions describe a static rearrangement of the charges as a response to an applied potential energy $\delta U(x)$ with Fourier components δU_{q}. The number of redistributed charges is $\delta n(x)$ with Fourier components δn_{q}. From Fig. 14.1 δn_{q} is estimated to be

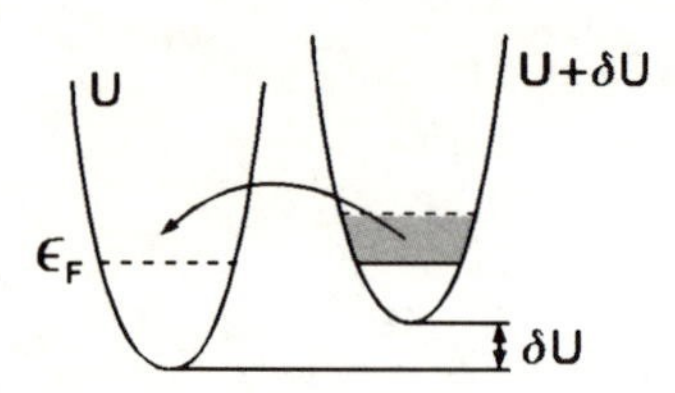

Fig. 14.1. Redistribution of free carriers as a consequence of a potential δU.

$$\delta n_{\mathrm{q}} = -g_{\mathrm{v}}(\epsilon_{\mathrm{F}})\delta U_{\mathrm{q}} \,, \tag{14.1}$$

where $g_{\mathrm{v}}(\epsilon_{\mathrm{F}})$ is the density of states at the Fermi level per unit volume. The Poisson equation relates δn_{q} to the induced potential $\delta\Phi$ with Fourier components $\delta\Phi_{\mathrm{q}}$

$$-q^2\delta\Phi_{\mathrm{q}} = -\frac{e}{\varepsilon_0}\delta n_{\mathrm{q}} = \frac{e}{\varepsilon_0}g_{\mathrm{v}}(\epsilon_{\mathrm{F}})\delta U_{\mathrm{q}} \,. \tag{14.2}$$

As a consequence of $\delta\Phi$ the resulting potential is not $\delta U/e$ but $\delta\Phi'$ or

$$\delta\Phi'_{\mathrm{q}} = \delta U_{\mathrm{q}}/e - \delta\Phi_{\mathrm{q}} \,. \tag{14.3}$$

The response function $\varepsilon(q)$ is defined as the ratio between the resulting potential $\delta\Phi'$ and the applied potential $\delta U/e$. Combining (14.2) and (14.3) we obtain

$$\delta\Phi'_{\mathrm{q}} = \left(1 + \frac{e^2 g_{\mathrm{v}}(\epsilon_{\mathrm{F}})}{q^2\varepsilon_0}\right)\frac{\delta U_{\mathrm{q}}}{e} \,. \tag{14.4}$$

The resulting response function is called *Thomas–Fermi* dielectric function and has the form

$$\varepsilon_{\mathrm{TF}}(q) = 1 + \frac{e^2 g_{\mathrm{v}}(\epsilon_{\mathrm{F}})}{q^2\varepsilon_0} = 1 + \frac{q_{\mathrm{TF}}^2}{q^2} \,. \tag{14.5}$$

The *Thomas–Fermi screening length* is

$$\boxed{\lambda_{\mathrm{TF}} = \frac{1}{q_{\mathrm{TF}}} = \left(\frac{2\epsilon_{\mathrm{F}}\varepsilon_0}{3n_0e^2}\right)^{1/2} ,} \tag{14.6}$$

where in the last part of the equation $g_{\mathrm{v}}(\epsilon_{\mathrm{F}})$ has been replaced by its value for free carriers $g_{\mathrm{v}}(\epsilon_{\mathrm{F}}) = 3n_0/2\epsilon_{\mathrm{F}}$. For a nondegenerate electron system screening is determined by the *Debye screening length*

$$\boxed{\lambda_{\mathrm{D}} = \frac{1}{q_{\mathrm{D}}} = \left(\frac{\varepsilon_0 k_{\mathrm{B}}T}{e^2n_0}\right)^{1/2} .} \tag{14.7}$$

To derive the generalized response functions $\varepsilon(q,\omega)$ or $\chi(q,\omega)$ we proceed similarly but allow δU and $\delta\Phi$ to be periodic in space and time

$$\begin{aligned} \delta U &\propto \mathrm{e}^{\mathrm{i}(qr-\omega t)}\mathrm{e}^{-\alpha t} \\ \delta\Phi &\propto \mathrm{e}^{\mathrm{i}(qr-\omega t)}\mathrm{e}^{-\alpha t}. \end{aligned} \tag{14.8}$$

For particles interacting by a Coulomb potential the relationship between $\delta\Phi_{\mathrm{q}}$ and δU_{q} can be worked out within the *random phase approximation* (RPA). (The concept of this approximation is described in Appendix L.6). The result is

$$e\delta\Phi_{\mathrm{q}} = \frac{e^2}{\varepsilon_0 \mathcal{V} q^2} \sum_k \frac{f_{\mathrm{F}}(k) - f_{\mathrm{F}}(k+q)}{\epsilon(k) - \epsilon(k+q) + \hbar\omega + \mathrm{i}\alpha\hbar} \delta U_{\mathrm{q}} , \tag{14.9}$$

where

$$f_{\mathrm{F}}(k) = \frac{1}{\exp[(\epsilon(k) - \mu)/k_{\mathrm{B}}T] + 1} \tag{14.10}$$

is the Fermi distribution function of the chemical potential μ. Using the relation between $\delta U, \delta\Phi'$, and $\delta\Phi$ as before the dielectric response function is obtained in the form

$$\varepsilon(q,\omega) = \varepsilon_{\mathrm{L}}(q,\omega) = 1 - \frac{e^2}{\varepsilon_0 q^2 \mathcal{V}} \chi^0(q,\omega) , \tag{14.11}$$

where $\mathcal{V}$ is the sample volume and $\chi^0(q,\omega)$ is the *generalized susceptibility* or *density–density response function* for neutral Fermionic particles given by

$$\chi^0(q,\omega) = \sum_k \frac{f_{\mathrm{F}}(k) - f_{\mathrm{F}}(k+q)}{\epsilon(k) - \epsilon(k+q) + \hbar\omega + \mathrm{i}\alpha\hbar} . \tag{14.12}$$

The concept of generalized susceptibility is explained in Appendix L.1. The factor in front of $\chi^0(q,\omega)$ in (14.11) is the Fourier transform of the Coulomb energy. The generalized response function $\varepsilon_{\mathrm{L}}(q,\omega)$ for $\alpha = 0$ is known as the *Lindhard* dielectric function. (Note that α in (14.12) is not a damping constant but rather a convergency parameter).

Since the Lindhard DF is the general form of the DF for charged particles it must reduce to the Drude DF and to the Thomas–Fermi DF for $q = 0, \omega \neq 0$ and $q \neq 0, \omega = 0$, respectively. The latter is easily proven by replacing the sum over all k states by an integral in k space as $\sum_k = \mathcal{V}/(2\pi)^3 \int \mathrm{d}^3k$ and by expressing the differences in the numerator and in the denominator of (14.12) by gradients. Finally the resulting derivative of the Fermi distribution function is replaced by a δ function $f'_{\mathrm{F}}(\epsilon) = -\delta(\epsilon - \epsilon_{\mathrm{F}})$. With this and for $\omega = \alpha = 0$ we obtain successively

$$f_{\mathrm{F}}(k) - f_{\mathrm{F}}(k+q) = -q f'_{\mathrm{F}}(k) = -q \frac{\partial f_{\mathrm{F}}}{\partial \epsilon} \nabla_{\mathrm{k}} \epsilon(k)$$

and

$$-\epsilon(k+q) + \epsilon(k) = -q \nabla_{\mathrm{k}} \epsilon$$

which finally yields from (14.11) the same result for $\varepsilon_{\mathrm{TF}}$ as (14.5)

$$\varepsilon_{\mathrm{TF}} = \frac{e^2}{\varepsilon_0 q^2 \mathcal{V}} \frac{\mathcal{V}}{(2\pi)^3} \int f'_{\mathrm{F}}(\epsilon) \mathrm{d}^3k = \frac{e^2 g_{\mathrm{v}}(\epsilon_{\mathrm{F}})}{\varepsilon_0 q^2} + 1 . \tag{14.13}$$

Similarly, the Drude DF $\varepsilon_{\mathrm{D}}(0,\omega)$ can be derived from (14.11). The explicit calculation is left as a problem.

14.2 Excitations of the Electronic System

A number of different excitations from the Fermi surface exist in a system of free carriers. Most important are the single pair excitations and the plasmons. In the former an electron is excited from the Fermi sea and a hole remains. The plasmons are collective modes with a longitudinal polarization as already described in Sect. 6.3.2. The spectral range of the various excitations (with a given q vector) is shown schematically in Fig. 14.2. The intensity of the excitation is represented by the dynamical form factor $S(q,\omega)$. Multiple pair excitations have a rather low probability.

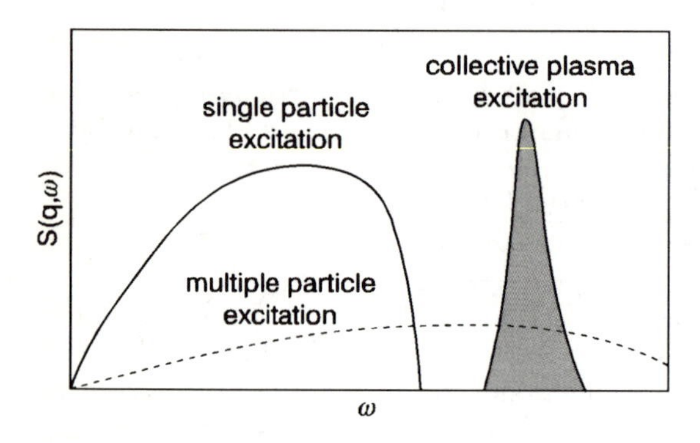

Fig. 14.2. Schematic representation of the spectral excitations $S(q,\omega)$ of a Fermi liquid.

14.2.1 Plasmons and Plasmon Dispersion

Plasmons are longitudinal excitations of the electronic system. Their eigenfrequency ω_{pl} was derived for the Drude model from the zeros of the dielectric function $\varepsilon_{\mathrm{D}}(\omega)$. In general the plasmon also has a dispersion $\omega_{\mathrm{pl}}(q)$. Similar to the Drude case it can be derived from the Lindhard DF by evaluating the zeros of $\varepsilon_{\mathrm{L}}(q,\omega)$. For small values of q (14.11) can be expanded and yields

$$\varepsilon(q,\omega) = \varepsilon_{\infty} - \frac{\omega_{\mathrm{p}}^2}{\omega^2} - \frac{6\epsilon_{\mathrm{F}}}{5m^*\omega_{\mathrm{p}}^2}q^2 + \dots \,. \tag{14.14}$$

For $\varepsilon(q,\omega) = 0$ the plasmon dispersion

$$\boxed{\omega_{\mathrm{pl}}(q) = \omega_{\mathrm{pl}}(0) + \frac{3\epsilon_{\mathrm{F}}q^2\omega_{\mathrm{pl}}(0)}{5m^*\omega_{\mathrm{p}}^2\varepsilon_{\infty}} + \dots} \tag{14.15}$$

follows. ω_{p} is the plasma frequency determined by the the particle concentration and particle effective mass as given in Sect. 6.3. Plasmons can originate from various sets of electrons such as conduction electrons, valence electrons, etc., provided the binding energy for these electrons is much smaller than the plasmon energy. If the energy range of the set includes an electronic gap the gap energy adds to the plasmon energy like

$$\hbar^2\omega_{\mathrm{pl}}^2 = \frac{\hbar^2 n e^2}{\varepsilon_0 m^* \varepsilon_\infty} + \epsilon_{\mathrm{g}}^2 \,. \tag{14.16}$$

14.2.2 Single Particle Excitation

Single particle excitations with dispersion $\epsilon(q)$ are possible for a wide continuum in energy and momentum. However, conservation rules and the Pauli principle impose characteristic limitations. Excitation of an electron with wave vector k from below the Fermi surface to a state with wave vector $k+q$ above the Fermi surface needs the energy

$$\hbar\omega(\boldsymbol{q}) = \frac{\hbar^2(\boldsymbol{k}+\boldsymbol{q})^2}{2m} - \frac{\hbar^2 k^2}{2m} = \frac{\hbar^2 \boldsymbol{k}\boldsymbol{q}}{m} + \frac{\hbar^2 q^2}{2m} \,. \tag{14.17}$$

From this and from the schematic drawing of the excitations in Fig. 14.3a the following limits for the excitations can be easily verified:

$$\begin{aligned} 0 \le \epsilon(q) &\le \frac{\hbar^2 k_{\mathrm{F}} q}{m} + \frac{\hbar^2 q^2}{2m} && \text{for } q < 2k_{\mathrm{F}} \\ -\frac{\hbar^2 k_{\mathrm{F}} q}{m} + \frac{\hbar^2 q^2}{2m} \le \epsilon(q) &\le \frac{\hbar^2 k_{\mathrm{F}} q}{m} + \frac{\hbar^2 q^2}{2m} && \text{for } q > 2k_{\mathrm{F}} \,. \end{aligned} \tag{14.18}$$

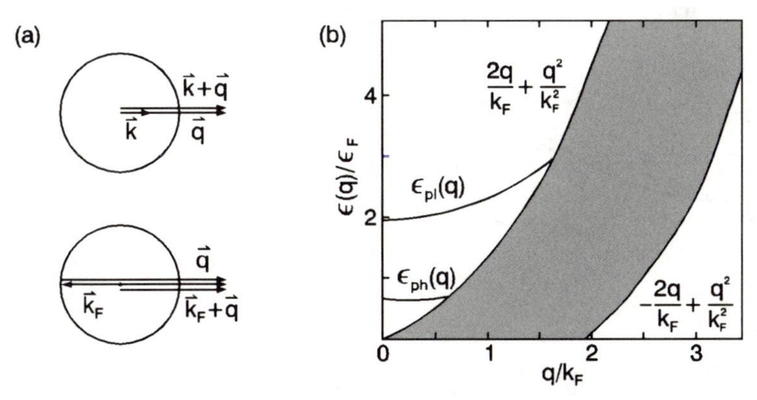

Fig. 14.3. Single particle excitations for a Fermi liquid; excitations from the Fermi sphere (**a**) and excitation diagram (**b**). $\epsilon_{\mathrm{pl}}(q)$ represents a plasmon dispersion and $\epsilon_{\mathrm{ph}}(q)$ the dispersion for an optical phonon. (The phonon is upshifted in energy for clarity). The hatched area in (b) covers the range of allowed single particle excitations.

The lower graph in Fig. 14.3a shows the limiting case where $k = k_F$ is oriented in the opposite direction to the excitation q. Figure 14.3b shows the acceptable excitation range in an energy-wave vector diagram as hatched area. The plasmon is only stable in the area which is not accessible to single particle excitation. In the hatched area the plasmon decays rapidly into an electron–hole pair. Similar arguments are valid for the phonons. As soon as the dispersion relation $\epsilon_{ph}(q)$ crosses the border line to single particle excitations the decay of the phonon into an electron–hole pair is an active channel and results in a strong broadening of phonon lines in Raman or neutron scattering.

14.2.3 Combination of the Dielectric Response

Since the susceptibilities discussed above or in Chap. 6 describe polarizations, they are additive and can contribute as a sum to the DF. An important and often observed example is the interaction of LO phonons and plasmons. The DF (for $q = \gamma = 0$, and $\tau = \infty$) is

$$\varepsilon(0,\omega) = 1 + \chi_\infty + \chi_p + \chi_{osc} = \varepsilon_\infty - \varepsilon_\infty \frac{\omega_p^2}{\omega^2} + \varepsilon_\infty \frac{\omega_L^2 - \omega_T^2}{\omega_L^2 - \omega^2} . \tag{14.19}$$

For $\varepsilon(0,\omega) = 0$ a combined phonon-plasmon mode is obtained with two characteristic frequencies $\omega_\pm$

$$\omega_\pm = \{\omega_p^2 + \omega_L^2 \pm [(\omega_p^2 - \omega_L^2)^2 + 4\omega_p^2(\omega_L^2 - \omega_T^2)]^{1/2}\}/2 . \tag{14.20}$$

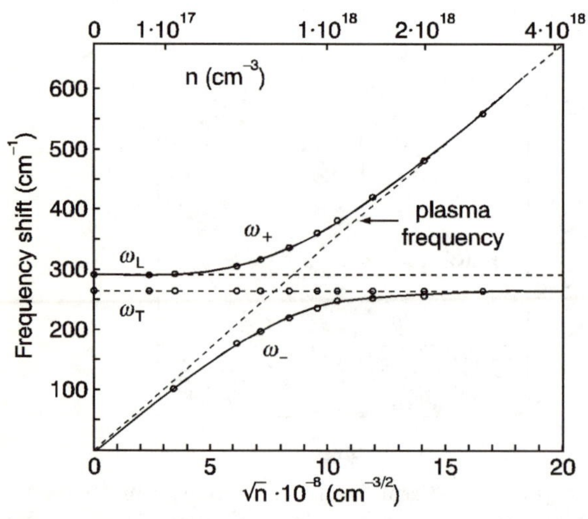

Fig. 14.4. Frequency of coupled phonon-plasmon modes ω_+ and ω_- in GaAs versus carrier concentration. The horizontal lines are the bare LO and TO phonon frequencies; after [14.2].

Since the plasma frequency shifts with the square root of the particle density n the behavior of the coupled phonon-plasmon mode scales with the latter. Figure 14.4 shows experimental results from Raman scattering on GaAs. The position of the modes is plotted versus the carrier concentration. From the figure the plasmon and the LO phonon remain independent as long as they are far enough apart in energy. With increasing carrier concentration the plasmon approaches the LO mode which starts to shift upwards. Finally, the lower branch of the combined mode merges with the TO mode and only the upper branch moves on.

Small particles with a DF ε_1 suspended in a matrix with DF ε_{M} are another example of a combined dielectric function. If the particles are sufficiently dilute, metallic and of the size of the light wavelength the resulting DF is given by

$$\boxed{\varepsilon_{\mathrm{MG}} = \varepsilon_{\mathrm{M}} \left(1 + 3f \frac{\varepsilon_{\mathrm{D}}(\omega) - \epsilon_{\mathrm{M}}}{\varepsilon_{\mathrm{D}}(\omega) - 2\varepsilon_{\mathrm{M}}} \right) ,} \tag{14.21}$$

where ε_{D} is the Drude DF and f the fill factor. The DF of (14.21) is known as the *Maxwell-Garnett* dielectric function. Such functions can be used to determine particle sizes from transmission and reflection experiments. For more details and applications of this function see [14.3].

Problems

14.1 Calculate the screening length for a nondegenerate electron gas.
(Purpose of exercise: observe the difference between degenerate and nondegenerate electron gas.)

14.2 Show that the Fourier transform of the Coulomb potential in three dimensions is $e/\varepsilon_0 q^2$.
Hint: Evaluate first the Fourier transform for the screened Coulomb potential $\Phi(r) = (e/\varepsilon_0 r)\exp(-a/r)$ and then reduce the screening to zero.
(Purpose of exercise: derive a very important relationship.)

14.3* Derive the Drude DF from the Lindhard DF for a Fermi liquid by specifying for $q = 0,\ \omega \neq 0$.
(Purpose of exercise: training in the evaluation of sums in k space.)

14.4* Discuss the possibility to find a broadening of lines in a light-scattering experiment on acoustic phonons in heavily doped Si. Take as an example Si with a hole concentration of $n = 10^{24}\,\mathrm{m}^{-3}$.
Hint: The light and heavy hole mass in Si are $0.16\,m_0$ and $0.5\,m_0$, respectively. The electron gas is degenerate so that the free electron model applies. The sound velocity of Si is 8×10^3 m/s.
(Purpose of exercise: understand the meaning of the single pair excitation regime.)

14.5 Discuss the relation between the frequency of the coupled phonon-plasmon mode and the carrier concentration for very high and very low carrier concentrations.

(Purpose of exercise: understand the limiting behavior of the curves in Fig. 14.4.)

14.6[a] Calculate the real part and the imaginary part of the density–density response function from (14.12) for $\alpha \to 0$ and discuss the physical meaning of the result.

Hint: Use the Dirac relation

$$\lim_{\alpha \to 0} \int \frac{f(x)\mathrm{d}x}{x - a + \mathrm{i}\alpha} = \mathrm{P} \int \frac{f(x)\mathrm{d}x}{x - a} - \mathrm{i}\pi \int f(x)\delta(x - a)\mathrm{d}x \, .$$

(Purpose of exercise: understand the meaning of the real and imaginary part of a response function.)

15. Spectroscopy with Electrons

Spectroscopy as discussed so far has been for electromagnetic radiation interacting with solids. According to the last chapter a beam of charged particles will also interact with the electronic system of the solid and will thus be subjected to an energy loss. This means energy transfer between the beam and the systems will take place and similar absorption and scattering spectra can be expected as described for the Raman effect and for optical absorption. In fact, spectroscopy with electrons extends the possible range of excitations dramatically. The much larger momentum of the electrons as compared to photons for the same energy enables not only the energy of the excitations but also their momentum to be measured. In addition, transport measurements for electrons or holes across junctions between two materials are an excellent method to obtain information on electron and even phonon densities of states. The two most important experimental techniques in this field are *electron energy loss spectroscopy* (EELS) and *tunneling spectroscopy.*

In EELS the energy loss of an electron is studied on its way through a crystal. The loss can be accomplished by any excitation of quasi-particles which couple to the moving electron. Formally the processes are similar to Raman scattering. A basic difference between the two comes from the request to generate an intermediate electron–hole pair in Raman spectroscopy. It is this pair which interacts with the quasi-particles of the solid whereas in EELS the probing electron interacts directly.

In early tunneling experiments the electron transport was studied only inside the joint solids. The current observed through a tunneling junction is then a measure for the change in density of states across the junction. Thus, a direct spectroscopy of the density of states is possible by measuring the current-voltage characteristics. In more recent experiments electrons are even allowed to tunnel over small distances outside the solid which allows for a scanning process during the experiment. This technique is known as *scanning tunneling spectroscopy* (STS).

15.1 Electron Energy Loss Spectroscopy

In EELS a beam of electrons is transmitted through or reflected from a crystal and the change in energy and direction is measured. The interaction between the electron beam and the electrons in the system is due to the Coulomb potential. Thus, the possible excitations are similar to those in light-scattering spectroscopy or optical absorption. On the other hand with electrons quasi-particles can be excited in the whole Brillouins zone as the wave vector for a 10-eV electron is about $2 \times 10^8\,\mathrm{cm}^{-1}$ whereas it is only $5 \times 10^5\,\mathrm{cm}^{-1}$ for a 10-eV photon.

It is interesting at this point to compare the spectral range in energy and momentum which is accessible to the various spectroscopic techniques. Results are shown in Fig. 15.1 for absorption of electromagnetic radiation, light scattering, x-ray scattering, inelastic neutron scattering and inelastic electron scattering. The figure clearly demonstrates the complimentary ranges for spectroscopy with radiation in the visible spectral range and with electrons. The lower limit on the energy scale for electron spectroscopy is determined by the energy resolution of the detection systems. The upper limit is given by the handling of high energy electrons and the condition that energy transferred in a scattering experiment should be much smaller than the energy of the exciting particle. Neutron scattering has a very good energy resolution for low energy scattering. This enables excitation energies of quasi-particles as low as 0.1 meV or $0.8\,\mathrm{cm}^{-1}$ to be detected.

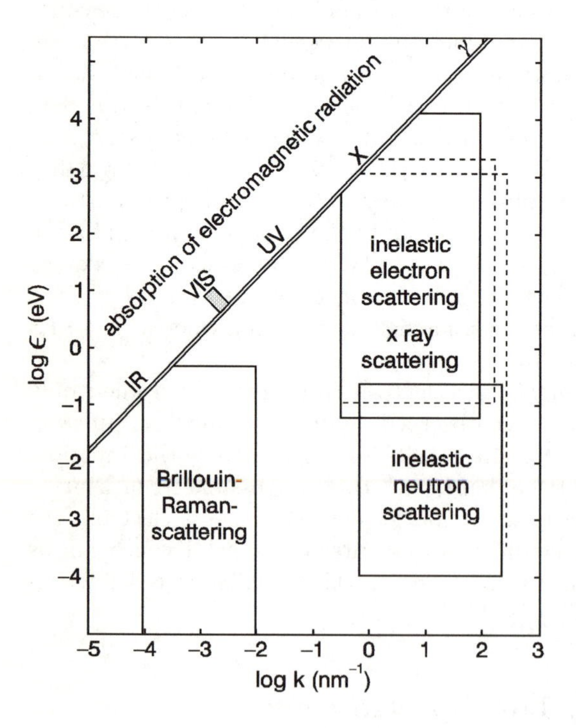

Fig. 15.1. Energy-wave vector diagram for spectroscopy with electromagnetic radiation and with particles; ($1\,\mathrm{nm}^{-1} = 10^7\,\mathrm{cm}^{-1}$), (IR: infrared, VIS: visible, UV: ultraviolet, X: x rays, γ: γ radiation)

15.1.1 Electron Energy Loss

Since EELS is based on an inelastic scattering process, all relationships concerning conservation of energy and momentum are valid as discussed in Chap. 9. There is in particular independent access to energy transfer and momentum transfer as demonstrated for Brillouin scattering experiments. In classical EELS experiments the electron energy was $< 100\,\text{eV}$ which limited the range of excitations to $\approx 10\,\text{eV}$ in energy and to $10^8\,\text{cm}^{-1}$ in wave number. With respect to transferred momentum this is already much larger than for experiments with light but still does not cover the whole Brillouin zone. In more recent EELS experiments the primary energy of the electrons is 100–200 KeV which corresponds to wave vectors of $200\,\text{nm}^{-1}$. This is much larger than the first Brillouin zone and means dispersion relations can be measured over the whole zone. Figure 15.2 shows the scattering geometry. The incident electron beam has energy and momentum ϵ_i and $\hbar k_\text{i}$, respectively. Scattering occurs with an energy loss $\Delta\epsilon \ll \epsilon$ and momentum change $q \ll k_\text{i}$ so that the scattering angle θ remains small. From energy and momentum conservation we find

$$
\begin{aligned}
q_\perp &= k_\text{s}\sin\theta \approx k_\text{i}\theta \\
q_\parallel &\approx k_\text{i} - k_\text{s} \approx k_\text{i}\frac{\Delta\epsilon}{2\epsilon_\text{i}} \ll q_\perp\,.
\end{aligned}
\tag{15.1}
$$

$q_\perp$ and $q_\parallel$ refer to the directions perpendicular and parallel to the electron beam. (15.1) implies that the angle θ is a good measure for the total transferred momentum.

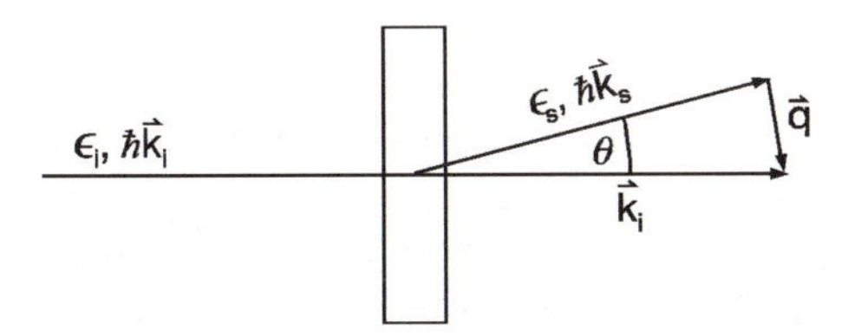

Fig. 15.2. Scattering geometry in electron energy loss spectroscopy.

For a formal description of the scattering intensities results can be used which were obtained in Chap. 9 for light scattering. Similar to (9.6) we define a scattering cross section $\mathrm{d}^2\sigma/\mathrm{d}\Omega\,\mathrm{d}\hbar\omega$ as the number of scattered electrons observed per unit of time, solid angle and energy interval for a given incident intensity (in electrons per unit of time and area). The evaluation of the scattering intensity needs a quantum-mechanical perturbation calculation. It is usually performed in the *Born approximation.* The resulting cross section is given by the dynamical form factor $S(q,\omega)$ as already used in (9.52) and as discussed in Sect. 17.4 for neutron scattering.

In the above approximation the scattering cross section is

$$\frac{\mathrm{d}^2\sigma}{\mathrm{d}\Omega\,\mathrm{d}\hbar\omega} = \left(\frac{\mathrm{d}\sigma}{\mathrm{d}\Omega}\right)_{\mathrm{el}} S(q,\omega) \,, \tag{15.2}$$

where $(\mathrm{d}\sigma/\mathrm{d}\Omega)_{\mathrm{el}}$ is the elastic contribution to the scattering process. The dynamical form factor for the electronic system is

$$S(q,\omega) = \frac{q^2\varepsilon_0\mathcal{V}}{e^2}\mathrm{Im}\left\{-\frac{1}{\varepsilon(q,\omega)}\right\} . \tag{15.3}$$

$\mathrm{Im}\{-1/\varepsilon(q,\omega)\}$ is the energy loss function. The factor in front is the scattering volume and the inverse of the Fourier transform of the Coulomb interaction

$$U_{\mathrm{q}}^{-1} = \frac{\varepsilon_0 q^2}{e^2} . \tag{15.4}$$

(15.3) is a special case of the *fluctuation–dissipation theorem* for particles subjected to a Coulomb interaction. For the elastic scattering the classical Rutherford cross section

$$\left(\frac{\mathrm{d}\sigma}{\mathrm{d}\Omega}\right)_{\mathrm{el}} = \frac{4}{r_{\mathrm{B}}^2 q^4} \tag{15.5}$$

is appropriate, where r_{B} is the Bohr radius.

The resulting relationship between cross section and loss function as given by (15.2), (15.3) and (15.5) is

$$\frac{\mathrm{d}^2\sigma}{\mathrm{d}\Omega\,\mathrm{d}\hbar\omega} = \frac{4\varepsilon_0\mathcal{V}}{e^2 r_{\mathrm{B}}^2 q^2}\mathrm{Im}\left\{-\frac{1}{\varepsilon(q,\omega)}\right\} . \tag{15.6}$$

It enables the imaginary part of $1/\varepsilon(q,\omega)$ to be determined from the scattering experiment. Finally, $\varepsilon(q,\omega)$ is obtained from a Kramers–Kronig analysis.

Since $\mathrm{Im}\{-1/\varepsilon(q,\omega)\}$ is the energy loss of the electrons while passing through the crystal it is interesting to compare this function with the imaginary part of the DF which describes the absorption of the beam. The loss function has the explicit form

$$\mathrm{Im}\left\{-\frac{1}{\varepsilon(q,\omega)}\right\} = \frac{-\varepsilon_{\mathrm{i}}(q,\omega)}{\varepsilon_{\mathrm{r}}^2(q,\omega)+\varepsilon_{\mathrm{i}}^2(q,\omega)} . \tag{15.7}$$

As long as $\varepsilon_{\mathrm{i}} \ll \varepsilon_{\mathrm{r}}$ it scales with ε_{i}, like the absorption $\alpha(\omega)$. This is not surprising since both originate from the same DF and from the same dynamical form factor. The possible excitations of the electronic system are the basic properties which enter both response functions.

The difference (or similarity) between loss function and ε_{i} can be studied by plotting the relevant relations for simple DFs. This is done in Fig. 15.3a,b for the Drude DF and for the Kramers–Heisenberg DF from Chap. 6. The two functions represent the response from free carriers and from dipole transitions, respectively. According to part (a) of the figure the shape of the loss

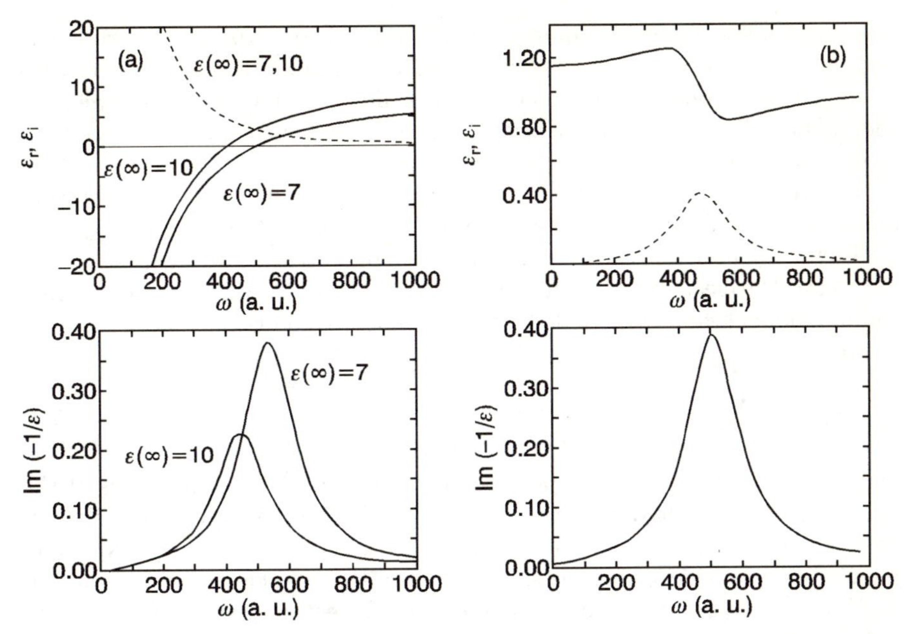

Fig. 15.3. Dielectric function and energy loss function for a free electron system according to (6.46) for $\omega_\mathrm{p} = 1400$, $\tau = 0.005$ and two different values for ε_∞, (**a**) and for a Kramers–Heisenberg oscillator function according to (6.36) for $\omega_\mathrm{T} = 500$, $\omega_\mathrm{p} = 200$, $\gamma = 200$ (**b**); frequencies are in arbitrary units; (– – –) ε_i, (—) ε_r

function is dramatically different from $\varepsilon_\mathrm{i}(\omega)$. It always has a maximum which depends strongly on ε_∞. The maximum appears for the frequency where ϵ_r crosses zero which is for $\omega = \omega_\mathrm{pl}$ in the case of the Drude DF. The response functions for the oscillator are different. Both the loss function and $\varepsilon_\mathrm{i}(\omega)$ exhibit a maximum as shown in Fig. 15.3b. Whereas the maximum for ε_i occurs at the transition energy ω_T of the dipole the maximum for the loss function occurs for $\omega^2 = \omega_\mathrm{T}^2 + \omega_\mathrm{pl}^2$, where ω_pl is determined by the number of oscillators in the system and by ε_∞ ($\omega_\mathrm{pl}^2 = \omega_\mathrm{p}^2/\varepsilon_\infty$). A typical example of the oscillator DF is the fundamental transition in a semiconductor with a wide gap and narrow bands. Such semiconductors are often found in organic materials.

The similarity of the loss function for the free carriers and for the oscillator as shown in Fig. 15.3 is striking. A simple calculation shows, however, that the functional dependences of $\mathrm{Im}\{1/\varepsilon(\omega)\}$ and $\varepsilon_\mathrm{i}(\omega)$ are different so that the Kramers–Kronig transformation gives the correct result for $\varepsilon_\mathrm{r}(\omega)$.

15.1.2 Spectrometers and Detectors

The first spectrometers for the determination of the energy loss of an electron beam in the solid were constructed for energies below 100 eV since only for this range a satisfactory resolution could be reached. Experiments were carried out in backscattering or for very thin foils because of the strong absorption of the electrons. Also the constructions did not allow the measurement of the angular distribution of the scattered electrons. Such spectrometers were discussed in Sect. 12.1.2. Spectrometers using more than 100-KeV electron energy allow transmission through samples with a thickness up to 100 nm and more. In the sense of solid-state physics this is bulk material.

Figure 15.4 sketches a spectrometer for high voltages. After emission from the source the electrons are deflected by 180° in an electrostatic monochromator and monochromatized to 0.05 eV. The energy resolution of the monochromator is

$$\Delta\epsilon = \epsilon \left(\frac{W}{2R} + \alpha^2 \right) , \tag{15.8}$$

where W, R, and α are the aperture of the entrance stop, the radius of the spectrometer and the divergency of the emerging electrons, respectively. After leaving the monochromator the electrons are accelerated to 170 KeV and hit the sample with this energy. The transmitted electrons can be selected with respect to their angular distribution by deflection plates. In the following stage the high energy electrons are decelerated to allow for an appropriate energy analysis. For the deceleration the same voltage is used as for the acceleration but with inverted sign. The application of the same voltage source for acceleration and deceleration is important for a good energy resolution since any noise in the voltage with frequency lower than the inverse transit time is eliminated. In this way a final energy resolution of 0.05 eV can be obtained. The analyzer is constructed equivalent to the monochromator and the detector is a channeltron, or, for high currents, a Faraday coup. The spectrometer needs a vacuum of 10^{-5} to 10^{-6} Pa.

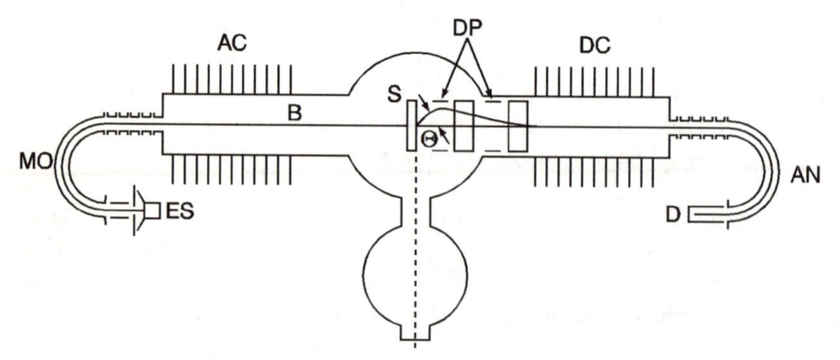

Fig. 15.4. Spectrometer for the analysis of electron energy loss; (ES: electron source, MO: monochromator, AC: accelerator, S: sample, DP: deflection plates, DC: decelerator, AN: analyzer, B: electron beam. D: detector); after [15.1].

15.1.3 Applications of Electron Energy Loss Spectroscopy

In the classical EELS experiments plasmons in metals were investigated for $q = 0$. Results for Mg after excitation with 2-KeV electrons are shown in Fig. 15.5. The series of peaks corresponds to multiple excitations. The double character of the peaks indicates two types of oscillations: a surface plasmon and a volume plasmon. The plasmon energies derived from the spectrum are 9.5 and 11.5 eV for the surface plasmon and for the volume plasmon, respectively. This yields from the latter a carrier concentration of $9\times10^{22}\,\mathrm{cm}^{-3}$ if ε_∞ is assumed to be one and the effective mass of the electrons is the free electron mass, in very good agreement with reported values of $8.9\times10^{22}\,\mathrm{cm}^{-3}$ for bulk Mg metal.

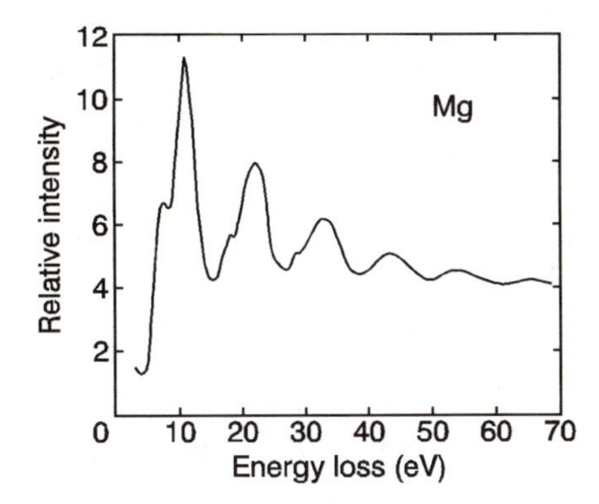

Fig. 15.5. Energy loss of electrons after excitation of volume plasmons and surface plasmons in Mg; after [15.2].

The wide range of applications for EELS can be seen from Fig. 15.6. Part (a) shows spectra of Al in the energy range between 1 eV and 2000 eV and part (b) the corresponding transitions. The various transitions are explained in the figure caption. The spectrum starts with low energy intraband excitations followed by the dominating scattering from surface and volume plasmons. Again multiple excitation is observed for the plasmons. With increasing energy transitions appear from the $2p$ and $2s$ valence bands to the conduction band above the Fermi energy. Finally, for energies as high as 1600 eV transitions from the $1s$ core levels can be observed.

In contrast to photoelectron spectroscopy where only occupied states can be investigated EELS allows the study of the unoccupied states of a solid. Figure 15.7 shows observed electron intensities from scattering experiments with TiC and VC. Transitions are from the conduction band into states above the Fermi edge. The dashed lines are calculated joint densities of states between the conduction band and higher bands.

The equivalency of the loss function obtained from EELS and from optical absorption is another important prerequisite of linear response theory. The agreement between two corresponding results is demonstrated in Fig. 15.8 for

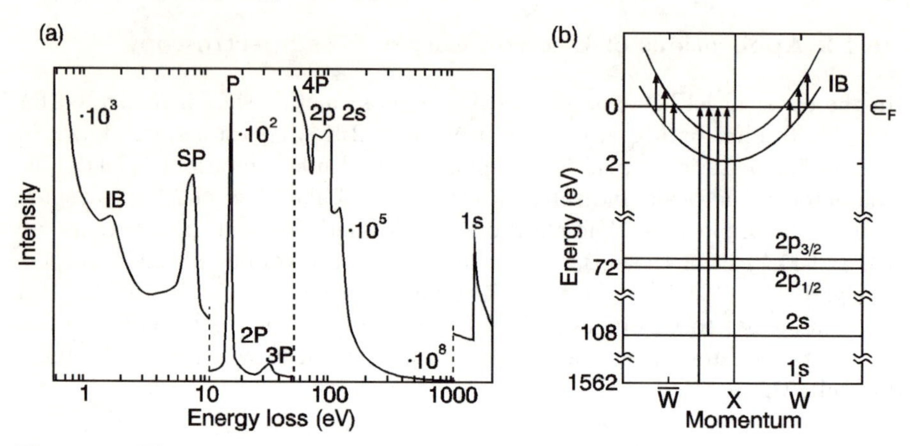

Fig. 15.6. Electron energy loss for Al between 1 eV and 2000 eV (**a**) and transitions in a schematic band picture (**b**); (IB: intraband, SP: surface plasmon, P: volume plasmon, $2p$, $2s$: valence band transitions, $1s$: core transitions); after [15.1].

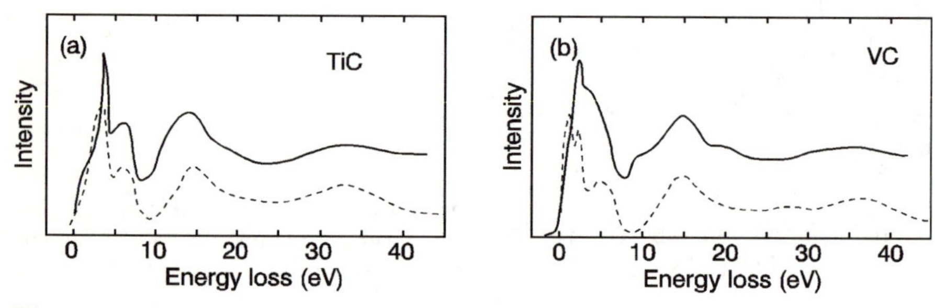

Fig. 15.7. Intensity of electrons scattered from TiC and VC versus distance from above the Fermi edge, as measured (—) and as calculated (– – –); after [15.1].

gold. The broad background in the two spectra originates from free carrier excitation. Even though EM radiation cannot excite plasmons due to the longitudinal polarization of the latter and the matrix element for plasmon excitation is very large for electrons the same DF results.

Finally, Fig. 15.9 shows the plasmon dispersion in highly oriented, *trans*-polyacetylene. Undoped polyacetylene is a quasi-one-dimensional insulator with an energy gap of 1.4 eV. The fully occupied valence band (π-band) and the empty conduction (π^*-band) are constructed from the p_z valence orbitals of carbon which are not involved in the sp^2 hybridization of the other three valence orbitals. Thus, the plasmon originates from excitations of the π electrons and corresponds to the example discussed in Fig. 15.3b. For wave vectors q_c larger than $10\,\mathrm{nm}^{-1}$ the plasmon decays into single particle excitations as discussed in Fig. 14.3 in Sect. 14.2.

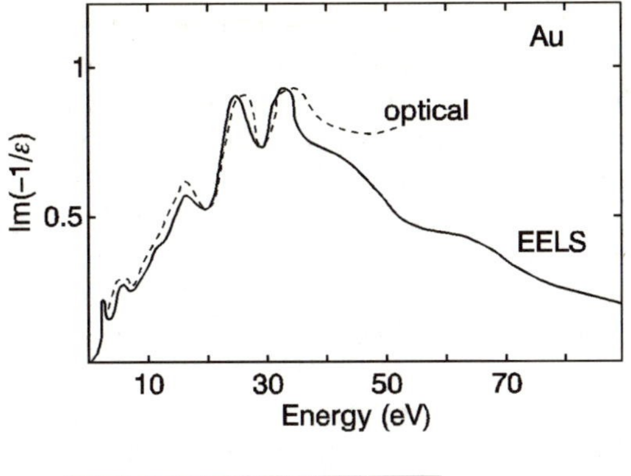

Fig. 15.8. Energy loss functions for gold as obtained from EELS (—), after [15.3], and and from optical measurements (− − −), after [15.4].

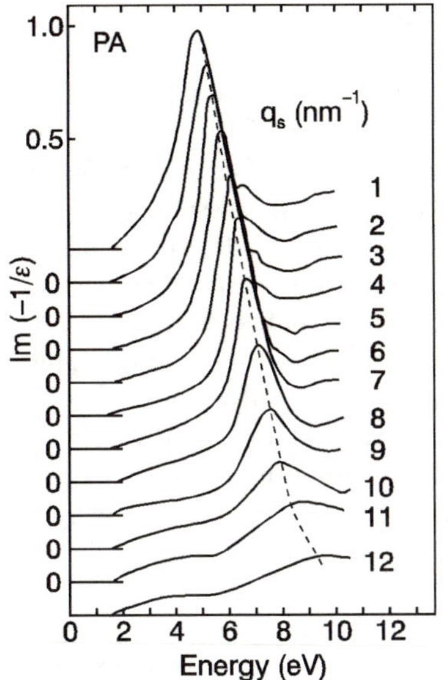

Fig. 15.9. Energy loss function for oriented polyacetylene for wave vectors of the excited π-band plasmon between 1 and $12\,\mathrm{nm}^{-1}$ as indicated; after [15.1].

15.2 Tunneling Spectroscopy

In the case of tunneling spectroscopy electrons intrinsic to the solid are used for an energy analysis. The electrons do not leave the solid or leave it only for submicroscopic distances from the surface. Subjected to an applied external field or potential they can propagate through areas of the crystal which are not accessible by drifting or by diffusion due to potential barriers. This propagation is established by a tunneling process. Tunneling is a typical quantum-mechanical transport phenomenon which allows the transmission of particles through a barrier with a certain probability, even if they do not have enough energy to cross the barrier. It relies basically on the probability distribution $|\psi(x)|^2$ of a particle which is continuous in space, even for abrupt changes of a potential $\Phi(x)$ which is supposed to confine the particle. For example the electron density of a metal is not zero just above its surface, but rather penetrates into the space above the surface and decays there exponentially. By approaching the surface with a second metal electrons can tunnel from one to the other in both directions. The potential barrier is in this particular case the work function of the metal. The probability of the tunneling process is enhanced if a potential difference between the two metals is applied which reduces or deforms the barrier. The lower and the narrower the barrier the larger the probability of tunneling. In addition efficient tunneling requires a high density of occupied electronic states on the one side of the barrier and a high density of unoccupied states on the other side of the barrier, both with respect to equal potential energy. Thus, for the case where the Fermi energy is located within one band the determination of the tunneling current versus applied voltage V is a good measure of the density of unoccupied states at the energy $\epsilon = eV$ above the Fermi edge.

Well known examples for the tunneling transport in solids are *Esaki diodes*. Esaki diodes are p-n junctions with the p-type and the n-type semiconductor doped to degeneracy. This means the respective Fermi energies are in the valence band and in the conduction band. Also metal-semiconductor junctions (Schottky diodes), where the semiconductor is doped to degeneracy show a tunneling characteristic. In both cases the carrier concentration must be high enough to make the tunneling current exceed the transport by carrier diffusion. Tunneling spectroscopy is of particular importance for the investigation of superconductors with structures of the type metal-insulator-superconductor (tunneling of quasi-particles) or with structures of the type superconductor-insulator-superconductor (tunneling of Cooper pairs). L. Esaki, I. Giaever, and B.D. Josephson received the Nobel prize in 1973 for their fundamental research work in connection with the problem of tunneling in solids.

The basic properties to which attention must be paid in tunneling experiments are the positions of the Fermi energies and the densities of states on the two sides of the tunneling junction.

15.2.1 The Tunneling Effect in Solids

The essential features of a tunneling process are shown in Fig. 15.10 for the particle transfer through a potential barrier $\Phi(x,V)$ between two metallic solids 1 and 2 with Fermi energies ϵ_{F_1} and ϵ_{F_2} and applied voltage V. From simple quantum mechanics the decay of the wave function within the barrier is given by the decay constant κ

$$\kappa = \sqrt{2m/\hbar^2}\sqrt{e\Phi(x,V) - \epsilon_x}\ , \tag{15.9}$$

where ϵ_x is the kinetic energy of the electrons on the side of the barrier with the higher energy. From κ the transmission factor D for the probability current through the barrier is obtained for the important case of small transmission from

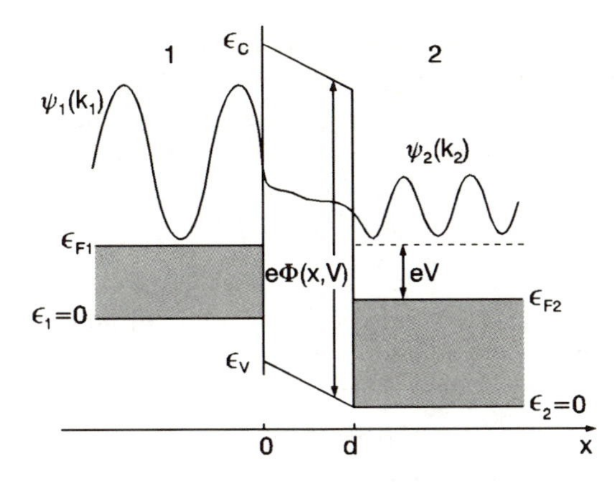

Fig. 15.10. Propagation and damping of a wave function across a rectangular potential barrier $\Phi(x, V)$ with an applied voltage V.

$$D(\epsilon_x, V) = g\mathrm{e}^{-2K} \qquad \text{with} \qquad g = \frac{16k_1k_2\kappa^2}{(k_1^2+\kappa^2)(k_2^2+\kappa^2)} \tag{15.10}$$

and

$$K = \int_0^d \kappa(x, \epsilon_x)\mathrm{d}x \; .$$

k_1 and k_2 are the wave vectors of the electrons on the one side and on the other side of the barrier, respectively [15.5]. K is an averaged decay constant times the barrier thickness d. κ and consequently also g and K depend strongly on the shape of the barrier. The value of (15.10) holds for the rectangular potential shown in Fig. 15.10. For the parabolic potential g was found to be 1 by G. Wenzel, H.A. Kramers, and L. Brillouin (WKB approximation). From the transmission factor D the current density $j_{1,2}$ from 1 to 2 and $j_{2,1}$ from 2 to 1 is obtained by integration of D over all states $k_1 = k$ weighted with the group velocity v_x and multiplied with the probability that the state is occupied on the one side of the barrier and empty on the other side of the barrier.

$$\begin{aligned} j_{1,2} &= \frac{2e}{(2\pi)^3}\int \mathrm{d}^2k_{yz}\mathrm{d}k_x v_x D f_\mathrm{F}(\epsilon)[1-f_\mathrm{F}(\epsilon+eV)] \\ j_{2,1} &= \frac{2e}{(2\pi)^3}\int \mathrm{d}^2k_{yz}\mathrm{d}k_x v_x D f_\mathrm{F}(\epsilon+eV)[1-f_\mathrm{F}(\epsilon)]. \end{aligned} \tag{15.11}$$

The factor 2 in front of the integral represents the spin degeneracy and $1/(2\pi)^3$ gives the number of states per unit volume in k space. The observed current density j is the difference between $j_{1,2}$ and $j_{2,1}$.

$$\boxed{j = \frac{2e}{(2\pi)^3}\int_0^\infty \mathrm{d}k_x v_x[f_\mathrm{F}(\epsilon) - f_\mathrm{F}(\epsilon+eV)]\int_0^\infty \mathrm{d}^2k_{yz}D(\epsilon_x, V) \; .} \tag{15.12}$$

Equation (15.12) represents the tunneling current under explicit inclusion of the transverse components of the momentum. Within the approximation where the transverse components are not explicitly taken into account the integration can be performed over the energy ϵ if v_x is replaced by $(2\epsilon_x/m^*)^{1/2}$ and d^3k by $k^2\mathrm{d}k$. m^* is the effective mass of the tunneling particle. Introducing the density of states on either side of the barrier yields the tunneling current from l to m as

$$j_{lm} = \frac{2e\pi^2\hbar^3}{m^{*2}\mathcal{V}^2}\int_0^\infty g_l(\epsilon)g_m(\epsilon)Df_{\mathrm{F}l}[1 - f_{\mathrm{F}m}]\mathrm{d}\epsilon\ ,$$

where l and m label either the left or the right side of the junction. From this finally the density of the tunneling current follows from

$$\boxed{j = \frac{2e\pi\hbar^3}{m^{*2}\mathcal{V}^2}\int_0^\infty g_1(\epsilon)g_2(\epsilon)D(\epsilon, V)[f_{\mathrm{F1}}(\epsilon) - f_{\mathrm{F2}}(\epsilon + eV)]\mathrm{d}\epsilon\ .} \tag{15.13}$$

This relationship can be specified for the various forms of the tunneling junctions. In the experiments the differential conductance

$$G(V) = \frac{\mathrm{d}I(V)}{\mathrm{d}V}$$

is often used to characterize the tunneling process.

15.2.2 The Tunneling Diode

The tunneling diode is the most traditional tunneling system in solids. The junction must be between a highly doped p-type and a highly doped n-type semiconductor as is the case in the Esaki diodes. High doping is required to make the junction sufficiently narrow. Only under such conditions the tunneling current exceeds the thermal current through the junction.

Figure 15.11a,b shows energy diagrams for p-n diodes; (a) is for the unbiased junction and (b) for the forward biased junction. The left part in each graph represents the p-type semiconductor, the right part the n-type semiconductor. In both cases the Fermi energy is within the band as a consequence of the strong doping. Both semiconductors are degenerated. For the unbiased junction fully occupied states on both sides are facing each other. If the junction is biased in forward direction with a small voltage V the Fermi energy on the right side is raised by eV. Now occupied states from the conduction band of the n-type semiconductor are facing empty states from the valence band of the p-type semiconductor. As a consequence a tunneling current (from left to right) is generated. To a first approximation this current is proportional to the relative shift of the two Fermi levels and thus raises linearly with V as shown in the current-voltage diagram Fig. 15.11c. A peak in the current is reached when the full fraction of the occupied conduction band faces the full fraction of the empty valence band at a voltage V_{peak}. For a further increase of the bias voltage the occupied part of the conduction band and the empty states of the valence band each start to face the gap region of the other partner of the junction. This reduces the tunneling current which finally reaches

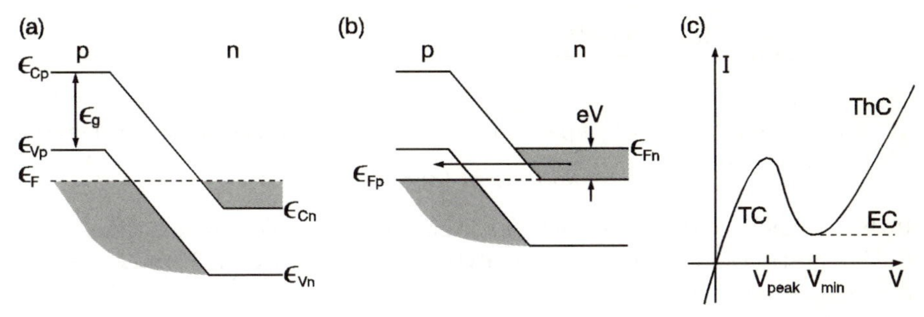

Fig. 15.11. Energy diagram (**a**), (**b**) and current-voltage characteristic (**c**) for the tunneling process in an Esaki diode. Part (a) is for the unbiased diode and part (b) for forward biasing with a voltage V. The arrow in (b) indicates the direction of tunneling for electrons. (TC: tunneling current, EC: excess current, ThC: thermal current.)

more or less zero. As a consequence the the current-voltage characteristic developes a minimum at some applied voltage V_{min}. In an idealized case the *valley current* in the minimum comes only from thermal excitation of carriers over the barrier. In reality some *excess current* at V_{min} originates also from tunneling into band tails or via impurity levels. Increasing the bias voltage beyond V_{min} raises the current exponentially as expected for a conventional p-n diode.

Reverse biasing will immediately lead to a current which increases linearly with the bias voltage. This follows straightforwardly from graphs like (a) and (b) in Fig. 15.11 with the bias voltage in the other direction. Hence a tunneling diode transmits current well in the reverse biased mode!

Observation of the maximum in the diode current is typical for a tunneling process and the width of the maximum is a good measure of the distances of the Fermi energies from the band edges. A quantitative description of the maximum can be obtained from (15.13). Tunneling transitions in both directions are only possible in the range of potentials between the lower edge of the conduction band ϵ_{Cn} of the n-type conductor and the upper edge of the valence band ϵ_{Vp} of the p-type conductor. Hence integration can be limited to this range. For 1 assigned to the p-type conductor, 2 to the n-type conductor, and V to the applied voltage (15.13) yields

$$j = C\int_{\epsilon_{\text{C}}}^{\epsilon_{\text{V}}} g_{\text{V}}(\epsilon)g_{\text{C}}(\epsilon)D[f_{\text{FC}}(\epsilon) - f_{\text{FV}}(\epsilon)]\text{d}\epsilon\,. \tag{15.14}$$

The applied voltage appears only in the Fermi function since $\epsilon_{\text{Fn}} - \epsilon_{\text{Fp}} = eV$. For a triangular shaped potential described by a transition width d and a constant field $\overline{E} = (\epsilon_{\text{g}} + \epsilon_{\text{Vp}} - \epsilon_{\text{Cn}})/ed$, D is given within the approximation used to derive (15.13) by

$$D = \exp\left(-\frac{4\epsilon_g^{3/2}(2m^*)^{1/2}}{3e\hbar\overline{E}}\right) .$$

Within the approximation of parabolic bands we use for the densities of states

$$g_C(\epsilon) \propto (\epsilon - \epsilon_C)^{1/2} \qquad \text{and} \qquad g_V(\epsilon) \propto (\epsilon_V - \epsilon)^{1/2} .$$

If the distances of the Fermi energies from the band edges are less than $2k_BT$ the integration can be performed analytically. In this case the Fermi functions are approximated by

$$f_{FC} \approx \frac{1}{2} - \frac{(\epsilon - \epsilon_{Fn})}{4k_BT} \qquad \text{and} \qquad f_{FV} \approx \frac{1}{2} + \frac{(\epsilon_{Fp} - \epsilon)}{4k_BT} .$$

The current density is obtained from the integration

$$\boxed{j = CD\frac{eV}{k_BT}(\epsilon_{Fn} + \epsilon_{Fp} - eV)^2 .} \tag{15.15}$$

Equations (15.14) and (15.15) show first an increase and then a drop in the current density in agreement with the schematic representation of Fig. 15.11.

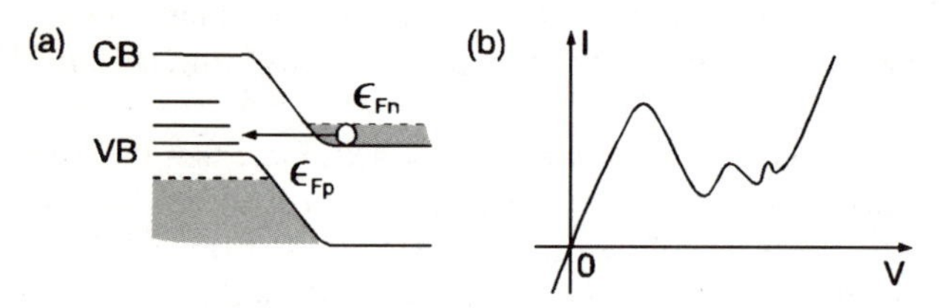

Fig. 15.12. Tunneling diode with impurity states in the gap (**a**) and corresponding current-voltage characteristic (**b**).

Tunneling is also possible into impurity states in the gap as shown schematically in Fig. 15.12. The current voltage characteristic exhibits several intermediate peaks of excess current in the valley before finally raising exponentially according to the Schockley equation. In this way electronic states in the gap can be analyzed. Less developed structures can be resolved by measuring the first or the second derivative of the current with respect to the voltage or vice versa.

A tunneling process can be supported from interaction with phonons. The current-voltage characteristic then exhibits a characteristic structures corresponding to phonon assisted tunneling transitions. An example is demonstrated in Figure 15.13 for a forward biased Si tunneling diode at 4.2 K. Such transitions will always occur in an indirect semiconductor like Si. They are a consequence of conservation of momentum for transitions from zone center carriers in the valence band of the p-type conductor to off zone center carriers in the conduction band of the n-type conductor (or vice versa). This and

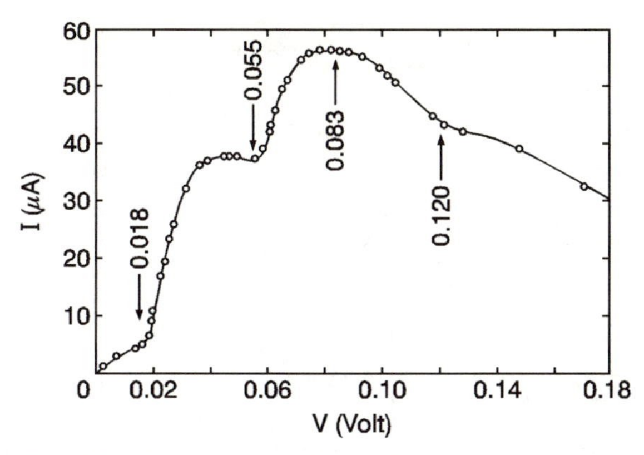

Fig. 15.13. Current-voltage characteristic of a Si tunneling diode at low temperatures. The structures in the current correspond to the energies of the phonons at the zone boundary; after [15.6]

the discussion about the impurity states from above shows that tunneling spectroscopy allows not only the determination of gap energies and positions of the Fermi level but also position and concentrations of impurity levels and phonon energies.

Tunneling characteristics for metal-semiconductor junctions (Schottky diodes) are similar to those for the Esaki diodes as long as the semiconductor is nondegenerate. However, instead of a maximum in the current only a turning point is observed in the forward biased current-voltage characteristic. The convention for forward bias in metal-semiconductor contacts is the configuration which reduces the barrier seen from the semiconductor. The energy level diagram for a (reversed biased) metal-semiconductor junction (with a degenerate n-type semiconductor) is shown in Fig. 15.14a. Φ_{b} is the work function of the metal.

For a forward biased Schottky diode of a metal with a nondegenerate semiconductor transport by diffusion yields an exponentially diverging differential conductivity $\mathrm{d}I/\mathrm{d}V$ with increasing voltage. A reverse biased diode exhibits a minimum in $\mathrm{d}I/\mathrm{d}V$ for a certain voltage, before it also diverges to infinity. If the transport across the junction occurs by tunneling the behavior is just the other way around. The differential conductivity goes to infinity for the reverse biased diode and passes a minimum for forward biasing. This case is shown in Fig. 15.14.

Calculations of the current-voltage characteristics for these junctions are difficult and need a detailed knowledge about the nature of the barrier. An exact solution was derived by *Conley* et al. [15.7] for a junction with a degenerate semiconductor and a barrier of the form

$$e\Phi(x,V) = \frac{e^2 N_{\mathrm{D}}[d(V)-x]^2}{2\varepsilon\varepsilon_0} + eV - \epsilon_{\mathrm{F}} \,, \tag{15.16}$$

and

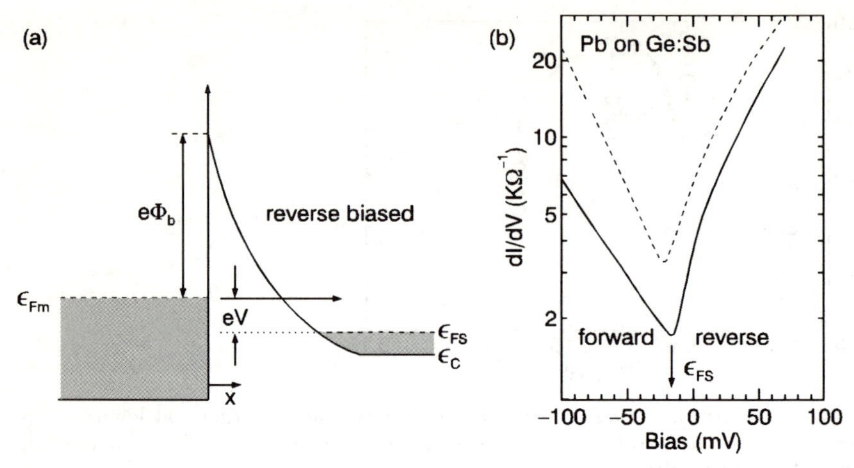

Fig. 15.14. Schematic representation of a Schottky diode (**a**) and experimental (—) and calculated (– – –) differential conductivity versus voltage (**b**) for $T = 4.2\,\mathrm{K}$ and $n = 7.510^{18}\,\mathrm{cm}^{-3}$; ($\Phi_\mathrm{b}$: barrier height, ϵ_Fm, ϵ_Fs: Fermi level of metal and semiconductor, respectively); after [15.8].

$$d(V) = \left(\frac{2\varepsilon\varepsilon_0(\Phi + \epsilon_\mathrm{Fs}/e - V)}{eN_\mathrm{D}} \right)^{1/2} .$$

$d(V)$ and N_D are the thickness of the barrier and the doping concentration in the semiconductor, respectively. $e\Phi(x, V)$ is measured from the Fermi level of the semiconductor. Figure 15.14b shows the result of the calculation for the differential conductivity in comparison to experimental results for a Pb-Ge Shottky diode. The minimum of $\mathrm{d}I/\mathrm{d}V$ at forward bias was found exactly for $eV = \epsilon_\mathrm{Fs}$ in the calculation. The minimum for $\mathrm{d}I/\mathrm{d}V$ observed in the experiment thus allows the distance of the Fermi energy from the band edge to be determined.

15.2.3 Tunneling Spectroscopy in Superconductors

Tunneling spectroscopy between metals is particularly useful for the analysis of the density of states in the superconducting phase. In this case the junctions consist of at least three components: two metallic contacts separated by a very thin insulating layer. The two metals can be the same or different. The insulating layer is usually a metal oxide film. At least one of the metals has to be in the superconducting state. Such structures are known as superconductor-insulator-normal conductor or SIN junctions. If both metals are in the superconducting state the structures are SIS or $\mathrm{S_1IS_2}$ junctions. In SIN junctions only quasi-particles (excited electrons) can tunnel. For SIS junctions both, quasi-particles and Cooper pairs can tunnel.

The density of states at the Fermi level of a metal changes characteristically at the transition to superconductivity. According to the theory of J.

Bardeen, L. Cooper, and R.J. Schrieffer (BCS theory) a gap of width 2Δ opens at ϵ_F and the density of states for the electrons (or holes) excited from the condensed pair state diverges right at the edges of the gap. This is shown schematically in Fig. 15.15. For $T = 0$ all electrons are in the paired ground states and the quasi-particle states are empty. For $T > 0$ some of the quasi-particle states above the superconducting gap are filled and corresponding holes appear for the states below the gap. As in the case of tunneling with semiconductors the current through a superconducting tunneling junction depends critically on the relative positions of the superconducting gaps and the relevant Fermi levels.

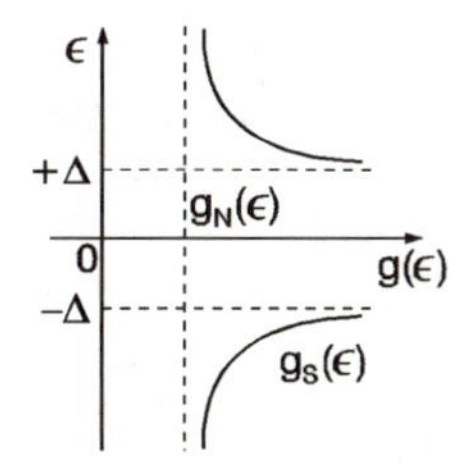

Fig. 15.15. Schematic density of states g_S for a BCS superconductor at $T = 0$. The energy is counted from the Fermi level, g_N is the density of states in the normal state.

Figure 15.16 shows the energy diagrams with the density of states for the quasi- particles corresponding to the various junctions, together with the relevant current-voltage characteristics. The presentation is for finite temperatures which means some of the Cooper pairs are broken up and quasi-particle states are occupied. The hatched areas in the density of states indicate occupied levels.

Part (a) of the figure represents the NIS structure with the left side of the junction in the normal state and the right side in the superconducting state. Except for the response from some thermally activated carriers an increase in the current is only observed for a voltage higher than a critical voltage $V_c = \Delta/e$. For the SIS structure shown in Fig. 15.16b the situation is similar. The Fermi levels of the two metals must be shifted at least by $V = 2\Delta/e$ before effective tunneling is possible. For finite temperatures the current below $2\Delta/e$ increases but the onset of the full tunnel current at $2\Delta/e$ remains sharp. Finally, for the case of S_1IS_2 junctions depicted in Fig. 15.16c a current can already start to flow for low voltages and reaches an intermediate maximum for $V = (\Delta_2 - \Delta_1)/e$. For a final continuous increase of the current $V \geq (\Delta_1 + \Delta_2)/e$ is required since only for this voltage the fully occupied states from the superconductor on the left side faces the empty states of the superconductor on the right side. The small peak in the current at $V = (\Delta_2 - \Delta_1)/e$ originates from tunneling into states which were thermally depopulated. For $T = 0$ this maximum does not exist. The figure shows that experimental results for the current-voltage characteristic allow the determination of the various energy gaps, even as a function of the temperature.

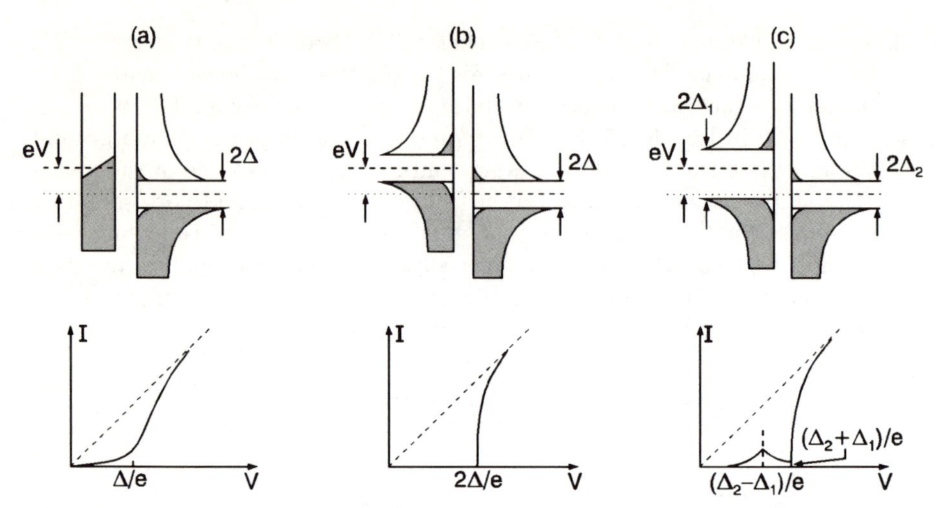

Fig. 15.16. Energy diagram (with energy plotted in y direction and density of states plotted in x direction) and current-voltage characteristic for various tunneling junctions at finite temperatures: metal-insulator-superconductor (a), superconductor-insulator-superconductor (**b**), superconductor 1-insulator-superconductor 2 (**c**).

A mathematical description of the tunneling processes of quasi-particles in superconducting junctions was first given by *Giaever* [15.9] and *Nicol* et al. [15.10]. The density of states shown in Fig. 15.15 is given within the BCS theory by

$$g_S(\epsilon) = \frac{g_N(0)|\epsilon|}{\sqrt{\epsilon^2 - \Delta^2}} , \tag{15.17}$$

where $g_N(0)$ is the density of states at the Fermi energy in the normal state and ϵ is counted from the Fermi level. The tunneling current is obtained from (15.13)

$$\boxed{I(V) = \frac{G_{NN}}{e g_N^2(0)} \int g_M(\epsilon + eV) g_S(\epsilon)[f_F(\epsilon + eV) - f_F(\epsilon)]d\epsilon ,} \tag{15.18}$$

where G_{NN} is the tunneling conductance (in units of A/V) in the normal state and $g_S(\epsilon)$ is the density of states from (15.17). (The transmission factor D is included in G_{NN}). g_M is the density of states for the metal on the left side of the junction. This metal can either be in the normal state (for a NIS junction) or in a superconducting state (for a SIS or S_1IS_2 junction). The integral in the equation is not solvable analytically but numerous good approximations exist.

For the NIS junctions the density of states from the superconductor dominates the energy dependence so that only g_S remains relevant in (15.18). In this case and for the voltage range $0 < V < 2\Delta/e$ the current density is

$$I(V) = \frac{2G_{\mathrm{NN}}\Delta}{e} \sum_{m=1}^{\infty} (-1)^{m+1} \mathrm{K}_1 \left(\frac{m\Delta}{k_{\mathrm{B}}T}\right) \sinh\left(\frac{meV}{k_{\mathrm{B}}T}\right) , \tag{15.19}$$

where K_1 is the first order modified Bessel function of the second kind. For the SIS structure and the same voltage range the current density is

$$\begin{aligned} I(V) &= \frac{2G_{NN}}{e} \exp\left(\frac{-\Delta}{k_{\mathrm{B}}T}\right) \sqrt{\frac{2\Delta}{eV+2\Delta}}(eV+\Delta) \\ &\quad \times \sinh\left(\frac{eV}{2k_{\mathrm{B}}T}\right) \mathrm{K}_0\left(\frac{eV}{2k_{\mathrm{B}}T}\right) , \end{aligned} \tag{15.20}$$

where K_0 is the zero order modified Bessel function of the second kind. For the current in a heterogeneous SIS junction approximations reveal for $T = 0$ a logarithmic singularity at $eV = \Delta_2 - \Delta_1$. This corresponds exactly to the peak in the current shown in Fig. 15.16 for finite temperatures.

For higher voltages the current increases linearly with applied voltage in all cases.

Figure 15.17 shows experimental results for the three different types of tunneling junctions from Fig. 15.16. T_{c} for Pb and Al is 7.19 K and 1.18 K, respectively. Part (a) and (b) can be described quantitatively with (15.19) and (15.20). For NIS structures very often the ratio between differential conductivity in the superconducting and in the normal state

$$\frac{(\mathrm{d}I/\mathrm{d}V)_{\mathrm{NS}}}{(\mathrm{d}I/\mathrm{d}V)_{\mathrm{NN}}} = \frac{G_{\mathrm{NS}}}{G_{\mathrm{NN}}}$$

is plotted as a function of the applied voltage. From (15.18) together with the density of states for the BCS superconductor (15.17) we obtain

$$\frac{G_{\mathrm{NS}}}{G_{\mathrm{NN}}} = \int \frac{|\epsilon|}{(\epsilon^2 - \Delta^2)^{1/2}} \left(-\frac{\partial f_{\mathrm{F}}(\epsilon + eV)}{\partial \epsilon}\right) \mathrm{d}\epsilon . \tag{15.21}$$

The derivative of the Fermi function $-f'(\epsilon)$ is sharply peaked at $\epsilon = 0$ with a width $3.5k_{\mathrm{B}}T$. For small values of T it approaches the δ function which allows a simple integration of (15.21). We obtain the very important result for the normalized conductance $\sigma(V)$

$$\begin{aligned} \sigma(V) = \frac{G_{\mathrm{NS}}}{G_{\mathrm{NN}}} &= 0 && \text{for } eV < \Delta \\ &= \frac{|eV|}{[(eV)^2 - \Delta^2]^{1/2}} && \text{for } eV > \Delta \text{ and } T \approx 0 . \end{aligned} \tag{15.22}$$

The equation shows that the normalized conductivity measures directly the density of states of the superconductor and the experimental results exhibit the shape of Fig. 15.15. An analysis of the fine structure of the current between $0 < eV < \Delta$ often allows additional information on defect states in

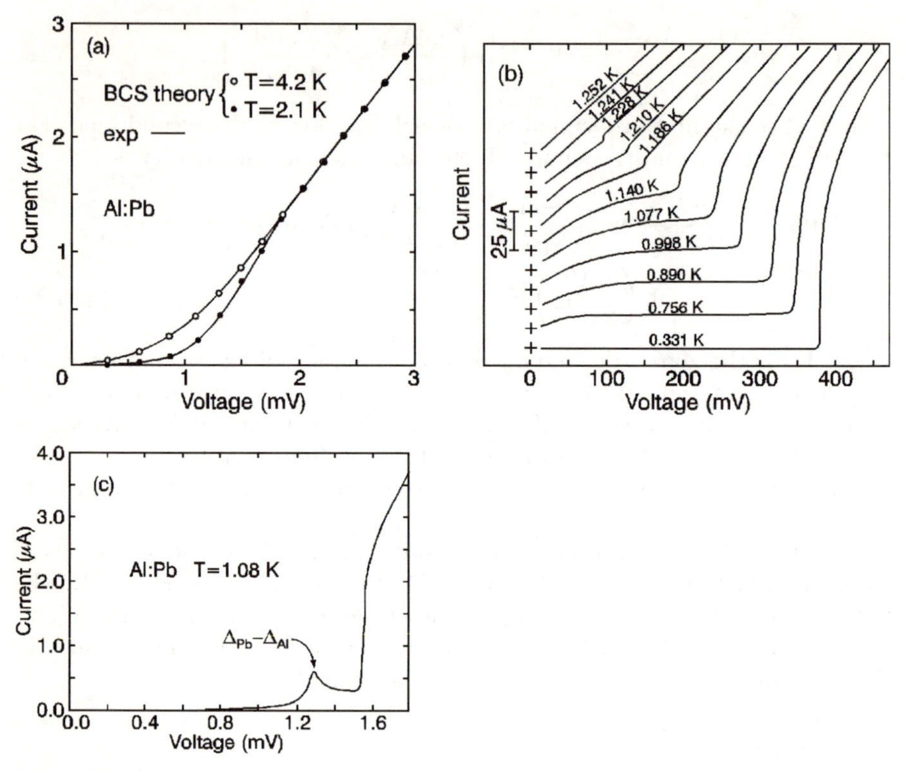

Fig. 15.17. Current-voltage characteristic for different tunneling junctions; NIS junction for Al-Pb; after [15.10] (**a**), SIS junction for Al-Al; after [15.11] (**b**), and S_1IS_2 junction (**c**); after [15.12].

the superconductor or on higher order tunneling processes to be drawn. Such experiments yield spectra which are very similar to those shown in Fig. 15.12 for the Esaki diode.

Another important application of tunneling spectroscopy refers to strong coupling superconductors where the BCS theory does not hold any more. For these materials G.M. Eliashberg and Y. Nambu have established an extended theory which differs from the BCS theory mainly by a better consideration of the phonon spectrum. An important parameter in the theory is the so called *Eliashberg function* $\alpha^2(\Omega)\varrho(\Omega)$ where α is an electron–phonon coupling constant and $\varrho(\Omega)$ is the density of states for the phonons. For superconductors with strong coupling a fine structure appears in the normalized tunneling conductivity $G_{\mathrm{NS}}/G_{\mathrm{NN}}$ for applied voltages larger than several Δ/e. These structures are superimposed on a BCS-like density of electronic states and can be extracted by normalizing the experiment on the latter. The structures are directly connected to the Eliashberg function and to the phonon density of states. *McMillan* and *Rowell* have worked out an iteration procedure which allows the extraction of the Eliashberg function from such experiments

[15.13]. Figure 15.18 shows a tunneling spectrum for Pb at 0.3 K (a) and the extracted Eliashberg function (b). The spectrum in (a) obviously has the form of Fig. 15.15 as expected from (15.22). If α depends only weakly on Ω the Eliashberg function is identical to the phonon density of states. This is indeed confirmed in part (b) of the figure where the extracted function is compared with a density of states for Pb obtained from inelastic neutron scattering. This means for strong coupling superconductors phonon densities of states can be determined. The technique described is known as *McMillan-Rowell* spectroscopy.

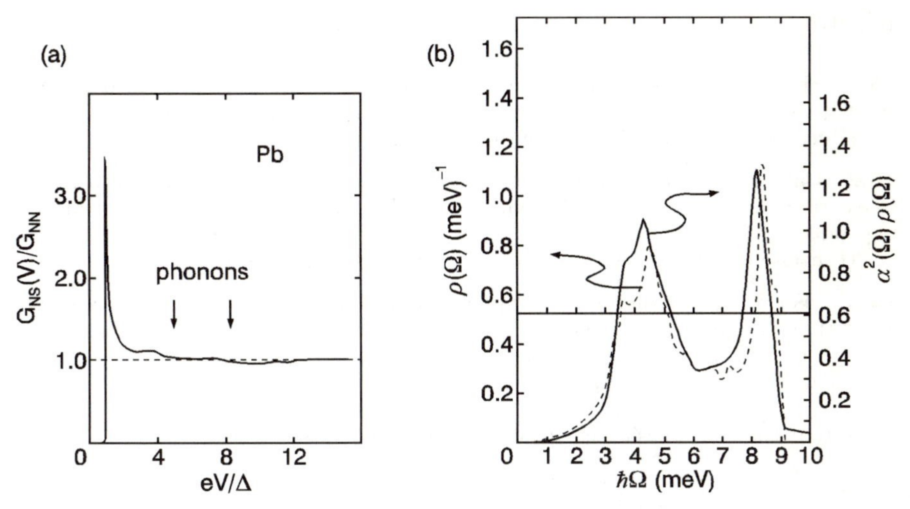

Fig. 15.18. Normalized tunneling conductance versus applied potential in units eV/Δ for Pb; after [15.14] (**a**), and extracted Eliashberg function (—) as compared to a phonon density of states from inelastic neutron scattering (– – –); after [15.15] (**b**).

The analysis of phonons by tunneling spectroscopy can also be applied to non-superconducting materials or weak coupling superconductors, since superconductivity can be induced using sandwich structures. A material N can become superconducting in junctions of the form SNS even though it would never be superconducting on its own. The tunneling junctions for such experiments consist usually of a metallic contact, an insulator, a normal conductor and a superconductor (CINS junctions). Spectroscopy with such junctions is known as *proximity* electron tunneling spectroscopy or PETS.

15.2.4 Scanning Tunneling Spectroscopy

Scanning tunneling spectroscopy has developed recently into a very active field of nanoresolution surface analysis. The technique employs a very fine tip which is approached to the surface until a tunnel current is recorded.

Atomic resolution can be obtained for the scanning process. Alternatively a current-voltage characteristic can be measured for the tunnel current, again with atomic resolution. The technique is used extensively to study energy gaps, impurity states, and quantum wells in semiconductor superlattices such as $Ga_{(1-x)}Al_xAs/GaAs$ [15.16]. Recently it has been applied also for the analysis of electronic levels splitted up as a consequence of quantum-size effects and for single-electron transport measurements in quantum wires of carbon nanotubes [15.17].

Problems

15.1 Show by expansion of the Lindhard DF that in the limit of $T \to 0$ and for small q the dispersion of the plasmon is given by $\omega_p(q) = \hbar\omega_p + (3\hbar\epsilon_F/5m_0\omega_p)q^2$.

(Purpose of exercise: training in handling a complicated DF.)

15.2 Calculate the frequency dependence of the energy loss function for a DF of free carriers and one oscillator. Convince yourself that it is different to the frequency dependence of ε_i.

(Purpose of exercise: get used to work with response functions.)

15.3* For a semiconductor with gap energy ϵ_g the damping constant for a triangular tunneling barrier is given by

$$\kappa(x) = \sqrt{2m^*(\epsilon_g/2 - e\overline{E}x)/\hbar^2} \; .$$

Show that the transmission coefficient for this barrier is

$$D = \exp\left(-\frac{4(2m^*)^{1/2}\epsilon_g^{3/2}}{3e\hbar\overline{E}}\right) \; .$$

(Purpose of exercise: recall the meaning of the transmission coefficient.)

15.4 Use the transmission coefficient from problem 15.2 to show that the current-voltage characteristic for parabolic bands has the form $j = aV(b - V)^2$.

(Purpose of exercise: solve a simple problem in tunneling theory.)

15.5 Using a personal computer discuss the shape of the current-voltage characteristic for a SIS junction and for a NIS junction.

(Purpose of exercise: training in handling a complex function on a personal computer.)

15.6 Show that the derivative of the Fermi function approaches a sharply peaked function of width $3.5k_BT$ at $\epsilon = 0$.

(Purpose of exercise: understand a very fundamental approximation used in (15.21).)

16. Spectroscopy with Positrons and Muons

Besides the stable electrons also several other leptons with rather short lifetime have been used successfully in solid-state spectroscopy. Most important are positrons and muons. If a positron is accommodated in a solid its lifetime depends critically on the environment and enables thus information to be drawn about the latter. Investigations of this type are known as *positron annihilation spectroscopy* (PAS).

Muons have a magnetic moment which can be used to detect local magnetic fields in crystals. Similar to the nuclei in the experiments of perturbed angular correlation these moments precess in a solid around the local field. The rotation frequency can be determined from the angular distribution of muon decay products. The technique is known as *muon spin rotation spectroscopy* (μSR).

16.1 Positron Annihilation Spectroscopy

Positron annihilation spectroscopy is one of the more recent developments in solid-state spectroscopy. Since positrons are the antiparticles to electrons and nearly identical to them except for the charge, one could expect a similar type of spectroscopy as discussed in Chap. 15. This spectroscopy would be very difficult because of the short lifetime of the positrons. It is in fact rather the decay of the positron in the crystalline environment which allows the study of the structure of the crystals and their electronic system. The technique is unique in the sense that energy and momentum of the electrons in the bands can be determined explicitly.

16.1.1 Positrons in Solids

Positrons are leptons produced during the radioactive decay of certain nuclei similar to the production of α or β particles. The positron is generated by the conversion of a proton into a neutron. The positive charge of the proton is carried away by the positron and a neutrino is generated simultaneously. Thus, the positron spectrum looks like a β spectrum. In contrast to classical particle beams from the radioactive decay the positron is highly unstable. Its lifetime in a metallic solid is only 10^{-10} s. The best known positron source is ^{22}Na with a positron emission energy of 545 KeV. These high-energy particles can be radiated into a solid and will be thermalized by inelastic scattering and excitations of the electronic system. A typical activity of a positron source is 0.1 Ci (1 Ci = 3.7 $\times 10^{10}$ decays/s) which means there will only be one positron in the crystal at the time, on average. For the thermalization a period of 10^{-11} s is sufficient which means the single positron will be fully thermalized during its lifetime in the crystal. Except for occasional thermal excitations it will be relaxed into its ground state with negligible values for energy and momentum in comparison to the values of the electrons.

The most likely decay channel for the positrons in the solid is an electron–positron annihilation process with a simultaneous emission of γ quanta. Energy, momentum, charge, angular momentum, and parity must be conserved for this process. Hence, at least two γ quanta must be emitted for electron–positron pairs with total spin 0 (para configuration) and at least three γ quanta must be emitted for the ortho configuration where the total spin of the pair is 1. In the first case the two emitted quanta have opposite spin and 511 KeV energy each. In the second case the quanta are subjected to an energy distribution and their total spin must be 1. The cross section σ for the 2-γ annihilation process of nonrelativistic positrons is known from nuclear physics to be $\pi r_0^2 c_0/v$ where r_0 is the free electron radius and v the particle velocity. The annihilation cross section for a 3-γ process is a factor 370 smaller. This means the only relevant process for spectroscopy with positron annihilation is the 2-γ decay channel.

The annihilation rate α for the positrons is independent of the particle velocity but proportional to the electron density $n_0 = |\psi(0)|^2$

$$\alpha = \sigma v |\psi(0)|^2 = \pi r_0^2 n_0 c_0 \, . \tag{16.1}$$

The sensitivity of the annihilation rate on the electron density is the reason why positron annihilation has attracted so much attention in solid-state physics. The particular importance of this new spectroscopic technique results from the fact that the angular distribution of the generated γ quanta carries the whole information on energy and momentum of the electrons participating in the annihilation process. The emission of the two γ quanta does not occur under an angle of 180° as one might expect from the very low momentum of the positron. Small deviations from the 180° emission are due to the contributions to the total momentum from the electrons involved in the annihilation process. Also, the energy of the two quanta is not exactly equal since there is a superimposed Doppler shift from the emitting electron. (The Doppler shift from the positron is more or less zero.) Finally, the lifetime τ of the positrons depends on the local electron density according (16.1) and thus on the local crystal structure where the electron–positron pairing takes place. Thus, the lifetime can be used to study this structure. Measurements of the lifetime can be particularly convenient if a γ quantum is emitted simultaneously with the positron. This quantum can be used as a time-setting for the positron annihilation.

According to the possibilities for spectroscopy with positron annihilation discussed above, three different types of experiment are possible. A corresponding setup is shown in Fig. 16.1. The process starts with the emission of a positron from the source S. The positron enters the sample and decays by pair annihilation. The two γ quanta $\gamma 1$ and $\gamma 2$ are emitted in opposite directions and measured in coincidence. The deviation from a 180° geometry is of fundamental importance. This type of experiment is known as *angular correlation of annihilation radiation* (ACAR).

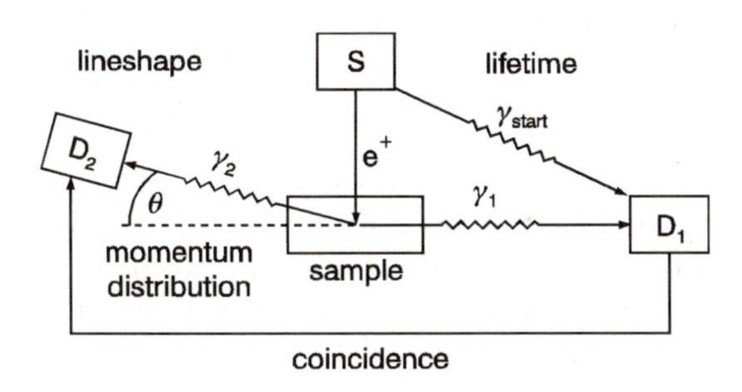

Fig. 16.1. Schematic representation of emission and detection processes for positron annihilation spectroscopy; (S: source, D1,D2: detectors, $\gamma 1$, $\gamma 2$: γ quanta produced by the annihilation process, $\gamma_{\rm start}$: signal quantum).

An analysis of the linewidth of the emitted γ radiation yields information on the Doppler broadening from the velocity distribution of the electrons in the direction of the emitted quanta. The lifetime for the positron in the sample is obtained from the time interval between the observation of the signal quantum $\gamma_{\rm start}$ to the detection of the quantum $\gamma 1$ from the annihilation process.

The probability $\Gamma(p)$ that the pair annihilation occurs under the emission of two γ quanta with total momentum $\boldsymbol{p}= \boldsymbol{p}_{\gamma 1}+ \boldsymbol{p}_{\gamma 2}$ is of fundamental importance for all experiments. From conservation of momentum and from the smallness of the momentum p_+ of the positron it is obvious from intuition that the pair momentum equals approximately the electron momentum $\boldsymbol{p} = \boldsymbol{p}_{\rm pair} \approx \boldsymbol{p}_{k_-} + \boldsymbol{p}_+ \approx \boldsymbol{p}_{k_-}$. To avoid confusion with positive holes in the bands here and for the rest of this chapter electrons and positrons are characterized by the symbols $-$ and $+$ instead of e and p, respectively.

A more fundamental but still highly simplified discussion of the pair-momentum distribution can be performed in a single particle picture. In this case the initial state is characterized by the product of the positron wave function and the single particle electron wave functions and $\Gamma(p)\mathrm{d}p$ has the form

$$\Gamma(p)\mathrm{d}p = \frac{\alpha}{(2\pi)^3 n_0} \Lambda(p)\mathrm{d}p \qquad \text{with} \tag{16.2}$$

$$\Lambda(p) = \sum_{k_-} \left| \int \mathrm{e}^{-\mathrm{i}pr/\hbar} \psi_+(r) \psi_{k_-}(r) \, \mathrm{d}r \right|^2 ,$$

where Λ is the *photonpair momentum density* or just the *momentum density* and ψ_+ and ψ_{k_-} are the wave functions for the positron and for the electrons in the crystal. The momentum density is the Fourier transform of the electron–positron overlap summed over all occupied states.

For low enough temperatures the positron is in its ground state with $p_+ = 0$ and the electrons occupy states up to $k = k_\mathrm{F}$. Thus, the sum in (16.2) extends only to the Fermi energy. To evaluate the contribution of electrons from the core levels the corresponding wave functions must be used in the sum. Such contributions are in general small since the positrons cannot penetrate very deep into the atom and the overlap with the core electrons remains small. This is a consequence of the positive charge of the positrons. For the contributions from valence electrons and conduction electrons Bloch functions are used.

For the evaluation of the integral at $T > 0$ the probability must be considered to find a particle with a particular wave vector. Since there is always only one positron in the crystal, the Boltzmann distribution f_B can be used for the latter despite of their Fermionic character. For the occupation of the electronic states the Fermi distribution f_F is required. The expression for Λ in (16.2) must be convoluted with f_F and f_B. This yields for the approximation where ψ_+ and ψ_{k_-} are Bloch functions

$$\Lambda(p,T) = \sum_{j,k_-,k_+} f_\mathrm{B} f_\mathrm{F} |A_j(p, k_+ k_-)|^2 \approx \sum_{j,k_-} f_\mathrm{F} |A_j(p, k_-)|^2 . \tag{16.3}$$

The index j refers to the electron band j and A_j are the Fourier transforms of the product from positron and electron wave functions

$$\begin{aligned} A_j(p, k_-) &= \int \mathrm{e}^{-\mathrm{i}pr/\hbar} \psi_+^*(r) \psi_{k_-}(r) \mathrm{d}r \\ &= \frac{1}{\mathcal{V}} \int \mathrm{e}^{-\mathrm{i}(p/\hbar - k_-)r} u_+^*(r) u_{k_-,j}(r) \mathrm{d}r . \end{aligned} \tag{16.4}$$

Because the functions $u_{k_-,j}$ and u_+ exhibit lattice periodicity the integration can be reduced to the volume $\mathcal{V}_\mathrm{u}$ of the unit cell. This renders the Fourier coefficients of (16.3) as

$$A_j(p, k_-) = \delta(\frac{p}{\hbar} - k_- - G) \frac{1}{\mathcal{V}_\mathrm{u}} \int_{\mathcal{V}_\mathrm{u}} u_{k_-,j}^*(r) u_+(r) \mathrm{d}r . \tag{16.5}$$

The relation shows explicitly that p means the momentum of the electrons except for a vector G of the reciprocal lattice.

16.1.2 Positron Sources and Spectrometers

Appropriate particle sources and spectrometers are requested for the experiments with positron annihilation. Table 16.1 lists the most important sources.

Table 16.1. Positron sources.

Isotope	Lifetime	Positron yield [%]	Positron energy [MeV]
^{13}N	9.96 min	100	1.20
^{18}F	1.83 h	97	0.635
^{22}Na	2.62 y	90.6	0.545
^{44}Ti	47 y	94	1.47
^{58}Co	71.3 d	15.0	0.475
^{68}Ge	275 d	88	1.88

The main parameters for the application are the decay yield and the lifetime for the emission. The most frequently used source is ^{22}Na which is obtained from a ^{24}Mg(d,α) reaction. ^{22}Na emits a γ quantum at 1.28 MeV simultaneously with the positron. This means the source is also applicable for lifetime experiments.

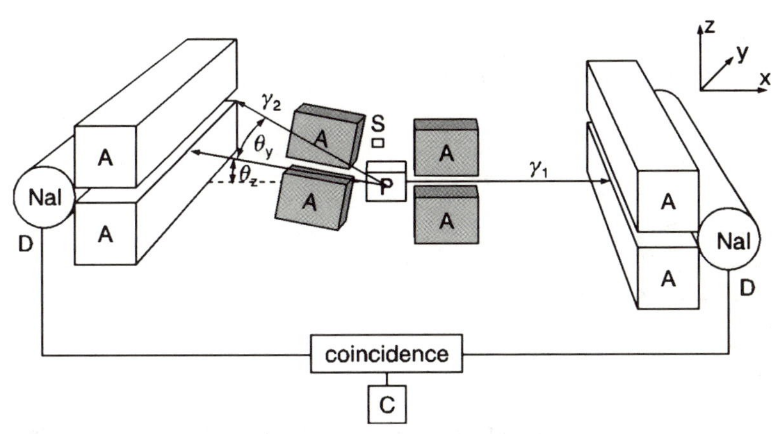

Fig. 16.2. Experimental arrangement for the determination of the angular correlated positron annihilation process with longslit geometry; (S: source, P: sample, D: detector, A: shielding, C: counter).

For the measurement of the ACAR two γ detectors are required which must be arranged in an 180° geometry as shown in Fig. 16.2. The positron source is shielded from the detectors and irradiates the sample. γ quanta resulting from the positron annihilation process and propagating in x direction are recorded with an angular resolution of 0.2–0.5 mrad in z direction. For the particular arrangement shown in Fig. 16.2 the signal is integrated in y direction over a wide angular range to keep the counting rate on a reasonable level. Thus, the geometry allows only a resolution of the momentum density in z direction. The emitted quanta are registered in coincidence by the two NaI detectors. In very elaborate instruments a *point geometry* is used for the detection and the registration occurs in the full yz plane. This allows a momentum resolution in two directions. The detectors are two-dimensional diode arrays as discussed in Sect. 5.4 or *Anger* cameras. The latter consist of a large plate of a NaI crystal with a set of photomultipliers mounted on the backside.

16.1.3 Experimental Results from Positron Annihilation Spectroscopy

None of the spectrometers discussed above can measure $\Gamma(p)$ explicitly. For the longslit geometry integration occurs over the x as well as over the y component of the distribution. For the point geometry integration is still over the x component. A possibility to determine all three components will be discussed later in this section. The observed count rates for the longslit and for the point geometry, respectively are

$$N(p_z)\Delta p_z = \Delta p_z \int \Gamma(p)\mathrm{d}p_x\mathrm{d}p_y \tag{16.6}$$

and

$$N(p_z, p_y)\Delta p_z\Delta p_y = \Delta p_z\Delta p_y \int \Gamma(p)\mathrm{d}p_x \,. \tag{16.7}$$

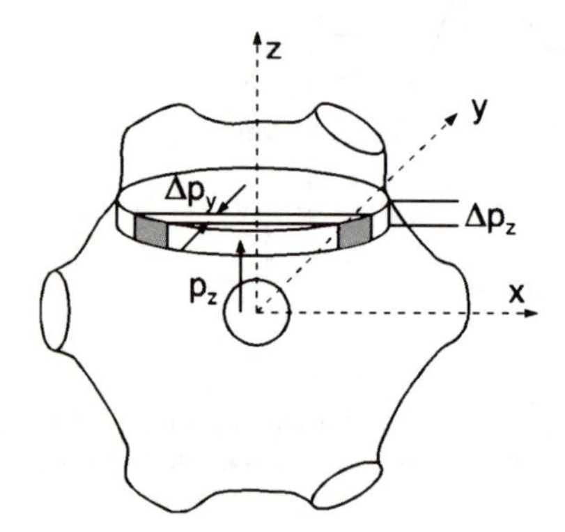

Fig. 16.3. Fermi surface of a metal and selected volume fractions probed by positron annihilation spectroscopy with a longslit and a point geometry.

Figure 16.3 shows the Fermi surface of a metal and the disk in k space which contributes to the annihilation signal. In the case of (16.6) integration is over the whole disc but only over the small distance Δp_z in z direction. In the case of point geometry the disc is further divided into small prisms along the y direction as shown in the figure. In both cases the maximum of N is obtained for $z = 0$ and N becomes zero for $p_z = p_\mathrm{F}$. For a spheric Fermi surface with radius p_F the relations between N and p are

$$\boxed{N(p_z) = C(p_\mathrm{F}^2 - p_z^2)} \tag{16.8}$$

for the longslit geometry and

$$\boxed{N(p_z, p_y) = C' \sqrt{p_F^2 - p_z^2 - p_y^2}} \tag{16.9}$$

for the point geometry, respectively. Within the approximation $p = p_{k_-}$ the observed momentum density distribution is therefore a parabola for the longslit geometry and a semi-sphere for the point geometry. Experimental results for Mg are shown in Fig. 16.4 as obtained for a longslit geometry. The count rate $N(p_z)$ is not exactly zero for $p_z = \hbar k_F$ since the signal from the conduction band electrons is superposed on small contributions from the core electrons.

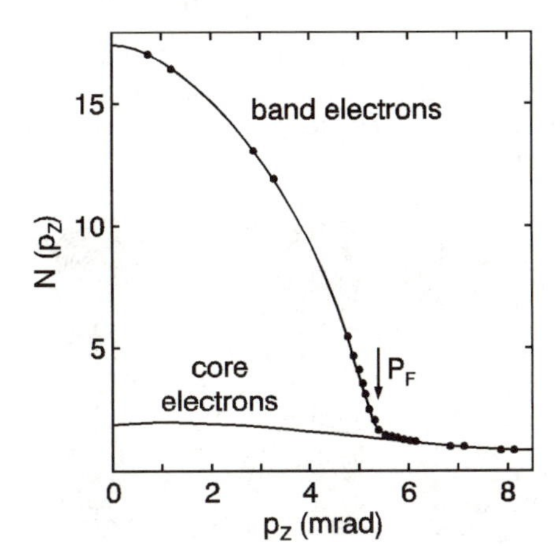

Fig. 16.4. Momentum density distribution $N(p_z)$ versus z component of the pair momentum for Mg as measured with a longslit geometry. p_z is defined by the angle θ_z in Fig. 16.2; after [16.1].

The momentum density distribution observed for a point geometry can be plotted as a two-dimensional diagram over p_y and p_z. Results are shown in Fig. 16.5a for Al. The momentum of the electrons is given in units of m_0c_0. In Fig. 16.5b a cut for $p_y = 0$ is presented together with the derivative of the distribution with respect to p_z. The result represents roughly a quarter of a circle as expected from (16.9). In detail deviations are observed from the simple circle. The deviations are more clearly seen from the plot of the derivative. They indicate a deviation of the Fermi surface from the one for a free electron system. The full drawn line is obtained from a band calculation.

Very often results for constant momentum density are plotted over the p_yp_z plane instead of the momentum density itself. This yields the *contour maps* which are more convenient for comparison between experiment and theory. Figure 16.6 displays contour maps for V_3Si. Part (a) is from a band structure calculation, part (b) are the experimental results and part (c) is for a Fermi surface model developed from the experiments.

The Doppler broadening of the γ radiation from the annihilation process originates from the motion of the electrons in the direction of the radiation

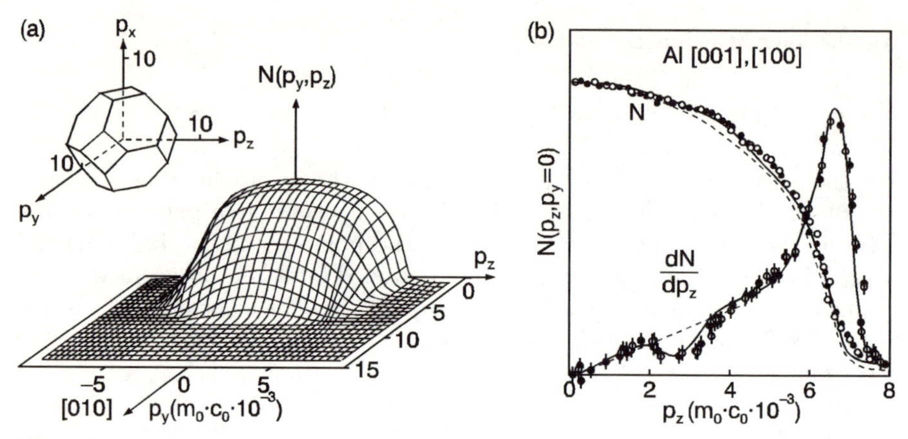

Fig. 16.5. Two-dimensional momentum density distribution for Al (**a**). The orientation of the crystal is determined by the orientation of the first Brillouin zone (insert to (a)). Part (**b**) shows the distribution along a cut for $p_y = 0$. m_0c_0 is the product of electron mass and light velocity; after [16.2]

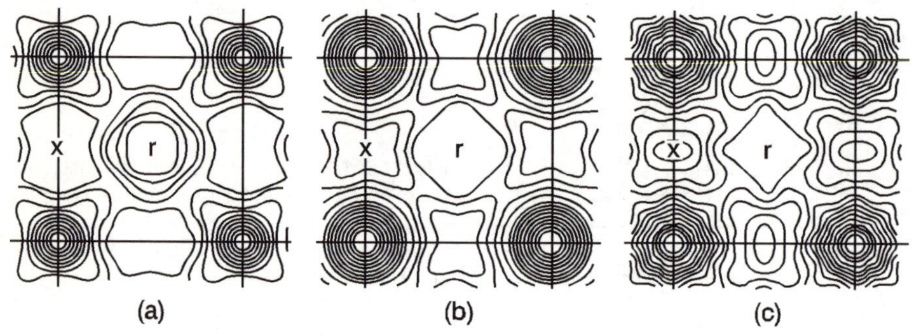

Fig. 16.6. Contour map (map of curves for constant momentum density) for V_3Si; from band structure calculation (**a**), from experiment (**b**), and from a model for the Fermi surface developed from the experiments (**c**); after [16.3].

and is of the order of KeV. The measurement of this quantity yields in principle the possibility to determine the third component p_x of the momentum distribution of the electrons. The Doppler broadening is obtained from a measurement of the line profile with a Ge diode. The diodes allow an energy resolution of 1.15 KeV which would correspond to about 4 mrad angular resolution for the momentum. Since the Doppler broadening does not give any information on the other components of the electron momentum it is comparable to measurements of the angular correlation in a longslit geometry with a resolution reduced by a factor of ten. Figure 16.7 shows a Doppler broadened γ line at 511 KeV in comparison to a γ line of ^{103}Ru at 497 KeV. The latter demonstrates the resolution of the detector. Because of the bad energy resolution measurements of the linewidths are not used so much for

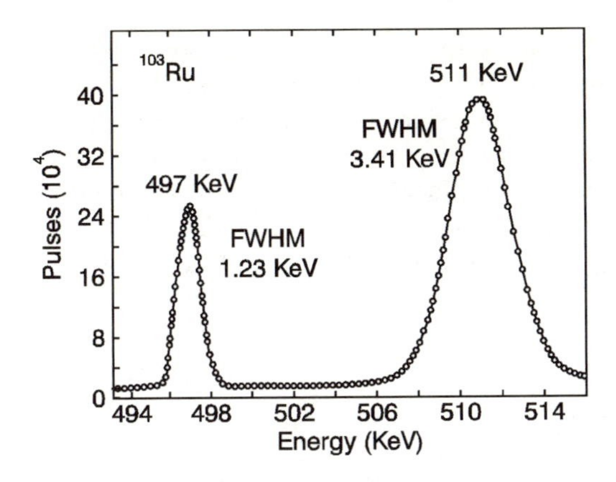

Fig. 16.7. Doppler broadening for a γ line from positron annihilation in ^{103}Ru in comparison to a line with a resolution-limited width; after [16.4].

the determination of momentum distribution but rather for the investigation of defects in crystals.

An alternative method for the investigation of defects in crystals is the determination of the positron lifetime. As already mentioned this is particularly successful if a γ quantum is emitted simultaneously with the positron. This quantum can be used as a start signal for the time recording. The measurement of the lifetime uses a so-called *fast-slow* coincidence as shown in Fig. 16.8a. The slow coincidence is responsible for the energy analysis. The two detectors with the photomultipliers are tuned to detect the start quantum γ_s and the two annihilation quanta γ_1 and γ_2. The coincidence of the two annihilation quanta determines the gate for the measurement of the time. The fast signal from γ_s of the left photomultiplier starts a time-amplitude

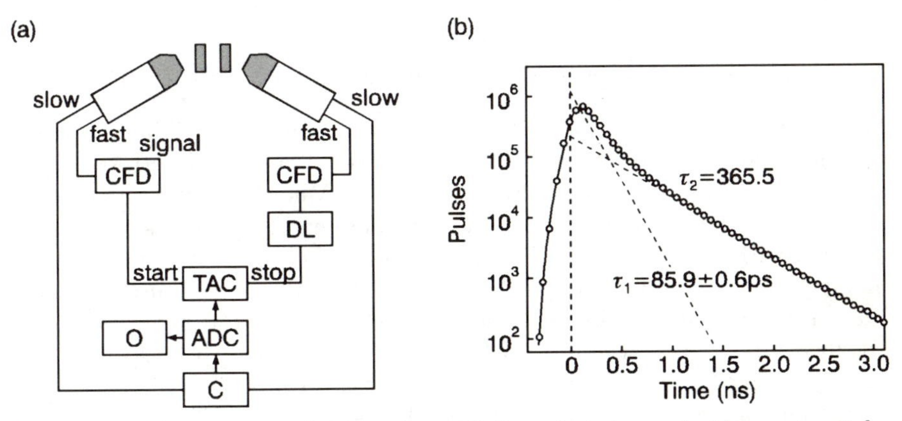

Fig. 16.8. Measurement of the positron lifetime; fast-slow coincidence circuit for the determination of the positron lifetime (**a**) and spectrum of the lifetime for Ni after irradiation with He (**b**); (CFD: constant fraction discriminator, DL: delay, TAC: time amplitude converter, ADC: analog-digital converter, C: coincidence, O: output); after [16.4].

ramp which is terminated by the fast signal of the annihilation quantum coming from the right photomultiplier. The CF discriminators allow a particularly accurate measurement of the time since the latter is determined for a constant fraction of the pulse height. In this way errors due to pulse height fluctuations are strongly reduced. The amplitude of the ramp is the lifetime stored in a memory. Figure 16.8b shows a lifetime spectrum for Ni after irradiation with He. The analysis of the result reveals two lifetimes of 85.9 ps and 356.5 ps, respectively.

16.2 Muon Spin Rotation

Muons are another type of particle with limited lifetime which are useful for spectroscopy of solids. Since muons are similar to protons in several aspects and since the spin of the muon plays an important role, similar questions and similar problems can be treated as for proton nuclear resonance. The information about the solid is obtained from the interaction of the spins with internal or external magnetic fields. The interaction appears as a rotation of the muon spin from where the name muon spin rotation is derived. The advantage of μSR spectroscopy over NMR spectroscopy results from the absence of any perturbing skin effect in the former and from the fact that only small external fields are required, or no external fields at all. A disadvantage is the lack of small-scale muon sources. Muon beams can be produced only from big accelerators. This spectroscopic technique is therefore limited to a few large laboratories.

Spectroscopy with μSR is mainly used for the investigation of diffusion processes in solids, of local magnetic fields, and of defects in the crystal lattice.

16.2.1 Muons and Muon Spin Rotation

Muons are available as positively charged and as negatively charged particles, μ^+ and μ^-, respectively. Experiments in solid-state spectroscopy are performed nearly exclusively with μ^+ particles. Muons are generated by the irradiation of light targets such as Be with high energy particles. They have spin 1/2 with a strong spin polarization antiparallel to the direction of their momentum in the muon beam. This polarization is retained after thermalization of the particles in the solid. The mass m_μ of the muon at rest is $183.98 \times 10^{-27}\,\text{g} = 1.84 \times 10^{-28}\,\text{VAs}^3/\text{m}^2$. This is about 10% of the proton mass. The magnetic moment is $\mu_\mu = \mu_\text{B}(m_0/m_\mu) = 4.49 \times 10^{-26}\,\text{J/T}$ and the magnetogyric ratio is $\gamma_\mu = 1.354 \times 10^8\,1/\text{sT}$.

The muon decays after $2.2{\times}10^{-6}\,\text{s}$ into a positron and two neutrons by a process of weak interaction. As a consequence of the parity violation for processes with weak interaction the positron emission is not isotropic. The angular distribution has the form

$$\boxed{W(\theta) = 1 + a(x)\cos\theta} \tag{16.10}$$

with

$$a(x) = \frac{2x - 1}{3 - 2x} \ .$$

x is the positron energy in units of the maximum positron energy. An averaged value for x would be 0.33. θ is the angle between spin orientation and positron emission. The muon is naturally polarized with its spin direction collinear to its momentum.

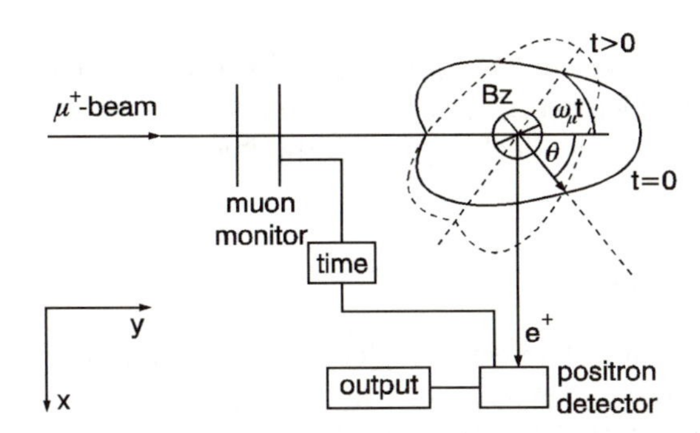

Fig. 16.9. Arrangement for the determination of muon spin rotation in a transverse field geometry; (B_z: magnetic field, ω_μ: rotation frequency, θ: orientation of emission at $t = 0$.

As soon as the muons are accommodated in the solid they precess in an external or internal field which causes a periodic signal at a fixed detector. A schematic setup for the observation of the spin rotation in a transverse field geometry is shown in Fig. 16.9. Transverse means the magnetic field is perpendicular to the direction of the incoming muons and to the direction of observation.

After passing a muon telescope (a muon monitor) at time $t = 0$ the muon penetrates into the crystal and starts to rotate in the local magnetic field with a frequency $\omega_\mu = \gamma_\mu B$. After a predefined time t the rate of positrons emitted from the muon decay process is recorded for the direction perpendicular to the field and to the incident muon beam. According to the precession of $W(\theta)$ a signal periodic with ω_μ is observed. The signal is damped because of the increasing dephasing of the spin precession with time. This signal is superposed on an exponentially decaying background originating from the finite lifetime of the muon. The dephasing of the spin rotation is completely analogous to the transverse relaxation of the spin excitations in the case of NMR spectroscopy. It is characterized by a dephasing factor $P_t(t)$.

For a longitudinal field geometry the positron detector is on the axis of the incident muon and the external field is collinear with this axis. For this geometry a precession is only possible if the local field has a transverse component. For an original orientation of the spins antiparallel to the field they will reorient by spin–lattice relaxation. This effect is again analogous to the longitudinal spin–lattice relaxation in NMR spectroscopy. The reorientation of the spins is described by another spin depolarization function $P_\mathrm{l}(t)$. This function is obtained experimentally from a difference of the signal with the positron detector in a 0° forward direction and in a 180° backward direction.

The positron count rates for the transverse and for the longitudinal geometry are

$$\begin{aligned} N_\mathrm{t}(t,\theta) &= N_0(\theta)\mathrm{e}^{-t/\tau}[1+a(x)P_\mathrm{t}(t)\cos(\omega_\mu t+\theta)] \,, \\ N_\mathrm{l}(t,\theta) &= N_0(\theta)\mathrm{e}^{-t/\tau}[1+a(x)P_\mathrm{l}(t)\cos\theta] \,. \end{aligned} \tag{16.11}$$

τ is the lifetime of the muon and $P_\mathrm{t}(t)$ and $P_\mathrm{l}(t)$ are the relaxation function and the depolarization function which describe the dephasing and reorientation of the spins. $N_0(\theta)$ is a normalization factor for the experiment if carried out under a certain angle θ.

16.2.2 Influence of Internal Fields

The relevant experimental results in µSR spectroscopy are the dependence of the rotation frequency on the internal magnetic fields and the spin relaxation or the spin polarization. For the internal contributions to B dipole fields are important to first order. This means B can be expressed as

$$B = B_\mathrm{ext} + B_\mathrm{dip} \,.$$

For zero external field the precession frequency is given by $\omega_\mathrm{dip} = \gamma_\mu B_\mathrm{dip}$ and the spin depolarization originates from the fluctuation in the local field. As in NMR spectroscopy it makes no difference whether this is due to a fluctuation of the local field in time or due to the diffusion of the muon through the crystal.

For slow fluctuations, which means for fluctuations with a correlation time τ_c large in comparison to $1/\omega_\mathrm{dip}$, the relaxation function and the depolarization function are given by

$$\begin{aligned} P_\mathrm{t}(t) &= P_\mathrm{t}(0)\exp\left\{-2\sigma^2\tau_\mathrm{c}^2\left[\exp\left(\frac{-t}{\tau_\mathrm{c}}\right)-1+\frac{t}{\tau_\mathrm{c}}\right]\right\} \,, \\ P_\mathrm{l}(t) &= P_\mathrm{l}(0)\left[\frac{1}{2}+\frac{2}{3}(1-2\sigma^2t^2)\exp(-\sigma^2t^2)\right] \,. \end{aligned} \tag{16.12}$$

σ is the van Fleck linewidth due to a static dipole–dipole interaction known from NMR as

$$\sigma^2 = \gamma_\mu^2\langle B_\mathrm{dip}^2\rangle \,. \tag{16.13}$$

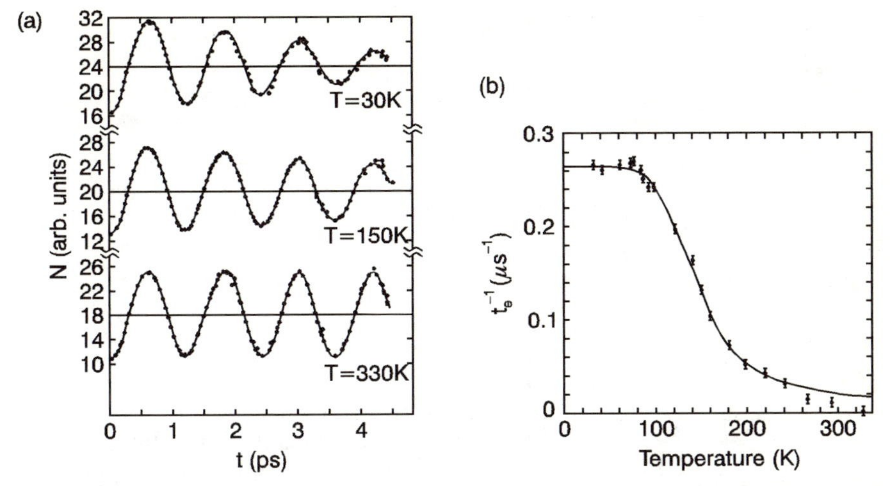

Fig. 16.10. Positron counting rates from Cu single crystals after irradiation of the crystal with muons at various temperatures and for a transverse field geometry (**a**) and inverse damping time $1/t_e$ versus temperature (**b**); after [16.5]

The relationship for $P_t(t)$ is known as the *Abragam formula*. The brackets mean averaging with respect to the square of the dipole field. $P_t(t)$ in (16.12) yields a Gaussian profile for small values of t ($t \ll \tau_c$) and an exponential decay for large values of t ($t \gg \tau_c$).

16.2.3 Experimental Results

Figure 16.10a shows positron count rates as a function of time for a muon decay in a copper crystal at different temperatures. The exponential decay of the signal due to the muon lifetime is compensated. For low temperatures the oscillations are clearly damped. For high temperatures the damping is not observed since the muons can diffuse more rapidly. By this process the dephasing of the spins is slowed down by motional narrowing as in the case of NMR. The inverse damping time is shown in Fig. 16.10b as a function of temperature. According to (16.12) this quantity is directly related to the coherence time τ_c. Since τ_c is on the other hand basically the time the muon stays at a certain crystal site such experiments are useful to study the elementary processes of diffusion for the muon.

In magnetic materials τ_c can be much larger than the time constant for the diffusion. This effect is due to fluctuations of the magnetic dipoles. In such cases the dephasing of the spin rotation directly determines the fluctuations of the magnetic moments. In this connection materials are of particular interest which exhibit a phase transition to a magnetic state established by an ordering process of free electrons. According to theoretical predictions a divergency of the longitudinal relaxation rate $1/T_1$ can be expected for $T = T_c$

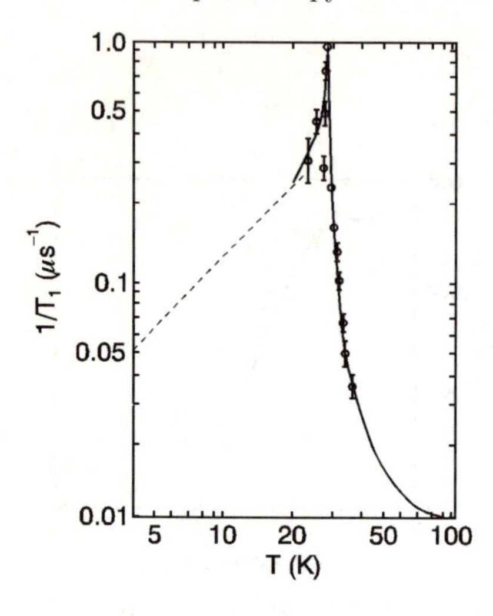

Fig. 16.11. Muon spin relaxation as a function of temperature for MnSi in a longitudinal field geometry; after [16.6].

in such cases. A corresponding measurement is not possible with NMR spectroscopy because of the sensitivity of T_c to the magnetic field. In contrast, μSR can demonstrate this behavior. An example is given in Fig. 16.11 of a phase transition in MnSi. The divergency of the relaxation with $1/(T - T_c)$ is evident.

Problems

16.1* Show explicitly that the integration for the calculation of the Fourier coefficients of the product of electron and positron wave function can be limited to the unit cell volume if the wave functions are given by Bloch functions.

(Purpose of exercise: training in simplification of an integral over periodic function.)

16.2 Show that the conservation of momentum results explicitly in the form $A_j(\boldsymbol{p}, \boldsymbol{k}_-) \propto \delta(\boldsymbol{p}/\hbar - \boldsymbol{k}_- - \boldsymbol{G})$.

(Purpose of exercise: derive mathematically a result which was obtained previously by intuition.)

16.3 Discuss an experimental arrangement for angular resolved positron annihilation which allows an energy resolution of 0.1 KeV in z direction.

(Purpose of exercise: get familiar with the geometric constraints in ACAR experiments.)

17. Neutron Scattering

Today neutron scattering is one of the most important experimental techniques in condensed matter analysis. Its application extends far beyond solid-state physics and plays an important role in fields such as structure of liquids, biophysics, biology, medicine, or environmental research.

The usefulness of neutron scattering originates from the comparatively simple and weak interaction of the neutrons with condensed matter and the well adapted "mechanical" properties of neutron beams to the microscopic and energetic structure of matter. The interaction is simple in comparison to the interactions relevant for light scattering or electron scattering because it is based on mechanical forces. The weakness of the interaction renders the system in the ground state for elastic scattering or in a well defined excited state for inelastic scattering. The hazard involved in degrading the samples is much lower as for example in the case of light scattering. The "mechanical" properties (energy and momentum) of the neutrons are appropriate since the de Broglie wavelength for thermal neutrons is of the order of interatomar distances and their energy compares well with the fundamental excitations in condensed matter. In addition the spin and magnetic moment of the neutron allow for special applications in magnetic materials and for the analysis of a random distribution and motion of isotopes.

It is useful at this point to compare the relationships between energy and wavelength for the various quasi-particles used for scattering experiments in condensed matter. According to de Broglie

$$\lambda = \frac{2\pi\hbar}{\sqrt{2M\epsilon_{\mathrm{kin}}}} \tag{17.1}$$

where M and ϵ_{kin} are the particle mass and kinetic energy, respectively. The following table compares the behavior of neutrons with electrons and photons. From it the advantage of the neutrons for atomistic resolution in space is evident. A disadvantage is the rather low resolution in energy which causes problems for the measurement of high energy phonons.

Neutron scattering has many applications in structural analysis where scattering is purely elastic. Like in the case of x-ray diffraction, this is not a central subject of this textbook. However, experimental and theoretical concepts for the elastic and inelastic scattering process are very similar and the same type of instrumentation is used. This means no strict differentiation

Table 17.1. Energies and wavelengths of quasi-particles.

	Neutrons	Electrons	Photons
wavelength λ [nm]	$0.028/\sqrt{\epsilon_{\text{kin}}\,[\text{eV}]}$	$1.2/\sqrt{\epsilon_{\text{kin}}\,[\text{eV}]}$	$1240/\epsilon\,[\text{eV}]$
energy for $\lambda = 1\,\text{nm}$	10 meV	1.5 eV	1.24 KeV
wavelength for 26 meV	2 nm	50 nm	100 μm
speed for 26 meV	200 m/s	2×10^4 m/s	3×10^8 m/s
energy resolution	5×10^{-2}	5×10^{-2}	10^{-5}

will be pursued in the first part of this chapter which covers instrumentation and scattering processes in general.

The chapter can only give an introduction to the subject. For extended information additional literature must be consulted. A good reference for the description of the theoretical background is the book by *Lovesey* [17.1]. Summaries of instrumentation and experimental procedures are in [17.2] and [17.3]. A more simple but still valuable summary is given in [17.4]. An instructive description of spallation sources can be found in [17.5].

17.1 Neutrons and Neutron Sources

Even though neutrons are the most abundant particles in the universe they are not easily obtained as a particle beam with high concentration. For the purpose of spectroscopy high flux reactors are the most widely used source. As an alternative spallation sources are gaining considerable importance because they can provide a higher neutron flux for the experiments in a certain spectral range.

17.1.1 Neutrons for Scattering Experiments

The elementary properties of the neutron are:

mass	1.675×10^{-27} Kg,
radius	6×10^{-16} m,
spin	1/2,
magnetic moment	$-0.9\times10^{-26}\,\text{Am}^2 = 1.9\,\mu_{\text{n}}$,
charge	0.

Interestingly the lifetime of the free neutron is only 932 seconds while it is fortunately practically infinite in condensed matter.

Table 17.2. High flux neutron sources available for neutron scattering experiments: reactors (upper part of table, capacity given in MW), spallation sources (lower part of table, capacity given in μA).

Name	Location	Capacity [MW]/[μA]	Thermal flux [10^{14}/cm^{2}s]	Cold source Number	Instruments Number
HFR	ILL,Grenoble	57	15	2	25
HFBR	Brookhaven	60	9	1	11
HFIR	Oak Ridge	100	10		10
NBSR	NIST,Washington	20	4	1	14
JRR-3	JAERI,Tokai	20	2	1	21
NRU	Chalk River	130	4		8
ORPHEE	Saclay	14	3	2	24
KENS-1	KEK,Tsukuba	4	3	1	17
ISIS	RAL,Chilton	200		1	13
SINQ	Villingen	1000	1.5	1	20
LANSCE	Los Alamos	100		1	5

It is useful to divide the neutrons in the following categories according to their energy:

ultra cold and very cold	:	below 0.5 meV ($T < 20\,\mathrm{K}$)
cold	:	0.5–2 meV ($10 < T < 50\,\mathrm{K}$)
thermal	:	2–100 meV ($30 < T < 1100\,\mathrm{K}$)
epithermal	:	$> 0.1\,\mathrm{eV}$, ($T > 1100\,\mathrm{K}$)

Neutrons with much higher energy are produced during the process of nuclear fission in the reactor and have to be cooled (moderated) to thermal energy to be useful for spectroscopy.

17.1.2 Thermal Neutron Sources

The most widely used sources for thermal neutrons are high flux reactors. The fission of ^{235}U produces in the reactor a total of 10^{18}–10^{19} neutrons per second with an energy of several MeV each. These neutrons must be slowed down in a moderator to thermal energy of about 26 meV for scattering experiments. High flux reactors can develop a thermal flux of more than 10^{15} neutrons/cm^{2}s. Unfortunately only a fraction of about 10^{-8} from this can be made available for scattering experiments. The dramatic loss of neutrons is due to the large spacial distribution of the neutrons in the reactor core, to the limitations when extracting the neutrons from the core into a beam, to the required selection in the process of monochromatization and to all other beam handling requirements.

Table 17.2 lists some of the most important reactors for scattering experiments together with their power, thermal flux and instrumentation. The latter refers to cold sources and various spectrometers.

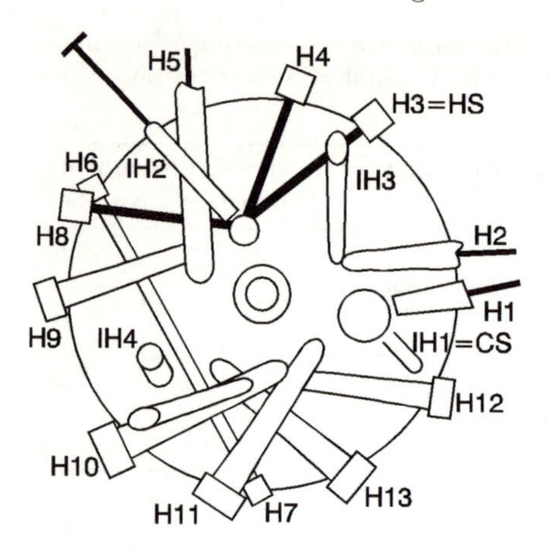

Fig. 17.1. Arrangement of beam tubes in the high flux reactor in Grenoble. The tubes assigned CS and HS are cold and hot sources, respectively. Black lines from H1 and H2 are connected to neutron guides for remote scattering experiments. IH3 and IH4 are vertical tubes.

The extraction of the neutrons from the interior of the reactor proceeds by a beam hole with a corresponding collimator called the *beam tube*. Figure 17.1 shows the arrangement of the beam tubes and beam holes in the high flux reactor in Grenoble. The beam hole must be well sealed with filters and several shields to prevent high energy neutrons and γ radiation escaping from the reactor and propagating in the beam. Appropriate filters are beryllium plates due to the low wavelength cutoff for Bragg scattering in Be at $\lambda = 4\,\text{Å}$.

From the beam hole the neutrons are guided to the experiment by the beam tube which allows the transfer of neutrons over distances of several meters without substantial loss. A beam tube consists of a Ni tube or a tube with Ni-plated walls which reflect the neutrons of a particular wavelength. The smaller the cross section of beam tube and collimator the less divergent the neutron beam, but the lower the flux available for the experiment.

Specially constructed beam tubes can act as *waveguides* for the neutrons and allow transportation of the neutron beam over 50–100 m without substantial loss. Neutrons with high k vectors can be transported if the walls are plated with *supermirrors* instead of Ni.

Alternative facilities to supply beams with a high neutron flux are *spallation sources*. Spallation sources are not restricted to special nuclei since they do not rely on a chain reaction. In contrast to the reactors energy is consumed to initiate and sustain the spallation process. This has the advantage of an easier control of the reaction. The energy is supplied in the form of a proton beam of several hundred μA and several 100 MeV energy which hits a target and initiates the spallation reaction in the nuclei. As soon as the current is turned off the reaction stops. The protons penetrate into the nuclei of the target and transfer energy to the protons and neutrons of the nuclei. The latter heat up and start to release neutrons (and also protons). Neutrons (or protons) directly emitted from the collision process (cascade particles) contribute to the heating of other nuclei. As in reactors the neutrons are released with a very high energy of several MeV and must be moderated. Moderation is easier in spallation sources since in contrast to the case of nuclear reactors no sophisticated balance for the concentration of thermal neutrons is required. The number N of neutrons released per incident proton can be estimated from

$$N = a(A + 20)(\epsilon - b) , \tag{17.2}$$

where A is the atomic weight of the target, ϵ is the energy of the proton and a and b are constants referring to constructive details. For a target of lead with 10 cm diameter and 60 cm length a and b are $0.1\,(\text{GeV})^{-1}$ and 0.12 GeV, respectively. This yields about ten neutrons per proton for 600-MeV primary protons. Close to the target a flux of thermal neutrons of the order of 2×10^{10} neutrons/cm^{2}s can be reached. This flux is guided to the outside of the moderator tank by beam tubes as in the case of a reactor. By special constructive details a very fast moderation can be obtained. This allows spallation sources to be operated in a pulsed mode with pulse times as short as 30 μs.

17.1.3 Cold and Hot Neutron Sources

For both the nuclear reactor and the spallation source the concentration of low energy neutrons can be considerably increased in a certain spectral range by inserting cooling units into the center of the source. These units operate usually with liquid hydrogen or liquid deuterium. The energy distribution in the cold neutron beam is shifted to lower values. The peak wavelength in the distribution of the thermal neutrons is at about 0.12 nm. This wavelength can be shifted to more than 1 nm for supercold neutrons. The downshift can be an advantage for the analysis of low-energy excitations since the resolution in energy scales with the energy of the neutrons.

In some cases hot sources can be of advantage. Some reactors are therefore supplied with hot cells where the neutrons can be equilibrated at high temperatures. The hot source of the high flux reactor in Grenoble consists of a graphite block heated to 2400 K which results in a downshift of the peak wavelength of the neutrons to 0.08 nm. Figure 17.2 shows the energy distribution of neutrons (plotted as a function of their wavelength) at several beam tubes for the reactor in Grenoble.

17.2 Neutron Spectrometer and Detectors

As in all other spectroscopic techniques the probe beam emerging from the sample must be analyzed with respect to its energy and intensity. Since the direction of the scattered neutrons is an important parameter and the incident beam needs a high degree of monochromatization neutron spectrometers require a considerable technical effort. In addition the weak interaction of the neutrons with matter makes their detection much more laborious as compared to light or electron detection.

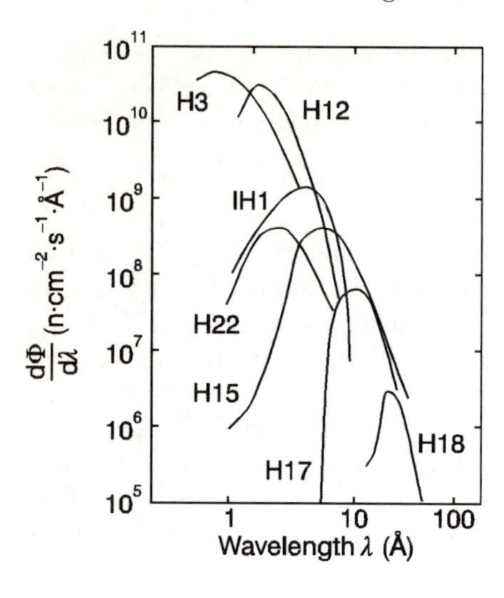

Fig. 17.2. Spectral distribution of neutrons at various beam tubes and guide facilities at the high flux reactor in Grenoble; (H3: hot, H12, H22: thermal, IH1: cold, H15-H18: very cold guides); after [17.6].

17.2.1 Neutron Spectrometer

Neutron spectrometers consist of two elaborate components: A facility to select monochromatic neutrons and the analyzing system including the detector. For the monochromator two concepts are in use. Bragg scattering from a high quality and large single crystal allows the selection of different neutron energies by choosing an appropriate geometry for elastic scattering of the neutrons from the beam hole. Bent crystals allow a focusing of the neutron beam. Alternatively, a mechanical chopper can be used for selection. Mechanical choppers consist of a rotor with axially or tangentially mounted blades and operate by selection of neutron velocity. If the chopper has a set of axial blades it rotates parallel to the flight direction of the neutrons and the blades must be twisted. Only neutrons with an appropriate transit time across the chopper can pass. Thus, the chopper generates a continuous beam of neutrons with a quasi-monochromatic energy distribution (mechanical velocity chopper). If the rotation axis of the chopper is perpendicular to the flight direction of the neutrons a monochromatic and pulsed beam is obtained (Fermi chopper).

One possibility to obtain full information on the wave vector and energy of the scattered neutrons is analysis with a *three-axis spectrometer* shown in Fig. 17.3a. The spectrometer has three axes where six angles can be selected independently. With the beam hole and the collimator a beam of neutrons is extracted from the reactor core and monochromatized with crystal MC. This crystal is usually bent to focus the diffracted beam to the sample. The diffracted beam strikes the sample S and the analyzer crystal is arranged to receive the beam after a second, inelastic scattering process in the sample. By rotating the analyzer and the detector the energy of the scattered neutrons can be investigated. In practice scans are usually performed for constant value of q or constant energy (see also Fig. 17.5 on p. 367).

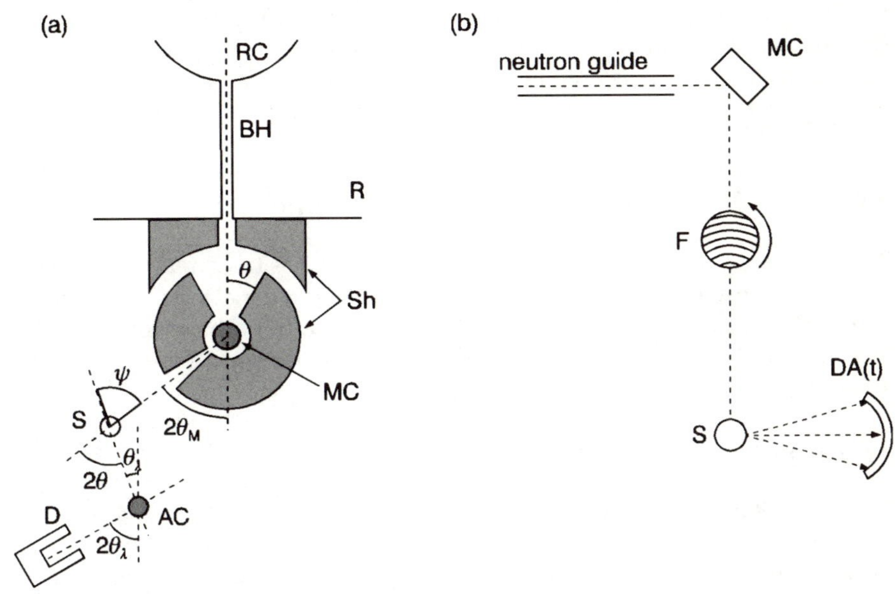

Fig. 17.3. Schematic representation of a three-axis neutron spectrometer (**a**) and a time-of-flight spectrometer (**b**); (RC: reactor core, BH: beam hole with collimator, MC: monochromator crystal, Sh: shielding, S: sample, AC: analyzer crystal, D: detector, DA: detector array, F: Fermi chopper);

Alternative to the use of an analyzer crystal the energy resolution can be obtained from a *time-of-flight* experiment. The setup is sketched in Fig. 17.3b. After a first monochromatization by the monochromator crystal MC the Fermi chopper generates a pulsed neutron beam and sets a time scale for each pulse. As a consequence of the inelastic scattering process in the sample S the emerging neutrons arrive at the detector after different times of flight. From the distances between chopper, sample and detector this time is easily converted to energy. Using a detector array scattering from quasi-particles with different wave vectors can be recorded simultaneously as discussed below.

Very high (angular) resolution can be obtained with spectrometers operating in a backscattering geometry.

17.2.2 Neutron Detectors

The scattered neutrons are detected from fission products after a reaction with either boron or ^{3}He. These detectors rely on the nuclear reactions

B(n, α)^{7}Li and
^{3}He(n,p)T .

The charged particles from the reaction are emitted with an energy of the order of MeV and can be detected in a proportional counter. To make the detection of the scattered neutrons more efficient one or two dimensional arrays of detectors with several hundred units are in use.

17.3 The Process of Neutron Scattering

The scattering of thermal neutrons is the basic process to extract information from the solids. The scattering process can be elastic or inelastic. In the latter case excitation of quasi-particles occurs in the solid. In both cases the process is well described by a scattering cross section similar to Raman or electron scattering.

17.3.1 The Scattering Cross Section

Neutrons penetrate material easily because of their weak interaction. The interaction length with the nuclei of the material investigated is only of the order of 10^{-13} cm. This means the wavelength of the neutron is always much larger then the size of the scatterer which implies isotropic or s-wave like scattering. As a consequence the scattering can be characterized by a single parameter b called the *scattering length.* The square of the scattering length determines the scattering cross section (see below). The distance between the nuclei is, on the other hand, much larger than their size which means scattering from all nuclei is just additive. The scattering length does not necessarily scale with the size of the nucleus. The coherent scattering lengths for carbon and titanium are, for example, 0.665×10^{-12} cm and -0.344×10^{-12} cm, respectively. In contrast, the covalent radii of the two atoms are $r_{\text{cov}}(\text{C}) = 0.77 \times 10^{-8}$ cm and $r_{\text{cov}}(\text{Ti}) = 1.32 \times 10^{-8}$ cm. This lack of scaling with the size of the scatter is in contrast to x-ray diffraction where the scattering cross section increases continuously with the atomic order and thus with the covalent radius of the atoms. In Fig. 17.4 scattering of x rays and neutrons is compared for several elements and isotopes. The diameters of the circles in part (a) represent scattering amplitudes for x-ray diffraction. The circles in the first row of part (b) are scaled to the scattering length for neutron diffraction. Even different isotopes of the same chemical element can have very different scattering lengths. This is demonstrated in part (b) for the isotopes of H, Ti, Fe, and Ni. The dramatic difference in their scattering lengths means that even a chemically homogeneous or coherent material looks very inhomogeneous or incoherent for the penetrating neutrons. As a consequence in the experiments and for the theoretical analysis coherent and incoherent scattering must be differentiated.

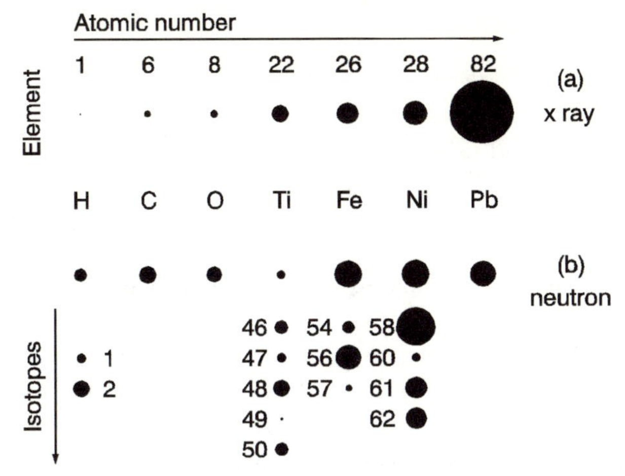

Fig. 17.4. Scattering efficiency of x rays (**a**) and thermal neutrons (**b**) for several elements and their isotopes.

17.3.2 Coherent and Incoherent Scattering in the Born Approximation

The concept of coherent and incoherent neutron scattering is very important and can be understood from a basic description of the scattering process within first-order perturbation theory or the *Born approximation.*

As in the case of optical absorption the golden rule allows the calculation of the probability $P_{k,k'}$ for the transition of an initial state to a final state. The latter are in the present case characterized by the wave vector k for the incident and k' for the scattered neutron. With a density of final states $\varrho_{k'}$ we have

$$P_{k,k'} = \frac{2\pi}{\hbar} \left| \int \mathrm{d}^3 x \psi_{k'}^* \mathbf{U} \psi_k \right|^2 \varrho_{k'} \ . \tag{17.3}$$

ψ_k and $\psi_{k'}$ are the wave functions for the incident and for the scattered neutron and $\mathbf{U}$ is the interaction potential (expressed as an operator). The absolute value of the matrix element in (17.3) is the *scattering amplitude.* The scattering cross section is defined as the ratio between the probability $P_{k,k'}$ of (17.3) and the incident neutron flux Φ.

For the simplest case of normalized plane waves for the incident and scattered neutrons

$$\psi_k = \frac{1}{L^{3/2}} \mathrm{e}^{\mathrm{i}\boldsymbol{kr}} \quad \text{and} \quad \psi_{k'} = \frac{1}{L^{3/2}} \mathrm{e}^{\mathrm{i}\boldsymbol{k'r}} \ ,$$

the flux per neutron is

$$\Phi = \frac{v_\mathrm{n}}{L^3} = \frac{\hbar k}{M_\mathrm{n} L^3} \ . \tag{17.4}$$

The scattering cross section is

$$\frac{\mathrm{d}\sigma}{\mathrm{d}\Omega} = \frac{P_{k,k'}}{\Phi} .$$

Inserting from (17.3) and (17.4) and using

$$\varrho_{k'} = \left(\frac{L}{2\pi}\right)^3 \frac{M_\mathrm{n} k'}{\hbar^2} \tag{17.5}$$

for the density of states per unit solid angle for the scattered plane waves yields

$$\boxed{\frac{\mathrm{d}\sigma}{\mathrm{d}\Omega} = |\langle \psi_{k'}|\mathbf{U}|\psi_k\rangle|^2 ,} \tag{17.6}$$

where the matrix element on the right-hand side of the equation was defined as

$$\langle \psi_{k'}|\mathbf{U}|\psi_k\rangle = \frac{L^3 M}{2\pi\hbar^2} \int \psi_{k'}^* \mathbf{U} \psi_k \mathrm{d}^3 x \qquad \text{(in m)} . \tag{17.7}$$

An explicit evaluation of the scattering cross section is instructive for an interaction potential expressed by a δ function for an assembly of scatterers at rigid positions $\boldsymbol{r}_l$ with individual scattering length b_l. For this (repulsive) interaction the *Fermi pseudopotential* can be used which is given by

$$U(\boldsymbol{r}) = \frac{2\pi\hbar^2}{M_\mathrm{n}} \sum_l b_l \delta(\boldsymbol{r} - \boldsymbol{r}_l) \qquad \text{(in J)} . \tag{17.8}$$

For this potential the elastic scattering cross section is obtained from (17.6)

$$\frac{\mathrm{d}\sigma}{\mathrm{d}\Omega} = \sum_{ll'} \overline{b_{l'}^* b_l}\, \mathrm{e}^{\mathrm{i}\boldsymbol{q}(\boldsymbol{r}_l - \boldsymbol{r}_{l'})} , \tag{17.9}$$

where $\boldsymbol{q} = \boldsymbol{k} - \boldsymbol{k}'$ is the scattering vector. The average is taken over all particles, isotopes, and spin orientations participating in the scattering process.

If l and l' refer to different sites there is no correlation between b_l and $b_{l'}$. This means

$$\overline{b_{l'}^* b_l} = \overline{b_{l'}^*}\, \overline{b_l} = |\bar{b}|^2 \qquad \text{for} \qquad l \neq l'$$

but

$$\overline{b_{l'}^* b_l} = \overline{|b_l|^2} = \overline{|b|^2} \qquad \text{for} \qquad l = l' .$$

This means in the latter case the square of the scattering lengths of the different components of the assembly of scatterers must be considered. We call the components with $l \neq l'$ in the sum of (17.9) the coherent contributions and the components with $l = l'$ the incoherent contributions to the scattering cross section. The general form for the averaged product of the scattering lengths is therefore

$$\overline{b_{l'}^* b_l} = |\bar{b}|^2 + \delta_{ll'}(\overline{|b|^2} - |\bar{b}|^2) . \quad (17.10)$$

With this we can express the scattering cross section as the sum of a coherent and an incoherent component

$$\begin{aligned} \frac{\mathrm{d}\sigma}{\mathrm{d}\Omega} &= \left(\frac{\mathrm{d}\sigma}{\mathrm{d}\Omega}\right)_{\mathrm{coh}} + \left(\frac{\mathrm{d}\sigma}{\mathrm{d}\Omega}\right)_{\mathrm{incoh}} \\ &= |\bar{b}|^2 \left|\sum_l \mathrm{e}^{\mathrm{i}\boldsymbol{q}\boldsymbol{r}_l}\right|^2 + N\overline{|b - \bar{b}|^2} . \end{aligned} \quad (17.11)$$

The equation shows that only the coherent part can lead to interferences for plane waves emerging from different nuclei.

The second part of (17.11) shows that σ_{incoh} is proportional to the mean square deviation of the scattering lengths for the different nuclei involved. This deviation can be very large. In the case of hydrogen $\sigma_{\mathrm{coh}} = 1.8\,\mathrm{bn}$, for example, and $\sigma_{\mathrm{incoh}} = 80\,\mathrm{bn}$. 1 bn (barn) is the unit for cross sections and amounts to $10^{-24}\,\mathrm{cm}^2$.

Table 17.3 lists scattering lengths b and total scattering cross sections σ as well as absorption cross sections for a selected number of elements and their isotopes. Note that the scattering lengths are complex for strongly absorbing nuclei and the real part of b can be negative. The total and the coherent cross section are $\sigma = 4\pi\overline{|b|^2}$ and $\sigma_{\mathrm{coh}} = 4\pi|\bar{b}|^2$, respectively.

17.3.3 Inelastic Neutron Scattering and Scattering Geometry

The scattering cross section from (17.9) is easily reformulated for an inelastic scattering process. In this case the k vectors for the incident and scattered neutrons do not cancel in the expression of (17.6) which means a factor k'/k is retained. Also, the system changes from an initial state i to a final f with energies ϵ_{i} and ϵ_{f}, respectively. This process is controlled by energy conservation which is guaranteed by multiplying the matrix element in (17.6) with a δ function. The differential scattering cross section (scattering per unit angle and per unit energy range) can be expressed by

$$\frac{\mathrm{d}^2\sigma}{\mathrm{d}\Omega\mathrm{d}\epsilon'} = \frac{k'}{k}|\langle k', \mathrm{f}|\mathbf{U}|k, \mathrm{i}\rangle|^2 \delta(\hbar\Omega + \epsilon_{\mathrm{i}} - \epsilon_{\mathrm{f}}) , \quad (17.12)$$

where $\hbar\Omega$ is the energy difference of the neutron before and after the inelastic collision.

As in the case of Raman scattering or electron energy loss processes the scattering geometry is determined by the conservation of energy and momentum

Table 17.3. Coherent scattering lengths $\overline{b}$, total scattering cross sections σ and absorption cross section σ_a for selected atoms; (av) refers to a natural isotope mixture; after [17.1].

Atom	Mass AMU	Abundance %	$\overline{b}$ $[10^{-12}\,\text{cm}]$	σ $[10^{-24}\,\text{cm}^2]$	σ_a $[10^{-24}\,\text{cm}^2]$
H	1	100	−0.374	81.67	0.333
	2	0.015	0.667	7.63	0.0005
He	av		0.326	1.21	0.001
	3	0.0001	0.574	5.6	5333.
	4	100	0.326	1.21	≈ 0
B	av		0.535−i0.021	5.01	767.
	10	20	0.0−i0.11	0.98	3837.
	11	80	0.666	5.8	0.006
C	av		0.665	5.564	0.004
	12	98.9	0.665	5.564	0.004
	13	1.11	0.62	5.5	0.001
N	av		0.93	11.5	1.9
	14	99.63	0.937	11.5	1.9
	15	0.37	0.644	5.21	< 0.001
Al	27	100	0.349	1.506	0.231
Si	av		0.415	2.173	0.171
	28	92.23	0.411	2.119	0.177
	29	4.67	0.47	2.9	0.1
	30	3.1	0.458	2.65	0.107
Au	197	100	0.763	7.81	98.65

$$\begin{aligned} \hbar\boldsymbol{k} &= \hbar\boldsymbol{k}' + \hbar\boldsymbol{q} \\ \hbar\Omega &= \frac{\hbar^2}{2m}(k^2 - k'^2)\,. \end{aligned} \tag{17.13}$$

Positive values for Ω mean generation of quasi-particles or Stokes scattering. There are two important differences to light scattering and electron scattering:

a) The difference in the magnitude of k and k' can be very large since the neutron can exchange a considerable part of its energy. This means the fraction k'/k is important in neutron scattering.

b) The scattering vector $\boldsymbol{q}$ is not equal to the q vector of the excited quasi-particle but can in addition contain any reciprocal lattice vector $\boldsymbol{G}$. Equation (17.13) is therefore only correct if $\boldsymbol{G}$ is chosen to be zero. To be sufficiently general and in agreement with most of the literature on neutron scattering we will assign the scattering vector $\boldsymbol{k}' - \boldsymbol{k}$ in the following as

$$\boldsymbol{k}' - \boldsymbol{k} = \boldsymbol{\kappa} = \boldsymbol{G} + \boldsymbol{q}\,.$$

In other words, the analysis of quasi-particle excitation can be performed close to any diffraction spot of the reciprocal lattice. The scattering geometry is shown schematically in Fig. 17.5. If the scattered neutron is observed in direction OA″ any quasi-particle vector $\boldsymbol{q}$ shown satisfies the momentum conservation but only for one selected vector is the energy conservation simultaneously valid. $|\boldsymbol{k}| - |\boldsymbol{k}'|$ determines the energy of the excited quasi-particle. If $|\boldsymbol{k}'|$ is scanned for constant direction of observation the quasi-particles satisfying the conditions for detection change. In this case the scanning is along the dashed line in Fig. 17.5 and a wide range of energies given by $|\boldsymbol{k}'|$ will satisfy the scattering conditions. If k' is larger than k only antiStokes scattering is possible. In practice scanning for the spectrum of the scattered neutrons is either for a constant value of $\boldsymbol{q}$ (energy scan) or for a constant value of Ω (wave vector scan). In both cases a well defined synchronous motion of the analyzing crystal and of the detector is required but selective excitations in energy and momentum can be studied.

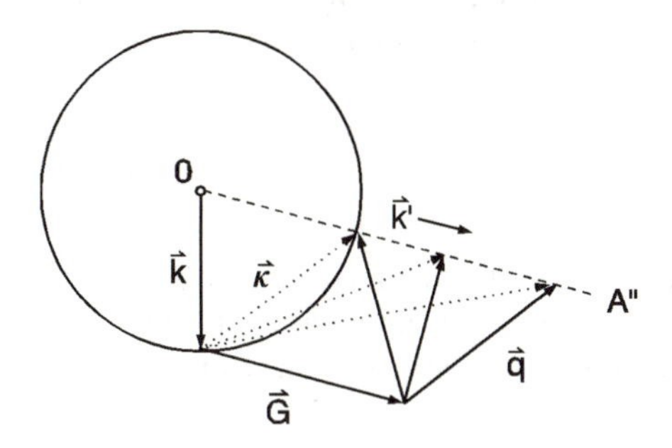

Fig. 17.5. Scattering geometry for inelastic neutron scattering. $\boldsymbol{G}$ and $\boldsymbol{q}$ are vectors of the reciprocal lattice and of the excited quasi-particle, respectively.

17.4 Response Function and Correlation Function for Inelastic Neutron Scattering

Neutron scattering is conveniently described in the frame of linear response theory. Most of the results from the discussion of the generalized susceptibility in Chap. 14 and from EELS in Sect. 15.1 can be used. The situation is even more simple since the particles are uncharged and non-interacting in the present case.

The relationship for the inelastic scattering cross section at finite temperatures is immediately obtained from (17.12) in the form

$$\frac{\mathrm{d}^2\sigma}{\mathrm{d}\overline{\Omega}\mathrm{d}\epsilon'} = \frac{k'}{k}\sum_{\mathrm{i,f}} W(\epsilon_\mathrm{i})|\langle k',\mathrm{f}|\mathbf{U}|k,\mathrm{i}\rangle|^2\delta(\hbar\Omega + \epsilon_\mathrm{i} - \epsilon_\mathrm{f}) , \tag{17.14}$$

where $W(\epsilon_\mathrm{i})$ is the the probability of the occupation of the initial state of energy ϵ_i. The cross section describes the probability of the scattering of a neutron from initial state $\langle k|$ and any system energy ϵ_i to a final state $\langle k'|$ with system energy ϵ_f for an interaction potential U and conservation of energy and momentum.

The matrix element and the δ function in (17.14) can be expressed by the dynamical form factor from Chap. 14. For a polynuclear interaction potential of the general form

$$U = \frac{2\pi\hbar^2}{M_\mathrm{n}} \sum_j U_j(\boldsymbol{r} - \boldsymbol{r}_j) \tag{17.15}$$

and a total of N independent scatterers (17.14) becomes

$$\frac{\mathrm{d}^2\sigma}{\mathrm{d}\Omega\mathrm{d}\epsilon'} = N\frac{k'}{k}|\overline{U_j}(\boldsymbol{\kappa})|^2 S(\boldsymbol{\kappa}, \Omega)\ . \tag{17.16}$$

The cross section is expressed by three components: a geometry factor Nk'/k, the Fourier component of the scattering potential $U_j(\boldsymbol{\kappa})$ and $S(\boldsymbol{\kappa}, \Omega)$ which describes the intrinsic structural and dynamical properties of the scatterer. For a particle density

$$n(\boldsymbol{r}, t) = \frac{1}{N}\sum_j \delta[\boldsymbol{r} - \mathbf{r}_j(t)] \tag{17.17}$$

$S(\boldsymbol{\kappa}, \Omega)$ can be expressed by the Fourier transform of the density–density correlation function between nucleus j at time 0 and nucleus j' at time t. As in (L.24) the dynamical form factor becomes in this representation

$$S(\boldsymbol{\kappa}, \Omega) = \frac{1}{2\pi\hbar}\int \mathrm{d}t \exp(-\mathrm{i}\Omega t)\langle n_{\boldsymbol{\kappa}} n_{-\boldsymbol{\kappa}}(t)\rangle\ . \tag{17.18}$$

The above concept can be used to calculate the scattering cross section for phonons in a crystal. In this case the interaction potential is periodic and the cross section for coherent Stokes scattering for one phonon with frequency Ω_j becomes

$$\begin{aligned}\left(\frac{\mathrm{d}^2\sigma}{\mathrm{d}\Omega\mathrm{d}\epsilon'}\right)_{\mathrm{st}} &= \frac{N\sigma_\mathrm{c}\hbar k'}{8\pi M_\mathrm{n} k}\exp[-2W(\boldsymbol{\kappa})]|\boldsymbol{\kappa}\boldsymbol{\sigma}_j|^2 \\ &\quad\times\frac{f_\mathrm{E}(\Omega_j\boldsymbol{q})}{\Omega_j(\boldsymbol{q})}\delta[\hbar\Omega - \hbar\Omega_j(\boldsymbol{q})]\ ,\end{aligned} \tag{17.19}$$

where $\boldsymbol{k}' = \boldsymbol{k} - \boldsymbol{q}_j - \boldsymbol{G}$ and ϵ' are the wave vector and the energy of the scattered neutron, σ_c is the coherent cross section, $2W(\boldsymbol{\kappa})$ is the Debye–Waller factor

$$2W(\boldsymbol{\kappa}) = \frac{3\hbar^2\kappa^2 k_\mathrm{B}T}{M(k_\mathrm{B}\Theta_\mathrm{D})^2}\ ,$$

and $\boldsymbol{\sigma}_j$ is the polarization of the phonon with frequency Ω_j. M and Θ_D are the mass of the nucleus and the Debye temperature, respectively. Integration of (17.19) over the energy ϵ' yields the total scattering cross section

$$\left(\frac{\mathrm{d}\sigma}{\mathrm{d}\overline{\Omega}}\right)_{\mathrm{st}} = \frac{N\sigma_{\mathrm{c}}\hbar k'}{8\pi M_{\mathrm{n}} k} \exp[-2W(\boldsymbol{\kappa})|\boldsymbol{\kappa}\boldsymbol{\sigma}_j|^2 \frac{f_{\mathrm{E}}(\Omega_j \boldsymbol{q})}{\Omega(\boldsymbol{q}_j)} \times \left\{1 - \frac{\hbar}{2\epsilon'}\boldsymbol{k}'\nabla\Omega_j(\boldsymbol{q})\right\}^{-1} . \tag{17.20}$$

Evaluation of (17.20) yields for ambient conditions and phonons in metals a cross section of the order of 10^{-3} bn.

17.5 Results from Neutron Scattering

Inelastic neutron scattering is extensively used to study lattice vibration or magnetic excitations. Figure 17.6 exhibits the dispersion of acoustic phonons in lead as measured with a three-axis spectrometer. Shown are longitudinal (L) and transverse (T) components in different crystallografic directions. No optical phonons are observed since lead has only one atom per primitive unit cell. The maximum frequency is about 10^{12} Hz or 3.5 meV. This is typical for metals. Frequencies for optical phonons in inorganic nonmetallic crystals are about one order of magnitude larger.

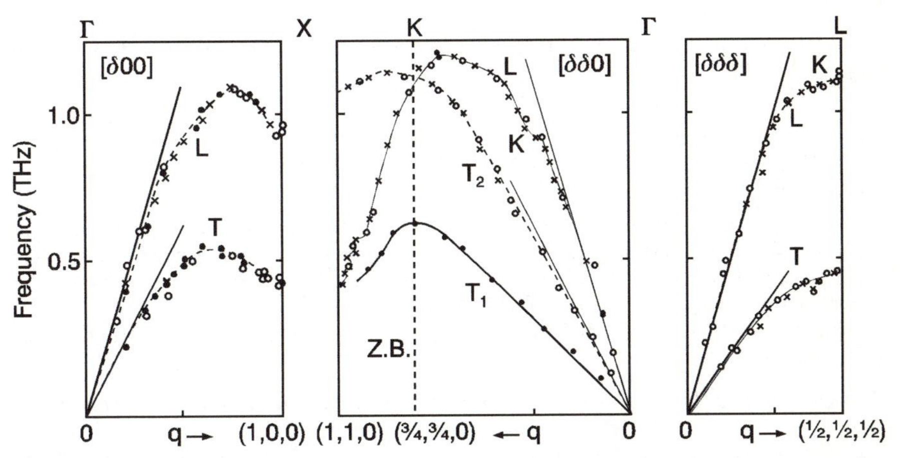

Fig. 17.6. Phonon dispersion in lead for different crystallographic directions; after [17.7].

Figure 17.7 shows a time-of-flight spectrum recorded for copper. The central line represents the elastically scattered neutrons. Neutrons from an antiStokes scattering process arrive earlier, while those from a Stokes process arrive later. Note the nonlinear relation between time and energy. The spectral structure represents a phonon density of states.

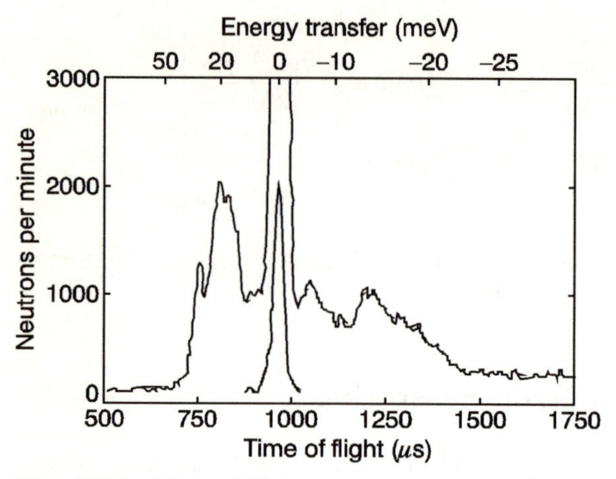

Fig. 17.7. Time-of-flight neutron spectrum for copper. The resolution in energy is represented by the widths of the central line; after [17.5].

Problems

17.1 Show that the density of states for a plane wave in a volume L^3 is given by (17.5).

(Purpose of exercise: recall the meaning of the density of states for the outgoing wave.)

17.2 Show that the total scattering cross section for a Fermi potential with a single scattering center is

$$\sigma = 4\pi b \, .$$

(Purpose of exercise: Use the simplest version of a scattering potential in the Born approximation.)

17.3 For a time-of-flight spectrometer with thermal neutrons the distances from chopper to sample and sample to detector are 3 and 4 m, respectively. Estimate the time of the arrival of the neutrons at the detector after inelastic scattering with optical phonons in classical semiconductors.

(Subject of training: Understand the time to energy conversion in time-of-flight spectrometers.)

17.4 Calculate the cross section for scattering of thermal neutrons by transverse acoustic phonons in Si. Consider two different points in the reciprocal lattice.

(Purpose of exercise: Using (17.20) convince yourself which parameters of the scattering experiment are most important for the scattering intensity.)

17.5* Calculate the dynamical form factor $S(\Omega)$ for the damped harmonic oscillator.

(Purpose of exercise: Apply a sophisticated mechanism to a simple example to understand the meaning of the dynamical form factor.)

18. Spectroscopy with Atoms and Ions

Scattering of atoms and ions from solid surfaces is another example of the development of an excentric experiment into a very useful and important technique for material analysis. These types of experiments are the youngest in the family of spectroscopy with particles. Also, the masses of the particles used as a probe are up to two orders of magnitude larger than those for the heavy particles discussed so far. Since the penetration of the atoms and ions into the solid material can be very small very thin films or surfaces can be investigated. Hence the booming activities in surface science has naturally contributed to the development and to applications of this new analytical technique.

Basic elements in the field of spectroscopy with atoms and ions are instrumentation, energy loss of the particles within the target material, backscattering of atoms, and secondary ion emission. Since the range of particle masses and the range of particle energies used is very wide an extended set of instrumentation is required. The range of the former extends from protons to heavy atoms like iron or gold, the range of the latter from several KeV to several tens of MeV. According to this wide range of energies a wide range of penetration depths can be obtained as demonstrated in Fig. 18.1. For very high energy particles as they are used in Rutherford backscattering experiments penetration can be as high as 100 nm. This is already considered as the bulk of the material. On the other hand, for heavy ions and energies in the KeV range penetration is only of the order of 1 nm which allows pure surface analysis.

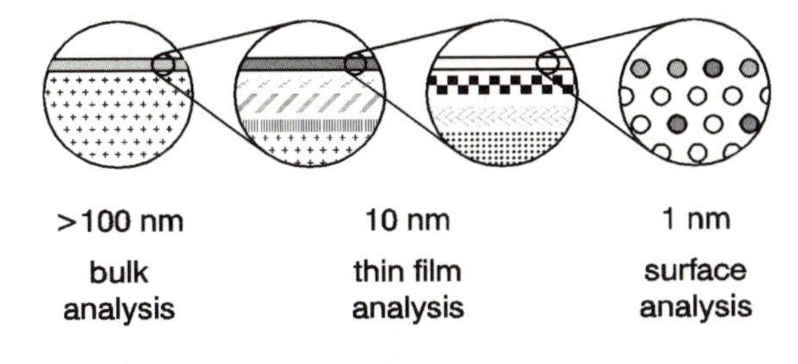

Fig. 18.1. Range of particle penetration and analysis for spectroscopy with atoms and ions.

Two classes of experiments can be distinguished at present. If the energy of the incident particle is very high, in the range of several MeV, the information about the atoms in the target is drawn from the backscattered particles or from particles recoiled from the target. Such experiments are known as *Rutherford backscattering* (RBS) and *elastic recoil detection* (ERD). In both cases energy resolution is low and low-mass particles are used for the beam. Alternative experiments employ low energy for the incident atoms but a wide range of masses and very high energy resolution for secondary emitted ions. These experiments are summarized as *secondary ion mass spectroscopy* (SIMS). The following sections will give an introduction to the instrumentation and to the applications of the different techniques available at present in the two classes.

18.1 Instrumentation for Atom and Ion Spectroscopy

The instrumentation for atom and ion beam spectroscopy must cover a wide range of specifications. For some techniques very high particle energies in the range of several tens of MeV are required but energy resolution for the beam after its interaction with the sample may not be so crucial. Other techniques like SIMS uses very high mass resolution and sophisticated ion sources but much lower energies for the primary ion beam. Instrumentation consists of the following basic components:

- ion beam sources including ion extractor
- accelerator and beam handling facilities
- energy analyzer and particle detector.

18.1.1 Ion Beam Sources

For the supply of the ion beam three concepts are in use: sources from the gas phase, sources using surface ionization, and liquid metal sources. The most common type of the gas phase source is the *duoplasmatron* where a gas discharge is used to generate the ions. The source operates at a rather high gas pressure of 10^{-1}–1 Pa so that a hot and dense gas plasma is generated. The duoplasmatron can be constructed with a hot or with a cold electrode. It is particularly useful where high beam currents are required. Figure 18.2 displays a schematic drawing. Constructive elements are the magnetic coils and the intermediate electrode to confine the ions, the orifice for extraction of the ion beam and the extractor. As the name says there are two discharge regimes, one between the cathode and the intermediate electrode and a second between the latter and the anode. The orifice has a typical diameter of 300 μm and current densities are of the order of 20–150 mA/cm^2. Cold cathode duoplasmatrons are more suitable for the supply of reactive ions like Cs^+ or O^+.

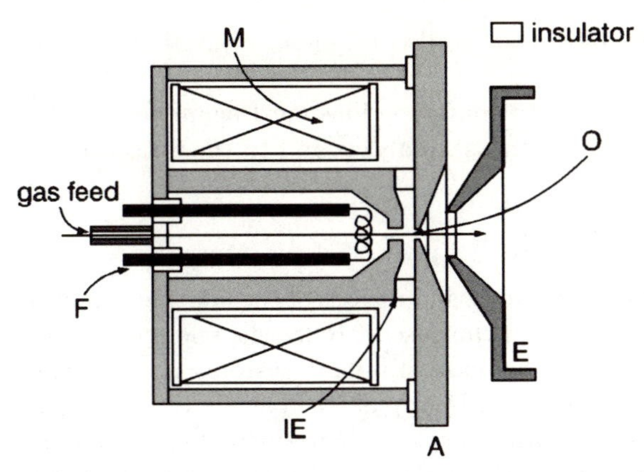

Fig. 18.2. Schematic drawing of an ion gun using a hot cathode plasma discharge; (M: magnetic coils, F: filament (cathode), IE: intermediate electrode with channel, A: anode, O: orifice, E: ion extractor).

The other frequently used type of ion gun is based on surface ionization. Surface ionization means the spontaneous (or more exactly the thermal) ionization of adatoms on a metal surface if the ionization potential of the adatom is much smaller than the work function of the metal. Thus, high work function metals like tantalum, tungsten, or rhenium are appropriate as supporting surfaces and alkali metals are good for the ionization process. Cs^+ ions emitted from a porous tungsten plug are commercial sources. They can be operated with a very small spot size and small energy distribution of only 0.2 eV for the emitted ions. This means a high beam current density and a high beam brightness up to 500 mA/cm^{2}ster as compared to 100–200 mA/cm^{2}ster for a duoplasmatron.

18.1.2 Accelerators and Beam Handling

After the ions are extracted from the source they must be accelerated to the required energy and guided to the target. For acceleration up to about 50 KeV, as required for SIMS experiments, conventional electronics is appropriate. For higher voltages, up to about 400 KeV, voltage multiplications similar to the original suggestion of Cockcroft and Walton are required. Accelerators up to this voltage can still be operated on air. Higher voltages need a van de Graaff generator and the power supply must be insulated by high-pressure gases like SF_6. Acceleration of the ions can be performed in a one stage or in a tandem system. In the latter the particles are first accelerated as negative ions. Half way down the acceleration path the applied voltage and the charge of the ions are reversed and their acceleration continues as positive ions.

Before the particle beam hits the target it must be cleaned of impurities and components with an undesired degree of ionization. For many applications like SIMS monochromatization is also required. This is done with magnetic filters.

After interaction with the target the outgoing ions must again be collected and finally guided to the energy (or mass) analyzer and to the detector.

18.1.3 Analyzers and Detectors

If the constraints on resolutions for energy and masses of the emitted atoms or ions are not so high, as for example in the case of RBS, the energy analyzer and detector can be the same. In this case standard systems are semiconductor surface barrier detectors followed by a pulse height analyzer. Surface barrier detectors are reverse biased p-n junctions where the region of the junction (depletion layer) is very close to the surface of the crystal. High-energy particles penetrating into this junction excite electron–hole pairs which are detected by the following electronic. For light particles a very good linear relation exists between the deposited energy and the number of generated electron–hole pairs. This energy is 3.6–3.7 eV per electron–hole pair for protons and He^+ ions if the particle energy is high enough. High enough means the electronic component must dominate the stopping power. Thus, a 2-MeV He^+ ion creates a charge of about 8×10^{-14} A s and the result of the pulse height analysis is the spectrum of the particles. Resolution in energy is rather low, of the order of 10–20 KeV. For particles with higher masses deviation from linearity occurs. This is known as *pulse height defect* and makes analysis more difficult.

For higher mass and better energy resolution detection must be performed in combination with a mass spectrometer. Very good resolution is obtained with time-of-flight (TOF) systems, even for very high energy ions. An example is shown in Fig. 18.3. The carbon foil at the spectrometer entrance emits

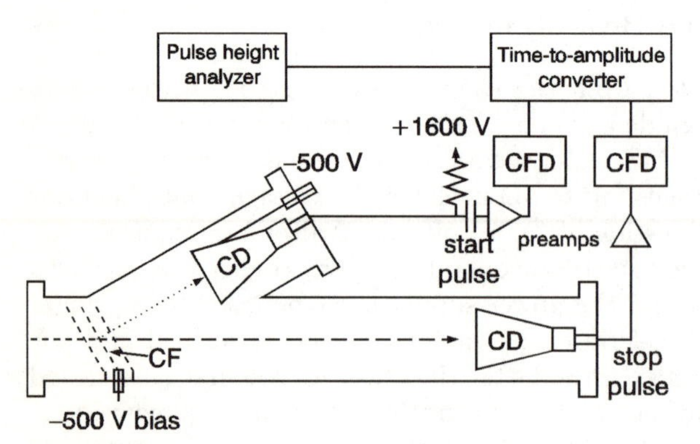

Fig. 18.3. Time-of-flight mass spectrometer for high energy ions; (CF: 2 μg/cm^2 carbon foil, CD: channeltron detector, CFD: constant fraction discriminator), (– – –) ion beam, (. . . .) electrons.

electrons as soon as it is hit by an incoming ion. The electrons are immediately detected with the channeltron which sets the start pulse. The second channeltron sets the stop pulse when the ion is registered. Channeltrons are very fast electron detectors. They were described in Sect. 12.1.3. The high speed of operation is needed since the time of flight for a 2-MeV He^+ ion is only about 500 ps per meter. The fast detection is routed to a fast electronic operating with constant fraction discriminators and time-to-amplitude converters as described in Sect. 16.1.3. This type of electronics allows for a time resolution of 150 ps.

If the particle energy is not so high as, e.g., in SIMS experiments timing of the ions is easier and can be carried out with more precision. A system often employed uses reflection of the ions from a repeller after they have traveled a certain distance and detection after returning the same distance. The energy distribution in the particle beam is deduced from the different times of flight for ions with slightly different energies. Such *reflectron* TOF spectrometers allow for a very good resolution in energy. In contrast to the TOF spectrometer shown in Fig. 18.3 the time ramp in a TOF spectrometer for SIMS is usually triggered by pulsing the ion gun.

If the very high resolution of the TOF spectrometer is not required conventional electrical quadrupole or magnetic sector field mass spectrometers are used. A typical application for these spectrometers are SIMS systems dedicated to depth profile analysis.

18.2 Energy Loss and Penetration of Heavy Particles in Solids

As soon as the ions from the probe beam hit the surface of a solid they start to collide with the atoms and loose energy. The interaction is based on the Coulomb potential. For low particle energies the energy loss is predominately by collision with atoms. For higher energies the efficiency of this process decreases as $1/\epsilon$ and drops to about 1% for particles with 200 KeV/amu where amu stands for *atomic mass units*. For particles with velocities $v > 0.1v_0$ energy loss is predominantly by collision with individual electrons. v_0 is the Bohr velocity (velocity of the 1s electron in hydrogen). It corresponds to the velocity of a 25-KeV proton. If the particle velocity is much larger than the Bohr velocity electrons are striped off so that the degree of ionization of the particle increases.

A quantitative description of the energy loss of the beam particles is performed in terms of the *stopping power* or *specific energy loss* $\mathrm{d}\epsilon/\mathrm{d}x$. Considered with respect to the density of particles n in the target a *stopping cross section* is defined by

$$\sigma = \frac{1}{n}\frac{\mathrm{d}\epsilon}{\mathrm{d}x} . \tag{18.1}$$

This cross section depends mainly on the atomic numbers Z_0 and Z_1 of the beam particles and target atoms, respectively, and on the beam particle energy. The cross sections for frequently used atomic beams like protons or ^{4}He are tabulated for various targets in corresponding handbooks [18.1].

As long as the beam particles do not penetrate into the target atoms the energy loss is well described by a classical Rutherford scattering based on the repulsive Coulomb interaction. Due to the statistical fluctuation in this scattering process the energy of the particles is spread out along its path in the target. This phenomenon is called *energy straggling.*

If the particle energy is high enough to allow for penetration into the target atoms deviation from Rutherford scattering is observed and finally nuclear reactions may occur. The spectroscopy of nuclear reaction products or of emitted γ radiation is another possibility for analyzing the material (*nuclear reaction analysis*, NRA). In any case experimentalists must be alert to nuclear reaction hazards if they use very high energy particle beams.

18.3 Backscattering Spectrometry

Instead of being completely stopped in the target the beam particle can be backscattered after an inelastic collision with an atom of the target. Analysis of backscattered particles is the basis of Rutherford backscattering spectroscopy (RBS). Alternatively, recoiled particles ejected from the target can be analyzed which is the basis of elastic recoil detection spectroscopy (ERD). The two different scattering processes are sketched in Fig. 18.4.

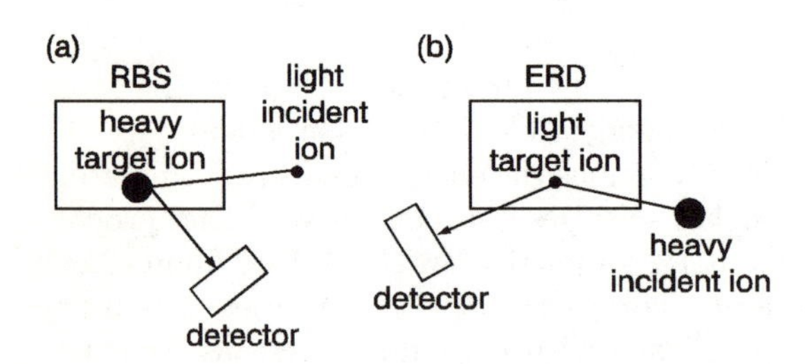

Fig. 18.4. Experimental geometry for Rutherford backscattering (**a**) and elastic recoil detection (**b**).

As long as the beam particles do not penetrate into the target atoms the kinematic of the scattering process is well described by the classical law for the scattering of two particles with masses M_0 and M_1 and a scattering angle θ. From the number and energy of the backscattered atoms the concentration and distribution of the atoms in the target can be determined, at least to the penetration depth of the particle beam.

A possible geometry for backscattering from a thin film is shown in Fig. 18.5a. The film is assumed to consist of two elements A and B which are homogeneously distributed and have relative concentrations m and n. The

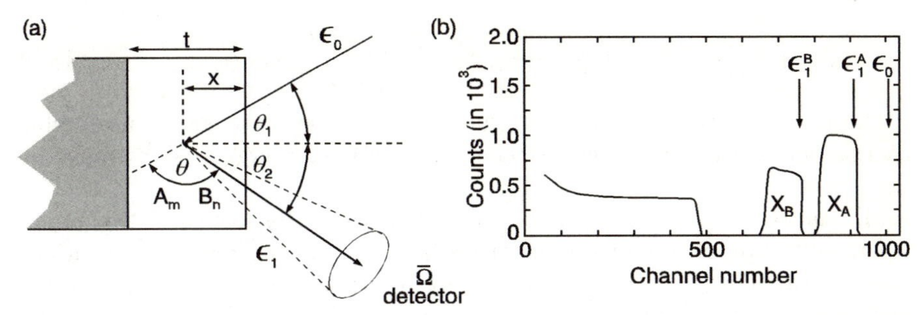

Fig. 18.5. Backscattering from a thin film consisting of two elements A and B with concentrations m and n (**a**) and schematic backscattering spectrum with integrated count rates X_A and X_B (**b**).

corresponding backscattering spectrum is shown schematically in Fig. 18.5b. It is plotted as counts per channel number where the latter is proportional to the particle energy measured after backscattering. The spectrum consists of several flat-topped peaks and a low energy continuum. Each peak corresponds to a particular element in the film. The width of the peaks is a consequence of the energy loss of the particles on their way through the film until they hit their partner for the backscattering process. The elements can be identified from the kinetic factor K of the scattering process defined, for the ith element, by

$$K_i = \frac{\epsilon_i}{\epsilon_0} , \tag{18.2}$$

where ϵ_0 and ϵ_i are the energies of the incident particle and the particle after backscattering, respectively. The latter is taken from the high energy edge of the flat-topped response peak as shown in Fig. 18.5b. To identify the scattering element i a relationship is needed between its mass M_i and K_i. For an incident particle with mass M_0 and a scattering angle $\theta = 180 - (\theta_1 + \theta_2)$ this relation is obtained from energy and momentum conservation.

$$\boxed{K_i = \frac{\epsilon_i}{\epsilon_0} = \left[\frac{(M_i^2 - M_0^2 \sin^2\theta)^{1/2} + M_0 \cos\theta}{M_0 + M_i}\right]^2 .} \tag{18.3}$$

The area density $n_{ai} = n_i d$, where d is the thickness of the layer, can be determined from the integrated count rate X_i by

$$\boxed{n_{ai} = \frac{X_i \cos\theta}{\Phi \overline{\Omega} \sigma_i(\epsilon_0, \theta)} ,} \tag{18.4}$$

if the other parameters of the scattering experiment such as the number of incoming particles per second, Φ, the detector collection angle $\overline{\Omega}$ and the scattering cross section $\sigma_i(\epsilon_0, \theta)$ are known well enough. If the scattering is of Rutherford type (pure Coulomb scattering) then $\sigma_i(\epsilon_0, \theta)$ can be calculated from

$$\sigma_{\mathrm{iR}} = \left(\frac{Z_0 Z_i e^2}{16\pi\varepsilon_0 \epsilon_0} \right)^2 \frac{4\left[(M_i^2 - M_0^2 \sin^2\theta)^{1/2} + M_i \cos\theta\right]^2}{M_i \sin^4\theta (M_i^2 - M_0^2 \sin^2\theta)^{1/2}} \,. \tag{18.5}$$

If the films are thick so that the incident particles have lost considerable energy before they are backscattered an averaged energy $\epsilon < \epsilon_0$ must be used.

Deviation from Rutherford scattering is observed for low particle energies where screening effects of the electrons in the target atoms become important and for high particle energies where interpenetration of the scattering partners occurs and nuclear forces start to contribute to the interaction.

The resolution for the separation of two different masses is obtained from a combination of (18.2) and (18.3) by

$$\delta M_i = \frac{\delta\epsilon}{\epsilon_0} \left(\frac{\mathrm{d}K_i}{\mathrm{d}M_i} \right)^{-1} , \tag{18.6}$$

where $\delta\epsilon$ is the energy resolution of the spectrometer. The relation indicates a much better mass resolution for high energy particles if the resolution of the analyzer is kept constant. Also, the evaluation of $\mathrm{d}K_i/\mathrm{d}M_i$ reveals a better resolution for heavier beam particles.

If the energy resolution for the scattered particles is high enough and the broadening from the straggling phenomenon is low enough even the thickness of the layered structures can be determined from the width of the response peak.

18.3.1 Rutherford Backscattering Spectroscopy

For Rutherford backscattering spectroscopy the description given above holds in principle, except that particle energies are often rather high and thus beyond the Rutherford limit. Most of the RBS experiments are performed either with ^{1}H or with ^{4}He. Two scattering geometries are in use and assigned as the *IBM configuration* and the *Cornell configuration*, respectively. In the former the incident beam, the surface normal of the target and the scattered beam are in the same horizontal plane (as shown in Fig. 18.5). In the latter the incident beam and the surface normal of the target are horizontal as well but the scattered beam is detected immediately below the incident beam. This means the three directions determining the scattering process are not coplanar.

Figure 18.6 shows the RBS spectra from a 50-bilayer film of Fe/Mo on a sapphire substrate. The total thickness of the film was 380 nm. The upper spectrum was excited with 1.9-MeV He atoms with a resolution of $\delta\epsilon =$ 15 KeV. The responses from the Mo and from the Fe atoms overlap but can be disentangled to yield an area density of $n_{\mathrm{aMo}} = 114 \times 10^{16}$ atoms/cm^2 and $n_{\mathrm{aFe}} = 142 \times 10^{16}$ atoms/cm^2 for the two atoms. As shown in the lower spectrum the resolution can be improved by increasing the energy of the incident He particles to 3.8 MeV. At this energy the scattering of Mo and Fe

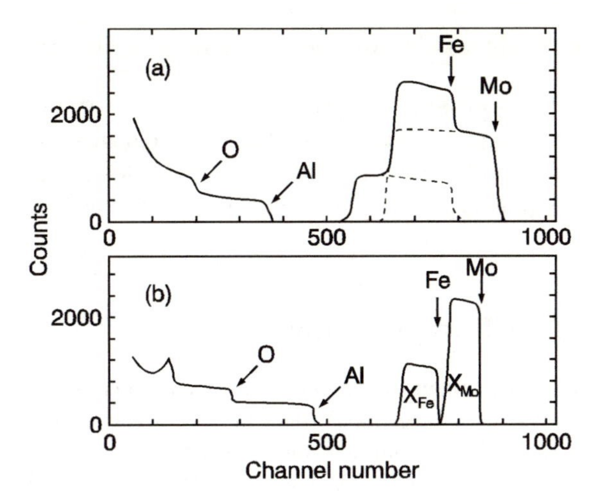

Fig. 18.6. Backscattering spectrum for 50 bilayers of FeMo on Al_2O_3 after excitation with 1.9-MeV ^{4}He (**a**) and after excitation with 3.8-MeV ^{4}He (**b**); adapted from [18.1].

is still Rutherford. Note that the channel numbers are the same for the two experiments but the energies assigned to the channels are about a factor of two larger in the second spectrum. The peak at channel 130 in the second spectrum is from a resonance in the He-O scattering cross section.

18.3.2 Elastic Recoil Detection Spectroscopy

The alternative method of depth profiling is elastic recoil detection spectroscopy. It is particularly useful for the analysis of light atoms in the target with masses up to $Z \approx 9$. Since the mass of the incident particle must be larger than the mass of the ejected particle heavier atoms are often used as a projectile and the beam energies are of the order of 1 MeV/amu. As far as the kinematic of the energy and momentum transfer, the stopping power, straggling, or depth profiling is concerned the same fundamental processes are valid for ERD and for RBS, respectively, except that in the case of the former the kinematic factor and the energy loss must be considered for the ejected particle.

The intermediate steps in an ERD process are best described by an example. A 28-MeV Si^{+6} projectile hits the surface of a Si_3N_4 sample which contains hydrogen. The Si probe looses energy from ϵ_0 to ϵ_0' determined by the stopping power $S_{Si} = 480\,eV/Å$. As it hits a hydrogen atom the latter is recoiled with an energy $\epsilon_H = K\epsilon_0 \approx 2.8\,MeV$ since $K \approx 0.1$ for forward scattering. On traveling through the target the hydrogen atom looses energy $\epsilon_H \to \epsilon_H'$ according to a stopping power $S_H \approx 3\,eV/Å$. Finally it is detected and energy analyzed by a surface barrier detector. In front of the detector a particle filter (range foil) is usually mounted to prevent the primary particles from hitting the detector and thus spoiling the signal. The range foil must have a very high stopping power for the heavy primary particle and a low stopping power for the light-ejected particle. Appropriate range foils are thin mylar or aluminum sheets.

Composition analysis and depth profiling can be performed with ERD in a similar way as with RBS. Optimum conditions for the resolution in mass separation and depth profile can be obtained for a sophisticated selection of energy and mass of the primary particle and of the material and thickness of the range foil.

18.4 Secondary Ion Mass Spectroscopy

Secondary ion mass spectroscopy has been developed as an extremely sensitive tool for the elemental analysis of material. As the name indicates secondary ions are generated from a primary particle beam hitting the sample surface. The ions are subsequently analyzed with respect to their mass and energy. The difference to the systems described previously comes from the application of lower energies for the primary particles. These energies are of the order of 0.5–50 KeV and allow the use of mass spectrometers for the analysis of the emitted ions, thus leading to a dramatic increase in resolution. Excitation sources and elemental analyzes are not restricted to a special group of materials. The full periodic system from hydrogen to uranium can be used as a primary ion beam and the full periodic system can be analyzed in the target. The method is extremely sensitive. Detection limits can be as low as 10^{-9} and 10^4–10^5 sputtered ions are enough for a signal.

There are two major disadvantages of the method. The first is the inherent destructive character of the technique which is not so serious because of the extremely small amount of masses needed for the analysis. The second disadvantage is more serious. A quantitative analysis can be very difficult as the emission characteristics do no only differ for the various elements but also depend on their chemical environment and on the matrix in general.

The basic process in SIMS is the sputtering of target atoms. The incident ions interact with the atoms and transfer energy by nuclear collision. The atoms are displaced and a small fraction receives enough energy to leave the sample as a neutral or as a charged particle. Computer simulation and experiments revealed an emission depth of the order of 0.3 nm. This means the primary application of SIMS is surface analysis.

Recent developments in SIMS are concerned with high resolution imaging of sample surfaces. This imaging can be performed with ion optics or by scanning techniques. Lateral resolution down to 20 nm is possible.

Another branch of the development is directed towards a discrimination between static SIMS and dynamic SIMS. The intention in the first case is to reduce the destructive nature of the method to a minimum. This means doses of primary ions are kept extremely low, of the order of 10^{13} ions/cm^2. In the case of dynamic SIMS the idea is opposite. High doses of the order of 10^{17} ions/cm^2 and more are used. The sample is eroded with time to a well defined depth. This allows a depth profile for distributed elements to be determined. Again depth resolution can be very high, of the order of 1–2 nm. Together with the lateral resolution this allows a 3-D imaging of the distribution of elements. The dynamical range of the depth profiling is very large. Concentrations of $5\times10^{22} - 5\times10^{12}$ atoms/cm^3 can be studied. Figure 18.7 gives an example. The depth profile of ^{11}B in Si is shown after 10^{16} atoms/cm^2 of ^{11}B were implanted with 70 KeV. The figure demonstrates a dynamical range of six orders of magnitude for the detection of the boron atoms in the silicon matrix.

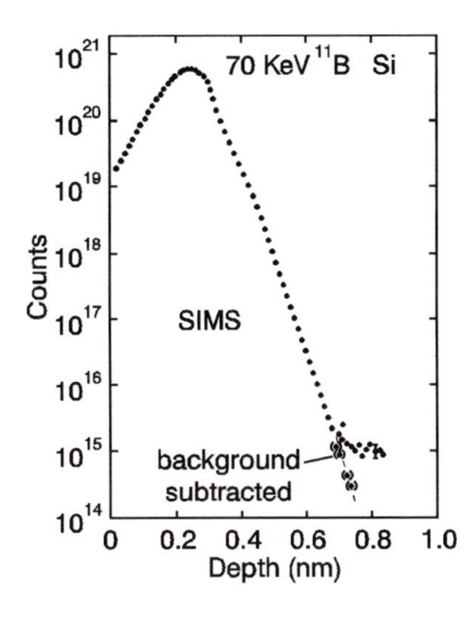

Fig. 18.7. Depth profile for boron in silicon after random implantation with 70 KeV. The primary ions are 12-KeV O_2^+; after [18.2].

Problems

18.1 Show that the Rutherford velocity is equal to the velocity of a 25-KeV proton.

(Purpose of exercise: remember that v_0 is a critical value in atomic scattering spectroscopy.)

18.2 Show that the kinetic factor K for Rutherford backscattering is given by (18.3).

(Purpose of exercise: derive the most fundamental relation for backscattering spectroscopy.)

18.3* Calculate the kinematic factor K for an elastic recoil detection experiment for a heavy mass M_0 with energy ϵ_0 and a light mass M_i at rest.

(Purpose of exercise: extend the information from the text to another fundamental technique in ion scattering spectroscopy.)

Appendix

A. To Chapter 1, Introduction

Table A.1. Fundamental physical constants.

Name		Value, in SI	SI units[a)]	Other units
Planck constant/2π	$\hbar$	1.0546E–34	VAs^2	6.582 E–16 eVs
Boltzmann constant	k_B	1.381E–23	VAs/K	8.616E–5 eV/K
Elementary charge	e	1.602E–19	As	4.77E–10 cgs
Dielectric constant of vacuum	ε_0	8.854E–12	As/Vm	
Permeability of vacuum	μ_0	1.257E–06	Vs/Am	
Mass of free electron	m_0	9.110E–31	VAs^3/m^2	9.110E–28 g
Mass of protons	M_p	1.673E–27	VAs^3/m^2	1.673E–24 g
Mass of neutron	M_n	1.675E–27	VAs^3/m^2	1.675E–24 g
Mass of muon	m_μ	1.840E–28	VAs^3/m^2	1.8400E–25 g
Bohr radius $4\pi\hbar^2\varepsilon_0/m_0e^2$	r_B	0.529E–10	m	
Rydberg $m_0e^4/32\pi^2\varepsilon_0^2\hbar^2$	Ry	2.180E–18	VAs	13.607 eV
Radius of electron $e^2/4\pi\varepsilon_0 m_0 c_0^2$	r_0	2.82E–15	m	
Bohr magneton	μ_B	9.273E–24	Am^2	VAs/T
Nuclear magneton	μ_n	5.051E–27	Am^2	VAs/T
Magnetic moment of proton	μ_p	1.409E–26	Am^2	VAs/T
Magnetic moment of muon	μ_μ	4.49 E–26	Am^2	VAs/T
Magnetic moment of neutron	μ_{ne}	–9.663E–27	Am^2	VAs/T
Magnetogyric ratio of				
electron	γ_e	1.76E+11	m^2/s^2V	1/sT[b)]
proton	γ_p	2.675E+8	m^2/s^2V	1/sT
muon	γ_μ	1.354E+8	m^2/s^2V	1/sT
Hyperfine constant for H	K	9.412E–25	VAs	$\doteq$1420 MHz
Fine structure constant $e^2/4\pi\varepsilon_0\hbar c_0$	α	7.297E–3	–	(= 1/137.039)
1 Curie	Ci	3.7E+10	1/s	

a) V, A, m, s are used as SI units for convenience to application.
b) Note: multiplication of γ with B (in Tesla) yields angular frequencies, not Hz.

Table A.2. Relations between energy units.

	Ws[a)]	eV	cm^{-1}	K	cal	Ry[b)]
Ws	1	6.242E18	5.031E22	7.245E22	0.239	4.587E17
eV	1.602E–19	1	8060	1.116E4	3.829E–20	0.0735
cm^{-1}	1.988E–23	1.240E–4	1	1.44	4.75E–24	9.12E–6
K	1.381E–23	8.617E–5	0.694	1	3.299E–24	6.333E–6
cal	4.186	2.611E19	2.103E23	3.031E23	1	1.920E18
Ry	2.18E–18	13.607	1.096E5	1.579E5	5.208E–19	1

[a)] The energy in erg is obtained by multiplication of the value in Ws with 10^7.
[b)] The energy in Hartree is obtained by multiplication of the value in Ry with 1/2.

B. To Chapter 2, Electromagnetic Radiation

B.1 The Maxwell Equations

The Maxwell equations in SI units are

$$1.\quad \text{curl}\,\boldsymbol{H} = \boldsymbol{j} + \frac{\partial \boldsymbol{D}}{\partial t}\,, \qquad 2.\quad \text{curl}\,\boldsymbol{E} = -\frac{\partial \boldsymbol{B}}{\partial t}\,, \tag{B.1}$$

$$3.\quad \text{div}\,\boldsymbol{D} = \varrho\,, \qquad 4.\quad \text{div}\,\boldsymbol{B} = 0\,, \tag{B.2}$$

where the five vectors $\boldsymbol{E}$, $\boldsymbol{D}$, $\boldsymbol{H}$, $\boldsymbol{B}$, $\boldsymbol{j}$ and the polarization $\boldsymbol{P}$ are connected by the additional relations

$$\boldsymbol{D} = \varepsilon\varepsilon_0\boldsymbol{E}, \qquad \boldsymbol{B} = \mu\mu_0\mathbf{H}, \qquad \boldsymbol{j} = \sigma\boldsymbol{E}, \qquad \boldsymbol{P} = (\varepsilon - 1)\varepsilon_0\boldsymbol{E}\,.$$

Elimination of $\boldsymbol{E}$ or $\boldsymbol{B}$ from (B.1) and (B.2) leads for zero charge ($\varrho = 0$) and nonconducting material ($\sigma = 0$) to the wave equations for the electric and the magnetic field:

$$\Delta\boldsymbol{E} = \frac{\varepsilon\mu}{c_0^2}\frac{\partial^2\boldsymbol{E}}{\partial t^2}\,, \qquad \Delta\boldsymbol{B} = \frac{\varepsilon\mu}{c_0^2}\frac{\partial^2\boldsymbol{E}}{\partial t^2}\,. \tag{B.3}$$

A related equation can be obtained for conducting material with conductivity σ.

$$\Delta\boldsymbol{E} = \frac{\varepsilon\mu}{c_0^2}\frac{\partial^2\boldsymbol{E}}{\partial t^2} + \mu\mu_0\sigma\frac{\partial\boldsymbol{E}}{\partial t}\,. \tag{B.4}$$

It becomes particularly simple if the ohmic current σE strongly exceeds the displacement current.

B.2 Potentials for the Electromagnetic Field.

In the static case $\boldsymbol{E}$ and $\boldsymbol{B}$ can be derived from a scalar potential Φ and from a vector potential $\boldsymbol{A}$ by

$$\boldsymbol{E} = -\text{grad}\,\Phi\,, \qquad \boldsymbol{B} = \text{curl}\,\boldsymbol{A}\,, \tag{B.5}$$

where $\Phi(\boldsymbol{r})$ is determined except for an arbitrary constant and $\boldsymbol{A}$ is determined except for an additive function grad χ where χ is an arbitrary function of $\boldsymbol{r}$.

Rather than solving the Maxwell equations for $\boldsymbol{E}$ and $\boldsymbol{B}$ for a given distribution of charges $\varrho(\boldsymbol{r})$ and currents $\boldsymbol{j}(\boldsymbol{r})$ it is often more convenient to calculate the potentials Φ and $\boldsymbol{A}$ directly from these distributions and then evaluate $\boldsymbol{E}$ and $\boldsymbol{B}$ with the help of the relationships in (B.5). The evaluation of Φ and A is possible using the Poisson equations

$$\Delta\Phi = -\varrho/\varepsilon\varepsilon_0, \qquad \Delta\boldsymbol{A} = -\mu\mu_0\boldsymbol{j} \tag{B.6}$$

or their integral forms

$$\boxed{\Phi(\boldsymbol{r}) = \frac{1}{4\pi\varepsilon\varepsilon_0}\int\frac{\varrho(\boldsymbol{r'})}{|\boldsymbol{r}-\boldsymbol{r'}|}\mathrm{d}^3x', \quad \boldsymbol{A}(\boldsymbol{r}) = \frac{\mu\mu_0}{4\pi}\int\frac{\boldsymbol{j}(\boldsymbol{r'})}{|\boldsymbol{r}-\boldsymbol{r'}|}\mathrm{d}^3x'\,,} \tag{B.7}$$

where d^3x' is the three-dimensional volume differential of r'-space.

B.3 Expansion of the Potential in Multipole Moments

The integral for the scalar potential in (B.7) can be evaluated by an expansion in spherical harmonics. The potential is then represented as

$$\boxed{\Phi(\boldsymbol{r}) = \sum_{l=0}^{\infty}\sum_{m=-l}^{l}\frac{4\pi}{2l+1}P_{lm}\frac{Y_{lm}(\theta,\phi)}{r^{l+1}}\,,} \tag{B.8}$$

where P_{lm} are the multipole moments of the charge distribution given by

$$P_{lm} = \int Y^*_{lm}(\theta',\phi')r'^l\varrho(\boldsymbol{r'})\,\mathrm{d}^3x'\,. \tag{B.9}$$

The lowest order multipole moments can be expressed by the total charge Q (monopole moment, $l = m = 0$), the the dipole moment $\boldsymbol{P}_D$ ($l = 1, m = 0, \pm 1$)

$$\boldsymbol{P}_D = \int \boldsymbol{r'}\varrho(\boldsymbol{r'})\,\mathrm{d}^3x', \tag{B.10}$$

and the quadrupole moment (quadrupole tensor)

$$(P_Q)_{ij} = \int (3x'_ix'_j - r'^2\delta_{ij})\varrho(\boldsymbol{r'})\mathrm{d}^3x'. \tag{B.11}$$

For rotational symmetry of the charges around the z axis the quadrupole tensor has only one independent component $(P_Q)_{33}$:

$$(P_Q)_{33} = P_Q = \int (3z'^2 - r'^2)\varrho(\boldsymbol{r'})\mathrm{d}^3x'\,. \tag{B.12}$$

Note that this equation immediately transforms to (2.20) for a discrete charge distribution. The multipole moments can also be used to expand the potential into a Taylor series in cartesian coordinates. In this case we obtain the relation

$$\Phi(\boldsymbol{r}) = \frac{Q}{r} + \frac{\boldsymbol{P}_D \boldsymbol{r}}{r^3} + \frac{1}{2}\sum_{i,j}(P_Q)_{ij}\frac{x_i x_j}{r^3} + \dots \, . \tag{B.13}$$

B.4 Time-Retarded Potentials

If the charge distribution and the current density is time-dependent the equations from Appendix B.2 become more complicated. This is due to the finite propagation velocity of light even in vacuum. The field which exists at point $\boldsymbol{r}$ at time t was generated from a charge distribution $\varrho(\boldsymbol{r}', t')$ which appears retarded to t by the propagation time $\tau = (\boldsymbol{r} - \boldsymbol{r}')/c$. Thus, the expressions in (B.7) can still be used but $\varrho(\boldsymbol{r}')$ and $\boldsymbol{j}(\boldsymbol{r}')$ have to be replaced by their time-dependent and retarded quantities $\varrho(\boldsymbol{r}', t - |\boldsymbol{r} - \boldsymbol{r}'|/c)$ and $\boldsymbol{j}(\boldsymbol{r}', t - |\boldsymbol{r} - \boldsymbol{r}'|/c)$. The geometrical configuration is shown in Fig. B.1. Also, in general the electric field is no longer obtained from the gradient of the potential in the simple form of (B.5). For the time-dependent charges and currents these equations change to

$$\boldsymbol{E} = -\text{grad}\,\Phi - \frac{\partial \mathbf{A}}{\partial t} \qquad\qquad \boldsymbol{B} = \text{curl}\,\boldsymbol{A}. \tag{B.14}$$

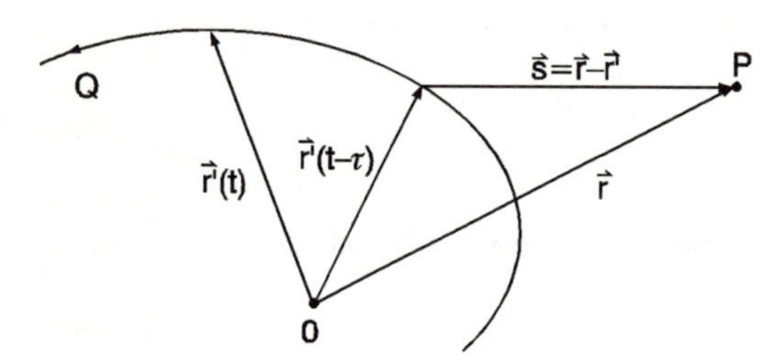

Fig. B.1. Configuration for the retardation of the potential at point P with respect to the position of the charge Q.

Their new forms implicate that we are no longer quite as free to choose the additive functions for the vector potentials $\boldsymbol{A}$ as grad $\chi(\boldsymbol{r}, t)$. In fact we can still choose any such function but we have to recalibrate Φ simultaneously with $-\partial\chi/\partial t$ as it is immediately proved by inserting the renormalized potentials into the (B.14). It is on the other hand straightforward to show that the fields from (B.14) satisfy the second and fourth Maxwell equation. From the first and third equation which contain the source terms ϱ and $\boldsymbol{j}$ both types of sources turn out to be needed for the evaluation of the potentials Φ and $\boldsymbol{A}$. Only for the special choice of the calibration function χ for which the potentials Φ and $\boldsymbol{A}$ satisfy the Lorentz condition

$$\text{div}\,\boldsymbol{A} + \frac{1}{c^2}\frac{\partial \Phi}{\partial t} = 0\,, \tag{B.15}$$

Φ and $\boldsymbol{A}$ are decoupled and can be evaluated independently as it is implicated in (B.7) by using the time-retarded forms of ϱ and $\boldsymbol{j}$. This is known as the Lorentz gauge. In this special gauge the differential equations for $\boldsymbol{A}$ and Φ have the simplified and decoupled form

$$\Delta \boldsymbol{A} - \frac{1}{c_0^2}\frac{\partial^2 \boldsymbol{A}}{\partial t^2} = -\mu\mu_0 \boldsymbol{j} \qquad \text{and} \qquad \Delta \Phi - \frac{1}{c_0^2}\frac{\partial^2 \Phi}{\partial t^2} = -\frac{\varrho}{\varepsilon\varepsilon_0} \,. \tag{B.16}$$

B.5 Radiation from an Arbitrarily Accelerated Charge

The radiation from a charge moving with an arbitrarily varying velocity can also be calculated from the formalism described in the above section by using the relations in (B.7) with retarded charges and currents. The field observed at point $\boldsymbol{r}$ and time t is determined by the position of the charge at the retarded time $t' = t - |\boldsymbol{r} - \boldsymbol{r}'(t_r)|/c_0 = t - s/c_0$. The symbols have the same meaning as in Fig. B.1 from the above section. This problem is discussed in standard textbooks for electrodynamics like *Jackson* [2.3] or *Feynman* [2.2]. The evaluation yields for Φ and $\boldsymbol{A}$ the relation

$$\begin{aligned} \Phi(\boldsymbol{r},t) &= \frac{Q}{4\pi\varepsilon\varepsilon_0|\boldsymbol{r}-\boldsymbol{r}'|}\frac{1}{(1-v(t')/c_0)} \quad \text{and} \\ \boldsymbol{A}(\boldsymbol{r},t) &= \frac{\mu\mu_0 Q}{4\pi|\boldsymbol{r}-\boldsymbol{r}'|}\frac{\boldsymbol{v}(t')}{(1-v(t')/c_0)} \,. \end{aligned} \tag{B.17}$$

These are the *Lienard–Wiechert* retarded potentials for a single charge Q with coordinates $\boldsymbol{r}'(t')$. $\boldsymbol{v}$ is the particle velocity $\mathrm{d}\boldsymbol{r}'(t')/\mathrm{d}t'$ at the retarded time t'. The Lienard–Wiechert potentials enable the calculation of the electric and magnetic fields as a function of $\boldsymbol{r}$ and $\boldsymbol{r}'(t)$. For large distances between the charge and the field point the vector $\boldsymbol{s}$ giving the distance and direction between charge and field point at the retarded time may be replaced by the vector $\boldsymbol{r}'(t) - \boldsymbol{r}$ at the actual time. Then the emitted fields are given by

$$\boldsymbol{E}(\boldsymbol{r},t) = \frac{\mu\mu_0 Q}{4\pi|\boldsymbol{r}-\boldsymbol{r}'(t)|}(\boldsymbol{e}_s \times (\boldsymbol{e}_s \times \boldsymbol{a})) \tag{B.18}$$

and

$$\boldsymbol{B}(\boldsymbol{r},t) = \boldsymbol{e}_s \times \boldsymbol{E}/c \,. \tag{B.19}$$

$\boldsymbol{a}$ is the acceleration $\mathrm{d}^2\boldsymbol{r}'/\mathrm{d}t^2$ of the charge and $\boldsymbol{e}_\mathrm{s}$ is the unit vector in $\boldsymbol{s}$ direction. Within the above approximation $\boldsymbol{a}$ is taken at the actual time t and $\boldsymbol{s}$ is assumed $\parallel \boldsymbol{r}$. The approximation is only good for non-relativistic particles. From (B.18) a complete polarization of the electric field in the plane given by the vectors $\boldsymbol{a}$ and $\boldsymbol{s}$ is evident. This result is very important for the emission of synchrotron radiation as will be discussed in the next chapter.

Since

$$|\boldsymbol{e}_s \times (\boldsymbol{e}_s \times \boldsymbol{a})| = a \sin\theta \,,$$

where θ is the angle between $\boldsymbol{s}$ and the acceleration $\boldsymbol{a}$ the emission is strongest in the direction perpendicular to $\boldsymbol{a}$. The pointing vector $\boldsymbol{I}$ for an actual distance s in direction $\boldsymbol{e}_s$ is

$$\boldsymbol{I} = \frac{\mu\mu_0 Q^2 \sin^2\theta}{16\pi^2 c s^2} a^2 \boldsymbol{e}_s \ . \tag{B.20}$$

B.6 Fourier Transformations

We review some more results on FTs and properties of Fourier pairs. This will be limited to the integral version of the Fourier theorem. For more details the book of *Brigham* [2.4] or other special textbooks on FT may be consulted.

The Fourier theorem is very often written in the form

$$E(t) = \frac{1}{2\pi} \int \mathrm{e}^{\mathrm{i}\omega t}\, \mathrm{d}\omega \int E(t')\mathrm{e}^{-\mathrm{i}\omega t'}\, \mathrm{d}t' \ , \tag{B.21}$$

where the second integral represents the Fourier transform $E(\omega)$ of the time function $E(t)$. There is unfortunately some discrepancy in the various textbooks with respect to the factor $1/2\pi$. It may be either included into $E(\omega)$ or into the back-transformation to $E(t)$. Alternatively it may be split into two factors of $1/\sqrt{2\pi}$, each of them in front of the two integrals. The advantage of the latter formulation is the symmetry for the transformation with respect to the back-transformation. The way we were using the theorem in Sect. 2.3 is even more convenient. The factor $(1/2\pi)$ was included into the differential $\mathrm{d}\omega$ yielding the differential $\mathrm{d}f$.

Proving the Fourier theorem as it is written in (B.21) is straightforward by interchanging the two integrations and considering the integral representations for the δ function from Appendix B.7.

Fourier pairs exist for a wide variety of functions. A sufficient condition for their existence is the absolute integrability of the function but even functions which are not integrable like $E(t) = \mathit{constant}$ can have a Fourier transform. More relaxed constraints are Dirichlet's conditions which require monotonic behavior in a finite number of intervals and left and right continuity at the positions of discontinuity. For a more detailed discussion of the existence problem special books on Fourier transformation must be checked.

Fourier pairs are subjected to several constraints which are quite useful to remember and in general very easy to prove. We summarize a few of them in Table B.1. Note that the intensity spectrum $E(f)E(f)^*$ is real, even, and positive in all cases where the time spectrum is real.

Another interesting list of relations holds between Fourier pairs of composed functions. These relations are likewise easy to prove and helpful to remember. Some of them are listed in Table B.2 where $E(t)$ and $E(f)$ are assumed to be Fourier pairs.

Finally, Table B.3 gives a listing of Fourier pairs for some representative functions.

B.7 The δ Function

The δ function is one of the most important mathematical tools in spectroscopy and spectroscopy-related calculations. The reason for this is evident

Table B.1. Properties of Fourier transforms.

Time function $E(t)$	Frequency function $E(f)$
Real	Real part even, imaginary part odd
Imaginary	Real part odd, imaginary part even
Real part even, imaginary part odd	Real
Real part odd, imaginary part even	Imaginary
Real and even	Real and even
Real and odd	Imaginary and odd
Complex and even	Complex and even
Complex and odd	Complex and odd

Table B.2. Fourier pair relations for composed functions.

Type	Time function	Frequency function
Summation	$E_1(t) + E_2(t)$	$E_1(f) + E_2(f)$
Time scaling	$E(at)$	$(1/\|a\|)E(f/a)$
Time shift	$E(t - t_0)$	$E(f)e^{-i2\pi f t_0}$
Modulation	$E(t)e^{i2\pi f_0 t}$	$E(f - f_0)$
Multiplication	$E_1(t)E_2(t)$	$E_1(f) * E_2(f)$(convolution)

from our findings in Sect. 2.3 that the FT of an idealized and continuously propagating plane wave is a δ function with the two components $\delta(f - f_0)$ and $\delta(f + f_0)$. In simple terms the δ function is zero everywhere except at the position where the argument is zero. At this single point the function becomes infinite to an extent that its integral is one as long as the point

Table B.3. Fourier pairs for some selected functions.

Function	$E(t)$	Range	$E(f)$
Constant	A	$-\infty$ to ∞	$A\delta(f)$
Cosine	$2\cos 2\pi f_0 t$	$-\infty$ to ∞	$\delta(f - f_0) + \delta(f + f_0)$
Rectangular pulse	A	$\|t\| \leq T_0$	$2AT_0(\sin 2\pi T_0 f)/(2\pi T_0 f)$
	0	elsewhere	
Truncated cosine	$A\cos 2\pi f_0 t$	$\|t\| \leq T_0$	$A^2 T_0[Q(f + f_0) + Q(f - f_0)]$
	0	elsewhere	$Q(f) = (\sin 2\pi T_0 f)/(2\pi T_0 f)$
Exponential decay	$\exp -\gamma\|t\|$	$-\infty$ to ∞	$(2\gamma^2)/[\gamma^2 + (2\pi f)^2]$
Gauss line	$\gamma/\sqrt{\pi}\exp(-\gamma t^2)$	$-\infty$ to ∞	$\exp(-\pi^2 f^2/\gamma^2)$
Damped wave, compl.	$\exp(-\gamma t/2)\exp i\omega_0 t$	$-\infty$ to ∞	$1/[\gamma/2 + i2\pi(f - f_0)]$
Damped wave, real	$\exp(-\gamma t/2)\cos\omega_0 t$	$-\infty$ to ∞	$2/(\gamma + 2i\omega_-) + 2/(\gamma + 2i\omega_+)$ with $\omega_{-,+} = \omega \mp \omega_0$

of divergence is included in the integration. Actually this can be taken as a definition for the δ function. More exactly speaking the δ function is a distribution to be described by a series of functions. The limit of the series must have the properties described above. Accordingly, we have many choices for representing a δ function.

B.7.1 Representations of the δ Function

In the following we give four convenient representations of the δ function.

a) $g_a(x)$ is a manifold of rectangular pulses of the form

$$g_a(x) = 1/a \quad \text{for} \quad -a/2 < x < a/2 \quad \text{and zero elsewhere.}$$

In the limit $a \to 0$ this function becomes infinite for $x = 0$ and zero everywhere else. Since its integral is always one it represents $\delta(0)$.

b) The manifold of functions

$$g_a(x - x_0) = \sqrt{\frac{a}{\pi}} \mathrm{e}^{-a(x-x_0)^2}$$

with variable a represent $\delta(x - x_0)$ in the limit $a \to \infty$ since in this case it is zero everywhere except at $x = x_0$ where it becomes infinite and its integral is normalized to 1.

c) Similarly, $\delta(x - x_0)$ is represented by the manifold of the functions

$$g_a(x - x_0) = \frac{\sin a(x - x_0)}{\pi(x - x_0)} \quad \text{for} \quad a \to \infty$$

and the functions

d) $$g_a(x - x_0) = \frac{a/\pi}{a^2 + (x - x_0)^2} \quad \text{for} \quad a \to 0.$$

As a consequence of these representations it is straightforward to show some integral representations for the δ function.

$$\begin{aligned}
\int \cos 2\pi f t \, \mathrm{d}t &= \delta(f) = 2\pi\delta(\omega) \\
\int \mathrm{e}^{-2\pi \mathrm{i} f t} \, \mathrm{d}t &= \delta(f) \\
\int \mathrm{e}^{2\pi \mathrm{i} f t} \, \mathrm{d}f &= \delta(t) \quad \text{equivalent to} \quad \int \mathrm{e}^{\mathrm{i}\omega t} \, \mathrm{d}\omega = 2\pi\delta(t) \, ,
\end{aligned} \tag{B.22}$$

and finally

$$\mathrm{i}\pi \int f(x)\delta(x - a)\mathrm{d}x = \mathrm{P} \int \left(\frac{f(x)\mathrm{d}x}{x - a} \right) - \lim_{\eta \to 0} \int \frac{f(x)\mathrm{d}x}{x - a + \mathrm{i}\eta} \, . \tag{B.23}$$

The latter equation is known as the *Dirac relation.*

B.7.2 Some Properties of the δ Function

We summarize some of the most important properties of the δ function. The limits of the integration are always considered to include the position of the divergence. Otherwise the integrals are zero.

1) $\int_{-b}^{b} g(x)\delta(x-a)\,\mathrm{d}x = g(a)$.
2) $\delta(ax) = (1/|a|)\delta x$; consequently: $\delta(x-a) = \delta(a-x)$.
3) $\int g(x)\delta'(x-a)\mathrm{d}x = -g'(a)$.
4) $\delta[g(x)] = [1/g'(x)]\delta(x-a)$; provided $g(a) = 0$;
5) or, more generally,
   $\delta[g(x)] = \sum_{i=1}^{n} \delta(x-x_i)/|g'(x_i)|$ where $g(x)$ is a scalar function with zeros at $x_1, \ldots, x_n$ and derivatives $g'(x_i)$ nonzero at these positions.
6) Immediately derived from (5) is
   $\int f(\boldsymbol{r})\delta[g(\boldsymbol{r})]\mathrm{d}^3x = \int g(\boldsymbol{r})[f(\boldsymbol{r})/|\nabla g(\boldsymbol{r})|]\mathrm{d}S(\boldsymbol{r})$.

C. To Chapter 3, Light Sources with General Application

C.1 Moments of Spectral Lines

The nth moment of a spectral distribution $I(\omega)$ is defined as

$$\sigma^n = \int \omega^n I(\omega)\,\mathrm{d}\omega. \tag{C.1}$$

The first moment is the *mean* or *expectation* value ω_0. The second moment with respect to the expectation value is the *variance* defined as

$$\sigma^2 = \int (\omega-\omega_0)^2 I(\omega)\,\mathrm{d}\omega. \tag{C.2}$$

The square route of the variance is the mean deviation from the expectation value.

C.2 Convolution of Spectral Lines

A spectral line as observed from an experiment may have a different shape as compared to the emitted spectrum. This is due to deviations from ideal transmission by the instrument. If the input to the spectrometer is a δ function of the form $\delta(\omega-\omega_0)$ the output may be a function $G(\omega-\omega_0)$. In this case the observed shape $g_{\mathrm{con}}(\omega)$ of a spectral line with lineshape $g(\omega)$ is given by the convolution between the *transfer functions* $G(\omega)$ and $g(\omega)$

$$g_{\mathrm{con}}(\omega') = \int G^*(\omega-\omega')g(\omega)\,\mathrm{d}\omega\,. \tag{C.3}$$

The line width resulting from the convolution can be calculated analytically in some simple cases. A straightforward calculation shows in particular that the convolution of two Gaussian functions with FWHM γ_1 and γ_2 results in a Gaussian function of width γ with

$$\gamma = \sqrt{\gamma_1^2 + \gamma_2^2}\,. \tag{C.4}$$

Similarly, for two Lorentzian functions the resulting line width is

$$\gamma = \gamma_1 + \gamma_2\,. \tag{C.5}$$

The convolution between a Gaussian line and a Lorentzian line is a Voigtian line which cannot be represented analytically. If γ_L and γ_G are the widths for the Lorentzian and for the Gaussian lines, respectively, the resulting line width is approximately given by

$$\gamma = 0.5346\gamma_\mathrm{L} + \sqrt{0.2156\gamma_\mathrm{L}^2 + \gamma_\mathrm{G}^2}. \tag{C.6}$$

An approximate description for the Voigtian line was given in [3.16] as a superposition of four generalized Lorentzian lines.

Remember the difference between convolution and correlation. The correlation between $G(\omega)$ and $g(\omega)$ is

$$g_\mathrm{cor}(\omega') = \int G^*(\omega + \omega')g(\omega)\,\mathrm{d}\omega. \tag{C.7}$$

C.3 Fano Lines

The physical meaning of the parameters for the Fano lines can be obtained from the original work of *Fano* [3.15] in which the energy loss of electrons scattered in He was studied.

The condition where Fano's theory applies is shown in Fig. C.1. Transitions from a ground state g are possible to either a single discrete state s with energy ϵ_s and described by a wave function φ_s or to a continuum of states ϵ' described by wave functions $\psi_{\epsilon'}$. The discrete state and the continuum are interacting which is described by a Hamiltonian $\mathbf{H}_v$ with matrix elements $V_{s\epsilon'} = \langle\psi_{\epsilon'}|\mathbf{H}_v|\varphi_s\rangle$. From this interaction the discrete state is renormalized

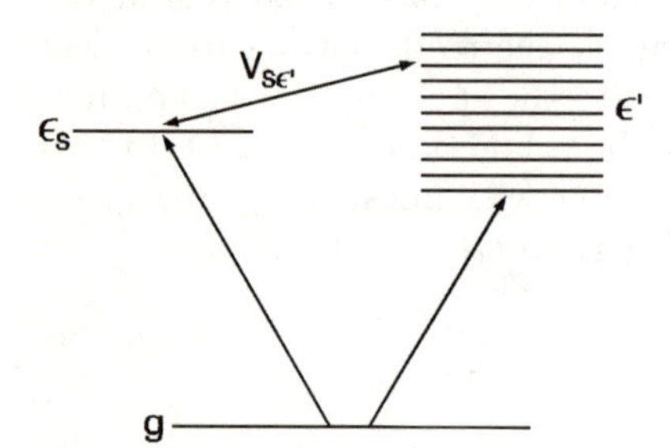

Fig. C.1. Energy schedule for interference effects between excitations from a ground state g to a single discrete level s and to a continuum ϵ'.

to a new state described by a wave function Φ_s and energy $\overline{\epsilon}_s = \epsilon_s - F(\epsilon)$ given by

$$\begin{aligned} \Phi_s &= \varphi_s + \mathrm{P}\int \mathrm{d}\epsilon' \frac{V_{s\epsilon'}}{\psi_{\epsilon'}} \qquad \text{and} \\ F(\epsilon) &= \mathrm{P}\int \mathrm{d}\epsilon' \frac{|V_{s\epsilon'}|^2}{\epsilon - \epsilon'} , \end{aligned} \tag{C.8}$$

where P indicates the principal value of the integral. The total interacting system is described by wave functions Ψ_ϵ. The lineshape for any transition in the total system originating from a perturbation T can be described by the renormalized and reduced energy $\overline{\epsilon}$

$$\overline{\epsilon} = \frac{\epsilon - \epsilon_s - F(\epsilon)}{\pi |V_{s\epsilon'}|^2} = \frac{\epsilon - \epsilon_s - F(\epsilon)}{\Gamma/2} , \tag{C.9}$$

where $\Gamma = 2\pi |V_{s\epsilon'}|^2$ is the spectral width of the interacting discrete state, and by the lineshape parameter Q

$$Q = \frac{\langle \Phi_s | T | g \rangle}{\pi V_{s\epsilon'} \langle \psi_{\epsilon'} | T | g \rangle} . \tag{C.10}$$

Expressing the lineshape as the ratio between the transition probability in the total system to the transition probability for the unperturbed continuum yields (3.8) in the form

$$\frac{|\langle \Psi | T | g \rangle|^2}{|\langle \psi_{\epsilon'} | T | g \rangle|^2} = \frac{(Q + \overline{\epsilon})^2}{1 + \overline{\epsilon}^2} . \tag{C.11}$$

A more detailed description with further references is given in [3.3].

C.4 Stimulated Emission of Laser Radiation

In a system consisting of two states with inverted electronic population (state 1 with lower energy and state 2 with higher energy) three processes with respect to light interaction are possible: light-induced absorption, light-induced (stimulated) emission, and spontaneous emission. Each of them is determined by the corresponding Einstein coefficient B_{12}, B_{21}, and $A_{21} = 1/\tau_{\mathrm{sp}}$, respectively, where $1/\tau_{\mathrm{sp}}$ is the lifetime versus spontaneous emission. For reasons of thermodynamic equilibrium the following two relations hold between the Einstein coefficients:

$$B_{21} = B_{12} \quad \text{and} \quad A_{21} = \frac{8\pi h}{\lambda^3} B_{21} , \tag{C.12}$$

where λ determines the transition energy. The Einstein coefficients for the stimulated processes may be defined as the number of excitation processes n per unit volume and time for a given spectral energy density $L_{\mathrm{f}}(f)$ as, e.g., given in (3.4). This means

$$\frac{\mathrm{d}n_{1,2}}{\mathrm{d}t} = -B_{21} n_{1,2} L_{\mathrm{f}}(f) , \tag{C.13}$$

where 1,2 refers to the absorption and to the stimulated emission process, respectively. The units of B_{21} are $\mathrm{m^3/J\,s^2}$. For the definition of the gain coefficient α it is convenient to consider the increase in photon concentration per one incident photon and per unit of length. This increase is identical to the number of net excitations $(1/c_0)\mathrm{d}(n_2 - n_1)/\mathrm{d}t$. Thus,

$$\alpha = \frac{\mathrm{d}(n_2 - n_1)}{\mathrm{d}x} = \frac{\mathrm{d}n}{c_0 \mathrm{d}t} = -\frac{B_{21} n h f g(f)}{c_0} , \tag{C.14}$$

where the spectral intensity distribution for the single photon is described by a normalized lineshape function $g(f)$. Expressed by the lifetime versus spontaneous emission this yields with the help of (C.12)

$$\boxed{\alpha = n\sigma(f) = n \frac{\lambda^2}{8\pi\tau_{\mathrm{sp}}} g(f) .} \tag{C.15}$$

σ has the dimension of $\mathrm{m^2}$ and defines a cross section for the absorption or emission process. From the gain coefficient the total net gain is obtained for one round trip of length $2L$ by

$$G_n(f) = \mathrm{e}^{2L\alpha - \gamma} , \tag{C.16}$$

where γ summarizes all radiation losses during the round trip. The spontaneous emission is considered to be small and contributes to the noise of the otherwise coherent radiation.

D. To Chapter 4, Spectral Analysis of Light

D.1 Multiple Beam Interference for a Plane Parallel Plate

The beam configuration shown in Fig. 4.13 enables the following relationships between the amplitudes for the reflected and transmitted partial beams to be obtained:

$$E_{\mathrm{r},1} = \sqrt{R} E_0, \quad E_{\mathrm{r},2} = \sqrt{R}(1 - R) E_0, \quad E_{\mathrm{r},n+1} = R E_{\mathrm{r},n} \quad \text{for } n \geq 2 \tag{D.1}$$

and

$$E_{\mathrm{t},1} = (1 - R) E_0, \quad E_{\mathrm{t},n+1} = R E_{\mathrm{t},n} \quad \text{for } n \geq 1 , \tag{D.2}$$

where $R(\theta_i) = I_{\mathrm{r}}/I_0$ is is the reflection coefficient for an arbitrary angle of incidence θ_i. The phase difference ϕ between two consecutive partial beams is as given in (4.22). The first reflected partial beam has a phase shift $\phi = \pi$ with respect to the incident light because of the reflection at the interface to a material with a higher optical density. Thus, the phase factor between E_0 and E_1 is $\mathrm{e}^{\mathrm{i}\pi} = -1$. This factor must not be considered for any of the other partial beams, neither in reflection nor in transmission. With this the two sums for the partial beams can be written explicitly in the form

$$E_\mathrm{r} = -\sqrt{R}E_0 + (1-R)\sqrt{R}E_0\mathrm{e}^{\mathrm{i}\phi} + (1-R)\sqrt{R}RE_0\mathrm{e}^{\mathrm{i}2\phi} + \ldots \tag{D.3}$$

and

$$E_\mathrm{t} = (1-R)E_0 + (1-R)RE_0\mathrm{e}^{\mathrm{i}\phi} + \ldots . \tag{D.4}$$

Separating $E_0\sqrt{R}(1-R)\mathrm{e}^{\mathrm{i}\phi}$ in (D.3) and $E_0(1-R)$ in (D.4) yields

$$E_\mathrm{r} = E_0\sqrt{R}\left(\frac{(1-R)\mathrm{e}^{\mathrm{i}\phi}}{1-R\mathrm{e}^{\mathrm{i}\phi}} - 1\right) \tag{D.5}$$

and

$$E_\mathrm{t} = E_0\frac{(1-R)}{1-R\mathrm{e}^{\mathrm{i}\phi}} . \tag{D.6}$$

Evaluating E^*E in both cases yields (4.23) and (4.24).

E. To Chapter 6, The Dielectric Function

E.1 Reflection and Transmission at an Interface for Arbitrary Incidence (Fresnel Equations)

If the direction of incidence for a light beam is not perpendicular to the (plane and flat) surface (6.18) and (6.19) become more complicated. In this case a different behavior is obtained for the polarization parallel (E_p) or normal (E_n) to the plane of incidence. If N and N' and θ and θ' are refractive indices and angles between beam and surface normal on the side of incidence and refraction, respectively, the ratios between the amplitudes of the propagating fields and the incident field E_i are

$$\begin{aligned} r_\mathrm{p} &= \frac{E_\mathrm{r,p}}{E_\mathrm{i,p}} = \frac{N'\cos\theta - N\cos\theta'}{N'\cos\theta + N\cos\theta'} = \frac{\tan(\theta-\theta')}{\tan(\theta+\theta')} \\ r_\mathrm{n} &= \frac{E_\mathrm{r,n}}{E_\mathrm{i,n}} = \frac{N\cos\theta - N'\cos\theta'}{N\cos\theta + N'\cos\theta'} = \frac{\sin(\theta'-\theta)}{\sin(\theta+\theta')} \end{aligned} \tag{E.1}$$

for the reflected wave and

$$\begin{aligned} t_\mathrm{p} &= \frac{E_\mathrm{t,p}}{E_\mathrm{i,p}} = \frac{2N\cos\theta}{N\cos\theta' + N'\cos\theta} = \frac{2\sin\theta'\cos\theta}{\sin(\theta+\theta')\cos(\theta-\theta')} \\ t_\mathrm{n} &= \frac{E_\mathrm{t,n}}{E_\mathrm{i,n}} = \frac{2N\cos\theta}{N\cos\theta + N'\cos\theta'} = \frac{2\sin\theta'\cos\theta}{\sin(\theta+\theta')} \end{aligned} \tag{E.2}$$

for the transmitted wave. These equations are known as the *Fresnel formulae*. For normal incidence ($\theta, \theta' = 0$) and $N = 1, N' = N$ equation (6.18) is immediately obtained.

The coefficients R_p and R_n for the reflected intensities are the square of the magnitude of r_p and r_n. The transmitted intensities are obtained from $1-R$.

In general N' is complex. In this case the angle for the refracted light also becomes complex since the generalized Snellius law of refraction

$$N \sin\theta = N' \sin\theta' \tag{E.3}$$

requires complex angles θ'. As long as $n'^2 \gg \kappa'^2$ the reflection coefficients for parallel and normal polarization can be obtained (for $N = 1, N' = N$) from

$$\begin{aligned} R_\mathrm{p} &= R_\mathrm{n} \frac{(N - \sin\theta \tan\theta)^2 + \kappa^2}{(N + \sin\theta \tan\theta)^2 + \kappa^2} \quad \text{and} \\ R_\mathrm{n} &= \left| \frac{(N - \cos\theta)^2 + \kappa^2}{(N + \cos\theta)^2 + \kappa^2} \right| . \end{aligned} \tag{E.4}$$

For most non-metallic materials the anticipated condition is rather well satisfied, even in the range of fundamental absorption. It holds in particular for conventional semiconductors, where n is of the order of 3 and κ of the order of 1.

E.2 Reflection and Transmission Through Plane and Parallel Plates

In many spectroscopic experiments transmission or reflection is studied from a triple layer of parallel plates with three different refractive indices N_0, N_1, and N_2, respectively. We will give in the following a few summarizing and useful formulae for three special cases: a plane crystal plate in air, a thin film on a reflecting mirror, and a thin film on a thick but non-absorbing layer.

The general formulae for the total field transmission and field reflection including multiple reflection are

$$\begin{aligned} t\mathrm{e}^{\mathrm{i}\theta} &= \frac{t_1 \mathrm{e}^{\mathrm{i}\theta_1} t_2 \mathrm{e}^{\mathrm{i}\theta_2} \mathrm{e}^{-\mathrm{i}\Delta}}{1 + r_1 \mathrm{e}^{-\mathrm{i}\theta_1} r_2 \mathrm{e}^{\mathrm{i}\theta_2} \mathrm{e}^{-\mathrm{i}\Delta}} \\ r\mathrm{e}^{\mathrm{i}\theta} &= \frac{r_1 \mathrm{e}^{\mathrm{i}\theta_1} + r_2 \mathrm{e}^{\mathrm{i}\theta_2} \mathrm{e}^{-\mathrm{i}\Delta}}{1 + r_1 \mathrm{e}^{-\mathrm{i}\theta_1} r_2 \mathrm{e}^{\mathrm{i}\theta_2} \mathrm{e}^{-\mathrm{i}\Delta}} , \end{aligned} \tag{E.5}$$

where the phase shift Δ is complex and for normal incidence of the form

$$\Delta = \delta - \mathrm{i}\beta = 2\pi n_1 d/\lambda - \mathrm{i}2\pi\kappa_1 d/\lambda$$

with

$$\beta = \alpha_1 d/2 = 2\pi\kappa_1 d/\lambda .$$

The quantities $t_\mathrm{j}\mathrm{e}^{\mathrm{i}\theta_\mathrm{j}}$ and $r_\mathrm{j}\mathrm{e}^{\mathrm{i}\theta_\mathrm{j}}$ are the complex field transmission and reflection coefficients at the various interfaces. $j = 1$ and 2 refer to the interface 0/1 and 1/2. The field transmission and reflection coefficients are given by the Fresnel formulae (E.1) and (E.2) for normal incidence and complex N. Note that from t in (E.5) the reflectivity and the transmittance are obtained as

$$\begin{aligned} t^2 &= \frac{t_1^2 t_2^2 e^{2\beta}}{1 + r_1^2 r_2^2 e^{4\beta} + 2 r_1 r_2 e^{2\beta} \cos\phi} \\ r^2 &= \frac{r_1^2 + r_2^2 e^{4\beta} + 2 r_1 r_2 e^{2\beta} \cos\phi}{1 + r_1^2 r_2^2 e^{4\beta} + 2 r_1 r_2 e^{2\beta} \cos\phi} \end{aligned} \tag{E.6}$$

with

$$\phi = 2\delta + \theta_1 - \theta_2 \ .$$

The response without interference is obtained from averaging over ϕ from 0 to 2π. This yields finally

$$\begin{aligned} \langle T \rangle &= \frac{t_1^2 t_2^2 e^{2\beta}}{1 - r_1^2 r_2^2 e^{4\beta}} \\ \langle R \rangle &= \frac{r_1^2 + r_2^2 e^{4\beta} - 2 r_1^2 r_2^2 e^{4\beta}}{1 - r_1^2 r_2^2 e^{4\beta}} \ . \end{aligned} \tag{E.7}$$

These equations are general enough to cover the above-mentioned experimental geometries.

For the free crystal plate $r_1^2 = r_2^2$ and $t_2^2 = t_1^2(1 + \kappa_1^2/n_1^2)$ which immediately gives (6.20) and (6.21).

For the film on a metal surface $r_2 = 1$ and $t_2 = 0$. Accordingly, $\langle T \rangle = 0$ and

$$\langle R \rangle = \frac{e^{-2\alpha d} + R_{\mathrm{film}}(1 - 2e^{-2\alpha d})}{1 - R_{\mathrm{film}} e^{-2\alpha d}} \ , \tag{E.8}$$

where R_{film} and α are the reflectivity of the film/air interface and the absorption coefficient of the film, respectively. If R_{film} and $2\alpha d$ are small compared to 1 $\langle R \rangle$ can be used to determine αd.

Finally, for a film on a thick transparent substrate (as, e.g., on Si) the reflectivity can be evaluated from (E.7) with $N_0 = 1$, $N_1 = n_1 + i\kappa_1$ and $N_2 = n_2$. The total transmittance cannot be evaluated from the given formulae because it is necessary to consider a four layered structure.

E.3 Experimental Techniques in Ellipsometry

For the experimental determination of the ellipsometric angles two types of ellipsometers are in use: the rotating analyzer ellipsometer and the zero ellipsometer. Since in both cases the relative settings of several polarizers and compensators are essential, it is recommended to refer all positions to a well defined coordinate system according to Fig. 6.8.

In the case of the zero ellipsometer the settings for the polarizer and for the compensator are such that after reflection at the sample surface linear polarized light is obtained. This light is blocked with the analyzer until a signal minimum is obtained at the detector. From the positions of the polarizer (P), the compensator (C), and the analyzer (A) the complex quantity r_{s} from (6.60) is obtained by

$$r_\mathrm{s} = -\tan A \frac{\tan C + r_\mathrm{c} \tan(P - C)}{1 - r_\mathrm{c} \tan C \tan(P - C)} \tag{E.9}$$

with

$$r_\mathrm{c} = T_\mathrm{c} \mathrm{e}^{\mathrm{i}\delta_\mathrm{c}} \, .$$

T_c and δ_c are the ratio of amplitude attenuation for the slow and the fast propagation mode in the compensator and δ_c is the corresponding phase difference selected.

A similar equation can be obtained for the rotating analyzer ellipsometer. In this case the position $A = \omega t$ increases continuously with the time and a periodically fluctuating intensity of the form

$$I_\mathrm{D} = K(1 + \alpha \cos 2A + \beta \sin 2A) \tag{E.10}$$

is observed. K is a constant determined by the light intensity used in the experiment. The coefficients α and β are obtained from a Fourier analysis and yield r_s from

$$r_\mathrm{s} = \frac{1 + \alpha}{\beta \pm \mathrm{i}(1 - \alpha^2 - \beta^2)^{1/2}} \frac{\tan C + r_\mathrm{c} \tan(P - C)}{1 - r_\mathrm{c} \tan C \tan(P - C)} \, . \tag{E.11}$$

With this value for r_s from both types of ellipsometers the DF is obtained by (6.61).

F. To Chapter 7, Spectroscopy in the Visible and Near Visible Spectral Range

F.1 Matrix Elements and First-Order Perturbation Theory

In quantum mechanics physical states and physical quantities are described by wave functions and operators, respectively. The eigenvalues of the operators are the values observable in an experiment. Each operator **A** has eigenfunctions $u_k(x), k = 1, 2, \ldots j, \ldots l, \ldots$ which can be used to construct the matrix elements

$$A_{jl} = \int u_j^* \mathbf{A} u_l \mathrm{d}x = \langle j | \mathbf{A} | l \rangle \, . \tag{F.1}$$

The diagonal matrix elements are the eigenvalues of the operators. The non-diagonal matrix elements are zero because of the orthogonality of the wave functions. If **A** is the Hamilton operator **H** the corresponding eigenfunctions can be obtained from the Schrödinger equation and the eigenvalues are the stationary states of the system.

The non-diagonal matrix elements become important if the system considered is subjected to a perturbation which transfers it from one state into another. This perturbation is in general time-dependent and characterized by an operator $\mathbf{A}'$ which is added to the original operator **A**. The original eigenfunctions $\langle j|$ are now no longer eigenfunctions of the system. After all, this is not possible because the new system is time-dependent and therefore needs time-dependent wave functions for its characterization. If again $\mathbf{A}$ is the Hamilton operator the new wave functions can be determined by solving the time-dependent Schrödinger equation for the operator $\mathbf{H}+\mathbf{H}'$. Since this is not possible in general approximations are necessary. One possibility is to perform a perturbation calculation. This anticipates that the perturbation $\mathbf{H}'$ is small as compared to the relevant energies of the system.

In perturbation calculations the new wave functions are expanded with respect to the wave functions of the unperturbed system and the time-dependence is considered by adding a factor $\exp(i\omega_{\text{fi}}t)$ where $\hbar\omega_{\text{fi}}$ is the energy difference between the initial and the final state. Within first-order perturbation theory the absolute square of the coefficient of the first term in the development $a_{\text{fi}}^{(1)}(t)$ determines the probability that a transition has occurred. This probability is expected to increase linearly in time. From inserting the new wave functions into the time-dependent Schrödinger equation the off diagonal matrix elements constructed from the total Hamiltonian and the original wave functions turn out to give the main contributions to $a_{\text{fi}}^{(1)}(t)$. Because of the orthogonality of the original wave functions these matrix elements are identical to the matrix elements constructed for the perturbation itself:

$$\langle \text{f}|\mathbf{H} + \mathbf{H}'|\text{i}\rangle = \langle \text{f}|\mathbf{H}'|\text{i}\rangle = H'_{\text{fi}}(t) \ . \tag{F.2}$$

For a perturbation $\mathbf{H}'$ periodic in time with frequency ω first-order perturbation theory yields for the coefficient of the first term in the development

$$a_{\text{fi}}^{(1)}(t) = -\frac{H'_{\text{fi}}(0)\{\exp[\text{i}(\omega_{\text{fi}} - \omega)t] - 1\}}{\hbar(\omega_{\text{fi}} - \omega)} \ , \tag{F.3}$$

where $H'_{\text{fi}}(0)$ is the time-independent part of the perturbation matrix element. The absolute square of this quantity is the transition probability $P_{\text{fi}}(t)$

$$P_{\text{fi}}(t) = |a_{\text{fi}}^{(1)}(t)|^2 = \frac{4|H'_{\text{fi}}(0)|^2\{\sin^2[(\omega_{\text{fi}} - \omega)t/2]\}}{\hbar^2(\omega_{\text{fi}} - \omega)^2} \ . \tag{F.4}$$

In the limit of very large t the frequency-dependent part on the right hand side of the equation approaches $\pi t/2$ times a δ function for the difference in the frequencies. This yields finally for the transition probability per unit of time

$$P_{\text{fi}} = \frac{P_{\text{fi}}(t)}{t} = \frac{2\pi|H'_{\text{fi}}(0)|^2}{\hbar}\delta(\hbar\omega_{\text{fi}} - \hbar\omega) \ , \tag{F.5}$$

known as the *golden rule* of quantum mechanics. The δ function represents the density of allowed final states. In the present problem it guarantees energy conservation for the transition.

F.2 Transitions Induced by Electromagnetic Radiation

Transitions between two energy eigenstates of a system can be performed by an incident EM radiation. In the sense of Appendix F.1 the radiation is a perturbation of the system. It is convenient to describe the radiation by its vector potential $\boldsymbol{A}$ and its scalar potential Φ from Appendix B.2. The Hamiltonian for the perturbed system then has the well known form

$$\mathbf{H} = \frac{1}{2m}(\mathbf{p} - e\mathbf{A})^2 + e\mathbf{\Phi} + \mathbf{U}\,. \tag{F.6}$$

For the evaluation of the expression in parentheses the quantum-mechanical definition for the momentum

$$\mathbf{p} = -\mathrm{i}\hbar\nabla$$

and the commutation relations

$$\mathbf{Ap} - \mathbf{pA} = -\mathrm{i}\hbar\nabla\mathbf{A}$$

must be considered. With this we obtain for the Hamiltonian

$$\mathbf{H} = \frac{\mathbf{p}^2}{2m} + \frac{\mathrm{i}\hbar e}{m}\mathbf{A}\nabla - \frac{\mathrm{i}\hbar e}{2m}\nabla\mathbf{A} + \frac{\mathrm{e}^2}{2m}\mathbf{A}^2 + e\mathbf{\Phi} + \mathbf{U}\,. \tag{F.7}$$

If there are no static space charges and no currents $\nabla\mathbf{A}$ and $\mathbf{\Phi}$ are zero in the Lorentz gauge [see Appendix B.4, (3.6)]. Also, since $\mathbf{A}$ is small the term proportional to $\mathbf{A}^2$ can be neglected as compared to the term linear in $\mathbf{A}$ which renders the latter as the only relevant perturbation in $\mathbf{H}$.

$$\boxed{\mathbf{H}' = -\frac{e}{m}\mathbf{Ap} = -\frac{e}{m}\mathbf{pA} = \frac{\mathrm{i}\hbar e}{m}\mathbf{A}\nabla\,.} \tag{F.8}$$

Since the perturbation is a plane wave we use for $\mathbf{A}$

$$\mathbf{A} = \mathbf{A}_0\mathrm{e}^{\mathrm{i}(kx-\omega t)}\,. \tag{F.9}$$

The matrix element for the time-independent part of the perturbation Hamiltonian $\mathbf{H}'(0)$ in (F.3) and (F.5) then has the form

$$H'_{\mathrm{fi}}(0) = \frac{\mathrm{i}e\hbar}{m}\langle \mathrm{f}|\mathbf{A}_0\mathrm{e}^{\mathrm{i}kx}\nabla|\mathrm{i}\rangle = \frac{\mathrm{i}e\hbar}{m}\int u_{\mathrm{f}}^*\mathrm{e}^{\mathrm{i}kx}\mathbf{A}_0\nabla u_{\mathrm{i}}\,\mathrm{d}^3x\,, \tag{F.10}$$

where the wave functions u_{f} and u_{i} are the explicit eigenfunctions of the unperturbed system.

For the evaluation of the transition probability it is convenient to replace the square of the vector potential by the intensity $I(\omega)$ of the radiation. This can be done using (2.8) and (2.5) and yields

$$A_0^2 = \frac{I(\omega)}{\varepsilon_0 c_0 \omega^2 n}\,. \tag{F.11}$$

From this the transition probability per unit time is

$$\boxed{P_{\mathrm{fi}} = \frac{1}{t}|a_{\mathrm{fi}}^{(1)}(t)|^2 = \frac{2\pi \mathrm{e}^2 I(\omega)}{m^2 \varepsilon_0 c_0 \omega^2} \left| \int u_{\mathrm{f}}^* \mathrm{e}^{ikx} \mathrm{grad}\, u_{\mathrm{i}}\, \mathrm{d}^3x \right|^2 \delta(\omega_{\mathrm{fi}} - \omega)\,.} \quad \text{(F.12)}$$

As long as the wavelength of the EM radiation is large compared to the periodicity of the electronic wave functions which means large compared to the dimension of the crystal unit cell the term e^{ikx} can be expanded in a Taylor series. The zero order contribution to the series is already a good approximation if the resulting matrix element is finite. In this case the transition probability per unit of time for the resonance conditions $\omega_{\mathrm{fi}} = \omega$ has the form

$$\begin{aligned} P_{\mathrm{fi}} &= \frac{2\pi \mathrm{e}^2 I(\omega)}{n m^2 \varepsilon_0 c_0 \omega_{\mathrm{fi}}^2} \left| \int u_{\mathrm{f}}^* \mathrm{grad}\, u_{\mathrm{i}}\, \mathrm{d}^3x \right|^2 \delta(\omega_{\mathrm{fi}} - \omega) \\ &= C|p_{\mathrm{fi}}|^2 I(\omega) \delta(\omega_{\mathrm{fi}} - \omega)\,, \end{aligned} \quad \text{(F.13)}$$

where

$$C = \frac{2\pi \mathrm{e}^2}{m^2 \varepsilon_0 c_0 \omega_{\mathrm{fi}}^2 \hbar^2}$$

and p_{fi} is the matrix element for the momentum operator

$$p_{\mathrm{fi}} = -\mathrm{i}\hbar \int u_{\mathrm{f}}^* \mathrm{grad}\, u_{\mathrm{i}} \mathrm{d}^3x\,. \quad \text{(F.14)}$$

Since the matrix element for the momentum operator is proportional to the matrix element for the dipole operator (see also Sect. F.3) the zero order approximation described above is called the *dipole approximation.* If the zero order transition matrix element is zero at least the next term in the expansion of $\exp \mathrm{i}kx$ must be considered. It has the form $\mathrm{i}kx$ and describes the quadrupole interaction. Accordingly, the higher terms $(\mathrm{i}kx)^n/n!$ in the expansion yield the matrix elements for the higher *multipole interactions.*

F.3 Matrix Elements in Dipole Representation

The relation between the momentum matrix elements and the dipole matrix elements can be obtained from the Heisenberg equation of motion for an operator $\mathbf{A}$.

$$\dot{\mathbf{A}} = -\frac{\mathrm{i}}{\hbar}[\mathbf{A}\mathbf{H}] = -\frac{\mathrm{i}}{\hbar}(\mathbf{A}\mathbf{H} - \mathbf{H}\mathbf{A})\,. \quad \text{(F.15)}$$

If $|\mathrm{f}\rangle$ and $|\mathrm{i}\rangle$ are eigenfunctions of $\mathbf{H}$ the matrix elements for the operator $\dot{\mathbf{A}}$ are

$$\begin{aligned} \dot{A}_{\mathrm{fi}} &= -\frac{\mathrm{i}}{\hbar}(AH - HA)_{\mathrm{fi}} = -\frac{\mathrm{i}}{\hbar}\{(AH)_{\mathrm{fi}} - (HA)_{\mathrm{fi}}\} \\ &= -\frac{\mathrm{i}}{\hbar}\{\langle \mathrm{f}|\mathbf{A}\mathbf{H}|\mathrm{i}\rangle - \langle \mathrm{f}|\mathbf{H}\mathbf{A}|\mathrm{i}\rangle\} = -\frac{\mathrm{i}}{\hbar}(\epsilon_{\mathrm{i}} - \epsilon_{\mathrm{f}})A_{\mathrm{fi}}\,. \end{aligned} \quad \text{(F.16)}$$

Using for $\mathbf{A}$ the operator $\mathbf{x}$ with $\dot{\mathbf{x}} = \mathbf{p}/m$ (F.16) yields

$$\frac{\mathbf{p}}{m} = \frac{-\mathrm{i}\hbar\nabla}{m} = \mathrm{i}\omega_{\mathrm{fi}}\mathbf{x}\,, \tag{F.17}$$

or for the matrix elements

$$p_{\mathrm{fi}} = \frac{\mathrm{i}m\omega_{\mathrm{fi}}}{e}ex_{\mathrm{fi}}\,. \tag{F.18}$$

$ex_{\mathrm{fi}} = M_{\mathrm{fi}}$ is the matrix element for the dipole moment. The relation (F.18) allows the use the dipole matrix element to calculate the transition probability in a similar way as the momentum matrix element

$$P_{\mathrm{fi}} = C'|M_{\mathrm{fi}}|^2 I(\omega)\delta(\omega_{\mathrm{fi}} - \omega) \tag{F.19}$$

with

$$C' = \frac{2\pi}{\varepsilon_0 c_0 \hbar^2}\,.$$

F.4 Quantum Mechanics of the Harmonic Oscillator

The Hamiltonian for the harmonic oscillator is

$$\mathbf{H} = -\frac{\hbar^2\nabla^2}{2M} + \frac{C\mathbf{x}^2}{2}\,, \tag{F.20}$$

with $C = M\Omega^2$. To find the solutions for the corresponding Schrödinger equation we introduce dimensionless coordinates

$$q = x\sqrt{\frac{M\Omega}{\hbar}} \tag{F.21}$$

and the abbreviation

$$d = \left(\frac{\Omega}{\hbar}\right)^{1/2}\,. \tag{F.22}$$

In this formulation the eigenvalues and the eigenfunctions are obtained by

$$\epsilon_\alpha = (\alpha + 1/2)\hbar\Omega \tag{F.23}$$

and

$$u_\alpha(q) = \left(\frac{d}{\pi^{1/2}2^n n!}\right)^{1/2} \exp(-q^2/2)H_\alpha(q)\,. \tag{F.24}$$

$H_\alpha(q)$ are the Hermit polynomials defined as

$$H_\alpha(q) = (-1)^\alpha \mathrm{e}^{q^2}\frac{\mathrm{d}^\alpha}{\mathrm{d}q^\alpha}\mathrm{e}^{-q^2}\,. \tag{F.25}$$

The first three polynoms have the explicit form

$$H_0(q) = 1, \quad H_1(q) = 2q, \quad H_2(q) = 4q^2 - 2\,.$$

Note that the wave functions in (F.24) are not normalized in q but normalized in the normal coordinate

$$Q = \frac{q}{d} = xM^{1/2} \,. \tag{F.26}$$

The eigenfunctions in (F.24) are certainly orthogonal in q but the integral over the product of two different eigenfunctions becomes nonzero if one of them is shifted from the origin by a constant value a. These integrals are called *Franck–Condon* integrals and can be evaluated analytically as

$$\begin{aligned} F_{\beta\alpha} &= \int u^*_\beta(q+a)u_\alpha(q)\mathrm{d}q \\ &= \mathrm{e}^{-a^2/4}\left(\frac{\alpha!}{\beta!}\right)^{1/2}\left(-\frac{a}{2^{1/2}}\right)^{\beta-\alpha}\mathrm{L}_\alpha^{\beta-\alpha}\left(\frac{a^2}{2}\right)\,, \end{aligned} \tag{F.27}$$

where $\mathrm{L}_n^m(x)$ are the associated Laguerre polynomials. They are derived from

$$\mathrm{L}_n^m(x) = \frac{1}{n!}\mathrm{e}^x x^{-m}\frac{\mathrm{d}^n}{\mathrm{d}x^n}\mathrm{e}^{-x}x^{m+n} \,. \tag{F.28}$$

For the evaluation of the Raman intensities the integrals $\langle\beta|\alpha\rangle$ and $\langle\beta|Q|\alpha\rangle$ are required with integration over Q. The first integral is of course $\delta_{\beta\alpha}$ since we consider unshifted wave function in this case. To show this and to evaluate the second integral it is useful to represent the Hermit polynomials by a generating function $S(q,s)$. The derivation is given in standard textbooks of quantum mechanics.

G. To Chapter 8, Symmetry and Selection Rules

G.1 Character Tables of Point Groups

Table G.1. Character tables for 32 point groups and $\boldsymbol{I}_h$. The top line in each subtable shows the point group and the symmetry elements as distributed among the classes.
($\epsilon = \exp(2\pi\mathrm{i}/3), \omega = \exp(2\pi\mathrm{i}/6)$, x, y, z refer to translations, X, Y, Z refer to rotations.)

triclinic

C_1	E
A	1

C_i	E	I	
A_g	1	1	X, Y, Z
A_u	1	−1	x, y, z

monoclinic

C_{1h}/C_s	E	σ_h	
A'	1	1	$x, y; Z$
A''	1	-1	$z; X, Y$

C_2	E	C_2	
A	1	1	$z; Z$
B	1	-1	$x, y; X, Y$

C_{2h}	E	C_2	I	σ_h	
A_g	1	1	1	1	Z
B_g	1	-1	1	-1	X, Y
A_u	1	1	-1	-1	z
B_u	1	-1	-1	1	x, y

orthorhombic

C_{2v}	E	C_2	σ_y	σ_x	
A_1	1	1	1	1	z
A_2	1	1	-1	-1	Z
B_1	1	-1	1	-1	x, Y
B_2	1	-1	-1	1	y, X

$D_2 = V$	E	C_2^z	C_2^y	C_2^x	
A	1	1	1	1	
B_1	1	1	-1	-1	z, Z
B_2	1	-1	1	-1	y, Y
B_3	1	-1	-1	1	x, X

$D_{2h} = V_h$	E	C_{2z}	C_{2y}	C_{2x}	I	σ_z	σ_y	σ_x	
A_g	1	1	1	1	1	1	1	1	
B_{1g}	1	1	-1	-1	1	1	-1	-1	Z
B_{2g}	1	-1	1	-1	1	-1	1	-1	Y
B_{3g}	1	-1	-1	1	1	-1	-1	1	X
A_u	1	1	1	1	-1	-1	-1	-1	
B_{1u}	1	1	-1	-1	-1	-1	1	1	z
B_{2u}	1	-1	1	-1	-1	1	-1	1	y
B_{3u}	1	-1	-1	1	-1	1	1	-1	x

trigonal

C_3	E	C_3	C_3^2	
A	1	1	1	zZ
$E(1)$	1	ϵ	ϵ^*	$(x, y); (X, Y)$
$E(2)$	1	ϵ^*	ϵ	$(x, y); (X, Y)$

C_{3i}/S_6	E	C_3	C_3^2	I	S_6^5	S_6	
A_g	1	1	1	1	1	1	Z
$E_g(1)$	1	ϵ	ϵ^*	1	ϵ	ϵ^*	(X, Y)
$E_g(2)$	1	ϵ^*	ϵ	1	ϵ^*	ϵ	(X, Y)
A_u	1	1	1	-1	-1	-1	z
$E_u(1)$	1	ϵ	ϵ^*	-1	$-\epsilon$	$-\epsilon^*$	(x, y)
$E_u(2)$	1	ϵ^*	ϵ	-1	$-\epsilon^*$	$-\epsilon$	(x, y)

C_{3v}	E	$2C_3$	$3\sigma_v$	
A_1	1	1	1	z
A_2	1	1	-1	Z
E	2	-1	0	$(x, y); (X, Y)$

D_3	E	$2C_3$	$3C_2$	
A_1	1	1	1	
A_2	1	1	-1	z, Z
E	2	-1	0	$(x, y); (X, Y)$

$\boldsymbol{D_{3d}}$	E	$2C_3$	$3C_2$	I	$2S_6$	$3\sigma_d$	
A_{1g}	1	1	1	1	1	1	
A_{2g}	1	1	−1	1	1	−1	Z
E_g	2	−1	0	2	−1	0	(X,Y)
A_{1u}	1	1	1	−1	−1	−1	
A_{2u}	1	1	−1	−1	−1	1	z
E_u	2	−1	0	−2	1	0	(x,y)

tetragonal

$\boldsymbol{C_4}$	E	C_4	C_2	C_4^3	
A	1	1	1	1	z,Z
B	1	−1	1	−1	
$E(1)$	1	i	−1	−i	(x,y)
$E(2)$	1	−i	−1	i	X,Y

$\boldsymbol{S_4}$	E	S_4	C_2	S_4^3	
A	1	1	1	1	Z
B	1	−1	1	−1	z
$E(1)$	1	i	−1	−i	(x,y)
$E(2)$	1	−i	−1	i	X,Y

$\boldsymbol{C_{4h}}$	E	C_4	C_2	C_4^3	I	S_4^3	σ_h	S_4	
A_g	1	1	1	1	1	1	1	1	Z
B_g	1	−1	1	−1	1	−1	1	−1	
$E_g(1)$	1	i	−1	−i	1	i	−1	−i	X,Y
$E_g(2)$	1	−1	−1	1	1	−1	−1	1	X,Y
A_u	1	1	1	1	−1	−1	−1	−1	z
B_u	1	−1	1	−1	−1	1	−1	1	
$E_u(1)$	1	i	−1	−i	−1	−i	1	i	x,y
$E_u(2)$	1	−i	−1	i	−1	i	1	−i	x,y

$\boldsymbol{D_4}$	E	$2C_4$	C_2	$2C_2'$	$2C_2''$	
A_1	1	1	1	1	1	
A_2	1	1	1	−1	−1	$z;Z$
B_1	1	−1	1	1	−1	
B_2	1	−1	1	−1	1	
E	2	0	−2	0	0	$x,y;X,Y$

$\boldsymbol{C_{4v}}$	E	$2C_4$	C_2	$2\sigma_v$	$2\sigma_d$	
A_1	1	1	1	1	1	z
A_2	1	1	1	−1	−1	Z
B_1	1	−1	1	1	−1	
B_2	1	−1	1	−1	1	
E	2	0	−2	0	0	$x,y;X,Y$

$\boldsymbol{D_{2d}/V_d/S_{4u}}$	E	$2S_4$	C_2	$2C_2'$	$2\sigma_d$	
A_1	1	1	1	1	1	
A_2	1	1	1	−1	−1	Z
B_1	1	−1	1	1	−1	
B_2	1	−1	1	−1	1	z
E	2	0	−2	0	0	$x,y;X,Y$

$\boldsymbol{D_{4h}}$	E	$2C_4$	C_2	C_2'	$2C_2''$	I	$2S_4$	σ_h	$2\sigma_v$	$2\sigma_d$	
A_{1g}	1	1	1	1	1	1	1	1	1	1	
A_{2g}	1	1	1	−1	−1	1	1	1	−1	−1	Z
B_{1g}	1	−1	1	1	−1	1	−1	1	1	−1	
B_{2g}	1	−1	1	−1	1	1	−1	1	−1	1	
E_g	2	0	−2	0	0	2	0	−2	0	0	(X,Y)
A_{1u}	1	1	1	1	1	−1	−1	−1	−1	−1	
A_{2u}	1	1	1	−1	−1	−1	−1	−1	1	1	z
B_{1u}	1	−1	1	1	−1	−1	1	−1	−1	1	
B_{2u}	1	−1	1	−1	1	−1	1	−1	1	−1	
E_u	2	0	−2	0	0	−2	0	2	0	0	(x,y)

hexagonal

$\boldsymbol{C_6}$	E	C_6	C_3	C_2	C_3^2	C_6^5	
A	1	1	1	1	1	1	zZ
B	1	−1	1	−1	1	−1	
$E_1(1)$	1	ω	$-\omega^*$	−1	$-\omega$	ω^*	$x,y;X,Y$
$E_1(2)$	1	ω^*	$-\omega$	−1	$-\omega^*$	ω	$x,y;X,Y$
$E_2(1)$	1	$-\omega^*$	$-\omega$	1	$-\omega^*$	$-\omega$	
$E_2(2)$	1	$-\omega$	$-\omega^*$	1	$-\omega$	$-\omega^*$	

$\boldsymbol{C_{3h}}$	E	C_3	C_3^2	σ_h	S_3	S_3^5	
A'	1	1	1	1	1	1	Z
$E'(1)$	1	ε	ε^*	1	ε	ε^*	x,y
$E'(2)$	1	ε^*	ε	1	ε^*	ε	(x,y)
A''	1	1	1	−1	−1	−1	z
$E''(1)$	1	ε	ε^*	−1	$-\varepsilon$	$-\varepsilon^*$	(X,Y)
$E''(2)$	1	ε^*	ε	−1	$-\varepsilon^*$	$-\varepsilon$	(X,Y)

$\boldsymbol{C_{6h}}$	E	C_6	C_3	C_2	C_3^2	C_6^5	I	S_3^2	S_6^5	σ_h	S_6	S_3	
A_g	1	1	1	1	1	1	1	1	1	1	1	1	Z
B_g	1	−1	1	−1	1	−1	1	−1	1	−1	1	−1	
$E_{1g}(1)$	1	ω	$-\omega^*$	−1	$-\omega$	ω^*	1	ω	$-\omega^*$	−1	$-\omega$	ω^*	X,Y
$E_{1g}(2)$	1	ω^*	$-\omega$	−1	$-\omega^*$	ω	1	ω^*	$-\omega$	−1	$-\omega^*$	ω	X,Y
$E_{2g}(1)$	1	$-\omega^*$	$-\omega$	1	$-\omega^*$	$-\omega$	1	$-\omega^*$	$-\omega$	1	$-\omega^*$	$-\omega$	
$E_{2g}(2)$	1	$-\omega$	$-\omega^*$	1	$-\omega$	$-\omega^*$	1	$-\omega$	$-\omega^*$	1	$-\omega$	$-\omega^*$	
A_u	1	1	1	1	1	1	−1	−1	−1	−1	−1	−1	z
B_u	1	−1	1	−1	1	−1	1	−1	1	−1	1	−1	
$E_{1u}(1)$	1	ω	$-\omega^*$	−1	$-\omega$	ω^*	−1	$-\omega$	ω^*	1	ω	$-\omega^*$	x,y
$E_{1u}(2)$	1	ω^*	$-\omega$	−1	$-\omega^*$	ω	−1	$-\omega^*$	ω	1	ω^*	$-\omega$	x,y
$E_{2u}(1)$	1	$-\omega^*$	$-\omega$	1	$-\omega^*$	$-\omega$	−1	ω^*	ω	−1	ω^*	ω	
$E_{2u}(2)$	1	$-\omega$	$-\omega^*$	1	$-\omega$	$-\omega^*$	−1	ω	ω^*	−1	ω	ω^*	

$\boldsymbol{D_6}$	E	$2C_6$	$2C_3$	C_2	$3C_2'$	$3C_2''$	
A_1	1	1	1	1	1	1	
A_2	1	1	1	1	−1	−1	z, Z
B_1	1	−1	1	−1	1	−1	
B_2	1	−1	1	−1	−1	1	
E_1	2	1	−1	−2	0	0	$x, y; X, Y$
E_2	2	−1	−1	2	0	0	

$\boldsymbol{C_{6v}}$	E	$2C_6$	$2C_3$	C_2	$3\sigma_v$	$3\sigma_d$	
A_1	1	1	1	1	1	1	z
A_2	1	1	1	1	−1	−1	Z
B_1	1	−1	1	−1	1	−1	
B_2	1	−1	1	−1	−1	1	
E_1	2	1	−1	−2	0	0	$x, y; X, Y$
E_2	2	−1	−1	2	0	0	

$\boldsymbol{D_{3h}}$	E	$2C_3$	$3C_2$	σ_h	$2S_3$	$3\sigma_v$	
A_1'	1	1	1	1	1	1	
A_2'	1	1	−1	1	1	−1	Z
E'	2	−1	0	2	−1	0	x, y
A_1''	1	1	1	−1	−1	−1	
A_2''	1	1	−1	−1	−1	1	z
E''	2	−1	0	−2	1	0	X, Y

$\boldsymbol{D_{6h}}$	E	$2C_6$	$2C_3$	C_2	$3C_2'$	$3C_2''$	I	$2S_3$	$2S_6$	σ_h	$3\sigma_d$	3σ	
A_{1g}	1	1	1	1	1	1	1	1	1	1	1	1	
A_{2g}	1	1	1	1	−1	−1	1	1	1	1	−1	−1	Z
B_{1g}	1	−1	1	−1	1	−1	1	−1	1	−1	1	−1	
B_{2g}	1	−1	1	−1	−1	1	1	−1	1	−1	−1	1	
E_{1g}	2	1	−1	−2	0	0	2	1	−1	−2	0	0	X, Y
E_{2g}	2	−1	−1	2	0	0	2	−1	−1	2	0	0	
A_{1u}	1	1	1	1	1	1	−1	−1	−1	−1	−1	−1	
A_{2u}	1	1	1	1	−1	−1	−1	−1	−1	−1	1	1	z
B_{1u}	1	−1	1	−1	1	−1	−1	1	−1	1	−1	1	
B_{2u}	1	−1	1	−1	−1	1	−1	1	−1	1	1	−1	
E_{1u}	2	1	−1	−2	0	0	−2	−1	1	2	0	0	x, y
E_{2u}	2	−1	−1	2	0	0	−2	1	1	−2	0	0	

cubic

$\boldsymbol{T}$	E	$4C_3$	$4C_3^2$	$3C_2$	
A	1	1	1	1	
$E(1)$	1	ϵ	ϵ^*	1	
$E(2)$	1	ϵ^*	ϵ	1	
F	3	0	0	−1	$x, y, z; X, Y, Z$

$\boldsymbol{T_d}$	E	$8C_3$	$3C_2$	$6S_4$	$6\sigma_d$	
A_1	1	1	1	1	1	
A_2	1	1	1	−1	−1	
E	2	−1	2	0	0	
F_1	3	0	−1	1	−1	X, Y, Z
F_2	3	0	−1	−1	1	x, y, z

$\boldsymbol{T_h}$	E	$4C_3$	$4C_3^2$	$3C_2$	I	$4S_6$	$4S_6^2$	$3\sigma_h$	
A_g	1	1	1	1	1	1	1	1	
$E_g(1)$	1	ϵ	ϵ^*	1	1	ϵ	ϵ^*	1	
$E_g(2)$	1	ϵ^*	ϵ	1	1	ϵ^*	ϵ	1	
F_g	3	0	0	−1	3	0	0	−1	X, Y, Z
A_u	1	1	1	1	−1	−1	−1	−1	
$E_u(1)$	1	ϵ	ϵ^*	1	−1	$-\epsilon$	$-\epsilon^*$	−1	
$E_u(2)$	1	ϵ^*	ϵ	1	−1	$-\epsilon^*$	$-\epsilon$	−1	
F_u	3	0	0	−1	−3	0	0	1	x, y, z

$\boldsymbol{O}$	E	$8C_3$	$3C_2$	$6C_4$	$6C_2'$	
A_1	1	1	1	1	1	
A_2	1	1	1	−1	−1	
E	2	−1	2	0	0	
F_1	3	0	−1	1	−1	$x, y, z; X, Y, Z$
F_2	3	0	−1	−1	1	

$\boldsymbol{O_h}$	E	$8C_3$	$3C_2$	$6C_4$	$6C_2'$	I	$8S_6$	$3\sigma_h$	$6S_4$	$6\sigma_d$	
A_{1g}	1	1	1	1	1	1	1	1	1	1	
A_{2g}	1	1	1	−1	−1	1	1	1	−1	−1	
E_g	2	−1	2	0	0	2	−1	2	0	0	
F_{1g}	3	0	−1	1	−1	3	0	−1	1	−1	X, Y, Z
F_{2g}	3	0	−1	−1	1	3	0	−1	−1	1	
A_{1u}	1	1	1	1	1	−1	−1	−1	−1	−1	
A_{2u}	1	1	1	−1	−1	−1	−1	−1	1	1	
E_u	2	−1	2	0	0	−2	1	−2	0	0	
F_{1u}	3	0	−1	1	−1	−3	0	1	−1	1	x, y, z
F_{2u}	3	0	−1	−1	1	−3	0	1	1	−1	

icosahedral

$\boldsymbol{I_h}$	E	$12C_5$	$12C_5^2$	$20C_3$	$15C_2$	I	$12S_{10}$	$12S_{10}^3$	$20S_6$	15σ	
A_g	1	1	1	1	1	1	1	1	1	1	
F_{1g}	3	$\frac{1+\sqrt{5}}{2}$	$\frac{1-\sqrt{5}}{2}$	0	-1	3	$\frac{1-\sqrt{5}}{2}$	$\frac{1+\sqrt{5}}{2}$	0	-1	X,Y,Z
F_{2g}	3	$\frac{1-\sqrt{5}}{2}$	$\frac{1+\sqrt{5}}{2}$	0	-1	3	$\frac{1+\sqrt{5}}{2}$	$\frac{1-\sqrt{5}}{2}$	0	-1	
G_g	4	-1	-1	1	0	4	-1	-1	1	0	
H_g	5	0	0	-1	1	5	0	0	-1	1	
A_u	1	1	1	1	1	-1	-1	-1	-1	-1	
F_{1u}	3	$\frac{1+\sqrt{5}}{2}$	$\frac{1-\sqrt{5}}{2}$	0	-1	-3	$\frac{\sqrt{5}-1}{2}$	$\frac{-\sqrt{5}-1}{2}$	0	1	x,y,z
F_{2u}	3	$\frac{1-\sqrt{5}}{2}$	$\frac{1+\sqrt{5}}{2}$	0	-1	-3	$\frac{-\sqrt{5}-1}{2}$	$\frac{\sqrt{5}-1}{2}$	0	1	
G_u	4	-1	-1	1	0	-4	1	1	-1	0	
H_u	5	0	0	-1	1	-5	0	0	1	-1	

G.2 Elements of Representation Theory

The following statements and relations about representations of groups hold and are in most cases easily proved by group theory.

1) Two representations are called independent or non-equivalent if their matrix groups are not conjugated. Conjugated means in this case the matrices are not related to each other by an orthogonal transformation.
2) Groups are composed of classes of conjugated elements. Since the elements of a class are related by orthogonal transformations, the characters for the elements of one class are equal.
3) For a d-dimensional representation defined by (8.9) already one or several subsets of coordinates $x_1, \dots x_{d_k}$ may be mapped onto themselves. In this case the subsets are already the basis for a d_k-dimensional representation and the matrices for the d-dimensional representation have the form

$$\boldsymbol{D}(R) = \begin{pmatrix} \boldsymbol{N}_{d_1}^{(1)} & \cdot & \cdot & \cdot & \cdot \\ \cdot & \boldsymbol{N}_{d_2}^{(2)} & \cdot & \cdot & \cdot \\ \cdot & \cdot & \cdot & \cdot & \cdot \\ \cdot & \cdot & \cdot & \cdot & \cdot \\ \cdot & \cdot & \cdot & \cdot & \boldsymbol{N}_{d_l}^{(l)} \end{pmatrix} . \tag{G.1}$$

The symbols $\boldsymbol{N}_{d_j}^{(j)}$ are d_j-dimensional matrices for d_j-dimensional representations and the points are zeros.
4) Applying all sorts of orthogonal transformations to any d-dimensional representation of a group $\boldsymbol{G}$ we can always find one form in which the elements in the matrix of (G.1) are arranged on the main diagonal in an optimized way. We call this a fully reduced representation of the group $\boldsymbol{G}$. The submatrices $\boldsymbol{N}_{d_j}^{(j)}$ are certainly representations of $\boldsymbol{G}$ in this case, but they cannot be further reduced and are therefore called the *irreducible* representations of $\boldsymbol{G}$. From this we conclude immediately that any d-dimensional representation $\Gamma = \boldsymbol{D}(R)$ can be represented by a sum of irreducible representations. Formally we can write

$$\Gamma = \boldsymbol{D}(R) = \sum_{\alpha} n_{\alpha} \boldsymbol{\Gamma}^{(\alpha)}(R) \ , \tag{G.2}$$

where n_α counts how often a particular irreducible representation $\Gamma^{(\alpha)}$ occurs in the total representation.

5) From (G.1) and (G.2) we learn immediately that the character $\chi^{(\Gamma)}(R)$ of the total representation Γ is the sum of the characters of the irreducible representations occurring in Γ

$$\chi^{(\Gamma)}(R) = \sum_{\alpha} n_{\alpha} \chi^{(\alpha)}(R) \ . \tag{G.3}$$

6) The matrices for the irreducible representations are orthogonal.

$$\sum_{R} D_{ik}^{\alpha *}(R) D_{lm}^{\beta}(R) = \frac{g}{(d_\alpha d_\beta)^{1/2}} \delta_{\alpha\beta} \delta_{il} \delta_{km} \ , \tag{G.4}$$

where d_α and d_β are the dimensions of the representations. Even more important is the orthogonality for the characters of two different irreducible representations as it is given by (8.10) in Sect. 8.2

$$\sum_{R} \chi^{*(\alpha)}(R) \chi^{(\beta)}(R) = g \delta_{\alpha\beta} \ . \tag{G.5}$$

This follows immediately from the orthogonality of the matrices and leads to the magic counting formula.

7) There are several sum rules.
The number of different irreducible representations equals the number of classes in the group.
The sum of the squares of the dimensions of the irreducible representations equals the order of the group.
The sum of the squared absolute values of the characters of an irreducible representation equals the order of the group. This follows immediately from (G.5) and is a convenient way to check whether a representation is reducible or irreducible.
Except for the trivial representation, the sum over equivalent matrix elements of an irreducible representation is zero.

8) The above sum rules and orthogonalities can be checked from the following table which presents the matrices for the irreducible representations explicitly for the point group $\boldsymbol{D_4}$.

Irreducible representations of the group $\boldsymbol{D_4}$

$\boldsymbol{D_4}$	E	C_4	C_2	C_4^3	C_2^x	C_2^y	C_2^{xy}	$C_2^{x'y'}$
$\Gamma^{(1)}$	1	1	1	1	1	1	1	1
$\Gamma^{(2)}$	1	1	1	1	-1	-1	-1	-1
$\Gamma^{(3)}$	1	-1	1	-1	1	1	-1	-1
$\Gamma^{(4)}$	1	-1	1	-1	-1	-1	1	1
$\Gamma^{(5)}$	$\begin{matrix} 1 & 0 \\ 0 & 1 \end{matrix}$	$\begin{matrix} 0 & -1 \\ 1 & 0 \end{matrix}$	$\begin{matrix} -1 & 0 \\ 0 & -1 \end{matrix}$	$\begin{matrix} 0 & 1 \\ -1 & 0 \end{matrix}$	$\begin{matrix} 1 & 0 \\ 0 & -1 \end{matrix}$	$\begin{matrix} -1 & 0 \\ 0 & 1 \end{matrix}$	$\begin{matrix} 0 & 1 \\ 1 & 0 \end{matrix}$	$\begin{matrix} 0 & -1 \\ -1 & 0 \end{matrix}$

G.3 Representation of Groups by Displacement Coordinates

In Sect. 8.3 we used a very simple procedure to evaluate the characters for the $3N$-dimensional representation of the normal coordinates. This calculation is based on the following consideration.

Instead of using the $3N$ normal coordinates we can also use the $3N$ cartesian displacements as the basis for the representation. This is possible since both representations are connected by an orthogonal transformation which leaves the traces of the representation matrices unchanged. The cartesian displacements are vectors and transform therefore like coordinates. The transformation behavior is best explained for a simple geometrical figure like the water molecule. Figure G.1 depicts the molecule with arbitrary displacement vectors $\boldsymbol{u}(1)$ to $\boldsymbol{u}(3)$. The displacement vectors have three coordinates each which yields nine coordinates altogether. The water molecule with displaced atoms certainly does not satisfy the SO for the $\boldsymbol{C_{2v}}$ point group. We can nevertheless apply them and check how the displacement vectors transform. If we apply, e.g., the rotation C_2 the displacement $\boldsymbol{u}(2)$ is mapped on $\boldsymbol{u}(3)$ and vice versa. In addition the coordinates are transformed by this operation. The displacement for atom 1 is mapped onto itself but its coordinates are still interchanged. Similar results are obtained for the other SO of $\boldsymbol{C_{2v}}$. So, the nine displacement coordinates are mapped onto themselves for the SO of the group. This is not surprising since we know they form a basis for a $3N$-dimensional representation of the group. The construction of the representation matrix is now rather easy. All coordinates transform with the 3×3 matrix $\boldsymbol{R}(\phi)$ of (8.2).

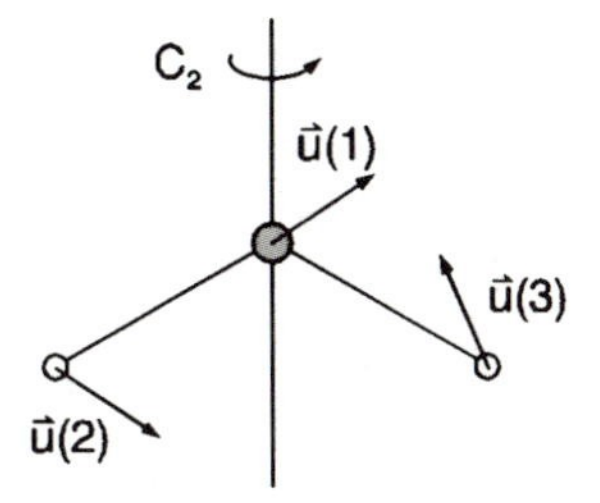

Fig. G.1. Water molecule with displacement vectors.

To get the correct mapping of the displacements from one atom to the other the 3×3 matrices must be properly arranged within the 9×9 transformation matrix for the total system of atoms. For the case of C_2 the transformation equation is easily verified to have the form

$$\begin{pmatrix} \boldsymbol{u}'(1) \\ \boldsymbol{u}'(2) \\ \boldsymbol{u}'(3) \end{pmatrix} = \begin{pmatrix} \boldsymbol{R}(\phi) & \boldsymbol{O} & \boldsymbol{O} \\ \boldsymbol{O} & \boldsymbol{O} & \boldsymbol{R}(\phi) \\ \boldsymbol{O} & \boldsymbol{R}(\phi) & \boldsymbol{O} \end{pmatrix} \begin{pmatrix} \boldsymbol{u}(1) \\ \boldsymbol{u}(2) \\ \boldsymbol{u}(3) \end{pmatrix}. \tag{G.6}$$

The displacement vectors $\boldsymbol{u}$ are written as matrices with one column and three rows, the symbols $\boldsymbol{R}$ and $\boldsymbol{O}$ are 3×3 matrices representing the orthogonal transformation and zeros, respectively. Thus, (G.6) is a 9×9 matrix equation. The matrix on the right-hand side is a nine-dimensional representation of $\boldsymbol{C}_{2v}$. Looking at the shape of the matrix in (G.6) it is immediately evident that only stationary atoms, i.e., atoms which are not moved by the SO, contribute to the trace of the representation matrix. Also, the traces for the individual matrices $\boldsymbol{R}$ do not depend on a particular orientation of the symmetry element under consideration. Thus, generally (8.2) is appropriate for their evaluation.

G.4 Vibrational Species of Rhombohedric $CaCO_3$

The space group for rhombohedric $CaCO_3$ is $\boldsymbol{D}_{3d}^6$ or $R\bar{3}c$ with two formula units per unit cell. The geometrical arrangement for the two molecules is shown in Fig. G.2. The crystallographic point group is $\boldsymbol{D}_{3d}$ with the symmetry elements $E, 3C_3, 3C_2, I, 2S_6, 3\sigma_d$ and the irreducible representations A_{1g}, A_{2g}, E_g, A_{1u}, A_{2u}, E_u. In the space group σ_d is a glide-mirror plane.

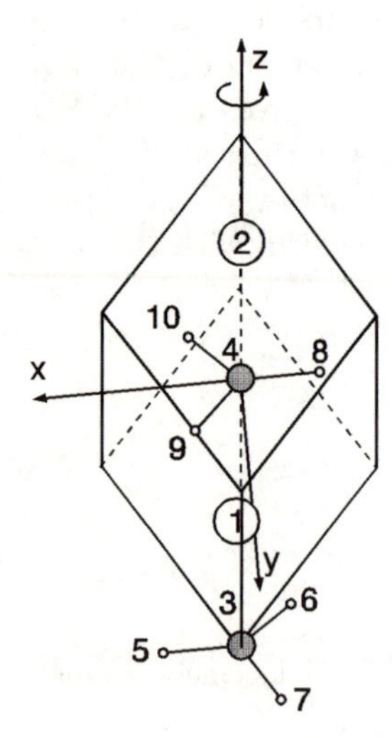

Fig. G.2. Unit cell for rhombohedric $CaCO_3$; (◯: Ca, •: C, ∘: O).

Table G.2. Effect of SO of space group $\boldsymbol{D}_{3d}^{6}$ on the position of the atoms in the unit cell of $CaCO_3$.

Symmetry element		Stationary Atoms	Non-stationary atoms Initial	Final
E		1 to 10		
$2C_3\|\|z$	120° 240°	1,2,3,4	8,9,10,5,6,7	10,8,9,7,5,6
$3C_2\|\|x$ or equivalent	$C_2(1)$	3,4,5,8	1,2,6,7,9,10	2,1,7,6,10,9
axis in	$C_2(2)$	3,4,6,9	1,2,5,7,8,10	2,1,7,5,10,8
xy plane	$C_2(3)$	3,4,7,10	1,2,5,6,8,9	2,1,6,5,9,8
I inversion	I	1,2	3,4,5,8,6,9,7,10	4,3,8,5,9,6,10,7
$2S_6\|\|z$ with reflection	$S_6(1)$	1,2	3,4,5,9,7,8,6,10	4,3,9,7, 8,6,10,5
on a plane $\|\|xy$	$S_6(2)$	1,2	3,4,5,10,6,8,7,9	4,3, 10,6,8,7,9,5
$3\sigma^g$ reflection on zy	$\sigma^g(1)$		1,2,3,4,5,8	2,1,4,3,8,5
or equivalent			6,10,7,9	10,6,9,7
plane with gliding	$\sigma^g(2)$		1,2,3,4,5,10	2,1,4,3,10,5
$(a+b+c)/2$			6,9,7,8	9,6,8,7
	$\sigma^g(3)$		1,2,3,4,5,9,6,8,7,10	2,1,4,3,9,5,8,6,10,7

At first the atoms which are stationary to the SO of the group of the unit cell must be determined. To do this it is convenient to number the atoms in the unit cell as shown in Fig. G.2 and to list their mapping in a table. The z axis is assumed to connect the two Ca atoms and the x axis is oriented symmetrically between the oxygen atoms 9 and 10. In Table G.2 the motions of the atoms are summarized. For the determination of the invariant atoms attention must be paid to atoms which transform to an equivalent position in the neighboring cell. Such atoms are counted as stationary. This happens, e.g., for the SO S_6 and concerns one of the Ca atoms.

Next the character table for the point group $\boldsymbol{D}_{3d}$ is used and Table G.3 is set up for the traces d_{R} of the three-dimensional orthogonal transformations and for the total representation.

Table G.3. Traces and characters for the vibrational analysis of $CaCO_3$.

R	E	$2C_3$	$3C_2$	I	$2S_6$	$3\sigma_d^g$
d_{R}	3	0	−1	−3	0	1
N_{c}	10	4	4	2	2	0
$\chi^{(3\mathrm{N})}$	30	0	−12	−6	0	0

Reducing $\chi^{(3\mathrm{N})}$ according to the irreducible representations of $\boldsymbol{D}_{3d}$ yields

$$\Gamma^{(\mathrm{tot})} = A_{1g} + 3A_{2g} + 4E_g + 2A_{1u} + 4A_{2u} + 6E_u \; . \tag{G.7}$$

From the character tables the species for the translations are A_{2u} and E_u. The same species represent the IR active modes. From the tables for the Raman tensors the A_{1g} and the E_g species are Raman active. This yields finally for the vibrational analysis

$$\Gamma^{(\mathrm{vib})} = A_{1g}(\mathrm{R}) + 3A_{2g} + 4E_g(\mathrm{R}) + 2A_{1u} + 3A_{2u}(\mathrm{IR}) + 5E_u(\mathrm{IR})\,. \quad (\mathrm{G.8})$$

The total number of vibrational degrees of freedom is, as it must be, $3N-3 = 27$.

H. To Chapter 9, Light-Scattering Spectroscopy

H.1 Raman Tensors for the 32 Point Groups

Table H.1. Raman-tensors for vibrational species.

monoclinic

		$\begin{vmatrix} a & d & . \\ d & b & . \\ . & . & c \end{vmatrix}$	$\begin{vmatrix} . & . & e \\ . & . & f \\ e & f & . \end{vmatrix}$
$Oz \parallel C_2$	C_2	A, z	B, x, y
$Oz \perp \sigma_h$	C_{1h}	A', x, y	A'', z
$Oz \parallel C_2$	C_{2h}	A_g	B_g

orthorhombic

		$\begin{vmatrix} a & . & . \\ . & b & . \\ . & . & c \end{vmatrix}$	$\begin{vmatrix} . & d & . \\ d & . & . \\ . & . & . \end{vmatrix}$	$\begin{vmatrix} . & . & e \\ . & . & . \\ e & . & . \end{vmatrix}$	$\begin{vmatrix} . & . & . \\ . & . & f \\ . & f & . \end{vmatrix}$
$Ox, Oy, Oz \parallel C_2^x, C_2^y, C_2^z$	D_2	A	B_1, z	B_2, y	B_3, x
$Oz \parallel C_2^z, Ox \parallel \sigma_y$	C_{2v}	A_1, z	A_2	B_1, x	B_2, y
$Oz \parallel C_2^x, Ox \parallel \sigma_y$	D_{2h}	A_g	B_1g	B_2g	B_3g

trigonal

		$\begin{vmatrix} a & . & . \\ . & a & . \\ . & . & b \end{vmatrix}$	$\begin{vmatrix} c & d & e \\ d & -c & f \\ e & f & . \end{vmatrix}$	$\begin{vmatrix} d & -c & -f \\ -c & -d & e \\ -f & e & . \end{vmatrix}$
$Oz \parallel C_3$	C_3	A, z	E, x	E, y
	C_3	A, z	E, x	E, y
		$\downarrow$	$\begin{vmatrix} c & . & . \\ . & -c & d \\ . & d & . \end{vmatrix}$	$\begin{vmatrix} . & -c & -d \\ -c & . & . \\ -d & . & . \end{vmatrix}$

$Oz \parallel C_3, Ox \parallel C_2$	D_3	A_1	E, x	E, y
	D_{3d}	A_{1g}	$E_g, 1$	$E_g, 2$
		$\downarrow$	$\begin{vmatrix} c & . & d \\ . & -c & . \\ d & . & . \end{vmatrix}$	$\begin{vmatrix} . & -c & . \\ -c & . & d \\ . & d & . \end{vmatrix}$
$Oz \parallel C_3, Ox \parallel \sigma_v$	C_{3v}	A_1, z	E, x	E, y

tetragonal

		$\begin{vmatrix} a & . & . \\ . & a & . \\ . & . & b \end{vmatrix}$	$\begin{vmatrix} c & d & . \\ d & -c & . \\ . & . & . \end{vmatrix}$	$\begin{vmatrix} . & . & e \\ . & . & f \\ e & f & . \end{vmatrix}$	$\begin{vmatrix} . & . & -f \\ . & . & e \\ -f & e & . \end{vmatrix}$	
$Oz \parallel C_4$	C_4	A, z	B	E, x	E, y	
	C_{4m}	A_g	B_g	$E_g, 1$	$E_g, 2$	
		$\downarrow$	$\downarrow$	$\downarrow$	$\begin{vmatrix} . & . & f \\ . & . & -e \\ f & -e & . \end{vmatrix}$	
$Oz \parallel S_4$	S_4	A	B, z	E, x	E, y	
		$\downarrow$	$\begin{vmatrix} c & . & . \\ . & -c & . \\ . & . & . \end{vmatrix}$	$\begin{vmatrix} . & . & d \\ d & . & . \\ . & . & . \end{vmatrix}$	$\begin{vmatrix} . & . & e \\ . & . & . \\ e & . & . \end{vmatrix}$	$\begin{vmatrix} . & . & . \\ . & . & e \\ . & e & . \end{vmatrix}$
$Oz \parallel C_4, Ox \parallel \sigma_v$	C_{4v}	A, z	B_1	B_2	E, x	E, y
		$\downarrow$	$\downarrow$	$\downarrow$	$\begin{vmatrix} . & . & . \\ . & . & e \\ . & e & . \end{vmatrix}$	$\begin{vmatrix} . & . & -e \\ . & . & . \\ -e & . & . \end{vmatrix}$
$Oz \parallel C_4, Ox \parallel C_2'$	D_{4v}	A_1	B_1	B_2	E, x	E, y
	D_{4h}	A_{1g}	B_{1g}	B_{2g}	$E_g, 1$	$E_g, 2$
		$\downarrow$	$\downarrow$	$\downarrow$		$\begin{vmatrix} . & . & e \\ . & . & . \\ e & . & . \end{vmatrix}$
$Oz \parallel S_4, Ox \parallel C_2'$	D_{2d}	A_1	B_1	B_2, z	E, x	E, y

hexagonal

		$\begin{vmatrix} a & . & . \\ . & a & . \\ . & . & b \end{vmatrix}$	$\begin{vmatrix} . & . & c \\ . & . & d \\ c & d & . \end{vmatrix}$	$\begin{vmatrix} . & . & -d \\ . & . & c \\ -d & c & . \end{vmatrix}$	$\begin{vmatrix} e & f & . \\ f & -e & . \\ . & . & . \end{vmatrix}$	$\begin{vmatrix} f & -e & . \\ -e & -f & . \\ . & . & . \end{vmatrix}$
$Oz \parallel C_6$	C_6	A, z	E_1, x	E_1, y	$E_2, 1$	$E_2, 2$
$Oz \parallel C_3$	C_{3h}	A'	$E'', 1$	$E'', 2$	E', x	E', y
$Oz \parallel C_6$	C_{6h}	A_g	$E_{1g}, 1$	$E_{1g}, 2$	$E_{2g}, 1$	$E_{2g}, 2$
		$\downarrow$	$\begin{vmatrix} . & . & . \\ . & . & c \\ . & c & . \end{vmatrix}$	$\begin{vmatrix} . & . & -c \\ . & . & . \\ -c & . & . \end{vmatrix}$	$\begin{vmatrix} d & . & . \\ . & -d & . \\ . & . & . \end{vmatrix}$	$\begin{vmatrix} . & -d & . \\ -d & . & . \\ . & . & . \end{vmatrix}$
$Oz \parallel C_6, Ox \parallel C_2'$	D_6	A_1	E_1, x	E_1, y	$E_2, 1$	$E_2, 2$
$Oz \parallel C_6, Ox \parallel C_2'$	D_{6h}	A_{1g}	$E_{1g}, 1$	$E_{1g}, 2$	$E_{2g}, 1$	$E_{2g}, 2$
$Oz \parallel C_3, Ox \parallel C_2$	D_{3h}	A_1	$E'', 1$	$E'', 2$	E', x	E', y
		$\downarrow$	$\begin{vmatrix} . & . & c \\ . & . & . \\ c & . & . \end{vmatrix}$	$\begin{vmatrix} . & . & . \\ . & . & c \\ . & c & . \end{vmatrix}$	$\downarrow$	$\downarrow$
$Oz \parallel C_6, Ox \parallel \sigma_v$	C_{6v}	A_1, z	E_1, x	E_1, y	$E_2, 1$	$E_2, 2$

cubic

		$\begin{vmatrix} a & . & . \\ . & a & . \\ . & . & a \end{vmatrix}$	$\begin{vmatrix} b+c\sqrt{3} & . & . \\ . & b\text{-}c\sqrt{3} & . \\ . & . & \text{-}2b \end{vmatrix}$	$\begin{vmatrix} c\text{-}b\sqrt{3} & . & . \\ . & c+b\sqrt{3} & . \\ . & . & \text{-}2c \end{vmatrix}$	$\begin{vmatrix} . & . & . \\ . & . & d \\ . & d & . \end{vmatrix}$	$\begin{vmatrix} . & . & d \\ . & . & . \\ d & . & . \end{vmatrix}$	$\begin{vmatrix} . & d & . \\ d & . & . \\ . & . & . \end{vmatrix}$
$Ox, Oy, Oz \parallel$	T	A	$E, 1$	$E, 2$	F, x	F, y	F, z
O_2^x, C_2^y, C_2^z,	T_h	A_g	$E_g, 1$	$E_g, 2$	$F_g, 1$	$F_g, 2$	$F_g, 3$
		$\downarrow$	$\begin{vmatrix} b & . & . \\ . & b & . \\ . & . & \text{-}2b \end{vmatrix}$	$\begin{vmatrix} \text{-}b\sqrt{3} & . & . \\ . & b\sqrt{3} & . \\ . & . & 0 \end{vmatrix}$	$\downarrow$	$\downarrow$	$\downarrow$
$Ox, Oy, Oz \parallel$	O	A_1	$E, 1$	$E, 2$	$F_2, 1$	$F_2, 2$	$F_2, 3$
or S_4	T_d	A_1	$E, 1$	$E, 2$	F_2, x	F_2, y	F_2, z
	O_h	A_{1g}	$E_g, 1$	$E_g, 2$	$F_{2g}, 1$	$F_{2g}, 2$	$F_{2g}, 3$

The first and the second columns in the table give the geometry and the point group, respectively, for which the Raman tensors are listed. The following columns give the vibrational species and the corresponding Raman tensors. A down arrow means the Raman tensor from the species above can be used. The coordinates x, y, z next to the Mulikan symbols assign the polarization of the mode.

H.2 Averaging of Raman Tensor Components

The average value of a tensor component is obtained by averaging over all orientations in space. This is done by evaluating the tensor component for an arbitrary orientation in space using Euler angles ψ, θ, ϕ. The definition of the Euler angles as they are used here is given in Fig. H.1. Orthogonal transformation of the Raman tensor $\chi_{mn,k}$ with a general rotation matrix $\boldsymbol{O}(\psi, \theta, \phi)$ gives the tensor for arbitrary orientation. Written in operator form this means

$$\chi' = \mathbf{O}\chi\mathbf{O}^T \, . \tag{H.1}$$

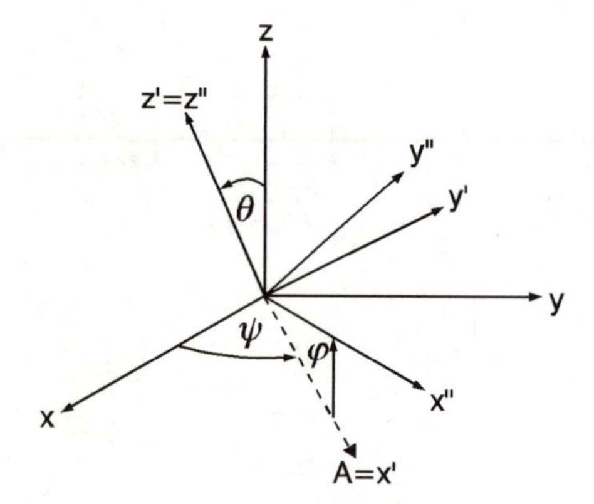

Fig. H.1. Euler angles for the rotation of coordinates by arbitrary angles.

The Matrix $\boldsymbol{O}$ can be obtained from several successive rotations of the following form

$$\boldsymbol{O} = \boldsymbol{O}_{z'}(\phi)\boldsymbol{O}_A(\theta)\boldsymbol{O}_z(\psi) \,, \tag{H.2}$$

with

$$\begin{aligned} \boldsymbol{O}_A(\theta) &= \boldsymbol{O}_z(\psi)\,\boldsymbol{O}_x(\theta)\boldsymbol{O}_z(-\psi) \\ \boldsymbol{O}_{z'}(\phi) &= \boldsymbol{O}_A(\theta)\,\boldsymbol{O}_z(\phi)\boldsymbol{O}_A(-\theta) \,. \end{aligned} \tag{H.3}$$

Inserting (H.3) into (H.2) yields

$$\boldsymbol{O}(\psi,\theta,\phi) = \boldsymbol{O}_z(\psi)\boldsymbol{O}_x(\theta)\boldsymbol{O}_z(\phi) \,. \tag{H.4}$$

This means the matrix for the general rotation is obtained from rotations around the z axis and around the x axis by the angles ψ, θ, and ϕ. The individual rotation matrices are given by (8.2). With the abbreviations

$$\begin{array}{lll} \cos\theta = c_1, & \cos\psi = c_2, & \cos\phi = c_3 \\ \sin\theta = s_1, & \sin\psi = s_2, & \sin\phi = s_3 \end{array}$$

we obtain from the matrix multiplication

$$\boldsymbol{O}(\psi,\theta,\phi) = \begin{pmatrix} c_2c_3 - c_1s_2s_3 & s_2c_3 + c_1c_2s_3 & s_1s_3 \\ -c_2s_3 - s_1s_2c_3 & -s_2s_3 + c_1c_2c_3 & s_1c_3 \\ s_1s_2 & -s_1s_2 & c_1 \end{pmatrix} . \tag{H.5}$$

We verify as an example the first relation in (9.40) for $\chi_{33,k}$. The index k is dropped for simplicity in the following. The relation to prove is

$$\begin{aligned} \overline{\chi_{33}'^2} &= a^2 + \frac{4}{15}\tau^2 \\ &= \frac{1}{5}(\chi_{11}^2 + \chi_{22}^2 + \chi_{33}^2) + \frac{2}{15}(\chi_{11}\chi_{22} + \chi_{22}\chi_{33} + \chi_{33}\chi_{11}) + \frac{4}{15}(\chi_{12}^2 \\ &\quad + \chi_{23}^2 + \chi_{31}^2) \,. \end{aligned} \tag{H.6}$$

The transformed component χ'_{33} is obtained from (H.1) and (H.5) by

$$\begin{aligned} \chi'_{33} = &\sum_{kl} O_{3k}O_{3l}\chi_{kl} \\ = &\ \chi_{11}s_1^2s_3^2 + \chi_{22}s_1^2c_3^2 + \chi_{33}c_1^2 \\ &+ 2\chi_{12}s_1^2s_3c_3 + 2\chi_{23}s_1c_1c_3 + 2\chi_{31}c_1s_1s_3 \,. \end{aligned} \tag{H.7}$$

The square of this expression has to be averaged over the Euler angles in the form

$$\begin{aligned} \overline{\chi_{33}'^2} &= \frac{1}{2\pi}\int_0^{2\pi} \mathrm{d}\phi \, \frac{1}{2\pi}\int_0^{2\pi} \mathrm{d}\psi \, \frac{1}{2}\int_0^{\pi} \sin\theta \mathrm{d}\theta \chi_{33}'^2 \\ &= \frac{1}{4\pi}\int_0^{2\pi} \mathrm{d}\phi \int_{-1}^{1} \mathrm{d}(\cos\theta)\chi_{33}'^2 \,. \end{aligned} \tag{H.8}$$

The explicit expression for $(\chi'_{33})^2$ is obtained in a straightforward way from (H.7). It consists of a sum of products of powers of sine and cosine functions with products of tensor components χ_{ik}. From the averaging all contributions c_3^m, s_3^n with uneven m, n can be dropped. The rest are simple integrals over powers of sine and cosine terms. The calculation is cumbersome but straightforward and finally yields

$$\begin{aligned}\overline{\chi_{33}'^2} = \; & \frac{1}{5}(\chi_{11}^2 + \chi_{22}^2 + \chi_{33}^2) + \frac{2}{15}(\chi_{11}\chi_{22} + \chi_{22}\chi_{33} + \chi_{33}\chi_{11}) \\ & + \frac{4}{15}(\chi_{12}^2 + \chi_{23}^2 + \chi_{31}^2) \, . \end{aligned} \tag{H.9}$$

I. To Chapter 10, Infrared Spectroscopy

I.1 Line Shape Function from the Fluctuation–Dissipation Theorem

The relationship between the absorption α and the autocorrelation function $\langle \boldsymbol{\mu}(0)\boldsymbol{\mu}(t)\rangle$ is based on a very general relationship between a linear response function and the autocorrelation function of a generalized displacement. It is known as the *fluctuation–dissipation theorem* [see (L.30) in Appendix L.5]. The generalized response function $T(\omega)$ is obtained from the relation between a generalized force $F(\omega)$ and the generalized displacement $X(\omega)$. The theorem relates the power spectrum of the displacement to the imaginary part of the linear response function. For more details on the fluctuation–dissipation theorem special textbooks or references must be considered [10.14, 10.15]. Since $\alpha(\omega)/\omega$ is related to the imaginary part of the dielectric function and the dipole moment is a special case of the generalized displacement the relationship between the two quantities is covered by the fluctuation–dissipation theorem. A detailed evaluation in [10.15] yields

$$\begin{aligned}&\frac{\alpha_{v_{ki}v_{kf}}(\omega + \Omega_k)}{(\omega + \Omega_k)(1 - \mathrm{e}^{-\hbar(\omega+\Omega_k)/k_\mathrm{B}T})} \\ &\quad = \frac{\pi n_\mathrm{d}}{3\hbar c_0 \varepsilon_0}\left[\int \mathrm{e}^{\mathrm{i}\omega t}\langle[\boldsymbol{\mu}]_{v_{ki}v_{kf}}(0)[\boldsymbol{\mu}]_{v_{ki}v_{kf}}(t)\rangle \mathrm{d}t\right] \, . \end{aligned} \tag{I.1}$$

The brackets on the right-hand side of the equation indicate an ensemble average. For an ergodic system this average is identical to the correlation function as defined in (L.23). Thus, the integral is the Fourier transform for the correlation function of the dipole moments. Fourier back transformation yields

$$\begin{aligned}&\langle[\boldsymbol{\mu}]_{v_{ki}v_{kf}}(0)[\boldsymbol{\mu}]_{v_{ki}v_{kf}}(t)\rangle \\ &\quad = \frac{3\hbar c_0 \varepsilon_0}{2\pi n_\mathrm{d}} \int_\mathrm{band} \frac{\mathrm{e}^{-\mathrm{i}\omega t}\alpha_{v_{ki}v_{kf}}(\omega + \Omega_k)}{(\omega + \Omega_k)(1 - \mathrm{e}^{-\hbar(\omega+\Omega_k)/k_\mathrm{B}T})}\mathrm{d}\omega \, . \end{aligned} \tag{I.2}$$

As in the case of optical absorption thermal averaging over the occupation numbers is required. This yields the absorption coefficient α_k for mode Ω_k on the right-hand side of the equation.

The correlation function on the left-hand side is evaluated by replacing it with the the transition matrix element from (L.23). This matrix element is approximated for $t = 0$ by (10.22) together with (10.23). For $\hbar\Omega_k > k_{\mathrm{B}}T$ the square of the matrix element becomes

$$|[\boldsymbol{\mu}]_{\mathrm{fi}}|^2 = \left|\frac{\mathrm{d}\boldsymbol{\mu}}{\mathrm{d}Q_k}\right|^2 \frac{\hbar}{2\Omega_k} \; .$$

With an appropriate transformation of the integration variable on the right-hand side we obtain finally

$$\left|\frac{\partial\boldsymbol{\mu}}{\partial Q_k}\right|^2 = \frac{12\Omega_k c_0 \varepsilon}{n_{\mathrm{d}}} \int \frac{\alpha_k(\omega)}{\omega} \mathrm{d}\omega \; .$$

J. To Chapter 11, Magnetic Resonance Spectroscopy

J.1 Transformation for Velocities Between Laboratory System and Rotating System

We consider the general case of a coordinate system with origin O' moving in a laboratory system with origin O as shown in Fig. J.1. The mathematical relationship between a vector $\boldsymbol{r}$ in the laboratory system and the same vector $\boldsymbol{r}'$ in the moving system is

$$\boldsymbol{r} = \boldsymbol{r}_0 + \boldsymbol{r}' = \boldsymbol{r}_0 + \sum x_i' \boldsymbol{e}_i' \; , \tag{J.1}$$

where $\boldsymbol{e}_i'$ and x_i' are the unit vectors and the components of $\boldsymbol{r}'$ in the moving system and $\boldsymbol{r}_0$ is the vector between O and O'. The absolute velocity (i.e., the velocity in the laboratory system) is

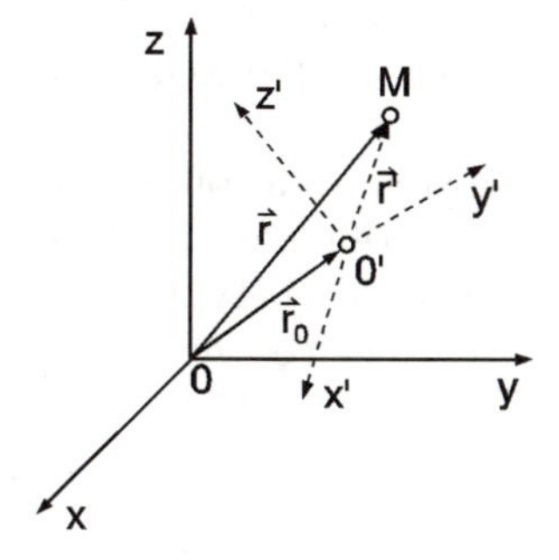

Fig. J.1. Geometry for the motion of a mass point M in the laboratory system O and in the moving system O'.

$$v_{\mathrm{l}} = \frac{\mathrm{d}\boldsymbol{r}}{\mathrm{d}t} = \frac{\mathrm{d}\boldsymbol{r_0}}{\mathrm{d}t} + \sum x_i' \frac{\mathrm{d}\boldsymbol{e}_i{}'}{\mathrm{d}t} + \sum \frac{\mathrm{d}x_i'}{\mathrm{d}t}\boldsymbol{e}_i{}' \,. \tag{J.2}$$

The last sum in (J.2) is the relative velocity $\mathrm{d}\boldsymbol{r}'/\mathrm{d}t$ in the moving frame. The time derivatives in the first sum describe the rotation of the frame O'

$$\frac{\mathrm{d}\boldsymbol{e}_{\mathrm{i}}'}{\mathrm{d}t} = \boldsymbol{\omega} \times \boldsymbol{e}_{\mathrm{i}}' \,. \tag{J.3}$$

Thus the general relation between the time derivatives of a vector $\boldsymbol{r}$ in the two systems is

$$\frac{\mathrm{d}\boldsymbol{r}}{\mathrm{d}t} = \frac{\mathrm{d}\boldsymbol{r_0}}{\mathrm{d}t} + \frac{\mathrm{d}\boldsymbol{r}'}{\mathrm{d}t} + \boldsymbol{\omega} \times \boldsymbol{r}' \,. \tag{J.4}$$

For the special case where O' coincides with O and the translational velocity $\mathrm{d}\boldsymbol{r}_0/\mathrm{d}t$ is zero relation (J.4) simplifies to

$$\frac{\mathrm{d}\boldsymbol{r}}{\mathrm{d}t} = \frac{\mathrm{d}\boldsymbol{r}'}{\mathrm{d}t} + \boldsymbol{\omega} \times \boldsymbol{r}' \,. \tag{J.5}$$

J.2 Exchange Interaction

The exchange interaction is a characteristic phenomenon of quantum mechanics. In a system of at least two nuclei a and b and two electrons 1 and 2 with distance r_{12} the interaction energy between the electrons is not only the Coulomb energy U_{C} but also the exchange energy of the form

$$K = \int \psi_{\mathrm{ab}}^*(x_1)\psi_{\mathrm{ab}}(x_2)\frac{\mathrm{e}^2}{r_{12}}\mathrm{d}x_1\mathrm{d}x_2 \tag{J.6}$$

with

$$\psi_{\mathrm{ab}}(x_i) = \psi_{\mathrm{a}}(x_i)\psi_{\mathrm{b}}(x_i) \,.$$

ψ_{a} and ψ_{b} are the wave functions for nucleous a and b with electrons 1 and 2, respectively. Thus, the wave function ψ_{ab} is given by the product of the wave function for the electron 1 in the orbit ψ_{a} times the fraction which this electrons occupies from orbit ψ_{b}. To obtain a sizable exchange interaction an overlap between the wave functions $\psi_{\mathrm{a}}(x_1)$ and $\psi_{\mathrm{b}}(x_2)$ must exist at least in a certain range of space. The higher the concentration of the exchanged electrons the higher the interaction energy.

A classical example of an exchange interaction by a spin–spin coupling is the Heisenberg model for ferromagnetism. The spin orientation and thus also the energy U of the system is given by an exchange interaction between spin S_i from atom i with spin j from atom j. In this case the exchange energy has the form

$$U = -2KS_iS_j \,. \tag{J.7}$$

K. To Chapter 13, Spectroscopy with γ Rays

K.1 Oscillator Models for Recoil-Free Emission of γ Radiation

The evaluation of the recoil-free fraction of γ quanta emitted from a nucleus needs the consideration of all phonon states. Let us label the phonons with s and their frequencies and occupation numbers with Ω_s and α_s, respectively. The fraction of recoil-free emission is then

$$f = \exp\left[-\frac{\hbar k^2}{3NM}\sum_s \frac{1}{\Omega_s}(\alpha_s + 1/2)\right] , \tag{K.1}$$

where N is the number of oscillators. This equation can be reformulated using the recoil energy from (13.1), replacing the sum by an integral, and introducing a density of states $\varrho(\Omega)$ for the phonons. This yields

$$f = \exp\left[-\frac{2\epsilon_{\mathrm{R}}}{3N\hbar}\int_0^{\Omega_{\max}} \frac{1}{\Omega}[n(\Omega) + 1/2]\varrho(\Omega)\mathrm{d}\Omega\right] , \tag{K.2}$$

where $n(\Omega)$ is given by the Bose–Einstein factor of (9.32). Equation (K.2) allows the calculation of the fraction of recoil-free emission for various lattice-dynamical models. In the simplest case we can use the Einstein oscillators with a density of states

$$\varrho_{\mathrm{E}}(\Omega) = 3N\delta(\Omega - \Omega_{\mathrm{E}}) . \tag{K.3}$$

Inserting the δ function into the integral of (K.2) yields

$$f = \exp\left[-\frac{\epsilon_{\mathrm{R}}}{k_{\mathrm{B}}\Theta_{\mathrm{E}}}\coth\left(\frac{\Theta_{\mathrm{E}}}{2T}\right)\right] , \tag{K.4}$$

where Θ_{E} is the temperature equivalent to the oscillator. For $T \gg \Theta_{\mathrm{E}}$ f can be approximated by

$$f = \exp\left[-\frac{\epsilon_{\mathrm{R}}}{k_{\mathrm{B}}\Theta_{\mathrm{E}}}\left(\frac{2T}{\Theta_{\mathrm{E}}} + \frac{\Theta_{\mathrm{E}}}{6T}\right)\right] . \tag{K.5}$$

The latter equation approaches (13.5) for high temperatures.

In the case of a Debye model for the lattice oscillators the density of states is

$$\begin{aligned} \varrho_{\mathrm{D}}(\Omega) &= \frac{9N}{\Omega_{\mathrm{D}}^3}\Omega^2 \quad \text{for} \quad \Omega < \Omega_{\mathrm{D}} \\ &= 0 \qquad\quad \text{for} \quad \Omega > \Omega_{\mathrm{D}} . \end{aligned} \tag{K.6}$$

Using this density of states in (K.2) yields

$$f(T) = \exp\left(-3\epsilon_{\mathrm{R}}\frac{1 + 4(T/\Theta_{\mathrm{D}})^2 I(\Theta_{\mathrm{D}}/T)}{2k_{\mathrm{B}}\Theta_{\mathrm{D}}}\right) ,$$

with

$$I(T/\Theta_{\mathrm{D}}) = \int_0^{\Theta_{\mathrm{D}}/T} \frac{x\mathrm{d}x}{\mathrm{e}^x - 1} \,.$$

$\Theta_{\mathrm{D}} = \hbar\Omega_{\mathrm{D}}/k_{\mathrm{B}}$ is the Debye temperature of the system.

L. To Chapter 14, Generalized Dielectric Function

L.1 Linear Response Theory

A general response function Φ_{AB} is defined by the *response* of a system to a perturbation $B = h(t)$ where the response is observed by the variable A. If the perturbation is small the response is linear and has the form

$$\overline{A(t)} = \langle A \rangle + \int_{-\infty}^{t} \mathrm{d}t' \Phi_{\mathrm{AB}}(t-t')h(t') = \langle A \rangle + \int_0^{\infty} \mathrm{d}\tau \Phi_{\mathrm{AB}}(\tau)h(t-\tau) \,. \quad \text{(L.1)}$$

The average $\langle\rangle$ is taken over the ensemble in space. We assume $\langle A \rangle = 0$ in the following and understand $\overline{A(t)} = \langle A \rangle$ as the change in A induced by the perturbation $h(t)$. $h(t)$ is assumed to be a real force so that $\Phi_{\mathrm{AB}}(t)$ is also real. The definition of the response function in (L.1) implies causality. The response at time t is determined by the perturbation acting the full time interval from $-\infty$ to t. Considering the Fourier transform of (L.1) and remembering that the Fourier transform for a convolution of two functions is the product of the transformed functions we have

$$\boxed{\langle A(\omega) \rangle = \Phi_{\mathrm{AB}}(\omega)h(\omega) = \chi_{\mathrm{AB}}(\omega)h(\omega) \,,} \quad \text{(L.2)}$$

with

$$\boxed{\chi_{\mathrm{AB}}(\omega) = \frac{\langle A(\omega) \rangle}{h(\omega)} = \int_0^{\infty} \mathrm{d}t \Phi_{\mathrm{AB}}(t) \mathrm{e}^{\mathrm{i}\omega t} \,.} \quad \text{(L.3)}$$

$\chi_{\mathrm{AB}}(\omega)$ is called the *generalized susceptibility* and (L.2) is a generalized form of (6.1) from Chap. 6. While $\Phi_{\mathrm{AB}}(t)$ is real $\chi_{\mathrm{AB}}(\omega)$ is complex of the form

$$\chi(\omega) = \chi_{\mathrm{r}}(\omega) + \mathrm{i}\chi_{\mathrm{i}}(\omega) \,, \quad \text{(L.4)}$$

where we have dropped the indices AB for simplicity. From the definition for $\chi(\omega)$

$$\chi(-\omega) = \chi^*(\omega)$$

follows immediately. Considering the real part and the imaginary part of χ separately we find

$$\chi_{\mathrm{r}}(-\omega) = \chi_{\mathrm{r}}(\omega), \quad \chi_{\mathrm{i}}(-\omega) = -\chi_{\mathrm{i}}(\omega) \,. \quad \text{(L.5)}$$

In other words, $\chi_{\mathrm{r}}(\omega)$ is an even and $\chi_{\mathrm{i}}(\omega)$ is an odd function of the frequency ω.

From the definition of the generalized susceptibility in (L.3) which implies causality between the perturbation of a system and its response two very important properties of the linear response functions can be derived. To discuss these properties we consider the response functions on the complex ω plane $\omega_r + i\omega_i$.

Firstly, $\chi(\omega)$ turns out to be always analytic (without poles) in the upper half of this plane.

Secondly, and even more important, a fundamental relation exists between the real part and the imaginary part of $\chi(\omega)$. We assume an arbitrary value of $\omega = \omega_0$ on the real axis and consider the path integral along C

$$I = \int_C \frac{\chi(\omega)}{\omega - \omega_0} \mathrm{d}\omega \, . \tag{L.6}$$

The path is a closed loop as displayed in Fig. L.1. It is chosen along the real axis from $-\infty$ to ∞ and closed by a half-circle in the upper complex plane. The pole of the integrand in (L.6) at $\omega = \omega_0$ is excluded by an infinitely small half-circle. Since the integrand is analytic along and within the integration path the value of the integral is zero. $\chi(\omega)$ is assumed to vanish for infinitely large values of ω so that the contribution of the half-circle in the upper complex plane to the integral vanishes[1]. The contribution of the infinitely small circle around ω_0 to the integral is $-i\pi\chi(\omega_0)$. This enables (L.6) to be rewritten in the form

$$I = \lim_{\varrho \to 0} \left\{ \int_{-\infty}^{\omega_0 - \varrho} \frac{\chi(\omega_r)\mathrm{d}\omega_r}{\omega_r - \omega_0} + \int_{\omega_0 + \varrho}^{\infty} \frac{\chi(\omega_r)\mathrm{d}\omega_r}{\omega_r - \omega_0} \right\} - i\pi\chi(\omega_0) = 0 \, . \tag{L.7}$$

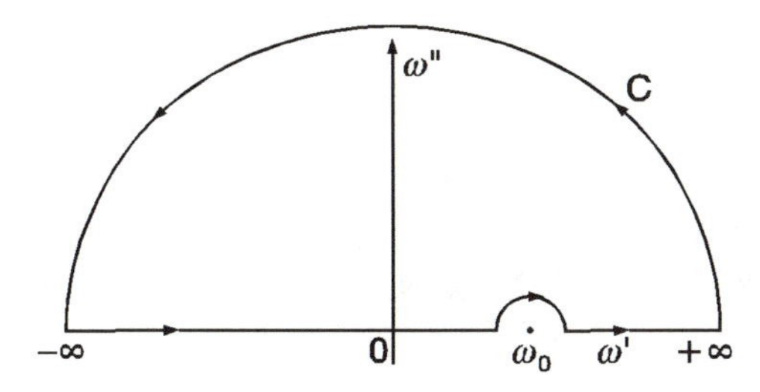

Fig. L.1. Path of integration to calculate the integral in (L.6).

The expression within the curly brackets is known as the *principal value* P of the integral from $-\infty$ to ∞. Relabeling ω_r as ω we obtain the complex *dispersion relation*

$$\boxed{\chi(\omega_0) = -\frac{i}{\pi} \mathrm{P} \int \frac{\chi(\omega)}{\omega - \omega_0} \mathrm{d}\omega \, .} \tag{L.8}$$

[1] If $\chi(\infty)$ is finite we can always consider $\chi(\omega) - \chi(\infty)$ as the response function which vanishes for $\omega = \infty$.

Separating real and imaginary parts yields the famous *Kramers–Kronig* relations

$$\begin{aligned}\chi_\mathrm{r}(\omega) &= \frac{1}{\pi}\mathrm{P}\int \frac{\chi_\mathrm{i}(\omega)}{\omega-\omega_0}\,\mathrm{d}\omega \\ \chi_\mathrm{i}(\omega) &= -\frac{1}{\pi}\mathrm{P}\int \frac{\chi_\mathrm{r}(\omega)}{\omega-\omega_0}\,\mathrm{d}\omega\,.\end{aligned} \tag{L.9}$$

The relations even hold when the response function has a pole of the form $\chi = \mathrm{i}A/\omega$ at $\omega = 0$. In this case the point $\omega = 0$ on the real axis must also be excluded from the integration and contributes a value of $-\mathrm{i}\pi A/\omega$ to the integral in (L.8).

Considering that $\chi_\mathrm{r}(\omega)$ and $\chi_\mathrm{i}(\omega)$ are even and odd, respectively, one arrives readily at the Kramers–Kronig relations from Sect. 6.1.3.

L.2 The Density–Density Response Function

To obtain a general analytical expression for the response function we restrict ourselves to the linear response of the particle density $n(\boldsymbol{r})$ and a scalar and velocity independent interaction energy $U(r,t)$ of a probe particle. The energy is assumed periodic in time with frequency ω and with Fourier components $U_\mathrm{q} = \mathcal{V}U(q,\omega)$. Since the perturbation is assumed small the response is not only expected to be linear but also selective in q. This means Fourier components can be studied instead of full space- and time-dependences.

The density for a discrete distribution of N particles is

$$n(\boldsymbol{r}) = \frac{1}{N}\sum_j \delta(\boldsymbol{r}-\boldsymbol{r}_j) \tag{L.10}$$

with the Fourier transform n_q

$$\begin{aligned}n_\mathrm{q} &= \int n(\boldsymbol{r})\mathrm{e}^{-\mathrm{i}\boldsymbol{q}\cdot\boldsymbol{r}}\,\mathrm{d}^3x \\ &= \frac{1}{N}\sum_j \int \delta(\boldsymbol{r}-\boldsymbol{r}_j)\mathrm{e}^{-\mathrm{i}\boldsymbol{q}\cdot\boldsymbol{r}}\,\mathrm{d}^3x = \frac{1}{N}\sum_j \mathrm{e}^{-\mathrm{i}\boldsymbol{q}\cdot\boldsymbol{r}_j}\,.\end{aligned} \tag{L.11}$$

The Fourier components of the total interaction energy have the form

$$H_\mathrm{int}(q,\omega) = U(q,\omega)n_\mathrm{q}\,.$$

The interaction Hamiltonian written as an operator with time-dependence given by the frequency ω is

$$\mathbf{H}_\mathrm{int} = \mathbf{n}_\mathrm{q}^{+}U(q,\omega)\mathrm{e}^{-\mathrm{i}\omega t}\mathrm{e}^{\alpha t}\,. \tag{L.12}$$

α is a small stability parameter to guarantee adiabatic boundary conditions. It can be reduced to zero in final results. With this interaction the system is transferred from the ground state $|0\rangle$ to an excited state $|l\rangle$ while the momentum of the probe is changed from p_pr to $p_\mathrm{pr} - q$.

The golden rule of quantum mechanics determines the probability for the transition

$$P(q,\omega) = \frac{2\pi}{\hbar}|U_{\mathrm{q}}|^2 \sum_l |(n_{\mathrm{q}})_{10}|^2 \delta(\hbar\omega - \hbar\omega_{10}) \; . \tag{L.13}$$

The sum extends over the complete set of eigenstates of the system, $\hbar\omega_{l0}$ are the transition energies, and the relevant matrix elements $(n_q)_{10}$ are given by the interaction Hamiltonian from (L.12)

$$\langle l|\mathbf{H}_{\mathrm{int}}|0\rangle = U(q,\omega)\langle l|\mathbf{n}_{\mathrm{q}}^{+}|0\rangle = U(q,\omega)(\mathbf{n}_{\mathrm{q}}^{+})_{10} \; . \tag{L.14}$$

According to (L.2) the Fourier components $\langle n(q,\omega)\rangle$ of the resulting particle density are related to the generalized susceptibility $\chi(q,\omega)$ by

$$\boxed{\frac{\langle n(q,\omega)\rangle}{U(q,\omega)} = \chi(q,\omega) \; .} \tag{L.15}$$

This function is called the *density–density response function*. The expectation value in (L.15) is a dimensionless quantity. Within the linear response theory it is proportional to the pertubation $U(q,\omega)$ and defined as

$$\langle n(q,\omega)\rangle = \langle\psi(r,t)|n_{\mathrm{q}}\mathrm{e}^{\mathrm{i}\omega t}|\psi(r,t)\rangle \; .$$

$\psi(r,t)$ are the exact wave functions including the probe particle. These wave functions must be evaluated from the time-dependent Schrödinger equation. For a linear response this can be done by first-order perturbation theory in a very similar way as demonstrated in Sect. F.1. The result yields the following expression for χ

$$\boxed{\chi(q,\omega) = \sum_l |(n_{\mathrm{q}})_{l0}|^2 \frac{2\hbar\omega_{l0}}{(\hbar\omega + \mathrm{i}\hbar\alpha)^2 - \hbar^2\omega_{l0}^2} \; .} \tag{L.16}$$

(For details of the calculation see [14.1], p. 97ff.)

The very famous result of (L.16) which expresses the linear response function in terms of the true eigenstates of the system was first reported by *R. Kubo* in 1956. The type of calculation is quite general and not restricted to a density–density response. Similar results have been obtained for a vector potential as a perturbation of charged particles. In this case a current–current response function is obtained.

For the explicit evaluation of the matrix elements $(n_q)_{l0}$ and transition energies $\hbar\omega_{10}$ of a particular system quantum-mechanical calculations are required.

An instructive example for the above model is the generalized susceptibility for noninteracting and neutral particles which follow Fermi statistics. The allowed states of the particles are given by the Fermi distribution function $f_{\mathrm{F}}(k)$ where $\hbar k$ is the momentum of the particles. The perturbation induces transitions from states $|k\rangle$ to states $|k+q\rangle$ with transition matrix elements $(n_q)_{10} = (n_q)_{k+q,k} = (n_q)_{qk}$ and transition energies

$$\hbar\omega_{l0} = \hbar\omega_{k+q,k} = \hbar[\omega(k+q) - \omega(k)] = \hbar\omega_{qk} \ .$$

The square of the matrix element $|(n_q)_{qk}|^2$ must be calculated explicitly. This has been done using a Hartree–Fock calculation or within a random phase approximation [14.1]. The correct result can also be obtained intuitively. The probability for a transition from state $|k\rangle$ to state $|k+q\rangle$ is given, at finite temperatures, by the product for the probabilities that state $|k\rangle$ is occupied and state $|k+q\rangle$ is empty. This product is $f_{\mathrm{F}}(k)[1 - f_{\mathrm{F}}(k+q)]$. Since the sum in (L.16) extends over all eigenstates k we obtain for a given vector q

$$\chi^0(q,\omega) = \sum_k f_{\mathrm{F}}(k)[1 - f_{\mathrm{F}}(k+q)] \frac{2\hbar\omega_{qk}}{(\hbar\omega + \mathrm{i}\hbar\alpha)^2 - \hbar^2\omega_{qk}^2} \ . \tag{L.17}$$

By separating the fraction into a sum of two fractions where the denominator is linear in energy and renormalizing the index which counts the k vectors in the second fraction in a suitable way (L.17) becomes

$$\chi^0(q,\omega) = \sum_k \frac{f_{\mathrm{F}}(k) - f_{\mathrm{F}}(k+q)}{\hbar\omega + \epsilon(k) - \epsilon(k+q) + \mathrm{i}\hbar\alpha} \ . \tag{L.18}$$

Note that the dimension of $\chi^0(q,\omega)$ is an inverse energy.

L.3 Dynamical Form Factor and Correlation Functions

The response of a system is closely linked to the correlations which exist between the positions and momenta of different particles. To describe such correlations it is common practice to introduce an additional function $S(q,\omega)$ which describes the dynamical structure of the system. This *dynamical form factor* is defined from (L.13) by

$$P(q,\omega) = \frac{2\pi}{\hbar} |U_q|^2 S(q,\omega) \ , \tag{L.19}$$

where

$$\boxed{S(q,\omega) = \sum_l |(n_q)_{l0}|^2 \delta(\hbar\omega - \hbar\omega_{l0}) \quad (\text{in J}^{-1}) \ .} \tag{L.20}$$

$S(q,\omega)$ is a very fundamental function of the solid since it describes all possible excitations for a particular probe. It is always real and positive and vanishes for $\omega < 0$. The latter is a consequence of the δ function in (L.20).

Like for all spectral excitations of solids sum rules exist for the dynamical form factor. One example is

$$\int_0^\infty \omega S(q,\omega) \mathrm{d}\omega = \frac{Nq^2}{2m} \ . \tag{L.21}$$

A realistic example for the above situation would be a scattering experiment with neutrons in liquid He^4. However, as we will see in Appendix L.5 the dynamical form factor is likewise obtained for charged particles and the concept is applicable to inelastic scattering of neutrons, electrons, light, and even to absorption, in a large variety of media such as crystals, amorphous solids, liquids, etc.

As we may expect from the above discussion the dynamical form factor is closely related to the density correlations and to the generalized susceptibility $\chi(q, \omega)$. The former relation is obtained by introducing the integral representation of the δ function from Appendix B.7 and representing the time evolution of the density excitations in the Heisenberg picture. This procedure is again very common in spectroscopy. From (L.20)

$$\begin{aligned} S(q,\omega) &= \frac{1}{2\pi\hbar}\int\sum_l |(n_q)_{l0}|^2 \mathrm{e}^{\mathrm{i}[\omega t-(\epsilon_l/\hbar-\epsilon_0/\hbar)t]}\,\mathrm{d}t \\ &= \frac{1}{2\pi\hbar}\int\sum_l \langle 0|\mathrm{e}^{\mathrm{i}\mathrm{H}t/\hbar} n_q(0)\mathrm{e}^{-\mathrm{i}\mathrm{H}t/\hbar}|l\rangle\langle l|n_q^+(0)|0\rangle \mathrm{e}^{\mathrm{i}\omega t}\,\mathrm{d}t \\ &= \frac{1}{2\pi\hbar}\int S(q,t)\mathrm{e}^{\mathrm{i}\omega t}\,\mathrm{d}t\,, \end{aligned} \tag{L.22}$$

where H is the total system Hamiltonian and $S(q, t)$ stands for

$$\sum_l \langle 0|\mathrm{e}^{\mathrm{i}\mathrm{H}t/\hbar} n_q(0)\mathrm{e}^{-\mathrm{i}\mathrm{H}t/\hbar}|l\rangle\langle l|n_q^+(0)|0\rangle\,.$$

Since $|l\rangle$ represents a complete set of eigenstates the sum over $|l\rangle\langle l|$ is one[2]. The first factor in the sum is the Heisenberg representation for the time-dependent operator $n_q(t)$. Thus, $S(q, t)$ describes the dynamic correlation between the particle density at the time $t = 0$ and at t by

$$\boxed{S(q,t) = \langle 0|n_q(t)n_{-q}(0)|0\rangle = \langle n_q(t)n^+q(0)\rangle\,.} \tag{L.23}$$

Formally the center part of this equation is a definition of the averaging process indicated at the right part of the equation. As it is written it refers to $T = 0$. In practice the symbols $\langle\rangle$ also include thermal averaging in the sense of (L.28). The expression on the right side of the equation is the *density–density correlation function.* Note the difference between the correlation function defined here on a quantum-mechanical basis and the classical correlation function we used in Sect. 3.5. Going back to (L.22) we obtain the final and very fundamental result

[2] Remember from quantum mechanics: If $|l\rangle$ is a complete set of eigenstates of a system $|\psi\rangle = \sum_l |l\rangle\langle l|\psi\rangle$.

$$\boxed{S(q,\omega) = \frac{1}{2\pi\hbar}\int \mathrm{e}^{\mathrm{i}\omega t}\langle n_q(t)n_{-q}(0)\rangle\,\mathrm{d}t\ .} \tag{L.24}$$

Expressed in words: the dynamical form factor is the Fourier transform of the density–density correlation or $S(q,\omega)$ is the spectral density of the state $n_q(t)|0\rangle$. Note that (L.24) is the quantum-mechanical analogon to the Wiener–Khintchin theorem (3.45).

The similarity between the definition of $S(q,\omega)$ and the generalized susceptibility $\chi(q,\omega)$ implies an immediate relationship between the two functions. Using (L.16) for the generalized susceptibility and (L.20) for the dynamical form factor it is indeed straightforward to show that

$$\chi(q,\omega) = \int_0^\infty \mathrm{d}\omega' S(q,\omega')\left\{\frac{1}{\omega-\omega'+\mathrm{i}\alpha} - \frac{1}{\omega+\omega'+\mathrm{i}\alpha}\right\}\ . \tag{L.25}$$

The density–density response function can be calculated immediately from the dynamical form factor. In contrast to $S(q,\omega)$, $\chi(q,\omega) = \chi_\mathrm{r}(q,\omega)+\mathrm{i}\chi_\mathrm{i}(q,\omega)$ is complex. The relations can be written in a more explicit form by comparing the real and the imaginary parts in (L.25) and using the Dirac relation (B.22) for $x = \omega - \omega'$. The result of the evaluation is

$$\boxed{\begin{aligned} \chi_\mathrm{r}(q,\omega) &= \mathrm{P}\int_0^\infty \mathrm{d}\omega' S(q,\omega')\left(\frac{2\omega'}{\omega^2-\omega'^2}\right) \\ \chi_\mathrm{i}(q,\omega) &= -\pi[S(q,\omega) - S(q,-\omega)]\ . \end{aligned}} \tag{L.26}$$

As required χ_r is an even function of ω while χ_i is odd. Since $S(q,\omega)$ is determined by the real transitions induced by the perturbation it describes the dissipative part of the interaction process. Thus the dissipative part involves the imaginary part of the response function[3]. As discussed in the next section $S(q,\omega)$ describes the energy transfer from the probe to the system whereas $S(q,-\omega)$ describes the energy transfer from the system to the probe. Hence for zero temperature $S(q,-\omega)$ is zero and the dynamical form factor becomes $-(1/\pi)\chi_\mathrm{i}(\omega)$.

L.4 The Fluctuation–Dissipation Theorem

At finite temperatures one can proceed in a similar way with the definition of the generalized susceptibility and the dynamical form factor. Two important differences must be considered, however. First, a finite number of initial states $|\mathrm{i}\rangle$ will be occupied. The probability $W(\epsilon_\mathrm{i})$ for the occupation is determined by the energy ϵ_i and given by the Boltzmann factor from statistical mechanics [see also the discussion in Chap. 7, equation (7.34)]

[3] Note: $\chi(q,\omega)$ is complex, even for vanishing α in (L.12). α is a convergency parameter and has nothing to do with a dissipation of energy from the probing particle to the system. This dissipation is described by χ_i.

$$W(\epsilon_{\rm i}) = \frac{1}{Z}{\rm e}^{-\beta\epsilon_{\rm i}} , \tag{L.27}$$

with $\beta = 1/k_{\rm B}T$ and $Z = \sum_{\rm i} {\rm e}^{-\beta\epsilon_{\rm i}}$. Second, energy exchange between probe and particle system will occur in both directions, from the probe to the system and vice versa, in a well defined balance. A good example of this behavior is the process of Stokes and antiStokes Raman scattering where in the first case energy from the photon probe is given to the lattice and in the second case from the lattice to the photon.

The dynamical form factor at finite temperatures is given by

$$S(q,\omega) = \frac{1}{Z}\sum_{\rm l,i} {\rm e}^{-\beta\epsilon_{\rm i}}|(n_q)_{\rm li}|^2\delta(\hbar\omega - \hbar\omega_{\rm li}) . \tag{L.28}$$

It describes the energy transfer from the probe to the system. A similar form factor $S(q,-\omega)$ describes the energy transfer from the system to the probe. The detailed balance between the two functions can be worked out to be

$$S(q,\omega) = {\rm e}^{\beta\hbar\omega}S(q,-\omega) . \tag{L.29}$$

An example where the principle of detailed balance applies was already discussed above when we investigated the generalized susceptibility $\chi^0(\omega)$ for a noninteracting system of Fermions at finite temperatures.

Inserting $S(q,-\omega)$ from (L.29) into the second part of (L.26) yields

$$\boxed{S(q,\omega) = -\frac{1}{\pi}[1 - {\rm e}^{-\beta\hbar\omega}]^{-1}\chi_{\rm i}(\omega) = \frac{1}{\pi}[1 + f_{\rm E}(\omega,T)]\chi_{\rm i}(\omega) ,} \tag{L.30}$$

where $f_{\rm E}(\omega,T)$ is the Bose–Einstein distribution.

Equation (L.30) is very famous and known as the *fluctuation–dissipation theorem.*

L.5 The Generalized Dielectric Function for Charged Particles

We proceed now to charged particles and evaluate the dielectric response function. The standard definition of the DF (Appendix B.1), written in Fourier components, is

$$\boldsymbol{D}(q,\omega) = \varepsilon(q,\omega)\varepsilon_0\boldsymbol{E}(q,\omega) . \tag{L.31}$$

Alternative but equivalent definitions using the charge density are more appropriate to study the linear response explicitly. The Poisson equations expressed by the electric displacement $D(q,\omega)$ or by the internal field $E(q,\omega)$ are

$$\mathrm{i}\boldsymbol{q}\boldsymbol{D}(q,\omega) = en_{\rm e}(q,\omega) \tag{L.32}$$

and

$$\mathrm{i}\boldsymbol{qE}(q,\omega) = \frac{e}{\varepsilon_0} n_{\mathrm{e}}(q,\omega) + \frac{e}{\varepsilon_0} \langle n(q,\omega) \rangle , \tag{L.33}$$

respectively. $en_{\mathrm{e}}(q,\omega)$ is the density of *external charges* (probes) and $e\langle n(q,\omega)\rangle$ is the expectation value for the resulting *induced charges* (response). Replacing the E field in (L.33) by using (L.31) and (L.32) we can express $1/\varepsilon(q,\omega)$ by the external charges n_{e} and obtain

$$\frac{1}{\varepsilon(q,\omega)} = 1 + \frac{\langle n(q,\omega) \rangle}{n_{\mathrm{e}}(q,\omega)} . \tag{L.34}$$

To obtain $\varepsilon(q,\omega)$ in terms of the density–density response function we introduce an interaction between the probing particles and the system particles as before. For the present case the Coulomb interaction with the Fourier components

$$U_{\mathrm{q}} = \frac{\mathrm{e}^2}{q^2 \varepsilon_0} n_{\mathrm{e}}(q,\omega) \tag{L.35}$$

is appropriate. With this we find from (L.34)

$$\frac{1}{\varepsilon(q,\omega)} = 1 + \frac{\mathrm{e}^2}{q^2 \varepsilon_0} \frac{\langle n(q,\omega) \rangle}{U_{\mathrm{q}}} . \tag{L.36}$$

With the definition of the density–density response function from (L.15) and considering $U(q,\omega) = U_{\mathrm{q}}/\mathcal{V}$ with U_{q} from (L.35) we obtain

$$\boxed{\frac{1}{\varepsilon(q,\omega)} = 1 + \frac{\mathrm{e}^2}{q^2 \varepsilon_0 \mathcal{V}} \chi(q,\omega) .} \tag{L.37}$$

To find the relation between $\varepsilon(q,\omega)$ and $S(q,\omega)$ we proceed as before and express the inverse of the DF in terms of the exact eigenstates of the electron system. Using (L.25) in (L.37) we find that

$$\frac{1}{\varepsilon(q,\omega)} = 1 + \frac{\mathrm{e}^2}{\varepsilon_0 q^2 \mathcal{V}} \int_0^\infty S(q,\omega') \left\{ \frac{1}{\omega - \omega' + \mathrm{i}\alpha} - \frac{1}{\omega + \omega' + \mathrm{i}\alpha} \right\} \mathrm{d}\omega' . \tag{L.38}$$

Hence the dynamical structure factor $S(q,\omega)$ represents the spectral density of $1/\varepsilon(q,\omega)$. Like $\chi(q,\omega)$, $\varepsilon(q,\omega)$ is complex and can be written as

$$\varepsilon(q,\omega) = \varepsilon_{\mathrm{r}}(q,\omega) + \mathrm{i}\varepsilon_{\mathrm{i}}(q,\omega) . \tag{L.39}$$

The imaginary part of $1/\varepsilon(q,\omega)$ is then obtained in analogy to (L.26)

$$\frac{\varepsilon_{\mathrm{i}}(q,\omega)}{|\varepsilon(q,\omega)|^2} = \frac{\mathrm{e}^2}{\varepsilon_0 q^2 \mathcal{V}} [S(q,\omega) - S(q,-\omega)] . \tag{L.40}$$

For $T = 0$ $S(q,-\omega) = 0$ so that $S(q,\omega)$ becomes

$$S(q,\omega) = \frac{q^2 \varepsilon_0 \mathcal{V}}{\mathrm{e}^2} \frac{\varepsilon_{\mathrm{i}}(q,\omega)}{|\varepsilon(q,\omega)|^2} = -\frac{q^2 \varepsilon_0 \mathcal{V}}{\mathrm{e}^2} \mathrm{Im} \left\{ \frac{1}{\varepsilon(q,\omega)} \right\} . \tag{L.41}$$

Since $S(q,\omega)$ describes the energy dissipated by the propagating probe, the expression $\mathrm{Im}\{-1/\varepsilon(q,\omega)\}$ is called the *energy loss function*.

Finally, we want to introduce screening effects into the expressions for the response functions, as we have done it for $q \neq 0, \omega = 0$ and $q = 0, \omega \neq 0$ in Sect. 14.1 and in Sect. 6.3. The simplest way to do this is to replace the unscreened potential U_q by a screened potential $U_\mathrm{q}/\varepsilon(q,\omega)$. With this the density–density response function for the electron liquid becomes

$$\chi_\mathrm{sc} = \frac{\mathcal{V}\langle n(q,\omega)\rangle}{U_\mathrm{q} n_\mathrm{e}(q,\omega)/\varepsilon(q,\omega)} = \varepsilon(q,\omega)\chi(q,\omega) \; . \tag{L.42}$$

Combining this with (L.37) yields for the DF

$$\varepsilon(q,\omega) = 1 - \frac{\mathrm{e}^2}{q^2\varepsilon_0\mathcal{V}}\chi_\mathrm{sc}(q,\omega) \; . \tag{L.43}$$

Within the RPA χ_sc is replaced by the density–density response function for noninteracting electrons $\chi^0(q,\omega)$ which yields the result of (14.11) in Sect. 14.1

$$\varepsilon(q,\omega) = 1 - \frac{\mathrm{e}^2}{q^2\varepsilon_0\mathcal{V}}\chi^0(q,\omega) \; , \tag{L.44}$$

where $\chi^0(q,\omega)$ is given by (14.12).

L.6 Random Phase Approximation

The *random phase approximation* (RPA) was developed as an approximation in calculations for screening and collective excitations in Fermi liquids in the 1950s. Since then its usefulness has led to a large number of applications in various problems of many particle systems. According to the concept of RPA contributions with random phases to a summation over the number of particles, usually occurring in product terms, can be neglected as compared to the total number of particles N. This simplifies expressions like

$$\sum_i \mathrm{e}^{\mathrm{i}(k-k')x_i} \approx N\delta_{k,k'} \; . \tag{L.45}$$

A useful example for the application of the RPA is the calculation of the screened generalized susceptibility for interacting particles. If the interaction energy is $U(q)$ the screened susceptibility evaluated within the RPA becomes

$$\chi_\mathrm{RPA}(q,\omega) = \frac{\chi^0(q,\omega)}{1 - U(q)\chi^0(q,\omega)} \; . \tag{L.46}$$

M. To Chapter 17, Neutron Scattering

M.1 Coherent and Incoherent Scattering for Hydrogen and Deuterium

The coherent and incoherent scattering cross sections for hydrogen are known to be very different whereas they are approximately equal for deuterium. It is very instructive to look at the reason for this.

As expected from the relation (17.11) in Sect. 17.3.2 incoherent scattering from a chemically homogeneous material can originate from scattering by isotopes l with different scattering length b_l. If the concentration of the lth isotope is c_l the relevant averaged quantities are

$$\bar{b} = \sum_l c_l b_l \quad \text{and} \quad \overline{|b|^2} = \sum_l c_l |b_l|^2 \, .$$

Scattering can also be incoherent for one and the same isotope since the scattering length depends on the relative orientation of the spins of the particles participating in the scattering process. If the nucleus has spin I the neutron can be scattered with a total spin $I + 1/2$ or $I - 1/2$ with corresponding scattering lengths b^+ and b^-. Since we have $2I + 2$ and $2I$ possible spin orientations for the first and second case, respectively, the probability for scattering in the two configurations is

$$\frac{2I+2}{(2I+2)+2I} = \frac{I+1}{2I+1} \quad \text{and} \quad \frac{2I}{(2I+2)+2I} = \frac{I}{2I+1} \, .$$

This yields for the relevant averaged scattering lengths

$$\bar{b} = \left(\frac{I+1}{2I+1}\right) b^+ + \left(\frac{I}{2I+1}\right) b^- \tag{M.1}$$

and

$$\overline{|b|^2} = \left(\frac{I+1}{2I+1}\right) |b^+|^2 + \left(\frac{I}{2I+1}\right) |b^-|^2 \, . \tag{M.2}$$

For neutron scattering by hydrogen we have $b^+ = 1.04 \times 10^{-12}\,\text{cm}$ and $b^- = -4.74 \times 10^{-12}\,\text{cm}$. This yields

$$\bar{b} = \frac{3}{4} b^+ + \frac{1}{4} b^- = -0.38 \times 10^{-12}\,\text{cm} \quad \text{and}$$
$$\overline{|b|^2} = \frac{3}{4}|b^+|^2 + \frac{1}{4}|b^-|^2 = 6.49\,\text{bn} \, .$$

From the definition of the total cross section and the coherent cross section we obtain finally for hydrogen

$$\sigma_\text{c} = 1.8\,\text{bn} \quad \text{and} \quad \sigma = 81.7\,\text{bn} \, .$$

For deuterium b^+ and b^- are 0.95×10^{-12} and 0.1×10^{-12}, respectively, which yields

$$\sigma_\text{c} = 5.6\,\text{bn} \quad \text{and} \quad \sigma = 7.6\,\text{bn} \, .$$

The incoherent cross section is the difference between the total cross section and the coherent cross section. The very large value of σ_inc for hydrogen is well established in neutron scattering experiments. In organic materials often the hydrogen is substituted by deuterium to avoid the dominance of incoherent scattering.

References

Chapter 1

1.1. A.A. Michelson: Phil. Mag. Ser. 5 **31**, 256 (1891)
1.2. P.B. Fellget: Dissertation, Cambridge University (1951)

Additional Reading

Spectroscopy

Holllas J.M.: *Modern Spectroscopy* (John Wiley & Sons, Chichester 1994)
Kuzmany, H.: *Festkörperspektroskopic. Eine Einführung* (Springer, Berlin, Heidelberg 1990)

Introduction to Solid-State and Semiconductor Physics

Ashcroft N.W., Mermin N.D.: *Solid State Physics* (Saunders College, Philadelphia 1976)
Born M., Huang K.: *Dynamical Theory of Crystal Lattices* (Oxford Univ. Press, Oxford 1985)
Challaway J.: *Quantum Theory of the Solid State* (Academic, New York 1976)
Ibach H., Lüth H.: *Festkörperphysik* 2nd edn. (Springer, Berlin, Heidelberg 1988)
Kittel C.: *Introduction to Solid-State Physics* 7th edn. (John Wiley, New York 1996)
Mihaly L., Martin M.C.: *Solid State Physics, Problems and Solutions* (John Wiley, New York 1996)
Seeger K.: *Semiconductor Physics, An Introduction* Springer Series in Solid-State Sciences, Vol.40 (Springer, Berlin, Heidelberg 1991)
Yu P.Y., Cardona M.: *Fundamentals of Semiconductors* (Springer, Berlin, Heidelberg 1996)
Ziman J.: *Principles of the Theory of Solids* (Cambridge Univ. Press, London 1964)

Chapter 2

2.1. F.H. Read: *Electromagnetic Radiation* (John Wiley & Sons, New York 1987)
2.2. R.P. Feynman, R.B. Leighton, M. Sands: In *Lectures on Physics*, Vol.2 (Addison-Wesley Publishing Company, Reading 1964)
2.3. J.D. Jackson: *Classical Electrodynamics* (John Wiley & Sons, New York 1974)
2.4. E.O. Brigham: *Fast Fourier Transform* (Prentice-Hall Inc., Englewood Cliffs, NJ 1974)

Additional Reading

Born M., Wolf E.: *Principles of Optics* (Pergamon Press, Oxford 1980)
Born M.: *Optik* (Springer, Berlin, Heidelberg, New York 1985)
Klein M. v., Furtak T.E.: *Optics* (John Wiley & Sons, New York 1986)
Read F.H.: *Electromagnetic Radiation* (John Wiley & Sons, New York 1987)
Reif F.: *Statistische Physik* (de Gruyter, Berlin 1985)
Smith F.G., Thomson J.H.: *Optics*, 2nd edn (John Wiley, Chichester 1988)

Chapter 3

3.1. R. Kubo: *Statistical Mechanics* (North Holland Publishing Company, Amsterdam 1965)
3.2. A. Eucken (ed.): Lanold-Björnstein, Atom und Molekularphysik, Part 1, *Atoms and Ions* (Springer, Berlin, Göttingen, Heidelberg 1950)
3.3. G. Abstreiter: In *Light Scattering in Solids IV*, M. Cardona and G. Güntherrodt (eds.), Topics in Applied Phys., Vol.54 (Springer, Berlin, Heidelberg 1984)
3.4. J. Schwinger: Phys. Rev. **75**, 1912 (1949)
3.5. C. Kunz (ed.): *Synchrotron Radiation*, Topics Curr. Phys., Vol.10 (Springer, Berlin, Heidelberg 1979)
3.6. C. Kunz: In *Vacuum Ultraviolet Radiation Physics*, E.E. Koch, R. Haensel, C. Kunz eds. (Vieweg, Braunschweig 1974)
3.7. U. Bonse: Festkörperprobleme, **23**, 77 (1983)
3.8. G. Huber: Physikalische Blätter **47**, 365 (1991)
3.9. R.L. Gunshor and A.V. Nurmikko: MRS Bulletin **20**, 15 (1995)
3.10. D.J. Bradley: In *Ultrashort light pulses,* 2nd edn., p.31, S.L. Shapiro (ed.) Topics Appl. Phys. Vol. 18, (Springer, Berlin, Heidelberg 1984)
3.11. W. Demtröder: *Grundlagen und Techniken der Laserspektroskopie* (Springer, Berlin, Heidelberg 1990)
3.12. Coherent Radiation Data Sheet 2017A 20M, Palo Alto, CA 1980
3.13. Y. Petroff: Z. Physik **B61**, 477 (1985)
3.14. T. Gray and C. Frederickson: Lasers and Optronics **9**, 40 (1990)
3.15. U. Fano: Phys. Rev. 124, 1866 (1961)
3.16. J. Puerta and P. Martin: Appl. Optics **20**, 3923 (1981)

Additional Reading

Bradley D.J.: In *Ultrashort Light Pulses*, 2nd edn., S.L. Shapiro (ed.) Topics Appl. Phys., Vol.18 (Springer Berlin, Heidelberg 1984)
Chen J., Ren H., Xie J., Ye M. (eds.): *Development and Applications of Free Electron Lasers* (Gordon and Breach, Amsterdam 1997)
Chow W.W., Koch S.W., and Sargent M.: *Semiconductor-laser Physics*, (Springer Berlin, Heidelberg 1994)
Demtröder W.: *Laser Spectroscopy, Basic Concepts and Instrumentation*, Springer Series in Chem. Phys., Vol.5 (Springer, Berlin, Heidelberg 1988)
Eberhardt W.: *Applications of Synchrotron Radiation*, Springer Series in Surface Sciences 35 (Springer Berlin, Heidelberg 1995)
Ewing J.J.: *Excimer lasers*, Physics Today **31**, 32 (1978)
Gudat W.: In *Synchrotron Strahlung in der Festkörperphysik*, IFF-Ferienkurs (Jülich 1987)
Ippen E.P., Shank C.V.: in *Ultrashort Light Pulses*, 2nd edn., S.L. Shapiro (ed.) Topics Appl. Phys., Vol.18 (Springer Berlin, Heidelberg 1984)

Kaiser W. (ed.): *Ultrashort Laser Pulses and Applications*, Topics Appl. Phys., Vol.60 (Springer, Berlin, Heidelberg 1988)
Koechner W.: *Solid-State Laser Engineering*, 2nd edn., Springer Series in Opt. Sci., Vol.1 (Springer, Berlin, Heidelberg 1988)
Milonni P.W. and Eberly J.H.: *Lasers* (John Wiley, NY 1988)
Rhodes C.K. (ed.): *Excimer Lasers*, 2nd edn., Topics Appl. Phys., Vol.30 (Springer, Berlin, Heidelberg 1984)
Saleh B.E.A. and Teich M.C.: *Fundamentals of Photonics* (John Wiley & Sons Inc. New York 1991)
Schäfer F.P. (ed.): *Dye Lasers*, 3rd edn., Topics Appl. Phys., Vol.1 (Springer, Berlin, Heidelberg 1990)

Chapter 4

4.1. M.V. Klein and T.E. Furtak: *Optics* (John Wiley & Sons, New York 1986)
4.2. D.A. Long: *Raman Spectroscopy* (McGraw-Hill, New York 1977)
4.3. J. Sandercock: in *Light Scattering in Solids*, p.9, M. Balkanski (ed.) (Flammarion, Paris 1971)

Additional Reading

Demtröder W.: *Grundlagen und Techniken der Laserspektroskopie* 2nd edn. (Springer, Berlin, Heidelberg 1990)
Demtröder W.: *Laser Spectroscopy. Basic Concepts and Instrumentation*, 2nd edn., Springer Series in Chem. Phys., Vol.5 (Springer, Berlin, Heidelberg 1988)
Klein M.V., Furtak T.E.: *Optics* (John Wiley & Sons, New York 1986)

Chapter 5

5.1. S.M. Sze: *Semiconductor Devices* (Wiley, New York 1981)

Additional Reading

Keyes R.J.: *Optical and Infrared Detectors*, 2nd edn., Topics Appl. Phys. Phys., Vol.19 (Springer, Berlin, Heidelberg 1980)
Kressel H.: *Semiconductor Devices for Optical Communication*, 2nd. edn., Topics Appl. Phys. 39 (Springer, Berlin, Heidelberg 1982)
Moss T.S., Burell G.J., Ellis B.: *Semiconductor Optoelectronics* (Butterworth, London 1973)
Papoulis A.: *Probability, Random Variables, and Stochastic Processes* (MacGraw-Hill, Auckland 1981)
Saleh B.E.A., Teich M.C.: *Fundamentals of Photonics* (John Wiley &Sons Inc., New York 1991)

Chapter 6

6.1. R.F. Wallis and M. Balkanski: *Many-Body Aspects of Solid-State Spectroscopy* (North-Holland, Amsterdam 1986)
6.2. A. Vasicek: *Optics of Thin Films* (North-Holland, Amsterdam 1960)
6.3. H. Ibach and H. Lüth: *Festkörperphysik*, 2nd edn. (Springer, Berlin, Heidelberg 1988)

6.4. J. Kittel: *Introduction to Solid-State Physics*, (Wiley Eastern, New Delhi 1977)
6.5. H. Raether: In *Ergebnisse der exakten Naturwissenschaften* p.84, Springer Tracts in Mod. Phys., Vol.38 (Springer, Berlin, Heidelberg 1965)
6.6. W.G. Spitzer, H.Y. Fass: Phys. Rev. **106**, 88 289 (1957)
6.7. J.R. Dixon: Proc. Int. Conf. on Physics of Semiconductors (Czech. Acad. Sci., Prague 1961)
6.8. R.M.A. Azzam, N.M. Bashra: *Ellipsometry and polarized light* (North-Holland, Amsterdam 1977)

Additional Reading

Brüsch P.: *Phonons: Theory and Experiment II* Springer Series in Solid-State Sciences, Vol.65 (Springer, Berlin 1986)
Heavens O.S.: *Optical Properties of Thin Solid Films* (Dover Publication, New York 1955)
Grosse P.: *Freie Elektronen in Festkörpern* (Springer, Berlin, Heidelberg 1979)
Seeger K.: *Semiconductor Physics, An Introduction* Springer Series in Solid-State Sciences, Vol.40 (Springer, Berlin 1991)
Yu P.Y., Cardona M.: *Fundamentals of Semiconductors* (Springer, Berlin, Heidelberg 1996)

Chapter 7

7.1. S.M. Sze: *Physics of Semiconductor Devices* (Wiley, New York 1969)
7.2. K. Seeger: *Semiconductor Physics* (Springer, Berlin, Heidelberg 1991)
7.3. P.Y. Yu, M. Cardona: *Fundamentals of Semiconductors* (Springer, Berlin, Heidelberg 1996)
7.4. T.S. Moss, G.J. Burell, B. Ellis: *Semiconductor Opto-Electronics* (Butterworth, London 1973)
7.5. P.J. Dean, D.G. Thomas: Phys. Rev. **150**, 690 (1966)
7.6. D.E. Aspenes: Spectroscopic Ellipsomety of Solids, in *Optical Properties of Solids, New Developments* (North-Holland, Amsterdam 1976)
7.7. D. Brust, J.C. Phillips, F. Bassani: Phys. Rev. Lett. **9**, 94 (1962)
7.8. M.L. Cohen, T.K. Bergstrasser: Phys. Rev. **141**, 789 (1966)
7.9. Y. Hamakawa and T. Nishino: Recent Advances in Modulation Spectroscopy, in *Optical Properties of Solids - New Developments* (North Holland, Amsterdam 1976)
7.10. P.W. Baumeister: Phys. Rev. **121**, 359 (1961)
7.11. J.C. Phillips: Solid State Phys. **18**, 55 (1966)
7.12. D.B. Fitchen, R.H. Silsbee, T.A. Fulton, E.L. Wolf: Phys. Rev. Lett. **11**, 275 (1963)
7.13. W. van Roosbroeck, W. Shockley: Phys. Rev. **94**, 1558 (1954)
7.14. C.H. Henry, P.J. Dean, J.D. Cuthbert: Phys. Rev. **166**, 754 (1968)
7.15. M. Gershenzon: Bell Syst. Techn. J. **45**, 1599 (1966)
7.16. D.G. Thomas, M. Gershenzon, F.A. Trumbore: Phys. Rev. **A133**, 269 (1964)
7.17. Y.E. Pokrovskii, K.I. Svituniva: Sov. Phys. Semicond. **4**, 409 (1970)
7.18. W.W. Chow, S.W. Koch, M. Sargent III: *Semiconductor Laser Physics* (Springer, Berlin, Heidelberg 1994)
7.19. W. Klöpfer: *Introduction to Polymer Spectroscopy* (Springer, Berlin, Heidelberg 1984)

Additional Reading

Johnson C.S., Pedersen L.G.: *Quantum Chemistry and Physics* (Dover Publ. Inc., New York 1986)
Keil T.H.: Phys. Rev. **A140**, 601 (1965)
Klingshirn C.F.: *Semiconductor Optics* (Springer, Berlin, Heidelberg 1995)
Kreibig U., Volmer M.: *Optical Properties of Metal Clusters*, Springer Series in Material Sceince, Vol.25 (Springer, Berlin, Heidelberg 1995)
Kressel H. (ed.): *Semiconductor Devices of Optical Communication* Topics Appl. Phys., Vol.39 (Springer, Berlin 1982)
Moss T.S., Burrel G.J., Ellis B.: *Semiconductor Optoelectronics* (Butterworth, London 1973)
Seraphin B.O.: *Optical Propertires of Solids* (North-Holland, Amsterdam 1976)
Schiff L.I.: *Quantum Mechanics* (McGraw-Hill, New York 1955)
Yen W.M. (ed.): *Laser Spectroscopy of Solids II*, Topics Appl. Phys., Vol.65 (Springer, Berlin, Heidelberg 1989)

Chapter 8

8.1. C. Herman: *Internationale Tabellen zur Bestimmung von Kristallstrukturen*, Vol.III, (Gebr. Borntraeger, Berlin 1935)
International Tables for X-Ray Crystallography, The International Union of Crystallography (The Kynoch Press, Birmingham, 1965)
8.2. A. Poulet, J.P. Mathieu: *Vibrational Spectra and Symmetry of Crystals* (Gordon and Breach, New York 1976)
8.3. L.D. Landau, E.M. Lifschitz: *Quantum Mechanics: non-relativistic theory* (Pergamon Press, London Berlin 1959)

Additional Reading

Behringer J.: *Factor Group Analysis Revisited and Unified*, in Solid State Physics, Springer Tracts in Mod. Phys., Vol. 68 (Springer, Berlin, Heidelberg 1973)
Decius J.C., Hexter R.M.: *Molecular Vibrations in Crystals* (McGraw-Hill Inc., New York 1977)
Hamermesh M.: *Group Theory and its Application to Physical Problems* (Reading, London 1962)
Ludwig W., Falter C.: *Symmetry in Physics*, Springer Series in Solid-State Sci., Vol.64 (Springer, Berlin, Heidelberg 1988)
Nye J.F.: *Physical Properties of Crystals* (Oxford Univ. Press, Oxford 1985)
Poulet A., Mathieu J.P.: *Vibrational Spectra and Symmetry of Crystals* (Gordon and Breach, New York 1976)
Sherwood P.M.A.: *Vibrational Spectra of Solids* (Cambridge Univ. Press, Cambridge 1972)
Vainstein B.K.: *Modern Crystallography I*, Springer Series in Solid-State Sci. Vol.15 (Springer, Berlin, Heidelberg 1981)

Chapter 9

9.1. W. Richter: In *Solid-State Physics*, p.121, Springer Tracts in Modern Phys., Vol.78 (Springer, Berlin, Heidelberg 1976)
9.2. H. Poulet, J.P. Mathieu: *Vibrational Spectra and Symmetry of Crystals* (Gordon and Breach, New York 1976)

9.3. S.P.S. Porto, J.A. Giordmain, T.C. Damen: Phys. Rev. **147**, 608 (1966)
9.4. C.H.Henry, J.J. Hopfield: Phys. Rev. Lett. **15**, 638 (1965)
9.5. T.R. Hart, R.L. Agarwal, B. Lax: Phys. Rev. **B1**, 638 (1979)
9.6. J.M. Worlock, J.F. Scott, P.A. Fleury: in *Light Scattering Spectra of Solids,* p.9, G. Wright (ed.) (Springer, Berlin, Heidelberg 1969)
9.7. H. Kuzmany, J. Kürti: unpublished
9.8. W. Richter: in *Solid-State Physics*, p.121, G. Höhler (ed.) Springer Tracts Mod. Phys., Vol.78 (Springer, Berlin, Heidelberg 1976)
9.9. A. Mooradian: in *Laser Handbook*, p.1309, F.T. Arecchi, E.O. Schulz du Bois eds., (North-Holland, Amsterdam 1972)
9.10. A. Pinczuk, L. Brillson, E. Burstein, E. Anastassakis: Phys. Rev. Lett. **27**, 317 (1971)
9.11. J.R. Sandercock: Opt. Commun. **2**, 76 (1970)

Additional Reading

Raman Spectroscopy

Cardona M. (ed.): *Light Scattering in Solids*, 2nd edn., Topics Appl. Phys. Vol.8 (Springer, Berlin, Heidelberg 1983)
Cardona M., Güntherodt G.: *Light Scattering in Solids II–V*, Topics Appl. Phys. Vols. 50, 51, 54, 66 (Springer, Berlin, Heidelberg 1982, 1982, 1984, 1989)
Poulet H., Mathieu J.P.: *Vibrational Spectra and Symmetry of Crystals* (Gordon & Breach, New York 1976)
Schrader B. (ed.): *Infrared and Raman Spectroscopy* (VHC Verlagsgesellschaft, Weinheim 1995)
Turell G.: *Infrared and Raman Spectra of Crystals* (Academic, New York 1972)

Brillouin Spectroscopy

Benedek G.B., Fritsch K.: Z. Physik **149**, 383 (1957)
Dil J.G.: Rep. Progr. Phys. **45**, 285 (1982)
Pine A.S.: in *Light Scattering in Solids I*, 2nd edn., p.253, M. Cardona (ed.) Topics in Appl. Phys., Vol.8 (Springer, Berlin, Heidelberg 1983)

Elastic Light Scattering

Hulst H.C. van: *Light Scattering by Small Particles* (Wiley, London 1957)
Mie G.: Ann. Phys. **25**, 377 (1908)

Chapter 10

10.1. J. Bohdansky: Z. Physik **149**, 383 (1957)
10.2. J. Hesse, H. Preier: Festkörperprobleme **15**, 229 (1975)
10.3. A. Mitsuishi, Y. Otsuka, S. Fujita, H. Yoshinaga: Jap.J.Appl.Phys. **2**, 574 (1963)
10.4. M.F. Kimmitt: *Far-Infrared Techniques* (Pion Limited, London 1970)
10.5. F.J. Low: Proc. Inst. Elec. Electron. Engrs. **54**, 477 (1966)
10.6. W. Kautzmann: *Quantum Chemistry* (Academic Press, New York 1957)
10.7. T. Ito, H. Shirakawa, S. Ikeda: J. Polym. Sc., Polym. Chemistry Ed. 13, 1943 (1975)
10.8. P.Knoll, H. Kuzmany: Phys. Rev. **B 29**, 2221 (1984)
10.9. M. Behmer: Dissertation, Universität Wien 1985

10.10. I. Melngailis: J. de Physique (suppl. No.11-12) **29**, C4-84 (1968)
10.11. S.M. Sze, J.M. Irvin: Solid State Electronics **11**, 599 (1968)
10.12. S.M. Kogan, T.M. Lifshits: phys. stat. sol. (a) **39**, 11 (1977)
10.13. P.L. Richards, M. Tinkham: Phys. Rev. **119**, 575 (1960)
10.14. L.D. Landau, E.M. Lifshitz: *Statistical Physics* (Pergamon Press, Oxford 1968)
10.15. R.G. Gordon: J. Chem. Phys. **43**, 1307 (1965)

Additional Reading

Borstel G., Falge H.J., Otto A.: *Surface and Bulk Phonon Polaritons Observed by Attenuated Total Reflection.* Springer Tracts Mod. Phys. **74**, 107 (Springer, Berlin Heidelberg 1974)
Chantry G.W.: *Submillimetre Spectoscopy* (Academic, London 1971)
Griffith P.R., de Haseth J.A.: *Fourier Transform Infrared Spectrometry* (John Wiley, New York 1986)
Grosse P.: *Freie Elektronen in Festkörpern* (Springer, Berlin, Heidelberg 1979)
Günzler H., Heise H.M.: *Infrared Spectroscopy: an Introduction* (Thieme, Stuttgart 1996)
Kimmitt M.F: *Far-Infrared Techniques* (Pion Limited, London 1970)
Rössler A.: *Infrared Spectroscopic Ellipsometry* (Akademie Verlag, Berlin 1990)
Sherwood P.M.A.: *Vibrational Spectroscopy of Solids* (Cambridge Univ. Press, Cambridge 1972)
Schrader B. (ed.): *Infrared and Raman Spectroscopy* (VCH Verlagsgesellschaft, Weinheim 1995)
Vinzent J.D.: *Fundamantals of IR Detector Operating and Testing* (John Wiley, New York 1990)

Chapter 11

11.1. D. Michel: *Grundlagen und Methoden der kernmagnetischen Resonanz* (Akademie Verlag, Berlin 1981)
11.2. C.P.Slichter: *Principles of Magnetic Resonance*, 3rd edn., Springer Series in Solid-State Sci., Vol.1 (Springer, Berlin, Heidelberg 1989)
11.3. G.E. Pake, T.L. Estle: *The Physical Principles of Electron Paramagnetic Resonance* (Benjamin, London 1973)
11.4. P.L. Scott, C.D. Jeffries: Phys. Rev. **127**, 32 (1962)
11.5. J. Kürti, G. Menczel: phys. stat. sol **102**, 639 (1980)
11.6. G. Emch, R. Lacroix: Helv. Phys. Acta **33**, 1021 (1960)
11.7. J. Kürti: unpublished
11.8. J.L. Hall, R.T. Schumacher: Phys. Rev. **127**, 1892 (1962)
11.9. M. Holz, B. Knüttel: Physikalische Blaetter **38**, 368 (1982)
11.10. M. Mehring: *NMR Basic Principles and Progress* Vol.11, (Springer, Berlin 1976)
11.11. B. Knüttel, with kind permission of the Bruker AG.

Additional Reading

Carrington A., McLachlan A.D.: *Introduction to Magnetic Resonance*, (J.W. Arrowsmith Ltd., Bristol 1979)
Mehring M.: *Principles of High Resolution NMR in Solids*, 2nd edn., (Springer, Berlin, Heidelberg 1983)

Michel D.: *Grundlagen und Methoden der kernmagnetischen Resonanz* (Akademie Verlag, Berlin 1981)
Pake G.E., Estle T.L.: *The Physical Principles of Electron Paramagnetic Resonance* (Benjamin, London 1973)
Panisod P.: *Nuclear magnetic resonance* in Microscopic Methods in Metals, p.365, U. Gonser (ed.) Topics Curr. Phys., Vol.40 (Springer, Berlin, Heidelberg 1986)
Slichter C.P.: *Principles of Magnetic Resonance*, 3rd edn., Springer Series in Solid-State Sci., Vol.1 (Springer, Berlin, Heidelberg 1989)

Chapter 12

12.1. H. Kuhlenkampff, L.S. Schmidt: Ann. Phys. **43**, 494 (1943)
12.2. Int. Tables of X-Ray Crystallography Vol. 4, J.A. Ibers, W.C. Hamilton (eds.) (The Kynoch Press, Birmingham 1974)
12.3. B.K. Agarawal: *X-Ray Spectroscopy. An Introduction*, 2nd edn., p.335, Springer Series in Opt. Sci., Vol.15 (Springer, Berlin, Heidelberg 1991)
12.4. P. Wobrauschek: J. Trace and Microprobe Techniques, **13**, 83 (1995)
12.5. S. Hüfner: *Photoelectron Spectroscopy*, Springer Series in Solid-State Sci. Vol.82, p.455 (Springer, Berlin, Heidelberg 1995)
12.6. P. Steiner, H. Höchst, S. Hüfner: in *Photoemission in Solids II*, L. Ley, M. Cardona (eds.) p.349, Topics Appl. Phys., Vol.27 (Springer Berlin, Heidelberg 1979)
12.7. D.W. Langer: Festkörperprobleme **13**, 193 (1973)
12.8. N.V. Smith, G.K. Wertheim, S. Hüfner, M.M. Traum: Phys. Rev. **B10**, 3197 (1974)
12.9. S. Hüfner, G.K. Wertheim, J.H. Wernick: Phys. Rev. **B8**, 4511 (1973)
12.10. S. Hüfner: *Photoelectron Spectroscopy*, Springer Series in Solid-State Sci., Vol. 82, p.7 (Springer, Berlin, Heidelberg 1995)
12.11. D.M. Poirier, J. Weaver: Phys. Rev. **B47**, 10959 (1994)
12.12. R.A. Pollak, L. Ley, S. Kowalczyk, D.A. Shirley, J.D. Joannopoulos, D.J. Chadi, M.L. Cohen: Phys. Rev. Lett. **29**, 1103 (1972)
12.13. A. Karpfen: J. Chem. Phys. **75**, 238 (1982)
12.14. K. Seki, U. Karlsson, R. Engelhardt, E.E. Koch: Chem. Phys. Lett. **103**, 343 (1984)
12.15. J. Weaver: J. Phys. Chem. Sol. **53**, 1433 (1992)
12.16. T.M. Hayes, J.B. Boyce: Solid State Phys. **37** 173 (1982)
12.17. J. Stöhr: *NEXAFS Spectroscopy* (Springer, Berlin, Heidelberg 1992)
12.18. S. Hüfner: *Photoelectron Spectroscopy*, Springer Series in Solid-State Sci., Vol. 82, p.437 (Springer, Berlin, Heidelberg 1995)

Additional Reading

Agrawal B.K.: *X-Ray Spectroscopy. An Introduction*, 2nd edn., Springer Series in Opt. Sci., Vol.15 (Springer, Berlin, Heidelberg 1989)
Briggs D., Seah M.P.: Practical Surface Analysis, Volume I, (John Wiley, Chichester 1992)
Cardona M., Ley L. (eds.): *Photoemission in Solids I*, Topics Appl. Phys., Vol.26 (Springer, Berlin, Heidelberg 1978)
Hayes T.M., Boyce J.B.: *Extended x-ray absorption fine structure spectroscopy*, Solid State Phys. Vol.37, 173 (1982)
Hüfner S.: *Photoelectron Spectroscopy*, Springer Series in Solid-State Sci., Vol. 82 (Springer, Berlin, Heidelberg 1995)
Smith N.V.: *Inverse Photoemission*, Rep. Progress Phys. **51**, 1227 (1988)

Chapter 13

13.1. U. Gonser: *Mössbauer spectroscopy*, U. Gonser (ed.) Topics Appl. Phys., Vol.5 (Springer, Berlin, Heidelberg 1975)
13.2. T. Moriya, H. Ino, F.E. Fujita, Y. Maeda: J. Phys.Soc. Japan **24**, 60 (1968)
13.3. A. Heiming, K.H. Steinmetz, G. Vogl, Y. Yoshida: J. Phys. F, Metal Phys. **18**, 1491 (1988)
13.4. W. Petry, G. Vogl: Material Science Forum **15-18**, 323 (1988)
13.5. Th. Wichert, E. Recknagel: In *Microscopic Methods in Metals*, p.317, U. Gonser (ed.) Topics Curr. Phys., Vol.40 (Springer, Berlin, Heidelberg 1986)
13.6. E. Matthias, D.A. Shirley, J.S. Evans, R.A. Naumann: Phys. Rev. **B140**, 264 (1965)
13.7. B. Lindgren, E. Karlsson, B. Jonsson: Hyperfine Interactions **1**, 505 (1976)
13.8. J. Christiansen, P. Heubes, R. Keitel, W. Klinger, W. Löffler, W. Sander, W. Witthun: Z. Phys. **B24**, 177 (1976)

Additional Reading

Catchen G.L.: Materials Research Society Bulletin, 37 (1995)
Gonser U. (ed.): *Mössbauer-Spectroscopy*, Topics Appl. Phys., Vol.5 (Springer, Berlin, Heidelberg 1975)
Gonser U. (ed.): *Mössbauer-Spectroscopy II*, Topics Curr. Phys., Vol.25 (Springer, Berlin, Heidelberg 1981)
Gonser U. (ed.): *Microscopic Methods in Metals*, Topics Curr. Phys., Vol.40 (Springer, Berlin, Heidelberg 1986)
Recknagel E., Schatz G., Wichert Th.: in *Hyperfine Interactions in Radioactive Nuclei*, J. Christiansen (ed.) p.133, Topics Curr. Phys., Vol.31 (Springer, Berlin, Heidelberg 1983)
Vogl G., Petry W.: Festkörperprobleme **25**, 655 (1985)
Wichert Th., Recknagel E.: *Perturbed Angular Correlation*, in Microscopic Methods in Metals, U. Gonser (ed.) p.317, Topics Curr. Phys., Vol.40 (Springer, Berlin, Heidelberg 1986)

Chapter 14

14.1. D. Pines, P. Nozieres: *The Theory of Quantum Liquids*, (Benjamin Inc., New York 1966)
14.2. M.V. Klein: in *Light Scattering in Solids*, 2nd edn., Topics in Appl. Phys., Vol.8, M. Cardona (ed.) (Springer, Berlin, Heidelberg 1983)
14.3. L. Genzel, U. Kreibig: Z. Physik **B37**, 93 (1980)

Additional Reading

Grosse P.: *Freie Elektronen in Festkörpern* (Springer, Berlin, Heidelberg 1979)
Pines D., Nozieres P.: *The Theory of Quantum Liquids*, (Benjamin Inc., New York 1966)
Landau L.D., Lifschitz E.M.: *Lehrbuch der Theoretischen Physik*, Vol.5 (Akademie Verlag, Berlin 1987)
Lovesey S.W.: *Theory of Neutron Scattering from Condensed Matter*, Vol.1 (Clarendon Press, Oxford 1984)
Mahan G.D.: Many Particle Physics, 2nd edn., (Plenum, New York 1990)
Ziman J.H.: *Principles of the Theory of Solids*, 2nd edn. (Cambridge University Press, Cambridge 1972)

Chapter 15

15.1. J. Fink: Advances in Electronics and Electron Physics **79**, 155 (1989)
15.2. C.J. Powell, J.B. Swan: Phys. Rev. **116**, 81 (1959)
15.3. C. Wehenkel: J. Phys. (Paris) **36**, 199 (1975)
15.4. L.R. Canfield, G. Hass, W.R. Hunter: J. Phys. (Paris) **25**, 124 (19964)
15.5. W.A. Harrison: Phys. Rev. **123**, 85 (1961)
15.6. L. Esaki, Y. Miyahara: Solid State Electron. **1**, 13 (1960)
15.7. J.W. Conley, C.B. Duke, G.D. Mahan, J.J. Tiemann: Phys. Rev. **150**, 466 (1966)
15.8. F. Steinrisser, L.C. Davis, C.B. Duke: Phys. Rev. **176**, 912 (1968)
15.9. I. Giaever: Phys. Rev. Lett. **5**, 147 (1960)
15.10. J. Nicol, S. Shpiro, P.H. Smith: Phys. Rev. Lett. **5**, 461 (1960)
15.11. B.L. Blackford, R.H. March: Can. J. Phys. **46**, 141 (1968)
15.12. J.K. Tsang, D.M. Ginsberg: Phys. Rev. **B22**, 4280 (1980)
15.13. W.L. McMillan, J.L. Rowell: Phys. Rev. Lett. **14**, 108 (1965)
15.14. I. Giaever, H.R. Hart, K. Megerle: Phys. Rev. **126**, 941 (1962)
15.15. W.L. McMillan, J.L. Rowell: In *Superconductivity 1*, R.D. Parks (ed.) (Dekker, New York 1969)
15.16. R.M. Feenstra: Semicond. Sci. Technol. **9**, 2157 (1994)
15.17. S.J. Tans et al.: Nature **386**, 474 (1997)

Additional Reading

Electron Energy Loss

Fink J.: Advances in Electronics and Electron Physics **79**, 155 (1989)
Fink J. et al.: J. Electron Spectroscopy **66**, 395 (1994)
Schülke W.: in *Handbook on Synchrotron Radiation*, Vol. 3, p. 565, G.S. Brown, D.E. Moncton eds. (North-Holland, Amsterdam 1991)

Tunneling Spectroscopy

Hansma P.K.: *Tunneling Spectroscopy: Capabilities, Applications and New Techniques* (Plenum, New York 1982)
Smoliner J.: Semicond. Sci. Technol. **11**, 1 (1996)
Soethout L.L., Van Kempen H. , Van de Walle: Advances in Electronics and Electron Physics **79**, 155 (1990)
Wolf E.L.: *Principles of Electron Tunneling Spectroscopy* (Oxford Univ. Press, Oxford 1985)

Chapter 16

16.1. P. Kubica, A.T. Stewardf: Phys. Rev. Lett. **34**, 852 (1975)
16.2. S. Berko, M. Haghgooie, J.J. Mader: Phys.Lett.A **63**, 335 (1977)
16.3. B.L. Blackford, R.H. March: Can. J. Phys. **46**, 141 (1968)
16.4. W. Trifthäuser: in *Microscopic Methods in Metals*, U. Gonser (ed.) p.249, Topics Curr. Phys., Vol.40 (Springer, Berlin, Heidelberg 1986)
16.5. V.G. Grebinnik, I.I. Gurevich, V.A. Zhukov, A.P. Manych, E.A. Meleshko, I.A. Muratowa, B.A. Nikolskii, v.I. Selivanov, V.A. Suetin: Soviet Phys. JETP **41**, 777 (1976)
16.6. R.S. Hayano, Y.J. Uemura, J. Imazato, N. Nishida, K. Nagamine, T. Yamazaki, Y. Ishikawa, H. Yasuoka: J. Phys. Soc. Japan **49**, 1773 (1980)

Additional Reading

Brandt W., Dupasquier A. (eds.): *Positron Solid State Physics* (North-Holland, Amsterdam 1983)
Chappert J., Yaouanc A.: Muon spectroscopy, in *Microscopic Methods in Metals* U. Gonser (ed.) p.297, Topics Curr. Phys., Vol.40 (Springer, Berlin, Heidelberg 1986)
Hautojärvi P. (ed.): *Positrons in Solids*, Topics Curr. Phys., Vol.12 (Springer, Berlin, Heidelberg 1979)
Seeger A.: *Positive muons as light isotopes of hydrogen*, in Hydrogen in Metals, G. Alefeld, J. Völkl (eds.) Topics Appl. Phys., Vol.28 (Springer, Berlin, Heidelberg 1978)
Trifthäuser W.: in *Microscopic Methods in Metals*, U. Gonser (ed.) p.249, Topics Curr. Phys., Vol.40 (Springer, Berlin, Heidelberg 1986)

Chapter 17

17.1. S.W. Lovesey: *Theory of Neutron Scattering from Condensed Matter* Vol.1 (Clarendon Press, Oxford 1984)
17.2. C.G. Windsor: Experimental Techniques in *Methods of Experimental Physics* Vol.23A (Academic Press, New York 1986)
17.3. W. Bührer: in *Introduction to Neutron Scattering*, 1st European Conference on Neutron Scattering, (Paul Scherrer Institute, Villingen 1996)
17.4. Bergmann–Schäfer: *Lehrbuch der Experimentalphysik* Vol.4 (Springer, Berlin, Heidelberg 1990)
17.5. M.K. Eberle: *The Spallation Neutron Source SINQ*, (Paul Scherrer Institut, Villingen 1994)
17.6. K. Ibel: *Guide to Neutron Research Facilities at the Institute Laue– Langevin*, (Institute Laue–Langevin, Grenoble 1994)
17.7. K.E. Larsson, U. Dahlborg, S. Holmryd: Ark. Fys. **17**, 369 (1960)

Additional Reading

Baruchel J., Hodeau J.L., Lehman M.S., Regnard J.R., Schlenker C.: *Neutron and Synchrotron Radiation for Condensed Matter Studies*, (Springer, Berlin, Heidelberg 1993)
Ibel K.: *Guide to Neutron Research Facilities at the Institute Laue–Langevin* (Institut Laue–Langevin, Grenoble 1994)

Chapter 18

18.1. J.R. Tesmer, M. Nastasi: Handbook of Modern Ion Beam Materials Analysis (Materials Research Society, Pittsburgh 1995)
18.2. K. Wittmack, J.B. Clegg: Appl. Phys. Lett. **37**, 285 (1980)

Additional Reading

Benninghoven A., Rüdenauer F.G., Werner H.W.: *Secondary Ion Mass Spectrometry*, (John Wiley, New York 1987)
Briggs D., Seah M.P.: *Practical Surface Analysis* Volume II (John Wiley, Chichester 1992)
Tesmer J.R., Nastasi M.: Handbook of Modern Ion Beam Materials Analysis (Materials Research Society, Pittsburgh 1995)

Subject Index

Springer and the environment

At Springer we firmly believe that an international science publisher has a special obligation to the environment, and our corporate policies consistently reflect this conviction.

We also expect our business partners – paper mills, printers, packaging manufacturers, etc. – to commit themselves to using materials and production processes that do not harm the environment. The paper in this book is made from low- or no-chlorine pulp and is acid free, in conformance with international standards for paper permanency.

Printing: Mercedesdruck, Berlin
Binding: Buchbinderei Lüderitz & Bauer, Berlin